Date Palm Genetic Resources and Utilization

Jameel M. Al-Khayri • Shri Mohan Jain
Dennis V. Johnson
Editors

Date Palm Genetic Resources and Utilization

Volume 1: Africa and the Americas

Springer

Editors
Jameel M. Al-Khayri
Department of Agricultural Biotechnology
King Faisal University
Al-Hassa
Saudi Arabia

Dennis V. Johnson
Cincinnati
Ohio
USA

Shri Mohan Jain
Department of Agricultural Sciences
University of Helsinki
Helsinki
Finland

ISBN 978-94-017-9693-4 ISBN 978-94-017-9694-1 (eBook)
DOI 10.1007/978-94-017-9694-1

Library of Congress Control Number: 2015934366

Springer Dordrecht Heidelberg New York London

Printed on acid-free paper

Springer is part of Springer Science+Business Media (www.springer.com)

Preface

Date palm (*Phoenix dactylifera* L.) trees bear edible fruits under adverse conditions of high temperature, limited water resources and high soil and water salinity. Dates have contributed to the emergence and survival of ancient societies in desert regions worldwide. Now they are being rediscovered as a nutritional health food due to richness in fructose, a low glycemic index sugar, vitamins and dietary fibers. Dates have diverse medicinal properties due to high content of phytochemicals including carotenoids, polyphenols (e.g. isoflavons, lignans and flavonoids), tannins and sterols. These compounds are known for antioxidant activity, cholesterol-lowering properties and chemoprevention of cancer, diabetes and cardiovascular diseases.

Date palm plantations worldwide are estimated to have over 150 million trees, distributed between 10° and 39°N Lat. but thrive mostly between 24° and 34°N Lat. in the Middle East and North Africa. Date production has steadily increased from (in thousands) 2,535 mt in 1980 to 7,549 mt in 2012 (FAOSTAT database apps.fao.org). Although modern agricultural practices have contributed to increase in production, expansion in agricultural land and utilization of natural resources has played an important role. Sustainable date palm cultivation in the twenty-first century will require implementation of a set of principles drawn from existing research results and farm-level experiences. Despite the existing body of research, realization of the full potential of the date palm is being hindered by the lack of comprehensive studies of key areas: (a) Comprehensive assessment of date palm genetic resources and their conservation, including the role of seedling palms; (b) Broad evaluation and identification of cultivars and the role of DNA fingerprinting; (c) Development of a set of best-cultivation practices to enhance yield and reduce post-harvest loss for both monoculture and mixed oasis farming systems; (d) Conservation and utilization of germplasm resources and agroforestry in light of climate change; (e) Appraisal of the role of tissue culture according to the scale of production; and (f) Evaluation of date palm products and marketing. This book fills the gaps in the preceding key areas for major date-producing countries worldwide. It provides an assessment of how to achieve these key principles across the date-producing countries with reference to the local environmental conditions and socioeconomic practices.

The book is published in two separate volumes; each consists of 15 chapters. Volume 1 addresses date-producing countries within Africa, North America and South America, whereas, Volume 2 pertains to Asia and Europe. Volume 1 begins with an introductory chapter highlighting modern scientific discoveries and the main book features. The second chapter discusses biodiversity of genetic resources of date palm in the world. Chapters 3, 4, 5, 6, 7, 8, 9, 10, 11, 12, and 13 provide an assessment specific to individual date-producing countries of Africa and the Americas. The African countries covered are: Egypt, Algeria, Sudan, Tunisia, Libya, Morocco, Mauritania, Niger, Cameroon, Djibouti, Chad, Mali Somalia, Ethiopia, Burkina Faso and Senegal. Chapters 14 and 15 cover the United States of America and the South American countries Chile and Peru. The African countries Kenya and Namibia, as well as Mexico in North America, are small date producers; however, they are not included in this volume due to unavailability of authors. The volume ends with appendixes summarizing available information on each country. The appendixes provide surveys of characteristics and distribution of major date palm cultivars, commercial sources of dates, offshoots and in vitro plants; and research institutes and scientific societies concerned with date palm research and production.

In this book volume, each chapter describes the current status and prospects on: cultivation practices and agroforestry; genetic resources focusing on threats and research progress in genetics, breeding, conservation, and germplasm banks; role and importance of cultivar classification and identification based on morphological and molecular techniques; importance of micropropagation technology and research progress toward scale-up production; and advances in dates processing, novel products and innovative utilizations and marketing. In addition, each chapter describes the main characteristics of important date palm cultivars including growth requirements, cultivar distribution shown in a country map, production statistics and economics, nutritional aspects as well as morphological descriptors supported with color photos. The chapter concludes with imperative recommendations for further development of date palm sector in the respective country.

This volume presents, for the first time, a comprehensive assessment of date palm genetic resources and utilizations in date palm-growing countries of Africa and the Americas. The book is a valuable resource for students, researchers, scientists, commercial producers, consultants, and policymakers interested in agriculture particularly in the date palm industry. Individuals and companies throughout the date industry would benefit greatly from the adoption of policies and practices discussed in this book to enhance production and expand industrialization and marketing of traditional and potentially new date products. The chapters were graciously contributed by prominent scientists of each country; the exception is the Mauritanian chapter which was written by a pronounced visiting scientist. All manuscripts were critically reviewed and revised to ensure accuracy and quality. We greatly appreciate all contributory authors for their contribution towards the success and quality of this book. We are grateful to Springer for giving us an opportunity to compile this book.

Al-Hassa, Saudi Arabia — Jameel M. Al-Khayri
Helsinki, Finland — Shri Mohan Jain
Cincinnati, OH, USA — Dennis V. Johnson

Contents

Contributors

Sahidou Abdoussalam Systems and Crops Diversification Program, International Crops Research Institute for Semi-Arid Tropics (ICRISAT-Niger), Niamey, Niger

Frederique Aberlenc-Bertossi IRD-Institut de Recherche pour le Dévelopement, Palm Developmental Biology Group, UMR DIADE, Montpellier, France

Jameel M. Al-Khayri Department of Agricultural Biotechnology, King Faisal University, Al-Hassa, Saudi Arabia

Souad Babahani Laboratory of Saharan Bio-Resources, Kasdi Merbah University, Ouargla, Algeria

Massimo Battaglia Istituto Agronomico per l'Oltremare, IAO-MAE, Florence, Italy

Shawky A. Bekheet Plant Biotechnology Department, National Research Center, Cairo, Egypt

Mohamed Ali Benabderrahim Arid and Oases Cropping Laboratory, Arid Area Institute, Elfjé, Medenine, Tunisia

Malika Bennaceur Department of Biology, Faculty of Nature Sciences and Life, University of Oran, Oran, Algeria

Faculty of Biological Sciences, Laboratory of Research in Arid Areas, University of Science and Technology Houari Boumediene, Algiers, Algeria

Salah Eddine Benziouche Agronomy Department, Faculty of Exact Sciences and Natural and Life Sciences, Mohamed Khider University, Biskra, Algeria

Carlo Bergesio Istituto Agronomico per l'Oltremare, IAO-MAE, Florence, Italy

Nadia Bouguedoura Faculty of Biological Sciences, Laboratory of Research in Arid Areas, University of Science and Technology Houari Boumediene, Algiers, Algeria

Sali Bourou Fruits Plants Program, Agricultural Research Institute of Cameroon, Garoua Station, Garoua, North Cameroon

Alessandro Camussi Department of Agri-Food Production and Environmental Sciences, Genetics Unit, University of Florence, Florence, Italy

Abdourahman Daher Institut des Sciences de la Vie/CERD-Centre d'Etude et de Recherche de Djibouti, Ministère de l'Enseignement Supérieur et de la Recherche, Djibouti, Djibouti

Mohammed Aziz Elhoumaizi Faculty of Sciences, University of Oujda, Oujda, Morroco

Sherif F. El-Sharabasy Central Laboratory for Date Palm Research and Development, Agricultural Research Center, Giza, Egypt

Hugo A. Escobar Facultad de Ciencias Agronómicas, Universidad de Tarapacá, Arica, Chile

Bashir Ghsera Department of Plant Protection, Faculty of Agriculture, El Fatah University, Tripoli, Libya

Hammadi Hamza Arid and Oases Cropping Laboratory, Arid Area Institute, Elfjé, Medenine, Tunisia

S. Mohan Jain Department of Agricultural Sciences, University of Helsinki, Helsinki, Finland

Abdullah A. Jaradat Department of Agronomy and Plant Genetics, USDA-ARS & University of Minnesota, Morris, MN, USA

Monia Jemni Département de Technologies de Dattes, Centre Régional de Recherche en Agriculture Oasienne de Degueche, Tozeur, Tunisia

Dennis V. Johnson Consultant, Cincinnati, OH, USA

Mohamed Khairi Faculty of Agriculture, Dongola University, Dongola, Sudan

Robert R. Krueger USDA-ARS National Clonal Germplasm Repository for Citrus and Dates, Riverside, CA, USA

Marta Mancini Slow Food Foundation for Biodiversity, Bra, Cuneo, Italy

Nabil Mohamed Institut des Sciences de la Vie/CERD-Centre d'Etude et de Recherche de Djibouti, Ministère de l'Enseignement Supérieur et de la Recherche, Djibouti, Djibouti

Abdesselem Mrabet Arid and Oases Cropping Laboratory, Arid Area Institute, Elfjé, Medenine, Tunisia

Ahmed Othmani Laboratoire de Culture In Vitro, Centre Régional des Recherches en Agriculture Oasienne de Degueche, Tozeur, Tunisia

Dov Pasternak Consultant, International Crops Research Institute for Semi-Arid Tropics (ICRISAT-Niger), Niamey, Niger

Milvia Luisa Racchi Department of Agri-Food Production and Environmental Sciences, Genetics Unit, University of Florence, Florence, Italy

Mohamed Ben Salah Research Regional Center on Oases Agriculture, Degache, Tunisia

Moulay Hassan Sedra Laboratory of Phytopathology, Genetics and Integrated Control, Regional Centre of Agricultural Research of Marrakech, National Institute of Agricultural Research (INRA), 533, Marrakech, Morocco

Sana Touil Department of Plant Protection, Centre Technique des Dattes, Kebilli, Tunisia

Rafael G. J. Valdivia Agronomist, Private Consultant, Santiago, Chile

Noé Woin Fruits Plants Program, Agricultural Research Institute of Cameroon, Garoua Station, Garoua, North Cameroon

Part I
Introduction

Chapter 1
Introduction: Date Production Status and Prospects in Africa and the Americas

Dennis V. Johnson, Jameel M. Al-Khayri, and S. Mohan Jain

Abstract Successful future development of date palm depends largely on evaluating, utilizing, and conserving genetic resources; assessing the value of present and potential cultivars; promoting the best cultivation, processing, and marketing practices; and broadening the number of fruit and tree products. The date palm was domesticated at the dawn of agriculture, new technologies were adopted for its cultivation, the palm was dispersed widely by humans and it has become a major world tree crop with the introduction of modern plantations. France, the USA, and international organizations have contributed to date palm development and more recently domestic programs in the larger producing countries. Creating a new international date palm organization to bring together research and development efforts and to serve as a resource center would be beneficial. Date palm cultivation has a long history across North Africa; Algeria, Egypt, Libya, Morocco, and Tunisia are major world fruit producers. Traditional date cultivation in oases is common, with recent and ongoing expansion of organized plantations. North Africa has potential for increased production given its proximity to existing European markets, but serious pest and disease problems must be overcome. Domestic and international research and development support and familiarity with the crop are assets. In the Sahel and Southern Africa, dates are minor crops but with potential for development to meet domestic demand. Advantages include favorable climatic and soil conditions and lesser pest and disease problems. Similar advantages exist in the Americas. Because they are not included in country chapters, summary accounts of date cultivation in Benin, Kenya, Mexico, and Namibia are provided.

D.V. Johnson (✉)
Consultant, Cincinnati, OH, USA
e-mail: djohn37@aol.com

J.M. Al-Khayri
Department of Agricultural Biotechnology, King Faisal University,
Al-Hassa 31982, Saudi Arabia
e-mail: jkhayri@kfu.edu.sa

S.M. Jain
Department of Agricultural Sciences, University of Helsinki, PL-27,
Helsinki 00014, Finland
e-mail: mohan.jain@helsinki.fi

J.M. Al-Khayri et al. (eds.), *Date Palm Genetic Resources and Utilization: Volume 1: Africa and the Americas*, DOI 10.1007/978-94-017-9694-1_1

Keywords Dispersal and distribution • Domestication • Food value • Modern plantations • Production statistics • Research and development

1.1 Introduction

The date palm (*Phoenix dactylifera* L.) is one of the most valuable domesticated fruit trees because of ritual significance in human societies, health benefits, productive capacity in harsh semiarid and arid environments, and the range of subsistence products from its fruits and other parts of the large palm. In planning this book, the editors strived for complementarity with their earlier book, *Date Palm Biotechnology* (Jain et al. 2011), as well as two books by other authors, *Dates: Production, Processing, Food, and Medicinal Values* (Manickavasagan et al. 2012) and *Dates: Postharvest Science, Processing Technology and Health Benefits* (Siddiq et al. 2014). Each of these three books is composed of individual chapters, on a broad range of topics. None of them deals with the date palm status of genetic resources and production in each of the date palm-growing countries.

Additionally in this book, four major themes were identified as exemplified in the book's title. Genetic resources of the date palm are of great importance in the future of the crop, for two reasons. First, the diversity present in the entire *Phoenix dactylifera* gene pool represents an invaluable resource for conventional and molecular breeding of future date palms of high fruit quality, high fruit yield, tolerance to abiotic stresses, and resistance to major pests and diseases. Genetic erosion and genetic sedimentation imperil the date palm. Second, the strong and increasing trend toward cultivation of known elite cultivars, so evident in nearly every new date palm plantation, is being carried on at the expense of a narrowing of the overall cultivar diversity as traditional cultivars are less likely to be propagated and unevaluated seedling date palms (Johnson et al. 2013) are eradicated as if they were weeds. Cultivar assessment, for all but a few well-known elite types, is at a very rudimentary stage of knowledge. Because the unassessed cultivars are often native to a single country, studies of promising cultivars must be organized at the national level, with active collaboration among neighboring countries which may harbor close relatives; in certain instances the same cultivar may be present in adjoining areas within a country, or in adjacent countries, and bears different vernacular names.

Cultivation practices exhibit considerable variability across the date-growing world. Certain accepted practices have been adapted to local environmental conditions, such as early picking and postharvest ripening because of climatic limitations preventing full maturity of the fruits on the trees. Best management practices have yet to be perfected for date palm growing under a basic set of environmental factors. The country-level practices described in this book provide a point of initiation to begin developing guidelines for growers to assure them with maximum return for their investment and labor. The novel products part of the book title was included to prompt individual countries to examine the products derived from date fruits not marketed as fresh or dried products. Descriptions of derived date fruit products in the various countries demonstrate the cottage-level and small- and large-scale industrialization in existence. These descriptions make possible a cross-fertilization of ideas from country to country to valorize date fruit production to the highest level possible, which represents a potential enhancement of income to producers and manufacturers. Although the novel products described in the book are almost exclusively focused on the fruits, other parts of the date palm tree have a range of other subsistence and industrial products such as those derived from the leaves, petioles, edible heart, and stem wood. These nonfruit date products merit a fresh in-depth assessment to update the earlier publication *Date Palm Products* (Barreveld 1993).

1.2 Agrobotany, Domestication, and Dispersal of Date Palm

1.2.1 Agrobotanical Description

A mature date palm can reach 30 m in height and is the tallest of the *Phoenix* species. The stem has a diameter of 40–50 cm, bearing suckering offshoots at its base during the early years of growth. The leaves are pinnate and erect with numerous stiff leaflets, and the rachis is armed with broad sharp spines on each side. An adult palm has a moderately dense crown of 100 or more leaves 3–6 m in length; each leaf has a lifespan of 3–7 years, and about 12 new leaves are produced each year. The date palm is dioecious, bearing male and female flowers on separate trees. Dioecy in date palm is an impediment to seed propagation and traditional crossbreeding experimentation as equal numbers of male and female plants are produced and their gender is not apparent in the field until flowering begins. Propagated by seed, it reaches sexual maturity at about 5–7 years of age, significantly sooner if by offshoots. Flowering is normally annual. Fruit development from pollination to maturity is reached in about 200 days and is variable in shape, color, and size, 4–7 × 2–3 cm, larger than in any other *Phoenix* species. Yields are highly variable, 20–100 kg per adult tree, and depend upon cultivar, environmental conditions, and cultivation practices. The Medjool cv. produces the largest commercial fruits. Fruits contain a single elongate seed, variable in shape and size, 20–30 × 5–8 mm, containing a useful oil. On average, a plantation has an economic life of about 50 years, although the trees will continue to produce fruit beyond that age, albeit at lower yields. When trees become very tall, they are more expensive to care for and harvest, which negatively affects the production economics.

The true date palm, *Phoenix dactylifera* L., is one of 14 recognized species of the genus (Barrow 1998). It is most closely related to the Cretan date palm (*P. theophrasti* Greu.) of the Eastern Mediterranean; to the Canary Islands date palm (*P. canariensis* Chab.), native to that island group; and to the sugar date palm (*P. sylvestris* (L.) Roxb.) which is native to South Asia. Where *Phoenix* species distributions overlap in the wild, natural hybrids occur; likewise *Phoenix* palms grown in botanical gardens readily hybridize (Gros-Balthazard 2013). This facility makes artificial crossbreeding of species easily achieved. The date palm has 36 chromosomes ($n=18$; $2n=36$).

1.2.2 Domestication

The date palm is one of the world's very early domesticates and a traditional Old World fruit. Date palm, olive tree (*Olea europaea* L.), common fig tree (*Ficus carica* L.), and wine grape (*Vitis vinifera* L.) together are a seminal quartet of ancient fruit crops closely associated with the earliest stages of agriculture. The first domestication of the date palm is believed to have taken place at least 6,000 years ago in Mesopotamia, the lands between the Tigris and Euphrates rivers, in what is today Iraq. The area is often referred to as the *cradle of agriculture*. The early history of date palm as a cultivated palm is known only from archaeological records and ancient cuneiform clay tablets. It is presumed that wild date palms were selected for propagation on the basis of their possessing desirable fruit characteristics.

Over an indeterminate period of time following the first planting of date palms, several simple but important innovations took place in Mesopotamia which fostered the development of date palm cultivation. In roughly presumed chronological order, the following took place:

(a) Female and male palms were distinguished.
(b) Organized plantings made with fixed spacing.
(c) Palms under cultivation segregated by gender.
(d) Irrigation and water management.
(e) Artificial pollination.
(f) Separation of offshoots for propagation.
(g) Stages of fruit development recognized and named.

From the latter are derived the five sequential stages of fruit growth which were named: hababauk, kimri, khalal, rutab, and tamar. Somewhere along the way, the first cultivar was given a name, which may have been Zahidi in the Basrah area of present-day Iraq, a cultivar prominent in that area to the present. The foregoing is summarized from Dowson (1921, 1923, 1982), Popenoe (1973), and Pruessner (1920).

Since the date palm was initially domesticated, new cultivars have repeatedly been created as a result of seedling propagation and the selection of those of superior fruit quality subsequently propagated by offshoots and assigned individual names. Cultivar names typically derive from either the particular fruit characteristics, the locality where the palm originated, or the name of the farmer who selected it.

1.2.3 Dispersal and Geographic Distribution

From its presumed homeland in Mesopotamia, over succeeding millennia, date palms were dispersed by humans to the east, west, and south and took hold wherever climatic conditions and water sources were favorable for fruit production. Initial dispersal was clearly by means of seed. Even when abandoned, dates persisted and became naturalized where conditions allowed. Munier (1973), in a general study of date growing, described dispersal of date palm in two major historic routes. One route is from Mesopotamia south into the Arabian Peninsula and eastward to undivided India. A second route begins in Egypt (itself the location of an independent domestication of date palm) across North Africa west to Morocco. The caravan trade and the Nile River may have been the means of spreading date palms southward into the Sahelian countries and Sudan in the early days. From North Africa, dates were carried to Spain by the Moorish invasion in the eighth century, or possibly earlier. Seedlings dates were grown successfully for fruit in southwestern Spain, although the climatic is marginal for carrying fruits to maturity on the tree. In the sixteenth century, the Spaniards successfully introduced date seeds to the Americas, initially to Peru and later to Mexico, and achieved modest fruit production. In the mid- to late 1800s, there are reports of seedling dates being grown in Southern Africa, Australia, and New Caledonia (Johnson 2010). In the past 100 or so years, as offshoots have successfully been transported over long distances, modern date palm plantations employing elite cultivars have emerged or are emerging in countries such as the UAE, Oman, Kuwait, the USA, Mexico, India, Israel, Jordan, Namibia, and Australia, as well as in several Sahelian countries in Africa.

1.3 The Rise of Modern Date Palm Plantations

Scientific literature on date palm published before the twentieth century consists mostly of brief accounts of date growing in a few countries or entries in reference volumes; an exception is the book *The Date Palm in India* (Bonavia 1885), which brings together information about date growing in the Persian Gulf and British India.

The US Department of Agriculture created a program in the 1890s to promote an irrigated commercial date industry in the American Southwest and in 1904 established a date experiment station in Mecca, California, later moved to Indio. Offshoots of recognized cultivars were imported from North Africa and the Middle East and provided to pioneer farmers. Commercial date production began in 1912. Needing a source of technical information, in 1924 the date growers created the Date Growers' Institute, which organized technical meetings each year and published an annual review (DGIAR 1924–1979). The USDA published a number of technical studies on date growing in the USA (e.g., Nixon and Carpenter 1978; Swingle 1904) as well as bulletins on date cultivation in other countries such as Egypt (e.g., Mason 1923). Three general books have helped to popularize dates in the USA (Dunham 1948; Paulsen 2005; Simon 1978). US scientific publications on date palm have made a major contribution to modern date growing. However, scientific research on date palm in the USA was greatly reduced after cessation of the date growers' annual meetings in 1979 and closure of the date experiment station a year later.

The French also made important contributions in the twentieth century to modern date cultivation in their North African colonies by organizing the first two international date palm conferences. In 1931 a weeklong meeting was held in Biskra, Algeria, with presentations on broad climatic, agronomic, and economic aspects of date production. One of the recommendations was to promote the Deglet Noor cultivar (Sem Dattier 1931). It was not until 1950 that a second such conference took place in Tunis, Tunisia. Among the recommendations made were to conduct scientific research on date palm agronomy and adopt uniform fruit export standards (Cong Int Datte 1951). French scientists produced numerous articles and technical reports on dates in Africa, broadening the knowledge base. Pereau-Leroy (1958) authored the first national study of dates on Morocco, and Munier (1973) the first general scientific study of date cultivation.

International backing for data palm development was provided by FAO in the second half of the twentieth century. This was initiated by sponsorship of three technical conferences on date palm: one in Tripoli, Libya, in 1959 and two in Baghdad, Iraq, in 1965 and 1975. No proceedings were published, but the presentations by leading date specialists from around the world generated background materials for the issuance by FAO of a series of important books on date palm (Dowson and Aten 1962; Dowson 1982; Barreveld 1993; Zaid 1999, 2002). Another contribution made to date palm science was *The Date Palm Journal* (FAO 1981–1988).

A recent contributor to date palm development is the International Center for Agricultural Research in the Dry Areas (ICARDA), recently relocated from Aleppo, Syria, to Beirut, Lebanon. One of its projects focuses on increasing date production among smallholder farmers in Iraq (Mazid et al. 2013), ravaged by war; another involves enhancing date palm production systems in the GCC (Gulf Cooperation Council) countries (ICARDA 2011). This project is being led by an ICARDA date palm specialist based in Oman. A study of date harvest and postharvest handling was a part of the project (Kader and Hussein 2009).

With the major contributions outlined above from the USA, France, FAO, ICARDA, and numerous lesser but important efforts, a firm scientific foundation has been established for the cultivation and processing of dates, upon which the future of the crop is being built.

1.4 Research and Development

Apart from the international development efforts described above which have fostered modernization of date palm production to varying degrees in individual countries, national programs also have played an important role. Major common problems shared by all date-producing countries are associated with cultivation practices, pests and diseases, soil and water conditions, harvest and postharvest practices, processing facilities, technical support and extension, the agricultural labor force, as well as domestic and international marketing. The best example of differences is pests and diseases. Remote areas of date growing in Southern Africa and the Americas have few such problems to contend with, whereas Morocco and Algeria are experiencing significant crop and tree losses given the occurrence of bayoud disease and the red palm weevil. Climatic factors present problems only in certain countries such as Pakistan and India where annual monsoon rain occurs before date fruits mature, but in Saudi

Arabia, for instance, climate presents no hindrance to date growing. Each date-growing country has a unique set of impediments to improving production, and the severity of each should set the agenda for domestic R & D.

This book provides information on the research centers carrying out research on date palm on various aspects, supported by mainly the federal government in academic institutions and private investment. A close link exists between the magnitude of date production and the funding to support date palm R & D. Egypt, Iran, Iraq, Saudi Arabia, and Pakistan, each ranking among the top six producers worldwide, have established research institutes devoted to date palm, which are funded by the federal governments. In other date palm-producing countries, including Algeria and Morocco, date palm research is carried out by the national agricultural research centers. In Syria and Palestine, date palm currently is considered as a minor crop, with annual production of 3,986 and 3,600 mt, respectively. The paucity of federal government funding becomes a major obstacle in developmental efforts. Therefore, these countries must fall back on larger date-producing countries for importing improved date palm cultivars, technologies to control pests and diseases, and agro-food-based products and other by-products. Any important factor in domestic support for date palm research depends on the priority of the national governmental support in creating date palm-based agro-food industry and local and international marketing boosting local economy.

Throughout the date palm world in general, insufficient attention has been given to programs of genetic improvement employing emerging biotechnology procedures. This neglect may be attributed to the limitation of research funds, inadequate advance research infrastructure, and scarcity of experts in many of the date palm-producing nations. Interrelated issues include the need for precise knowledge of the genetic resources present in a country, including that contained in seedling date populations, and the effective means to conserve genetic resources through germplasm gardens and cryopreservation.

Given the problems shared by date-growing countries, it would be of great benefit to create an international date palm center. The advantages of such a center include the following:

(a) Avoid expensive duplication of effort in addressing major pest and disease problems.
(b) Enhance opportunities for collaborative research both at the bilateral and international levels.
(c) Develop and disseminate information on best practices in date cultivation, harvesting, postharvest handling, and marketing by developing an interactive website. An excellent model for date palm exists in the Coconut Timeline (http://cocos.arecaceae.com/).
(d) Develop international industry descriptors and standards for fruit quality and packaging and marketing.
(e) Maintain a database of world date palm cultivars and their conservation status.
(f) Maintain information on date palm genetic diversity, genetic erosion, conservation, and utilization of germplasm.

(g) Develop programs on health benefits of dates and various commercial food products.
(h) Develop date palm functional genomics for studying useful genes leading to genetic improvement of date palm growing under climate change, enhanced fruit quality, and industrial products.

1.5 Date Fruit Production Statistics

During the planning of the present book, available data on date production were critically evaluated in order to determine the relative importance of the crop within a country based on national production in relation to global production quantities. The Food and Agriculture Organization of the United Nations (FAO) collects data from member nations and publishes annual agricultural production data on major crops. These data are the most comprehensive available and of primary concern in this book. In addition, FAO publishes data on the value of crops, as well as import and export quantities, but those data are of lesser importance given the focus of the book on production.

Domestic production statistics in most countries are incomplete or difficult to access. The exceptions are Saudi Arabia and Morocco which have published detailed accounts of their date palm cultivation which include some data on the numbers of palms being grown and area under cultivation, along with cultivar descriptions (INRA 2011; Min Agr Saudi Arabia 2006).

Aggregate world date fruit production for 2012 amounted to 7,548,918 mt, according to FAOSTAT (2012). Table 1.1 lists reported producing countries in descending order of magnitude. The production figures given in Table 1.1 reflect the extent of date growing around the world. These data are a mixture of production figures provided by the various countries and FAO estimates. Burkina Faso, Ethiopia, Mali, and Senegal, discussed in this book, are not included in FAO data, apparently because commercial date production hardly exists yet.

There are some serious problems with the FAO data that have a bearing on this book and need to be discussed; date production reported for five countries is incorrect. The production reported for China of 150,000 mt represents the production of the red date fruit, rather than the true date palm. Red date or jujube is scientifically *Zizyphus jujuba* (L.) H. Karst., a tree or shrub of the Rhamnaceae or buckthorn family. China has no production of true dates. A similar situation is found with respect to the reporting of 31,675 mt of date production in Turkey. In this instance, the figures represent production of the common fig. This misunderstanding arises from the same Turkish word being used to refer to dates and figs. Albania's reported production of 12,935 mt of dates is a reflection of date imports from other countries being considered as national production; the country's climate is unsuited to commercial date production. Swaziland's reported production of 330 mt is incorrect, because there is no commercial date cultivation within the country (McCubbin M, personal communication, 2014). The neighboring Republic of South Africa has some date

Table 1.1 Date fruit production by country, 2012, in mt.

Ranking	Country	Production (mt)
1	Egypt	1,470,000
2	Iran	1,066,000
3	Saudi Arabia	1,050,000
4	Algeria	789,357
5	Iraq	650,000
6	Pakistan	600,000
7	Sudan (former)	433,500
8	Oman	270,000
9	United Arab Emirates	250,000
10	Tunisia	190,000
11	Libya	170,000
12	China*	150,000
13	Morocco	113,397
14	Yemen	55,181
15	Israel	42,866
16	Kuwait	34,600
17	Turkey*	31,765
18	USA	28,213
19	Mauritania	22,000
20	Qatar	21,843
21	Chad	20,000
22	Niger	17,000
23	Bahrain	15,000
24	Somalia	13,000
25	Albania*	12,935
26	Jordan	10,417
27	Mexico	6,012
28	Spain	4,000
29	Syria	3,986
30	Palestine	3,600
31	Benin	1,300
32	Kenya	1,100
33	Cameroon	600
34	Namibia	400
35	Peru	400
36	Swaziland*	330
37	Djibouti	86
38	Colombia*	30

Source: FAOSTAT (2012)
Countries which erroneously are reported to have date production are indicated with an asterisk

growing focused on Medjool cv. in the Limpopo Region (McCubbin 2007), which does not appear in FAO statistics for the RSA. There are no FAO trade data on date imports and exports between the two countries, so it appears unlikely there is any connection. FAO has reported date production for Swaziland each year since 1999, all based on either an estimate or imputation methodology. The most plausible explanation is that the production reported as dates is for some other fruit. Finally, the statistic that Colombia produced 30 mt of dates is not from domestic production but rather from imports and exports. Climatic conditions in the country also are unsuited for commercial date growing. Inclusion of these five countries in world production totals resulted in an overstatement of 195,060 mt in 2012. India should be but is not included in Table 1.1 for unknown reasons, and data are either not collected in the country or, if they are, not submitted to FAO. Whatever the case, according to data from a newspaper estimate cited in Chap. 14, India's date production was estimated in 2010/2011 to be at least 120,000 mt, which would place it among the top 12 world producers.

The quality and accuracy of national date production data vary, irrespective of the significance of the crop to the individual country's agricultural sector. Among the factors at play are the following:

(a) The mix of modern formal plantation cultivation and traditional date growing where densities vary and other crops are grown in association such as is typical in oases. Statistical data on the former are much easier to gather and are more accurate.
(b) Planting densities in some countries are highly variable; hence it is difficult to take data on area under cultivation and derive the actual number of trees being grown.
(c) Data from farmers may be understated in countries where taxes are assessed on the basis of total production and likewise the quantities of dates consumed by growers and their families or used as livestock feed are poorly known.
(d) The level of proficiency of the national agency collecting agricultural production data is again highly variable.
(e) Data from one producing country may be double counted. For example, Mexico's small date production takes place just south across the border from the Yuma, Arizona, and Bard, California area, one of the two most important production regions of the USA. An indeterminate quantity of Mexican dates is known to be shipped to Yuma, Arizona, and sold through the large cooperative processing facilities in that city. Other such examples likely exist in some African countries.

Regarding production data by date cultivar at the national level, data are nonexistent. National date production statistics in even the most advanced agricultural countries such as the USA are aggregated and not collected by cultivar. As indicated in the country chapter in these volumes, the most prominent cultivars are identified, but their relative ranking as regards total national production is unknown. The most reliable production data by date cultivar is derived from farm or research station records of per tree productivity.

The lesson to be drawn from this discussion is that date palm production statistics for each country must be appraised for their accuracy and reliability within the context of the conditions which are known to exist in that country.

1.6 Food Value of Dates

In the human diet, date fruits are rich sources of sugars, protein, fiber, minerals, and certain vitamins. A quick energy source because of the high sugar content, dates have been called *nature's candy*; they are an excellent snack food and currently being recommended as such. When the fruits reach maturity (tamar), sucrose inverts into glucose and fructose. Dates contain reasonable amounts of vitamin A, thiamin, riboflavin, and niacin and are a good source of minerals such as potassium, calcium, and iron, as well as of dietary fiber. The burgeoning natural foods sector is promoting date consumption as a functional food and as possessing antioxidant qualities. Dates are included in the group of recommended healthy food choices for individuals suffering from type 2 diabetes. Each date cultivar has an individual profile of nutritional values. Table 1.2 gives the nutritional values for two common commercial cultivars: Deglet Noor and Medjool. Ahmed et al. (2014) provide a good current summation of research of the composition and nutritional value of date fruits, and Vayalit (2014) of the bioactive compounds and functional properties of the fruit.

1.7 Date Production in Africa and the Americas

Date cultivation in Africa is dominated by the countries across North Africa, from Morocco to Egypt. Deglet Noor from Algeria, Medjool from Morocco, and Zaghloul from Egypt are major commercial cultivars originating from North Africa. Egypt is predominant with date production that far exceeds that of all other countries of the continent combined and which made it, in 2012, the largest producer in the world. This group of countries includes the world's largest date producer, as well as the smallest, Djibouti, with reported production of 86 mt in 2012.

North Africa suffers from two major pest and disease problems which are present to varying degrees depending upon the country. The most serious threat is posed by bayoud disease, a fungus (*Fusarium oxysporum* f. sp. *albedinis*), which originates from southern Morocco and has spread eastward to Algeria and south into Mauritania. Bayoud is extremely difficult to control because it is a soilborne fungus. Control measures include restricting the transport of date palms out of infected areas and destruction of infected trees. Research into selecting and breeding trees with resistance has shown some encouraging results, but the disease represents a major threat to date palm growing worldwide. In recent decades, the red palm weevil (*Rhynchophorus ferrugineus*) has spread westward out of Asia through the region and crossed the Mediterranean Sea into southern Europe, largely by the

Table 1.2 Nutritional value of Deglet Noor and Medjool cv. dates (nutrient values and weights are for edible portion)

Nutrient	Unit	Value per 100 g	
		Deglet Noor cv.	Medjool cv.
Proximates			
Water	g	20.53	21.32
Energy	kcal	282	277
Protein	g	2.45	1.81
Total lipid (fat)	g	0.39	0.15
Carbohydrate	g	75.03	74.97
Fiber, total dietary	g	8.0	6.7
Sugars, total	g	63.35	66.47
Minerals			
Calcium, Ca	mg	39	64
Iron, Fe	mg	1.02	0.90
Magnesium, Mg	mg	43	54
Phosphorus, P	mg	62	62
Potassium, K	mg	656	696
Sodium, Na	mg	2	1
Zinc, Zn	mg	0.29	0.44
Vitamins			
Vitamin C, ascorbic acid	mg	0.4	0.0
Thiamin	mg	0.052	0.050
Riboflavin	mg	0.066	0.060
Niacin	mg	1.274	1.610
Vitamin B_6	mg	0.165	0.249
Folate, DFE	μg	19	15
Vitamin A, RAE	μg	0	.7
Vitamin A, IU	IU	10	149
Vitamin E	mg	0.05	–
Vitamin K	μg	2.7	2.7

Source: US Department of Agriculture National Nutrient Database for Standard Reference, Basic Reports 9087 and 9421, accessed 18 Sept 2014

shipment from North Africa of unknowingly infected date palms destined for ornamental use.

A considerable number of seedling dates exist in oases in North Africa, the Sahel and Southern Africa, providing subsistence and/or famine food for local people. Where seedling dates are present, there exists the potential for enhancing production with better cultivars and the adoption of best practices for cultivation. Seedling dates are being studied for their resistance to bayoud disease, a trait which could potentially be transferred to elite date cultivars.

North Africa is the presumed origin of date palm seed conveyed to Spain where plantings were established and later seed carried to the New World during the Spanish conquest of the sixteenth century. In coastal Peru and Baja California, in northwestern

Mexico, climatic conditions were conducive to date cultivation, and several oases of seedling dates were established and have persisted to the present day. In the New World, modern plantations from offshoots of elite cultivars from North Africa and the Middle East were established in the USA early in the twentieth century.

Three African counties (Table 1.1) are not discussed specifically elsewhere in this volume: Benin, Kenya, and Namibia. Mexico, a minor date producer, is likewise not covered. Below are brief accounts of the information available about date production in each of these countries.

1.7.1 Benin

According to an FAO estimate, Benin produced 1,300 mt of dates in 2012. The Alibori and Atakora departments in the northernmost part of the country have a climate suitable for date growing. No published information could be been found on date production in Benin.

1.7.2 Kenya

Random introductions of date palm to Kenya occurred beginning in the early twentieth century from seeds originating from Iraq, Bahrain, the USA, Aden, and Pakistan and test planted at various locations. Subsequent introduction of offshoots of named cultivars was brought in from Pakistan to Turkwell, Turkana District, which is at present the largest concentration of date palms in the country. Among the cultivation problems encountered were minimal offshoot production from imported trees, poor survival of separated offshoots, inadequate crop management practices, irrigation system failures, nonflowering male palms, and complications of communal land ownership (Wasilwa et al. 2007). Records of introductions appear to have been poorly kept; Gammell (1989) reported that only 6 of 15 cultivars imported from Pakistan in 1974 which showed promise could be reidentified some 3 years later. These were Mazwati, Chakri, Bagum Jangi, and Assil under their Pakistani names and Kaekerongole and Ome, local names given to unidentified introduced cultivars. Kenya has suitable climatic areas for date cultivation and could become a much larger producer if technical and social constraints can be overcome. In 2012, date production amounted to 1,100 mt.

1.7.3 Namibia

This southwest African country has climatic conditions appropriate for date cultivation and was the venue of an international symposium in 2000 to promote the crop within the country (Min Agr 2000); the Namibia Development Corporation owns and runs the plantations. Modern plantations were established from tissue-cultured plants, with emphasis on Medjool and Barhi cvs., with fruit chiefly for export. Based on general information from the Internet, there are at least three date plantations in Namibia; two sites named are Eersbegin and Naute. Unseasonal rainfall and alkaline soil conditions are mentioned as problems of date production. Apparently, there has not been a recent scientific study of the situation in Namibia. Based on production data, it is obvious that there have been difficulties. According to FAO, in 2003 Namibia's estimated production was 651 mt, but in 2008 and 2009, official production was 333 and 276 mt, respectively. Estimated production in 2012 amounted to 400 mt. Namibia certainly has the potential to market khalal stage Barhi and tamar stage Medjool fruits to the Middle East and Europe in the off-season.

1.7.4 Mexico

The early date palm oases established during the Spanish colonial period have persisted in certain locations in the Baja California Peninsula and provide small quantities of fruit for local consumption (Johnson et al. 2013). Modern date production in Mexico did not begin until the late 1960s when plantations were established in extreme northwestern Sonora State, immediately south across the international border from Yuma, Arizona. The new plantings were modeled after cultivation practices in the USA, which was also the source of offshoots. The major cultivar grown is Medjool. There is favorable potential for expansion of Mexican date growing given local water resources and lower labor costs than in the USA. Fruit production in 2012 amounted to 6,012 mt.

1.8 Conclusions and Prospects

The chapters in this volume describe date production in 16 African countries and 3 in the Americas. Although the date palm has been dispersed throughout the African continent, significant commercial production is yet to be realized in those countries to the south of the traditional producers along the Mediterranean Sea. Introduction of the date palm to the Americas has led to commercial production in the USA and Mexico. However, the magnitude of production is relatively small (the USA produced 28,213 mt in 2012) as measured against the countries in North Africa and the Middle East such as Egypt, Iran, and Saudi Arabia, each with production in excess of 1 million metric tons.

Three distinguishing characteristics of date production are represented in Africa and the Americas. One concerns the fact that in the dispersal of date palms to regions far distant from the core production areas (i.e., North Africa and the Middle East), major pest and disease problems such as bayoud disease and the red palm weevil are avoided. Neither Southern Africa nor the Americas have recorded these problems in date-growing areas. With vigilance, these two serious plagues of the date palm can be prevented. A second related issue is that areas free of these two organisms are safe sources of uninfected offshoots for expanding production anywhere in the world. The third issue concerns genetic resources. Climatic and soil conditions in Southern Africa and the Americas are unlike those found in the core production areas. Over generations of seedling dates propagating by themselves in remote areas, new forms have certainly arisen and may represent valuable germplasm resources for traditional and molecular breeding purposes.

Issues one and two above also give an opportunity to expand date cultivation in these remote areas and, in southern hemisphere locations, export dates to northern hemisphere markets in the off-season.

References

Ahmed J, Al-Jasass FM, Siddiq M (2014) Date fruit composition and nutrition. In: Siddiq M et al (eds) Dates: postharvest science, processing technology and health benefits. Wiley Blackwell, Chichester, pp 261–283

Barreveld WH (1993) Date palm products, Agricultural services bulletin 101. FAO, Rome

Barrow S (1998) A revision of phoenix. Kew Bull 53(3):513–575

Bonavia E (1885) The future of the date palm in India. Thacker Spink and Co., Calcutta

Cong Int Datte (1951) Congrès International de la Datte. Bull Infor 38, Office Tunisien de Standardisation, Tunis

DGIAR (1924–1979) Date growers' institute annual report, vol 1–54. Valley CA, Coachella

Dowson VHW (1921, 1923) Dates and date cultivation in the 'Iraq. Parts I,II,III. Heffer & Sons, Cambridge UK

Dowson VHW (1982) Date production and protection, vol 35, Plant production and protection paper. FAO, Rome

Dowson VHW, Aten A (1962) Dates – handling, processing and packing, vol 72, Agricultural development paper. FAO, Rome

Dunham WA (1948) It's a date. Publication Press, Pasadena

FAO (1981–1988) The date palm journal. Regional Project for Palm & Dates Research Centre in the Near East & North Africa, Baghdad

FAOSTAT (2012) Food and Agricultural Organization Corporate Statistical Database. FAO, Rome

Gammell T (1989) Date palms in Kenya. Acts Press, Naibobi

Gros-Balthazard M (2013) Hybridization in the genus Phoenix: a review. Emir J Food Agr 25(11):831–842

ICARDA (2011) Development of sustainable date palm production systems in the Gulf Cooperation Council countries. Project Report 1, International Center for Agricultural Research in the Dry Areas, Aleppo

INRA (2011) Atlas du palmier dattier au Maroc. Institut National de la Recherche Agronomique. In French and Arabic: Rabat, Morocco

Jain SM, Al- Khayri JM, Johnson DV (eds) (2011) Date palm biotechnology. Springer, Dordrecht

Johnson DV (2010) Worldwide dispersal of the date palm from its homeland. Acta Hort 882:369–375

Johnson DV, Al-Khayri JM, Jain SM (2013) Seedling date palms (Phoenix dactylifera L.) as genetic resources. Emir J Food Agric 25(11):809–830

Kader AA, Hussein AM (2009) Harvesting and postharvest handling of dates. ICARDA, Aleppo

Manickavasagan A, Essa MM, Sukumar E (eds) (2012) Dates: production, processing, food, and medicinal values. CRC Press, Boca Raton
Mason SC (1923) The Saidy date of Egypt: a variety of the first rank adapted to commercial culture in the United States, vol 1125, USDA Bull. USDA, Washington DC
Mazid A, Al-Hashimy MJ, Zwain A, Haddad N, Hadwan H (2013) Improved livelihoods of small-holder farmers in Iraq through integrated pest management and use of organic fertilizer. Socioeconomic and Agricultural Policy Report, Working Paper 11, ICARDA, Beirut
McCubbin MJ (2007) The South African date palm industry – strengths and weaknesses. Acta Hort 736:53–57
Min Agr (2000) Proceedings of the date palm international symposium, Windhoek, Namibia 22–25 February 2000. Namibia Ministry of Agriculture, Water and rural Development, Windhoek
Min Agr Saudi Arabia (2006) The famous date varieties in the Kingdom of Saudi Arabia. Ministry of Agriculture, Riyadh. In Arabic and English
Munier P (1973) Le palmier-dattier. Maisonneuve & Larose, Paris
Nixon RW, Carpenter JB (1978) Growing dates in the United States, vol 207, Agr Infor Bull. USDA, Washington DC
Paulsen ME (2005) The amazing story of the fabulous Medjool date. Mark Paulsen Press, Tualatin
Pereau-Leroy P (1958) Le palmier dattier au Maroc. Ministère de l'Agriculture, Rabat
Popenoe PC (1973) The date palm. Field Research Projects, Coconut Grove, Miami
Pruessner AH (1920) Date culture in ancient Babylonia. Am J Sem Lang Lit 36:213–232
Sem Dattier (1931) Semaine du dattier – compte rendu general. Imp Imbert, Algiers
Siddiq M, Aleid SM, Kader AA (eds) (2014) Dates: postharvest science, processing technology and health benefits. Wiley Blackwell, Chichester
Simon H (1978) The date palm: bread of the desert. Dodd Mead & Co., New York
Swingle WT (1904) The date palm: its utilization in the southwestern states, vol 53, Bur Plant Ind Bull. USDA, Washington DC
Vayalit PK (2014) Bioactive compounds, nutritional and functional properties of date fruit. In: Siddiq M et al (eds) Dates: postharvest science, processing technology and health benefits. Wiley Blackwell, Chichester, pp 285–303
Wasilwa LA, Esilaba AO, Moi T, Wasike VW, Kubaso S (2007) The current status of date palm production and utilization in Kenya. Afr Crop Sci Conf Proc 8:2209–2212
Zaid A (ed) (1999) Date palm cultivation. Plant production and protection paper 156. FAO, Rome
Zaid A (ed) (2002) Date palm cultivation. Rev. ed. Plant production and protection paper 156. FAO, Rome

Chapter 2
Biodiversity, Genetic Diversity, and Genetic Resources of Date Palm

Abdullah A. Jaradat

Abstract *Phoenix dactylifera* L. is composed of genetically discrete clones representing thousands of cultivars without the benefits of a dynamic mutation-recombination system; its genetic resources are the most important component of biodiversity in its natural habitats; these include modern cultivars, landraces, obsolete cultivars, breeding lines, and related wild species. Cultivated *Phoenix* is closely related to a variable aggregate of wild and feral palms distributed over a wide desert belt across the Middle East and North Africa. Genetic diversity and genetic structure of the species gene pool complex have been shaped and greatly altered by human and natural selection, clonal propagation, and spatiotemporal exchange of germplasm. The mixed sexual-clonal propagation system acted on the sexual traits, impacted the genetic structure of populations, and may have resulted in the accumulation of domesticated traits in the date palm. Traditional oases continue to play a vital role in the maintenance and enrichment of date palm biodiversity, genetic diversity, and genetic resources through multiple processes and dynamic conservation practices. However, with the advent of modern plantations emphasizing elite cultivars, a better understanding of the intraspecific genetic variation of date palm and its distribution in oasis agroecosystems is essential for the conservation and sustainable utilization of its biodiversity and genetic resources. In-depth assessment of the genetic vulnerability of date palm to biotic and abiotic stress requires knowledge of the extent and distribution of its genetic diversity, both of which depend on the species evolution and its unique breeding system, past genetic bottlenecks, and ecological, environmental, and anthropogenic factors.

Keywords Biodiversity • Differentiation • Genetic diversity • Genetic erosion • Genetic resources • Population structure

---

A.A. Jaradat
Department of Agronomy and Plant Genetics, USDA-ARS & University of Minnesota, 803 Iowa Ave., Morris, MN, USA
e-mail: abdullah.jaradat@ars.usda.gov, jarad001@umn.edu

J.M. Al-Khayri et al. (eds.), *Date Palm Genetic Resources and Utilization: Volume 1: Africa and the Americas*, DOI 10.1007/978-94-017-9694-1_2

2.1 Introduction

The scientific name of the date palm derives from the Greek meaning for color (*Phoenix*, purple or red) and shape (*dactylifera*, finger-like appearance) of its fruit. The cultivated date palm (*Phoenix dactylifera* L.) belongs to the Arecaceae which includes 183 genera and more than 2,400 species. It is widely distributed over a broad range of climatic conditions mostly between 44° N and 44° S. The family is comprised of two subfamilies, three tribes, and five subtribes (Table 2.1). *Phoenix dactylifera* L. is one of five domesticated species within the Arecaceae; the other four species are the areca or betel nut palm (*Areca catechu*), the peach palm (*Bactris gasipaes*), the coconut palm (*Cocos nucifera*), and the oil palm (*Elaeis guineensis*) (Dransfield et al. 2008). In spite of species differences, there are a few similarities as to some leaf and trunk characteristics. Primarily domesticated for its fruit, however, almost every part of the date palm provides one or more ecosystem services of economic or environmental value. Over the past centuries, some hundreds of uses have been found for different parts of the date palm. It is a source of food and shelter for the desert dwellers. The needs for the date palm are such that "It must have its feet in the running water and its head in the fire of the sky."

The natural distribution of date palms covers a belt that stretches from the Indus Valley in the Indian subcontinent in the east to the Atlantic Ocean of North Africa in the west; however, the main region of date palm cultivation and production is confined to a rainless belt of the deserts south of the Mediterranean and in the southern parts of the Middle East. As a horticultural fruit tree, the date palm is a dioecious perennial and bears fruit for 40–50 years or more.

All species in the genus *Phoenix* are dioecious, suggesting the existence of a common dioecious ancestor before speciation of the genus. The earliest records of its cultivation date from about 7,000 years ago in Eridu in southern Mesopotamia; however, cultivation probably began thousands of years earlier (Zohary and Hopf 2000). The long history of date palm domestication with an unknown origin, and the nature of its culture may have played an important role in the composition of its genome (Tengberg 2012). As a vegetatively propagated perennial fruit tree, the date palm is unique in that it is composed of genetically discrete clones representing highly heterozygous cultivars without the benefits of a dynamic mutation-recombination system. For breeding, selection, and genetic manipulation purposes, the date palm can be propagated by seed, offshoots, and tissue culture or other micro-propagation techniques (Zaid and de Wet 2005).

Table 2.1 Taxonomic relationships in the Arecaceae

Family	Subfamily	Tribe	Subtribe	Scientific name	Common name
Arecaceae	Aricoideae	Areceae	Arecinae	*Areca catechu*	Areca
		Cocoseae	Attaleinae	*Cocos nucifera*	Coconut
			Bactrinidae	*Bactris gasipaes*	Peach palm
			Elaeidinae	*Elaeis guineensis*	Oil palm
	Coyphoideae	Cryosophila	Phoeniceae	*Phoenix dactylifera*	Date palm

2.2 The Genus *Phoenix*

The genus *Phoenix* includes the date palm (*Phoenix dactylifera* L.) and is the only member of the tribe Phoeniceae of the family Arecaceae (Table 2.1). The *Phoenix* spp. have either single or multiple trunks; these range in height from rudimentary to 30 m or more. Several phenotypic traits distinguish *Phoenix* spp. from other palms, including feather-type leaves, modification of the basal leaflets into spines, the presence of a terminal leaflet, and a central crease or ridge on the leaflets. The dioecious inflorescences of *Phoenix* spp. come up among the leaves. The small, pale, yellowish flowers are borne singly; female flowers have three carpels, only one of which matures, whereas male flowers generally have six stamens. The fruit of *Phoenix* spp. is a drupe of variable sizes and a single grooved seed (Masmoudi-Allouchi et al. 2009).

The taxonomy of the genus *Phoenix* has not been well established in the literature until relatively recently (Dransfield et al. 2008). There has been disagreement between various taxonomic treatments and some confusion about species names and their validity. *Phoenix* spp. hybridize readily, which has led to the suggestion that the genus *Phoenix* is monotypic (Wrigley 1995). The ready interspecific hybridization of *Phoenix* spp. can lead to uncertainty, especially when several species are present in close proximity. In cultivation, *Phoenix* spp. are often mislabeled and in some cases are obvious hybrids (e.g., in the Canary Islands; González-Perez et al. 2004) or off-types (Wrigley 1995).

Phoenix dactylifera is the main Middle Eastern wild representative of its genus which comprises at least 12 species distributed over parts of Africa and south Asia. The only other wild date which occurs in the east Mediterranean basin is *P. theophrasti*, a narrow endemic confined to the island of Crete and several spots in coastal southwest Turkey. The cultivated date palm is closely related to a variable aggregate of wild and feral palms distributed over the southern warm and dry parts of the Middle East as well as the northeastern Saharan and north deserts of *Arabia Deserta* (interior of the Arabian Peninsula). These spontaneous dates show close morphological similarities and parallel climatic requirements with the cultivated clones. In addition, they are interfertile with the cultivars and are interconnected with them by occasional hybridization. Botanists place these wild dates with *P. dactylifera* L. The wild species, similar to the cultivated date palm, produce basal suckers but differ from them by their smaller and mostly inedible fruits; these contain relatively little pulp and are mostly indigestible. Sexual reproduction is the rule in the wild species, while cultivation has brought a shift to vegetative propagation which led to the fixation of desired highly heterozygous female clones (see Sect. 2.3.1). Apparently, there is little information available about the status of these species in the wild; their habitats are in many cases threatened by multiple anthropogenic and environmental factors (Chao and Krueger 2007; Munier 1981; Popenoe 1913). In addition, most of these species are lacking characterization and evaluation as potential genetic resources, and it is highly likely that the diversity of some species is threatened by genetic erosion (Krueger 1998, 2011).

Although 19 species of *Phoenix* have been named (Table 2.2), most taxonomic treatments list 12 as valid species names, with *P. acaulis*, *P. canariensis*, *P. dactylifera*, *P. paludosa*, *P. reclinata*, *P. rupicola*, and *P. sylvestris* as widely accepted species. The geographic distribution pattern of seed shapes points to human dispersal routes that spread cultivation of date palm from one or more initial domestication centers. Recent genetic data suggest that the cultivated date derives from wild populations of *P. dactylifera*; however, gene flow from cultivated to wild date palms could complicate the identification of true wild *P. dactylifera* populations. The geometric morphometry of modern and ancient seed differentiated *Phoenix* species (e.g., *P. dactylifera*, *P. canariensis*, *P. reclinata*, *P. sylvestris*, and *P. theophrasti*) at different taxonomic, geographic, and chronological levels (Terral et al. 2012).

2.2.1 The Wild Phoenix *Species*

Spontaneously growing dates occur in almost the entire area of date cultivation, and they frequently represent secondary escapees. However, in some areas in the Middle East and probably also in the northeastern Sahara desert, in *Arabia Deserta*, and

Table 2.2 Summary of palm species in the genus *Phoenix*

Species	Synonym(s)	Common name	Distribution	Notes
P. acaulis			North India	Stemless
P. anadamensis			Bay of Bengal	Semidwarf; single trunk
P. atlantica			Atlantic shores, Morocco	
P. caespitose	*P. arabica*		Somalia, Arabia	Edible fruits
P. canariensis	*P. cycadiflora*	Canary palm	Canary Islands	Edible fruits, single trunk
P. chevalierii		Barbary palm	SE Spain	
P. dactylifera		Date palm	Middle East, North Africa	Fruit, trunk, leaves
P. iberica			Spain	
P. loureiri	*P. formosana*		Indochina, Malaysia	Edible fruits, dwarf
P. paludosa			Indochina, Malaysia	Edible fruits, semidwarf
P. pusilla	*P. farinifera*		South India, Sri Lanka	Edible fruit, trunk, leaf
P. reclinata	*P abyssinica*	Senegal palm	South Arabia, Africa	Ornamental, fruits
P. roebelenii		Pygmy palm	South China	Ornamental
P. rupicola		Cliff palm	North India	Animal fodder
P. sylvestris		Rain palm	India, Pakistan	Sap for sugar
P. theophrasti		Cretan palm	Crete, coastal Turkey	

Adapted from Barrow (1998) and Krueger (1998, 2011)

Baluchistan, dates are genuinely wild and occupy primary habitats. Prominent among these locations is lowland Khuzestan at the southern base of the Zagros mountain range in Iran as well as the southern part of the Dead Sea basin in Jordan. In these warm, dry locations, wild *Phoenix dactylifera* palms with their characteristic small and mostly inedible fruits thrive in gorges, wet rocky escarpments, and seepage areas in wadi beds and near brackish springs where they constitute a conspicuous element in the vegetation (Fig. 2.1). This variable complex of wild forms, segregating escapees, and cultivated clones of *P. dactylifera* coexisted for millennia, and all of its components are genetically interconnected by occasional hybridization. Although it is impossible to delimit the preagriculture distribution of wild

Fig. 2.1 Feral date palms growing in wadi bed (**a**), rocky (**b**), and desert (**c**) habitats (Photos: Google Earth 2013)

Fig. 2.1 (continued)

Phoenix, there is little doubt that the wild *P. dactylifera* forms are indigenous to the warm and dry parts of the Middle East (Zohary and Hopf 2000).

Three wild *Phoenix* spp. grow on the fringe of traditional date cultivation in the Old World, and they may have enriched the gene pool of the cultivated fruit trees through spontaneous hybridization. These are *P. atlantica* which grows near the Atlantic shore of North Africa, and it is apparently involved in the formation of some Moroccan date cultivars; bushy *P. reclinata* occurs in south Arabia and in Africa south of the Sahara, and it probably hybridizes with *P. dactylifera* on the southern fringe of date cultivation; finally on its eastern border, the Indus Valley, the cultivated date comes in contact and occasionally hybridizes with the wild *P. sylvestris*, a nonsuckering tall *rain palm* not adapted to deserts but to wetter, tropical climates (Chevalier 1952; Dransfield et al. 2008; Krueger 1998, 2011).

Phoenix dactylifera was well suited for the shift from sexual to vegetative propagation since it produces easily transferable suckers or offshoots. It is often difficult to decide whether noncultivated material is genuinely wild or whether it represents weedy forms or secondary seedlings derived from cultivated clones. Also, it may be impossible to delimit the preagriculture distribution of the wild date palm; however, there is strong evidence suggesting that the wild *P. dactylifera* forms are indigenous to the warm and dry parts of the Middle East. Although 19 *Phoenix* species have been named, most taxonomic treatments considered 12 of these species as taxonomically valid. There is a general agreement that *P. acaulis, P. canariensis, P. dactylifera, P. paludosa, P. reclinata, P. rupicola,* and *P. sylvestris* are widely accepted as species (Barrow 1998; Krueger 1998, 2011).

Phoenix dactylifera is considered as the main Middle Eastern representative of its genus which comprises 12 recognized species distributed over Southwest Asia and Africa. The only other wild date palm which occurs in the Eastern Mediterranean region is *P. theophrasti*, a narrow endemic confined to the island of Crete and several small coastal areas in Southwest Turkey. As a narrow endemic, *P. theophrasti* is considered a threatened species due to its restricted distributional range and due to a gradual loss of its natural habitat. Three wild *Phoenix* species grow on the fringes of traditional date palm cultivation in the Old World, and they may have enriched the gene pool of the cultivated species through spontaneous hybridization. These are *P. atlantica* which grows near the Atlantic shores of North Africa, and it is apparently involved in the formation of some Moroccan date palm cultivars; however, *P. atlantica* may represent only feral populations of *P. dactylifera* and may not be considered as a valid species (Popenoe 1913; Rivera et al. 2008). *Phoenix iberica* is a wild species growing in wadis near the Mediterranean coast of Spain; it has glaucous leaves, stout stems, and small fruit with thin flesh; its fruits are intermediate between *P. theophrasti* and *P. sylvestris*. Some of its vegetative traits resemble those of the cultivars Medjool and Barhi. *Phoenix chevalierii* (Barbary palm) is an Iberian-Moroccan group of cultivars growing in SE Spain and are well known for their greener leaves as compared with *P. dactylifera* (Rivera et al. 2008). Bushy *P. reclinata* is found in southern Arabia and in Africa south of the Sahara Desert and it most probably hybridizes with *P. dactylifera* on the southern fringes of date palm cultivation (Tengberg 2012). Finally, *P. sylvestris*, a non-suckering tall, *rain palm* not adapted to deserts but to wetter, tropical climates which comes in contact with the cultivated date palm in the Indus Valley, the eastern border of its distribution. *P. sylvestris* may reach 20 m in height and is tapped for its sap to make sugar and other products, and its leaves are used in basketry; it occasionally hybridizes with *P. dactylifera* (Barrow 1998; Chevalier 1952; Zohary and Hopf 2000).

Other wild species include *Phoenix canariensis*, which is endemic in the Canary Islands and is adapted to more moderate climatic conditions and cooler temperatures than some of the other *Phoenix* spp. (González-Perez et al. 2004); it has a stout single trunk of about 20 m in height, and it is widely planted as an ornamental in the USA and the Mediterranean region; *P. rupicola*, the cliff date palm, has a thin trunk of about 7 m in height, is native to northern India, and is considered as a horticultural palm; *P. pusilla* is native to southern India and Sri Lanka and is about the same size and general appearance as *P. rupicola* but with a shorter trunk. The fruit and the pith of its trunk are edible; *P. reclinata*, the Senegal date palm, is a variable species native to tropical Africa, with thin, clustering trunks that may reach 10 m in height, and is widely planted as an ornamental; *P. paludosa*, the mangrove date palm, another ornamental palm, is similar in appearance to *P. reclinata* and is native to swampy areas in southeast Asia; *P. acaulis*, native to northern India and Burma, has short, clumping stems; *P. roebelenii*, the pygmy date palm, is native to Southeast Asia and is grown as an elegant ornamental and has both single and clustering trunks; and *P. loueiri* is native to northern India and southern China and is a poorly understood species with short stems that is often confused with *P. acaulis* (Barrow 1998; Chevalier 1952; Zohary and Hopf 2000).

2.2.2 *The Domesticated Species:* Phoenix dactylifera

The date palm has been cultivated since antiquity, but its wild progenitor(s) were undoubtedly used by humans long before actual cultivation began. At the present day, however, it is difficult to distinguish between true, wild forms, escapees, and hybrids between wild and domesticated forms of the date palm (Goldschmidt 2013). Nevertheless, exploitation of the date palm by humans probably began as simple gathering of the fruits, fronds, trunks, and other usable parts of the tree in ancient Mesopotamia (Nixon 1951; Zohary and Hopf 2000) which is considered as the center of origin of the species. Selection of trees with superior phenotypic traits may have originated in ancient times, along with clonal propagation by offshoot.

The domesticated date palm (*Phoenix dactylifera*) is the tallest of *Phoenix* spp. reaching heights of more than 30 m. It has clustering trunks smaller in diameter than *P. canariensis* but larger than other *Phoenix* spp. In cultivation, it usually appears with a single trunk, as the offshoots are usually removed at a suitable growth stage and used in vegetative propagation of the mother plant. The fruit is the largest of any *Phoenix* spp. reaching up to 100 × 40 mm in size (e.g., in the cultivar Anbarah originated in Medina, Saudi Arabia). Date palms evolved in a unique manner (Wrigley 1995). The date palm grows well in sand, but it is not arenaceous; it has air spaces in its roots and may grow well where soil water is close to the surface, but it is not aquatic; it grows well in saline conditions, but it is not a true halophyte and does better if planted in higher-quality soil and irrigated with freshwater; and its leaves are adapted to hot, dry conditions, but it is not a xerophyte and requires abundant irrigation water.

2.3 Center of Origin and Center of Diversity

The earliest records of date palm cultivation date from about 7,000 BP in Mesopotamia; however, several lines of evidence point to the fact that its culture began thousands of years earlier. Five-thousand-year-old date seeds have been found in Mohenjo Daro in the Indus river valley, and in 4,500 BP, the date palm was used in the construction of the temple of the Moon god in Ur of lower Mesopotamia. However, some authorities (Barrow 1998; Wrigley 1995; Zohary and Hopf 2000) suggest that the specific center of origin of *Phoenix dactylifera* within lower Mesopotamia is somewhat unclear, and a multidisciplinary approach, combining evidence from historical, cultural, archaeobotanical, and ecological sources, is needed to resolve this issue. Nevertheless, ecological observations suggested that the eastern parts of *Arabia Deserta* may be considered as a primary or secondary center of origin of the date palm. It is argued that conditions of a hot, dry land with wet, saline soils to which date palms are adapted are found precisely in this part of *Arabia Deserta*. This viewpoint is supported by paleobotanical data and archaeological, historical, and cultural information (Meyer et al. 2012; Terral et al. 2012).

2.3.1 The Domestication Syndrome

A domestication syndrome in plants may include a combination of a few but different traits, including seed retention (e.g., in grasses, legumes), increased fruit size, changes in branching and stature, changes in reproductive strategy, and changes in secondary metabolites. Perennial fruit trees, such as the date palm, exhibit significantly fewer domestication syndrome traits than annual crops, and their domestication may have occurred more slowly compared with annual crops because fewer sexual generations occur in a given period of time (Meyer et al. 2012). The domestication syndrome in date palm, as well as in other clonally propagated crops, has been poorly understood. Man-made selection obviously acted on the genetic diversity available in the wild progenitor of the date palm to increase yield, fruit size, and palatability, to reduce but not totally eliminate branching (offshoots), and to facilitate vegetative propagation. It is assumed that the mixed sexual-clonal propagation system (see Sect. 2.3.2) acted on the sexual traits, impacted the genetic structure of populations, and may have resulted in the accumulation of domesticated traits in the date palm (Masmoudi-Allouchi et al. 2009; Meyer et al. 2012; Nixon 1951). In addition, a critical change in habitat was decisive for the final domestication of the date palm. As a consequence, the genetic architecture of the domesticated date palm made it dependent on human interference and on the oasis as a man-made habitat (Popenoe 1913; Zohary and Hopf 2000).

Whether conscious or unconscious selection was behind date palm domestication, several steps were involved, including the shift from sexual to vegetative reproduction which was vital for the preservation of selected phenotypes but resulted in narrowing the genetic base and loss of biodiversity that may have endangered further breeding improvement (Zohary 2004). Vegetative reproduction reduced the duration of the juvenile growth phase and led to fruiting at an early age; it caused the breakdown of natural pollination system which imposed a serious limitation on fruit production of the date palm as a dioecious species. The wild-type mode of pollination was replaced by the more efficient method of artificial pollination that required a limited number of male trees (Barfod et al. 2011; Goldschmidt 2013).

A mixed sexual-clonal production system in the date palm prevented inbreeding depression, preserved a large amount of genetic diversity, and maintained a high adaptive potential which was used by early farmers for selection of horticulturally valuable date palm clones, whereas several mechanisms of the domestication process have led to spatiotemporal diversification of seed and fruit shape and of cultivars (Terral et al. 2012).

Very little is known about the productivity of the early domesticates of date palm because, historically, less attention has been paid to the domestication of fruit trees compared to field crops. However, it is postulated (Goldschmidt 2013) that there is an evolutionary continuum of productivity patterns among trees in their wild habitats, intermediate domesticates, and the most advanced domesticates in oases and that the alternate bearing may represent an intermediate step in the evolutionary pathway.

2.3.2 Domestication of Date Palm

The date palm was one of the first five fruit trees to be domesticated along with olive, grapevine, fig, and pomegranate as members of the *first wave* of domesticated fruit trees. Then, a few thousand years later, was followed by the domestication of almonds, apricots, and pistachios as members of the *second wave* of domesticates in the Old World (Zohary and Hopf 2000). These are either predominantly or obligatory outcrossing fruit tree species. Domestication of the date palm has led to the increase in fruit size and pulp quality and to a shift from sexual to vegetative propagation; the latter resulted in the immediate fixation of desirable fruit (and tree) traits in highly heterozygous female cultivars.

Date palm domestication, as was the case for most fruit trees, was a major positive change in the edibility of a wild, nonpalatable fruit brought about by a rare genetic event that would disappear in the wild without human intervention (Goldschmidt 2013; Zohary and Hopf 2000). Therefore, the study of its wild relatives is indispensible for the reconstruction of the domestication process and identification of the progenitors of the modern date palm (Goldschmidt 2013).

As mentioned earlier (see Sect. 2.3), the earliest records of domesticated date palm (i.e., under cultivation) in lower Mesopotamia go back to ~7,000 BP; however, there is ample evidence suggesting that date culture began thousands of years earlier. The ruins of Mohenjo Daro, along the Indus River in the Sind, yielded 5,000-year-old date palm seed, whereas during those times, the date palm was used in the construction of the temple of the Moon god in Ur in lower Mesopotamia. Also, the date palm is shown in the bas-reliefs at Nineveh, the capital of the Assyrian Empire, in northern Mesopotamia, and in the ruins of Palmyra, an ancient oasis in the heart of the Syrian Desert. From its center of origin, the date palm culture had spread, within a relatively short time after its domestication, to *Arabia Deserta, Arabia Felix* (southern Arabia), the Fertile Crescent, Egypt, North Africa, and the western parts of the Indian subcontinent.

2.3.3 Evolution Under Domestication

Crop evolution under domestication generally has led to increased productivity and adaptation of crop species, but narrowed their genetic basis (Dhanapal 2012; Goldschmidt 2013). Date palm evolved under domestication in a manner unlike most other fruit trees, especially those growing under its canopies in oasis agroecosystems. The tree has characteristics that adapted it to varied agroecological and edaphic conditions but differed from many other fruit trees that are found under the same conditions as indicated in Sect. 2.2.2. Diversity of the wild progenitor and wild relatives, the diverse ecologies of domesticated date palm, and the pressure of an age-old and intricate mix of natural and man-made selection resulted in complex and diverse evolutionary trajectories of the date palm under domestication.

The interaction between sexual reproduction and clonal propagation must have shaped the evolutionary dynamics of the date palm (Charlesworth 2013; Gschwend et al. 2012). Sexual reproduction was not totally replaced by clonal propagation and has been used by horticulturists since its domestication. This practice is more common today in traditional oasis agroecosystems and resulted in a mixed clonal-sexual reproductive system under which date palm populations become a mixture of interlinked subpopulations (Honnay and Jacquemyn 2008; Younis et al. 2008). Sex became a factor in the reproductive system when early horticulturists decided to select, save, transplant, and then clonally propagate *outstanding* volunteer seedlings they observed upon seed germination under mother trees. This practice allowed them to exploit the advantages of both sexual reproductive and clonal propagation systems (Jaradat 2011; Zaid and de Wet 2005).

Obviously, the evolution of many plant species, including the date palm, under strict clonal propagation is more dynamic than is previously thought (Honnay and Jacquemyn 2008; McKey et al. 2010). Somatic mutations, for example, are so frequent that strict genetic identity of clones is not guaranteed over many generations. These mutations create a necessary genetic variation that contributes to adaptive evolution of clonally propagated plant species. The effect of biased clonal propagation of female plants on the reproductive biology of the date palm led to the development of specialized cultural practices (i.e., hand pollination) to counter male limitations. In date palm, as well as in other dioecious plants, male limitations may have interacted with gender-dependent cost of sex. In date palm and several dioecious plants, the cost of flowering in males is lower than the cost of flowering and fruiting in females. Nevertheless, selection for clonal propagation has not always modified sexual fertility or the mating system in dioecious plants. Some date palm cultivars have evolved parthenocarpy, but because parthenocarpic fruits are smaller and grow more slowly than fruits of hand-pollination, this laborious cultural practice persisted for millennia and guaranteed the production of high-quality dates (Honnay and Jacquemyn 2008; McKey et al. 2010). In ancient Mesopotamia some 5,000 years ago, the maximum yield of artificially pollinated date palm trees was estimated at about 105 kg per tree which is in the range of normal productivity of today's date palm cultivars (Goldschmidt 2013).

The initial spread of date palm germplasm was probably made by seed, which was much easier to transport than offshoots before the domestication and use of the camel in transporting offshoots between scattered habitable desert locations. Vegetative propagation and increase of selected clones could have been practiced after seedling populations were established in what became an oasis. This gave rise to the many local cultivars that are found in oases scattered all over Middle Eastern and North African countries (Chao and Krueger 2007; Krueger 1998, 2011). The greater role of hybridization in date palm evolution may have been triggered by the discovery of its sexuality. Early horticulturists must have collected pollen from *elite* male plants and performed laborious artificial pollination of female plants. Natural solutions to the difficulties associated with artificial pollination were provided by the genetic shift from dioecism to hermaphroditism in grapes (*Vitis vinifera*) and parthenocarpy in figs (*Ficus carica*), but not in date palm, yet (Barfod et al. 2011; Goldschmidt 2013).

The discovery of the effects of xenia and metaxenia on the developing fruit must have led to further improvements in the size, shape, texture, color, and nutritional value of the date. The earliest record of artificial pollination in palm trees was found in the cuneiform texts of Ur in lower Mesopotamia some 4,500 BP and later from bas-reliefs in Ashurnasirpal palace in Nimrud about 3,000 BP. Seed morphological diversity, based on geometric measurements of modern and ancient date palm seed, showed a complex structure and appeared not to be structured according to the geographical origin of date palm cultivars; it reflected the scale of the dispersal routes along which early horticulturists and traders have spread cultivation techniques and cultivars through transport of seeds and clones over the last 5–6 thousand years and possibly through hybridization between genotypes from different regions in the Old World (Terral et al. 2012).

2.4 Biodiversity of Date Palm

Intraspecific genetic variation provides the basis for any evolutionary changes and is thus the most important component of biodiversity (Pauls et al. 2013). Biodiversity of the date palm, or any other crop, is comprised of genetic, organismal (i.e., plant), and ecological diversities, whereas genetic diversity represents heritable variation within and between wild or domesticated populations. On the other hand, phenotypic diversity represents the interaction effect between genetic diversity and the environment, and it is an apparent indicator of date palm diversity. The latter represents the basis for selection and conservation, as well as for date palm improvement for sustainable utilization. Biodiversity of date palm is a prerequisite for the proper functioning of the oasis ecosystem, which is a complex ecosystem characterized by horticultural, agronomic, ecological, economic, social, and cultural dimensions; it represents the climax of rigorous management of scarce water and land resources in alliance with the date palm. The date palm is the dominant component upon which the sustainable biophysical and socioeconomic structures of the oasis ecosystem are based (Jaradat 2011).

Throughout its long history, the date palm represented a powerful example of integrating sustainable use of renewable material resources, including soil, water, inputs, and outputs. All parts of the date palm, except perhaps its roots, are used for a purpose best suited to them. The date palm not only provides a concentrated energy food, it also creates a more amenable habitat and provides shade and protection from the desert wind and heat. In addition, the date palm yields a variety of products for use in agricultural production and for domestic utensils. It is said that there are as many uses for dates as there are days in the year. The reproductive phenological cycle varies greatly for various cultivars planted in the same or different oases. This allows for extended harvest season with dates suitable for fresh consumption, storage, and processing. One or a few elite cultivars usually dominate in certain oasis because of their high fruit quality or because of their early or late maturity; they fill a special market niche. For example, of some 180 cultivars producing over 230,000 mt annually, only 10 cultivars produce 80 % of dates in Oman (Alyahyai and Alkhanjari 2008).

Changes in biodiversity that alter the oasis agroecosystem function have economic impacts through the provisioning of goods and services to society. Large genetic diversity within the oasis agroecosystem can be expected to give rise to ecosystem stability; however, diversity is not the driver of this relationship; rather, ecosystem stability depends on the ability of the oasis to contain different species or functional groups (e.g., different cultivars of date palm, different species and cultivars of fruit trees, forage crops, annual grain crops, vegetable crops, semidomesticated crops, weedy relatives of crops, etc.) that are capable of differential responses to biotic and abiotic stresses and to different management practices. For example, (agro)biodiversity in the ancient Siwa Oasis in Egypt was described as being stable since the early nineteenth century (Battesti 2013); however, establishing a list of different landraces of date palm in that oasis proved to be far more difficult, even though only ~15 named date palm cultivars, mostly landraces, coexisted with other fruit trees and field crops in that oasis (Nabhan 2007). Oasis biodiversity at the species level has functional consequences because the number and kinds of species, besides *Phoenix dactylifera*, present determine the traits that influence a large number of processes within, and services provided by, the oasis agroecosystem.

The changes leading to the collapse of a few oases, such as Sijilmasa in southern Morocco and Timbuktu in Mali, can occur suddenly, although they often represent the cumulative result of a slow decline in biodiversity and reduced ecological resilience of the oasis. Currently, however, poor management is causing some oasis agroecosystems to pass ecological thresholds, leading to irreversible changes in the ecosystem and the loss of its services. The impact of modern irrigation techniques for human settlement in hyperarid regions, for example, is demonstrated by the large quantitative and qualitative changes in vegetation cover that have occurred in several Middle Eastern and North African oases over the past 50 years. The impact of this technology on natural vegetation is demonstrated by the disappearance of date palm orchards and associated biodiversity due to the depletion of desert aquifers (e.g., Wadi al-Ajal, Libya), abandonment, and declining oasis (e.g., Um Elma'a, southwestern Libya, Fig. 2.2) or to seawater intrusion (e.g., coastal regions of Ras al-Khaimah, UAE). Nevertheless, in-depth assessment of the genetic vulnerability of date palm to many threats (e.g., climate change, desertification, and salinity stress) requires knowledge of the extent and distribution of its genetic diversity, both of which depend on the species evolution and its unique breeding system, past genetic bottlenecks, and ecological, geographical, and anthropogenic factors (Pauls et al. 2013; Shabani et al. 2012). Therefore, a biodiversity-based model for sustainable agriculture in oasis agroecosystems may provide the most cost-effective and durable solution for the problems associated with or emanating from such factors.

Fig. 2.2 Abandoned oasis at Um Elma'a in the southwestern Libyan Desert (Photo: National Geographic)

2.5 Genetic Diversity of Date Palm

Genetic diversity is expressed as genetic differences between species, subspecies, cultivars, populations, or individual clones and may be measured at the morphological, physiological, biochemical, or molecular levels (see Sect. 2.5.1). Several measures are available to quantify genetic diversity within plant populations; these include, but are not limited to, the amount of polymorphism within populations, the allelic richness (i.e., the total number of alleles in the population), and the gene diversity or probability that two random copies of the gene will have dissimilar alleles and heterozygosity (i.e., the percentage of heterozygous genotypes in a population). So far, only a small part of the total genetic diversity, including date palm, has been characterized, evaluated, and used for crop breeding and improvement purposes (Dhanapal 2012). Sexual expression in palms is spatially separated at five discrete levels; these are within flowers (in between floral organs), within flower clusters (in between flowers), within inflorescences (in between partial inflorescences), within palms (in between inflorescences), and in between palms. The complexity of sexual expression in palms and its impact on genetic diversity only becomes clear when the spatiotemporal separation of male and female functions is considered as in the date palm (Masmoudi-Allouchi et al. 2009; Younis et al. 2008).

A meta-analysis of the relation between mating system, growth form, and genotypic diversity in clonal plant species, such as date palm, revealed the potential advantages of clonal growth; these advantages include: facilitation of resource uptake in heterogeneous environments, persistence under suboptimal environmental conditions, and increased attraction of pollinators by increased floral display (Honnay and Jacquemyn 2008). However, this strategy may incur fitness (i.e., fruit yield) costs, which are associated with the effects of large clonal individuals and floral displays on the patterns of pollen dispersal (excluding artificial pollination) and the rate of sexual reproduction (Iqbal et al. 2009).

Date palm populations are composed of genetically discrete clones representing highly heterozygous cultivars without the benefits of a dynamic mutation-recombination system. The strong artificial selection and clonal propagation of this unique clonally propagated perennial fruit tree in oasis agroecosystems greatly altered its original genetic structure. Selection, within and among date palm cultivars, is the primary force that shaped the levels and patterns of genetic diversity within and between date palm populations. Selection happens when certain individual trees in the population are more likely to survive to maturity and produce more offspring (either seedlings or offshoots) than other trees (Munier 1981; Nixon 1951).

Genetic diversity in date palm is not randomly or uniformly distributed in space or time. The amount of genetic diversity differs between oases and populations or between regions and localities; and several key historical, geographical, ecological, and anthropogenic factors determine its spatiotemporal distribution. The impact of these factors and their interaction is reflected on the level of population differentiation and especially on fruit quality traits (Alkhalifah et al. 2012; Battesti 2013). Genetic diversity and genetic structure of the gene pool complex of date palm (wild, feral, and domesticated) have been both shaped and greatly altered by natural and human selection, clonal propagation, and spatiotemporal exchange and movement of its germplasm. Traditional oases continue to play a vital role in the maintenance and enrichment of date palm genetic resources and their genetic diversity through multiple processes and dynamic conservation practices. However, a better understanding of the intraspecific genetic variation of the date palm and its distribution in the oasis ecosystems will be vital for the proper conservation and sustainable use of its genetic diversity. If properly designed and implemented, strategies for the study, productive conservation, and sustainable use of date palm (bio)diversity will minimize anthropogenic disturbance, interference, and impact; optimize ecosystem functions; and result in integrated protection of environmental resources of fragile oasis ecosystems (Jaradat 2011).

Traits related to vegetative and reproductive organs of date palm comprise a useful approach to analyzing its phenotypic diversity. Those related to leaflet length and spine, fruit, and seed size showed the highest discriminating power between cultivars (Ahmed et al. 2011). Accurate identification of, and discrimination between, cultivars requires the use of a large number of morphological markers and exploration of additional biochemical and molecular markers (See Sect. 2.5.1). Intracultivar stability of a few morphometric traits, including percent spined midrib, apical divergence angle, maximal pinnae width at leaf top, percent solitary spines, spine length at the middle and maximal spine angle, has been used for accurate cultivar identification. These traits helped cluster cultivars according to their fruit characteristics (Hammadi et al. 2009; Iqbal et al. 2009).

Estimates of genetic diversity, which is desirable for long-term date palm improvement and reduction of its vulnerability to biotic and abiotic stresses, can be compared to a minimum genetic distance which indicates that two cultivars are essentially the same (Elmeer et al. 2011). Accurate estimates of genetic diversity in traits of the trunk and crown, fruiting and fruit quality attributes, and their partitioning within and among the gene pool of date palm in its center of origin and center of diversity are important considerations for a successful date palm industry. This is particularly significant for the long-term survival of date palm plantations due to the long-life expectancy of each generation and, in particular, due to the high maintenance cost of mature female trees. Although the date palm may not be immediately threatened by genetic erosion in spite of documented isolated cases (Fig. 2.3), several reports indicated that the level of genetic diversity as to the number of cultivars in oases is declining due to interacting anthropogenic, biotic, and abiotic factors. In addition, the status of genetic diversity in North African countries is aggravated by the threat of destructive diseases such as the vascular fusariosis disease commonly known as *bayoud* and caused by *Fusarium oxysporum* f.sp. *albedinis* (Sedra 2013).

2.5.1 Analyses Techniques of Genetic Diversity

A rational and sustainable utilization of date palm genetic resources requires extensive characterization, evaluation, and documentation of the germplasm available at a local or regional level. Characterization is defined as assessment of the presence, absence, or degree of specific traits that are little influenced in their expression by varying environmental conditions, whereas evaluation is the assessment of plants at different levels of organization for potentially useful genetic traits, many of which may vary with changes in the micro- or macro-environment (e.g., pest or disease resistance, fruit quality, and flavor) (Gotor et al. 2008; IPGRI 2005).

Fig. 2.3 Loss of date palm cultivars (genetic erosion) due to salinity induced by the intrusion of sea water (**a**) and drought due to groundwater depletion caused by excessive flood irrigation and reduced recharge of underground aquifer (**b**) (Photos: AA Jaradat)

Discrimination between closely related cultivars and clones for genetic diversity studies is often extremely difficult. Identification of date palm cultivars is usually based on fruit morphology; however, morphological traits are often unreliable and may not precisely correlate with the genotype of the date palm. These traits often are influenced by environmental conditions or they vary with the developmental stage of the date palm. Nevertheless, a number of frond and leaflet morphological qualitative and quantitative traits of a selected number of elite cultivars have been reported to be stable and did not exhibit variation in response to environmental or management factors. Such morphological traits can be used as stable descriptors of date palm cultivars and for cultivar identification at any growth stage (Ahmed et al. 2011; Alyahyai and Alkhanjari 2008; Hammadi et al. 2009).

Biochemical markers (i.e., isozymes), due to their low level of polymorphism, appear to be of limited value in quantifying the genetic diversity of date palm populations. Allozyme electrophoresis can reveal genetic polymorphism allowing the direct study of genetic variation. However, there is a limit to the number of enzymes available for study (because of the requirements for detection) and therefore a limit to the proportion of the date palm genome which can be accessed. Due to their low level of polymorphism, allozymes appear to be of limited value in quantifying the genetic diversity of date palm populations. However, they have been extensively used prior to 2000 in genetic diversity of crops and fruit trees, but only a single study using allozyme markers was recently undertaken to assess levels of genetic diversity in 29 date palm cultivars belonging to three main date palm-growing regions (Ould Mohamed Salem et al. 2008). These researchers used data from starch and polyacrylamide gel electrophoresis of five polymorphic loci corresponding to four enzyme systems. Date palm was found to have high percentage of polymorphic loci, strong heterozygosity, and total genetic diversity. In addition, population differentiation between geographical groups was low. Multivariate analysis showed no well-defined structuring of cultivars in relation to their geographical origins. Twenty-seven of the 29 cultivars could be identified by one of the 28 multilocus genotypes observed in this study.

Genotyping is more reliable for cultivar identification and for genetic diversity analysis. Molecular markers are more precise and can accurately identify date palm cultivars and quantify their genetic diversity and phylogenic relationships. Molecular markers (i.e., DNA based) helped in the verification of accession identity and genetic contamination and also have been used to identify ecogeographic landraces within the domesticated or wild gene pool of several crop species (Dhanapal 2012; Gao et al. 2012). Several of these markers have been extensively used to study the genetic variation of date palm cultivars. These include randomly amplified polymorphic DNA (RAPD), amplified fragment length polymorphism (AFLP), and microsatellite markers (SSRs) (Ahmed et al. 2013; Elmeer and Mattat 2012; Fig. 2.4). The latter are locus specific, codominant, highly polymorphic, and highly reproducible (Akkak et al. 2009).

Each one of these methods has its own advantages and disadvantages and limitations; however, nuclear microsatellites seem to fulfill most requirements for an accurate analysis of date palm diversity and phylogeny (Khanam et al. 2012). The AFLP technology is sensitive enough to detect low levels of variation, allowing the

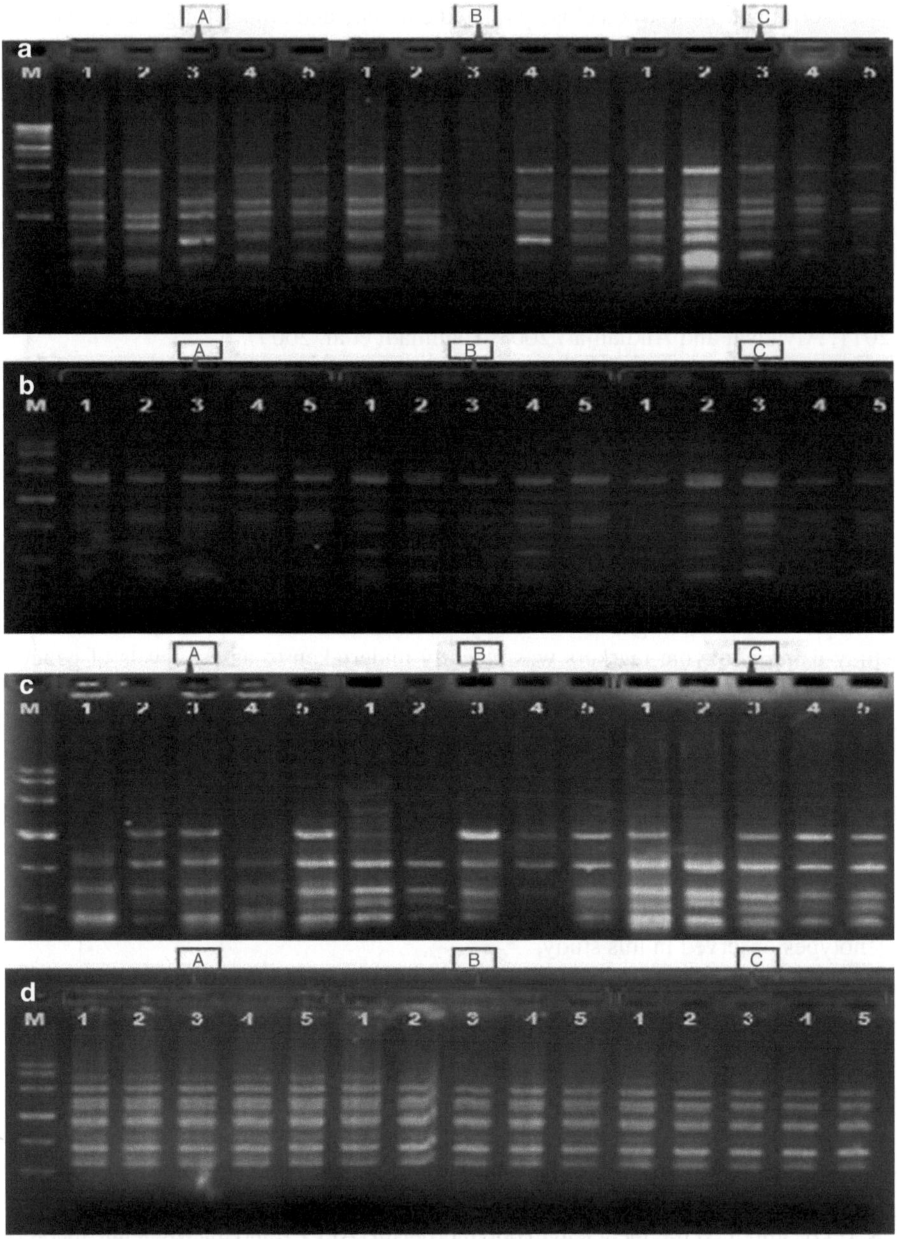

Fig. 2.4 Example of polymorphism in five date palm cultivars (1–5) grown in three locations (*A, B, and C, in Qatar*) using four primers (**a–d**) (Ahmed et al. 2013)

discrimination between highly related varieties (Diaz et al. 2003) and in quantifying genetic diversity and relatedness of date palm varieties, especially when large intervarietal polymorphism is encountered (Khieralla et al. 2011a, b). A recently completed and published sequence and transcriptomic analysis of date palm mitochondrial genome (Fig. 2.5) was described as being positioned at the root of known monocot mitochondrial genomes. It displayed several unique features such as having a very low level of repeat content, displayed abundant RNA editing events, and exhibited a high level of chloroplast sequence insertions as compared to other known angiosperm mitochondrial genomes (Fang et al. 2012).

2.5.2 *Population Genetic Structure of Date Palm*

Population structure refers to variation in size, maturity, and connectedness among the various populations of a species in a region; it takes into account the numbers and age range of individuals and their spatiotemporal distribution in plant

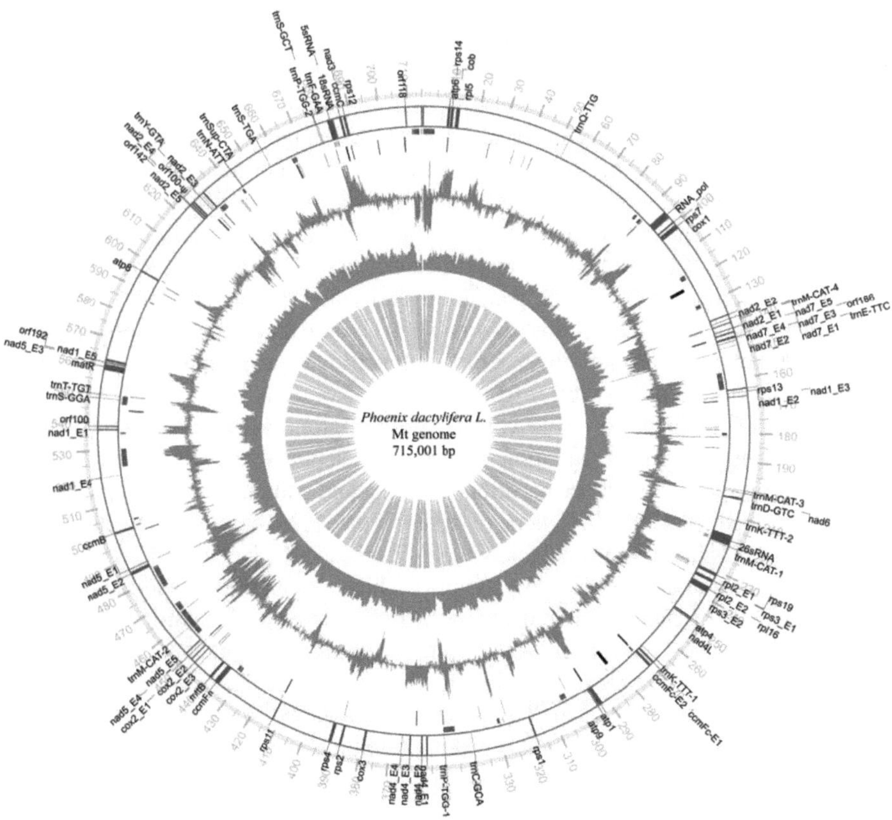

Fig. 2.5 A circular display including the physical map, coding sequences, and chloroplast-derived regions, among other components, to illustrate the complexity of date palm (*P. dactylifera* L.) mitochondrial genome (Fang et al. 2012) (Reproduced with permission, *PLoS ONE* (www.plosone.org/statis/license))

populations. Date palm must have been initially selected for certain tree and fruit attributes, and therefore, its population genetic structure must have been influenced by factors such as selection for high fruit yield, better quality, precocity, and regular, long-term, non-alternating productivity (Goldschmidt 2013). Available empirical evidence indicated that the genetic structure of date palm populations is governed by at least three main factors; these are (1) the environmental characteristics of the region of cultivation (i.e., oasis), (2) isolation by distance, and (3) the biological characteristics of the date palm tree. However, the history of cultivation and the cultural practices may have played strong roles in the structuring of genetic diversity of date palm populations. Factors other than geographic distances played an important role in affecting the gene flow and genetic structure of date palm populations worldwide. These factors include the exchange of vegetative propagules (e.g., offshoots, tissue culture plantlets) and pollen and seed dispersal, especially during the initial phase of date palm domestication and spread in the Old World.

The genetic architecture of a domesticated fruit tree resulted in almost total dependence of its population on human interference and man-made habitat (Goldschmidt 2013), that is, the oasis in the case of the date palm. Depending on decisions regarding the size and relative placement of their orchards, horticulturists and oasis dwellers can impact significantly the diversity and genetic structure of local date palm populations; orchards, within an oasis, a garden, or plantation, may be large or small, close together or widely dispersed, and composed of one or many closely related or different cultivars. However, habitat continuity within each of the above situations may improve the capacity to predict genetic differentiation. Depending on the reproductive biology being used (seedlings, offshoots, tissue culture), this structuring can have wide ranging effects on the genetic diversity and genetic structure of the date palm (Alkhalifah et al. 2012). It is reasonable to assume that large orchards or plantations will likely contain more diverse date palm cultivars and populations; therefore, their genetic structures may differ from those of small oasis and gardens.

The use of reliable and stable vegetative descriptors and microsatellite markers revealed significant genetic variation between soft, semidry, and dry varieties, with 7 and 93 % of molecular variance partitioned among and within subpopulations, respectively (Hammadi et al. 2011). Genetic diversity of date palm based on molecular markers in Sudan (Elshibli and Korpelainen 2008), Tunisia, (Dhieb et al. 2012; Karim et al. 2010; Mohamed and Rachid 2006), Egypt (Elkhishin et al. 2003), and Saudi Arabia (Askari et al. 2003) and based on fruit quality traits in the GCC countries (Jaradat and Zaid 2004) is mostly partitioned within rather than among populations, whereas heterozygosity and low genetic differentiation in populations of date palm were attributed to human impact and ecological determinants (Elshibli and Korpelainen 2009a).

Comparisons of genetic diversity and population genetic structure among different stages of life history provided important information on the effects of the different forces and (micro)evolutionary processes that may have influenced genetic diversity and genetic structure after population fragmentation. High genetic diversity may result from natural selection, which favors heterozygosity excess, coupled with a combination of a reproductive system and seed or pollen dispersal mechanisms that favor gene flow between populations (Elshibli and Korpelainen 2009b).

Genetic characterization of date palm genetic resources by microsatellite markers (Racchi et al. 2013) indicated that they were characterized by negative fixation indices due to heterozygosity excess; however, all cultivars were distinguishable and the genetic diversity and polymorphism were relatively large.

As a measure of population differentiation, the relative magnitude of gene differentiation (*GST* or *FST*) among subpopulations depends on total variation (*HT*); if the latter is small, then *GST* may become large even if the absolute gene differentiation is small (Fig. 2.6). Population genetic differentiation is usually shaped by several factors, including species life history and environmental attributes affecting dispersal of either propagules (e.g., offshoots) or adult trees (Dhanapal 2012; Elshibli and Korpelainen 2009a, b), where environmental and ecological distances proved to be better predictors of genetic differentiation between populations than simple geographic distances. The relationships between population differentiation based on molecular markers and each of heterozygosity and fixation index across three studies is illustrated in Fig. 2.6. A clear distinction can be seen between molecular markers representing sex- and autosomal-linked chromosomes (Cherif et al. 2012), and the larger variation in *FST* when *FIS* and the difference between *Ho* and *He* were small.

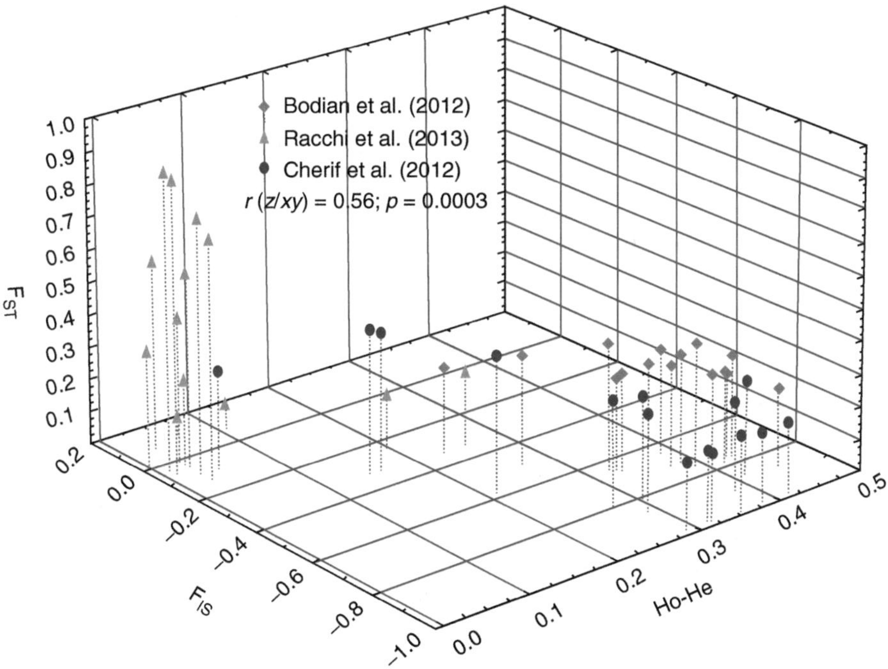

Fig. 2.6 Population differentiation (*FST*) as a function of the deviation of expected (*He*) from observed (*Ho*) heterozygosity and the fixation index (*FIS*) derived from three studies (Bodian et al. 2012; Cherif et al. 2012; Racchi et al. 2013) of date palm diversity based on microsatellite (SSR) marker analyses (Data from Cherif et al. 2012 was based on autosomal- and sex-linked chromosomes)

Knowledge of spatial patterns of genetic differentiation of date palm populations is key to understanding important processes, such as evolutionary mechanisms of differentiation and ecological or conservation consequences of loss of genetic diversity. Information derived from the mating system and long-distance gene flow in heart of palm (*Euterpe edulis* Mart.) (Gaitto et al. 2003) suggested that population dynamics can significantly affect evolutionary factors such as selection and genetic drift and consequently have important impacts on the genetic structure of date palm populations.

2.5.3 Genetics and Breeding System

Domestication and subsequent cultivation of the dioecious date palm was founded on a breeding system of mixed clonal-sexual mode of reproduction system which shaped its dispersal and evolution under domestication (Terral et al. 2012). Understanding of the genetic variability and proper cultivar identification of date palm constitute indispensable steps towards the development of breeding strategies (Ahmed et al. 2011). Selection for clonal propagation has not always modified sexual fertility or mating systems of fruit trees. In the date palm, some cultivars have evolved parthenocarpy, but because the resulting fruits are smaller and develop more slowly than those of artificially pollinated plants, the cultural practices of laborious hand pollination have been conserved and practiced for millennia (McKey et al. 2010). For example, high contemporary genetic diversity of date palm was attributed to continued multiplication by seed in Spain (Rivera et al. 2008) and Egypt (Battesti 2013), in addition to clonal reproduction. Historically, this mixed clonal-sexual reproduction prevented inbreeding depression, preserved a large amount of genetic diversity, and maintained a high adaptive potential used by horticulturists for selection of valuable genotypes; these underlying mechanisms of the domestication process has led to a spatiotemporal diversification of seed and fruit shapes and of cultivars (Terral et al. 2012).

Phylogenetic relationships within the Arecaceae suggested the presence of multiple origins of dioecy, often from hermaphroditism. Nevertheless, dioecy may have evolved several times in plants from bisexual (hermaphroditic or monoecious) ancestors by (1) a first mutation causing male sterility generating a gynodioecious population and (2) a second mutation resulting in a decreased female fertility in the population leading to functional dioecy (Cherif et al. 2012). Whether the evolution of dioecy results from selection for outcrossing or from sexual selection on separate components of male and female fitness has been a subjected of considerable debate (Elmeer and Mattat 2012; Weiblen et al. 2000). However, there is strong evidence suggesting that dioecy appears to have evolved most frequently via monoecy, perhaps through divergent adjustments of floral sex ratios between individual plants (Weiblen et al. 2000). The date palm may have a long history of dioecy (Charlesworth 2013). The evidence strongly suggested two clear-cut changes from monoecy to dioecy but four shifts from hermaphroditism to dioecy or polygamodioecy (Weiblen et al. 2000).

There is an economic value of determining sex in date palm seedlings and before flowering to increase the proportion of females. Understanding the evolution of sex chromosomes is impossible without first understanding the evolution of separate sexes, because the genetic changes involved in the latter lead with high probability to the chromosomes involved losing recombination, and it is the latter that led to the special properties of sex chromosomes. Molecular markers linked to the sex-determination region have been found in several dioecious plant species and are useful for determining the sex of the seedlings as well as immature plants. However, genetic maps with large number of markers are needed to confirm these findings (Charlesworth 2013).

The genetic component of sex determination in date palm is not well understood; however, several breakthroughs (e.g., Aldous et al. 2011; Cherif et al. 2012; Siljak-Yakovlev et al. 1996) established the date palm as a biological model with one of the most ancient sex chromosomes in flowering plants. The heterochromatin structure was first used (Siljak-Yakovlev et al. 1996) to determine sex chromosomes and identify male and female date palm plants at an early stage by inspecting root meristem cells. Sex-specific chromosomes have been identified for some date palm cultivars using RAPD and iSSRs (Younis et al. 2008); others (Elmeer and Mattat 2012) identified SSR markers of potential value in sex identification of date palm. Two iSSR markers, linked with flowering regulation, have been identified and were considered as potential candidate markers that may enhance our understanding of flower development and sex determination and identification in date palm (Masmoudi-Allouchi et al. 2009).

Using massively parallel sequencing on a cultivar of date palm with no documented inbreeding made it possible to detect a large number of parental allelic differences (Aldous et al. 2011). These researchers presented the first publically available draft of the nuclear genome for a member of the Arecaceae family. Three sex-linked SSR loci were identified and considered as reliable sex markers, demonstrated an XY chromosome system, revealed the existence of nonrecombining XY-like regions, and made it possible to trace date palm parental lineages (Cherif et al. 2012). Finally, the date palm mitochondrial genome showed obvious tissue-specific gene regulation patterns, among which male and female flowers, root and bud tissues exhibited higher gene expressions than other plant tissues. The gene expression profiles of male and female flowers, although clustered together in one cluster, were separated from each other. On the other hand, gene expression of seed and fruit formed a cluster totally separated from the rest (Fang et al. 2012).

2.6 Genetic Resources

The need to preserve and use plant genetic resources is well recognized, and the prospect of declining plant genetic diversity coupled with increased demand on these resources put them at the forefront of local, regional, and global discussions (Dhanapal 2012). Primitive forms and landraces of cultivated date palm, modern cultivars, obsolete cultivars, breeding lines, and related wild species are the most important component of its genetic resources and (agro)biodiversity (Dhanapal 2012). The value of date palm genetic resources is dependent upon the information utilized to promote their use (Gotor et al. 2008), while understanding the existing distribution and the extent of *Phoenix* spp. diversity is a prerequisite for its proper dynamic conservation and sustainable utilization. Plant genetic resources of date palm are defined as the *genetic material of date palm which is of value as a resource for the present and future generations of people*; their conservation is the link between genetic diversity of date palm and its utilization or exploitation by humans (IPGRI 2005).

Palm diversity is greatest in the tropics and subtropics, where palms are of great ecological and economic importance. However, palms in general are recognized as an increasingly threatened family, with a total of 222 species identified by the Palm Specialist Group of the International Union for Conservation of Nature (IUCN) Species Survival Commission (www.iucn.org/about/programmes/species) as highly threatened with extinction. Nine genera, in addition to *Phoenix* (i.e., *Chamaedorea, Sabal, Livistona, Dypsis, Arenga, Caryota, Pritchardia, Ptychosperma*, and *Licuala*) contain 257 species and represent only 6 % of the total generic diversity, account for 37 % of all cultivation records, and contain 22 highly threatened species. However, there are no highly threatened species in the genera *Phoenix, Livistona, Arenga,* and *Caryota.*

Traditional horticulturists and oasis dwellers in these countries practice de facto conservation of genetic resources and genetic diversity of date palm and other fruit trees and crops by maintaining traditional cultivars, especially in traditional oases (Alghamdi 2001; Battesti 2013; Nabhan 2007) (Table 2.3). They also develop or improve management practices, including the conscious selection of clones for various tree and fruit traits and selection of elite male cultivars for artificial pollination. Such practices may have far-reaching effects on the status of date palm genetic resources and their diversity (Allam and Cheloufi 2012; Elshibli and Korpelainen 2009b). These practices also go beyond traditional genetic resources conservation by improving and developing new and improved cultivars. Generally, these practices are not well documented, and their effectiveness in maintaining or creating new genetic combinations is not well known or documented. Usually, the choice of which cultivars or clones to grow is subject to each farmer's decision at

Table 2.3 Major oases in the Middle East and North Africa where date palm germplasm has been conserved, propagated, and exchanged for millennia

Country	Oasis
Algeria	Ouargla, Taut, Timimoun
Egypt	Bahraiya, Farafra, Khargah, Siwa
Libya	Ghadames, Kufrah
Oman	Buraimi, Maghta, Bahla
Saudi Arabia	Alqatif, Alahsa
Tunisia	Tozeur, Tamerza
Morocco	Tafilalt, Ourzzazzat

each planting and the factors influencing those choices are complex and not well understood (Jaradat 2011).

The oasis agroecosystem is a standard model for a spatially heterogeneous, three-story intercropping system of date palms, fruit trees, and annual crops. The composition and configuration of the three-story system creates different profiles of horizontal wind speed, relative air temperature, and relative air humidity. Date palms, fruit trees, and annual crops approximately intercept 20, 20, and 40 % of daily net radiation, respectively (Battesti 2013; Nabhan 2007). Highly adapted cultivars of date palm, fruit trees, and annual crops are managed through refined social practices and institutions. The indigenous knowledge associated with this diversity and its management is crucial to ensure a sustainable life in the oases. Agriculture in the oasis agroecosystem is mainly limited by the availability of suitable irrigation water; however, even with sufficient water, its use under the usually hot dry climate is often not sustainable, leading to soil salinization as a consequence of inappropriate irrigation and drainage techniques (Jaradat 2011).

Oases in the major center of origin and diversity of date palm (i.e., lower Mesopotamia, eastern Arabia, and Egypt) typically cover thousands of hectares, contain a large number of date palm and other fruit trees, and are composed of a mixture of adapted cultivars. However, oases away from the center of origin are smaller in size, may cover a few hectares, and contain a few date palm cultivars (Gebauer et al. 2007). Historically, these oases have been developed by transport of seed and, occasionally, offshoots from existing oases. The mode of propagation impacted the level of diversity and varietal composition in the new oases. In the case of elite selections, propagation by transported offshoots resulted in dissemination of genetically identical or nearly identical cultivars to various parts of a country or region. In the case of noncommercial or less desirable cultivars, seed dissemination resulted in the establishment of more localized and adaptive cultivars. Human selection of elite types traditionally was based on fruit characteristics and this would be the main selection pressure; however, some natural selection pressure in the new oases may have occurred due to resistance or susceptibility to biotic and abiotic stresses. Similarly, natural selection could have been applied on the nonelite cultivars that originated from seed.

The botanical composition and floristic inventory contained in even relatively small oases are rich and highly diverse. For example, 14–17 different date palm cultivars and a total of 107 different plant species have been recorded in three small oases in the northern mountains of Oman (Alyahyai and Alkhanjari 2008; Gebauer et al. 2007). In addition, the number of crops was very high in comparison with other small-scale cropping systems found under arid or semiarid conditions. In this mountainous region, a completely different form of agriculture has persisted for millennia. Date palms and annual crops are cultivated in oases that are watered either by springs or by *aflaj*, tunnel systems dug into the ground or carved into the rock to tap underground aquifers. Both systems require the oases to be located at the foot of cliffs and below plateaus, which accumulate the scarce rainfall of a large area (i.e., water harvesting) and never developed a serious salinity problem (Gebauer et al. 2007).

2.6.1 Field Genebanks

In addition to traditional conservation practices in home gardens, oases, and new plantations, there are a few documented date palm field genebanks (ex situ collections) in the world engaged in the conservation of date palm genetic resources and where characterization, evaluation, and in-depth molecular analyses are being carried out (Table 2.4). Although most of the germplasm diversity and genetic resources of date palm exist within its center of origin, an increasing proportion can be found in the wider center of diversity of the species. However, the latter is most probably composed of elite cultivars or breeding lines with limited genetic diversity (Krueger 1998, 2011). Consequently, there is an urgent need to establish more field genebanks throughout the center of diversity of date palm and for more concerted efforts to enhance the genetic diversity available in traditional oases; some of these oases continue to endure adverse anthropogenic and climatic threats (Pauls et al. 2013). Even with an increased number of ex situ collections of date palm germplasm as active field genebanks, these collections will be fewer and smaller than for most other crops. Specialists attribute this to the relatively limited geographic area in which cultivation of date palm is feasible, the nature of the date palm tree and the relatively narrow genetic base (Alghamdi 2001; IPGRI 2005; Krueger 1998). Nevertheless, these genebanks are critical sources of biodiversity and can serve as research laboratories for identifying genetic variation that can help to achieve reliable date production under anticipated biotic and abiotic stresses (Nawar and Mackay 2010).

In addition to conserving date palm germplasm, ex situ collections also increase the efficiency of date palm utilization by providing cultivars with genetic purity and furnishing favorable conditions for reduced incidence of biotic stresses and precise documentation of characterization and evaluation data. Moreover, such collections will be more capable of furnishing enough germplasm for genetic and management experimentation to be carried out (Gotor et al. 2008; Krueger 2011).

There are only a few known *classical* date palm field genebanks. Bettencourt et al. (1992) listed only 15, the largest of which were found in Algeria, India, Iraq, Nigeria, and the USA. Some of the date palm germplasm collections are far removed from the center of origin and maintain only a few cultivars or genotypes (e.g., Brazil, Dominican Republic, France, Taiwan), while others have been lost due to natural disasters (e.g., Miami, Florida, USA) (Krueger 2011). Except possibly for the Nigerian collection (Ataga et al. 2012), most date palm accessions maintained in field genebanks appear to be elite cultivars or breeding lines of low genetic diversity. The Near East and North Africa Plant Genetic Resources Network (NENA-PGRN 2011) published results of a survey in countries of that region as a part of a strategy for the conservation and sustainable use of plant genetic resources.

The study concluded that almost all countries have field genebanks and botanic gardens to conserve landraces of fruit trees and other perennial species. The study went on to state that most of the field genebanks have both landraces and varieties introduced from abroad but indicated that the richness in landraces of most fruit

Table 2.4 Examples of studies on date palm genetic diversity analyses, markers used, and measures or statistics estimated (compiled from references cited in this chapter)

Country	No. of accessions/ varieties	Marker(s)	Measures/ statistics	Reference
Algeria	1	Proteomics	Diversity	Sghaier-Hammami et al. (2009)
China	3	mtDNA	Sequence of mtGenome	Fang et al. (2012)
Egypt	4	RAPD, SSR	Polymorphism	Younis et al. (2008)
Egypt	8	RAPD, SSR	Diversity	Eissa et al. (2009)
Egypt	5	AFLP	Jaccard	Elkhishin et al. (2003)
GCC[a]	203	Fruit phenotypes	Genetic diversity	Jaradat and Zaid (2004)
Iraq	18	AFLP	Jaccard	Jubrael (2005)
Iraq	20	AFLP	Jaccard	Khieralla et al. (2011a)
Iraq	20		Heterozygosity, genetic distance	Khieralla et al. (2011b)
KSA[a]	7	RAPD	Nei's distance	Askari et al. (2003)
KSA	13	RAPD	Genetic similarity	Alkhalifah and Askari (2003)
KSA	8	cpDNA	Genetic similarity	Alquraini et al. (2011)
Libya	377	SSR	Polymorphism	Racchi et al. (2013)
Mauritania	29	SSR	Genetic diversity	Ould Mohamed Salem et al. (2008)
Mauritania	12	DNA	Discrimination	Ould Mohamed Salem et al. (2001)
Morocco	66	AFLP	Genetic similarity	Elhoumaizi et al. (2006)
Morocco	45	RAPD, SSR	Polymorphism	Sedra (2013)
Morocco	18	iSSR	AMOVA	Bodian et al. (2012)
Oman	?	Phenotypes	Phenotypic similarity	Alyahyai and Alkhanjari (2008)
Qatar	43	SSR	N/A	Elmeer and Mattat. (2012)
Qatar	N/A	SSR	Genetic diversity	Elmeer et al. (2011)
Spain	3	AFLP	Dice	Diaz et al. (2003)
Spain	N/A	RAPD	Discrimination	González-Perez et al. (2004)
Sudan	60	SSR	Genetic diversity	Elshibli and Korpelainen (2008)
Tunisia	27	SSR	Heterozygosity, fixation index	Hammadi et al. (2009); Karim et al. (2010)
Tunisia	3	Phenotypes	Variance analysis	Hamza et al. (2011)
Tunisia	5	N/A	Polymorphism	Ouarda et al. (2012)
Tunisia	1	In vitro hermaphroditism	N/A	Masmoudi-Allouchi et al. (2009)
USA[a]	12	EST-SSR	Polymorphism	Zhao et al. (2013)

[a]*GCC* Gulf Cooperation Council, *KSA* Kingdom of Saudi Arabia, *USA* United States of America

trees needs more collecting efforts for their inclusion in reliable genebanks. The study concluded that most of the old-field genebanks are suffering from a lack of adequate maintenance and regeneration. As for date palms, however, the study stated, without supporting statistics, that field genebanks have been established in Algeria, Egypt, Iran, Jordan, Morocco, Tunisia, and the UAE.

There are other germplasm collections that are either new since the compilation of Bettencourt et al. (1992) or possibly were unknown at the time of that compilation. These include a field genebank at:

- King Faisal University, Saudi Arabia
- International Center for Biosaline Agriculture, Dubai, UAE (Fig. 2.7)
- Wadi Quriyat Date Palm Research Station in Bahla, Oman

The Global Environmental Facility (GEF)-International Plant Genetic Resources Institute (IPGRI) project on participatory management of plant genetic resources in oases of North Africa was reported (NENA-PGRN 2011) as instrumental in developing in vitro propagation protocols for many local date palm cultivars; also it investigated several value-adding and technological options, which resulted in renewed interests in conserving some neglected date palm cultivars.

A preliminary consideration for the implementation of a dynamic conservation program of genetic resource of date palm is "what genotypes to incorporate in a field genebank?" If the objective is to supply germplasm for propagation (e.g., using tissue culture) for a local industry, such a field genebank would have a limited range

Fig. 2.7 Field genebank of 10 salt tolerant date palm cultivars (Abu Ma'an, Barhi, Djirbi, Fardh, Khnaizi, Khalas, Khasab, Lulu, Naghal, and Shahla) at the research farm of the International Center for Biosaline Agriculture, Dubai, UAE (Photo: M. Shahid 2013)

of genotypes compared to a field genebank that has the objective of general genetic resource conservation and utilization. The latter might incorporate diverse genetic resources with much larger genetic diversity and would contribute to widening the genetic base of new date palm plantations (Jaradat 2011; Krueger 2011).

In situ conservation of wild relatives of the date palm received very little attention and was not included in a global review of the status and trend of in situ conservation of crop wild relatives (Meilleur and Hodgkin 2004). Historically, crop wild relatives have been identified during the 1980s as a target for conservation due to an accelerating rate of species extinction and loss of genetic diversity (IPGRI 2005). Recently, however, wild crop relatives, including those of date palm were described as being undervalued, underutilized, and under threat, especially from global climate change (IPCC 2007). The continued survival and evolution of these genetic resources in their natural habitats is becoming increasingly important. Wild *Phoenix* populations should have sufficient evolutionary potential to respond to climate change (see Sect. 2.7.3). Most probably (Shabani et al. 2012; Turner et al. 2010) these wild relatives, along with the cultivated species, will need to be moved around the world more than ever in order to facilitate their adaptation in response to climate change. Nevertheless, these wild populations are buffered against negative effects of habitat disturbances and their spontaneous natural hybridization with the domesticated species in parts of its distribution range (e.g., Canary Islands, Atlantic coast of North Africa, and the Sind of the Indian subcontinent) may have contributed to the domesticated species richness through an influx and introgression of valuable genes for biotic and abiotic stresses (González-Perez et al. 2004).

2.6.2 In Vitro Conservation of Date Palm Genetic Resources

Traditional conservation of date palm genetic resources in field genebanks for a long period of time may pose serious problems caused by susceptibility to biotic and abiotic stresses (Bekheet et al. 2001; González-Benito et al. 2004), high cost of management, and the eventual loss of valuable and unique genetic material (Jaradat 2011). In addition, germplasm exchange using bulky offshoots developed in field genebanks was considered as difficult and may pose greater risk of disease and insect transfer (Munier 1981). Conservation of such genetic resource, as a highly heterozygous germplasm, is practically of limited value in conventional field genebanks. Therefore, the prospects of in vitro conservation of date palm genetic diversity for sustainable production have been recently recognized and were outlined in a number of seminal papers (Bekheet et al. 2001, 2005; González-Benito et al. 2004; Jain 2010; Othmani et al. 2009) and justified on the basis of future impact of adverse climate change on sustainable date production and the production of disease-free, true-to-type germplasm with low somaclonal variation and easy to handle for germplasm exchange (González-Benito et al. 2004).

In vitro techniques (Fig. 2.8) can be used for medium-term germplasm conservation (of up to 3 years) without subculturing. A combination of physical and chemical

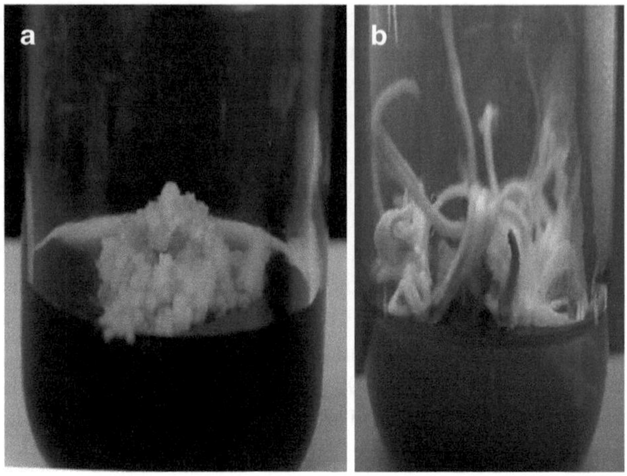

Fig. 2.8 Callus culture of date palm cv. Zaghloul stored for 9 months on medium containing 40 mg L^{-1} sorbitol under normal growth conditions (**a**) and shoot buds of date cv. Zaghloul stored for 12 months at 5 °C under dark conditions (**b**) (SA Bekheet 2013)

factors can be fine-tuned to achieve a successful level of conservation. Whereas; long-term conservation of germplasm can be achieved using cryopreservation methods. The development of in vitro technology has led to the production of a new category of germplasm which represents a value-added in terms of germplasm conservation and exchange (Cruz-Cruz et al. 2013). Such germplasm material is developmentally synchronized, miniaturized, and potentially genetically homogenous (Al-Bahrany and Al-Khayri 2012). Although such germplasm can be both rapidly and massively propagated and exchanged, however, its phytosanitary status may not be guaranteed, except in the case of shoot tip or meristem cultures (Bekheet et al. 2007), and its potential for genetic uniformity of elite cultivars at the expense of less-preferred ones should not be overlooked (Jaradat 2011).

Technologies involving in vitro conservation, cryopreservation, or cryo-storage and cold storage are increasingly becoming more practical and cost-effective methods of plant genetic resources conservation, including date palm and other horticultural fruit trees (Bekheet et al. 2001; Jain 2010; Othmani et al. 2009). Although the large-scale, routine application of cryopreservation is still limited to a small number of perennial fruit trees, the number of plant species, including date palm, for which cryopreservation techniques are already established and validated for genetically diverse species is increasing (Cruz-Cruz et al. 2013). Date palm germplasm has been conserved in vitro in different vegetative forms including shoot apices excised from in vitro plants, meristems, somatic and zygotic embryos, pollen, callus, and cell suspension cultures (Bekheet et al. 2007; González-Benito et al. 2004; Jain 2010).

Long-term preservation can be achieved using cryopreservation (i.e., at very low temperature in liquid nitrogen at −196 °C) where all cellular biological activities (e.g., cell division and metabolic processes) are virtually stopped. This procedure was successfully used (Bekheet et al. 2007) in preserving undifferentiated date palm tissue cultures and the subsequent regeneration of plants in the laboratory. Subsequent biochemical and molecular tests confirmed the genetic stability of the preserved date palm plant material.

2.6.3 Descriptor Lists

The descriptor list of date palm constitutes the basis for a standardized characterization system that provides an internationally agreed upon format and universally understood "language" of date palm genetic resources data (IPGRI 2005; Gotor et al. 2008). A descriptor list includes management, environmental, characterization, and evaluation descriptors of the tree and its different parts. A short list of tree, leaf, fruit, and seed descriptors is presented (Table 2.5) as a quick reference. Although most of these descriptors explain qualitative and phenological plant and fruit traits (Eissa et al. 2009; IPGRI 2005), an increasing number of molecular markers is being developed by several research organizations (e.g., Ahmed et al. 2013; Eltarras et al. 2002; Ponnuswami 2010 and see Jaradat 2013 for a review).

A varietal identification key, using multilocus genotyping, can be a powerful tool to correctly assign single plants to their definite cultivars, with implications for the conservation of date palm genetic resources and for breeding new cultivars (Racchi et al. 2013). Whereas; specific molecular markers can be used as descriptors to verify cultivar identity and place of origin, in addition to detecting and identifying intracultivar variation and polymorphisms between cultivars grown in different oases and plantations (Ahmed et al. 2013).

2.6.4 Cultivars

According to recent reports (e.g., Al-Bahrany and Al-Khayri 2012; Ahmed et al. 2013; Gebauer et al. 2007; see subsequent chapters in this volume for up-to-date information on date palm cultivars in individual countries), the number of recognized cultivars, as components of the global date palm genetic resources, range

Table 2.5 A short list of date palm tree, leaf, seed, and fruit morphological descriptors (IPGRI 2005; Eissa et al. 2009)

Plant part	Morphological variants
Tree	Trunk width, crown shape
Leaf	Length, width, color, midrib color, color of leaf base abaxial surface, curvature and curvature point, base width, blade length, blade-leaf ratio, blotches on leaf base abaxial surface
Petiole	Length, width, thickness, shape, petiole-leaf ratio
Pinnate	Number/leaf, density/unit length, length, width, shape, apex, rachis angle
Spine	Area length, spine area-leaf ratio, shape, length, base, type, color, rachis angle
Seed	Length, width, weight, volume, shape, color, apex, base, surface, micropyle position and elevation, transverse grooves, ventral furrow shape and ends
Fruit	Length, width, weight, volume, density, apex, shape, base, color at Khalal, color at maturity, skin texture at maturity, skin thickness, flesh thickness, flesh color, flesh texture, flavor, taste, pulp, maturity (early, medium, late)

from 2,000 to approximately 5,000; examples of cultivars with specific quality traits are presented in Table 2.6. Each cultivar is derived from a unique single seed, cloned, and multiplied by offshoots. More recently, however, an increasing number of date palm cultivars are being propagated through tissue culture. Reports are conflicting as to the number of date palm trees and cultivars in each of the major date-producing countries in the Middle East and North Africa (Jaradat 2011). According to a recently available account, there are approximately 120 million trees producing ~7 million mt of dates; these are almost equally divided between the Middle Eastern and North African countries with Egypt and Saudi Arabia as the leading countries in date production (FAOSTATS 2012). The top ten date-producing countries (in alphabetic order) are Algeria, Egypt, Iran, Iraq, Morocco, Oman, Saudi Arabia, Sudan, Tunisia, and the UAE.

Reports are conflicting as to the number of date palm trees and cultivars in each of the major Middle Eastern and North African date-producing countries. According to recently published accounts, there are approximately 120 million trees producing ~7 million mt of dates; these are almost equally divided between the Middle Eastern and North African countries with Egypt and Saudi Arabia as the leading countries in date production (FAOSTAT 2012). The 10-top date-producing countries (in alphabetic order), presumably cultivating the largest number of trees, but not necessarily the largest number of cultivars, are Algerian oasis harbor ~800 date palm cultivars, the largest number reported from North Africa and contributes ~6 % of world production. There are ~12 million date palm trees and ~255 cultivars, 55 of which are commercially produced in Egypt which is the largest date-producing country. Egypt contributes 17 % of world production from plantations along the

Table 2.6 Example of date palm cultivars having specific quality traits

Trait	Quality	Cultivar
Fruit color	Black glossy	Abbada
	Golden amber	Migraf
Fruit size	Large, long	Anbarah
Sweetness	Extreme	Halawi, Iteema
Fruit texture	Soft	Amir Haj
	Syrupy	Khastawi
	Semidry	Zahidi
	Dry, chewy	Thoori
	Crunchy	Zhaghloul
Maturity	Early	Manakbir
	Late	Barhi
Tolerance to moisture	Very tolerant	Dayri, Medjool, Tadla, Halawy, Maktoom
Texture	Firm	Badrya, Deglet Beidha, Deglet Noor, Thoory, Hoffa
	Caramel	Khadrawi
Fruit stalk	Long	Hoffa, Khosh Zebda
Medicinal properties		Ajwa
Overall quality	Superior	Amir Haj, Deglet Noor, Medjool

Nile, in the Nile Delta and from several oases in the western and southern deserts, and in the northern and southern parts of Sinai (Battesti 2013; Eissa et al. 2009; Elkhishin et al. 2003; Nabhan 2007).

Iran, the second largest date-producing country, contributes 13 % of world production with ~400 date palm cultivars mostly in the southern part of the country (FAOSTATS 2012). Prior to the 1980s, Iraq was the major date producer in the world with ~30 million date palm trees and ~700 cultivars, most of which are found in the southern part of the country along the Tigris-Euphrates River and Shatt Al-Arab waterway. During the past ~30 years, the country witnessed a huge loss (~20 million) in number of trees and date production (6 % of world production), some, if not most, of the lost trees represented unique clones and valuable genetic resources (Walsborn 2008). Morocco has ~5 million date palm trees and approximately 250 date palm cultivars and contributes ~1 % of world production (Elhoumaizi et al. 2006). Oman contributes 3.5 % of the world production from ~12 million date palm trees, 7.3 million of which are fruit bearing, and 230 date palm cultivars, 20 of which are grown commercially in a number of mountain and valley oases (Alyahyai and Alkhanjari 2008; Gebauer et al. 2007). Pakistan is one of the largest date producers, contributing ~10 % to the world production and has ~300 cultivars (Iqbal et al. 2011).

A total of 450 date palm cultivars were reported in Saudi Arabia contributing 12 % of the world production, whereas the Medina date market alone contains about 150 cultivars, the most popular and most expensive of which is Anbarah (Alghamdi 2001). Date palm agriculture in Sudan is mainly concentrated in the northern part of the country along the Nile; however, there are few oases scattered in northern Kordofan and northern Darfur. There are approximately five to six million date palm trees and about 400 cultivars of date palm in Sudan (Elshibli and Korpelainen 2009b).

Tunisia contributes 2 % of the world production from more than four million date palm trees and about 250 cultivars. Almost 60 % of the recently developed plantations in Tunisia are dominated by the elite variety Deglet Noor at the expense of other traditional, albeit less desirable, commercial cultivars (Hammadi et al. 2009; Hamza et al. 2011; Karim et al. 2010). Finally, the UAE contributes 11 % of world production and is considered as a leading country in date production; it has ~40 million date palm trees and a minimum of 200 cultivars, 68 of which are the most important commercially (Jaradat and Zaid 2004).

2.6.5 Germplasm Movement and Exchange

The overall partitioning of genetic diversity, based on results of phenotypic, biochemical, and molecular markers, and fruit quality traits (Jaradat 2013) suggests that date palm cultivars represent a complex gene pool within which historical movement of germplasm (as seed, tissue culture, offshoots, or any other propagules), recent introductions, and human selection are shaping its genetic structure.

Although a large portion of genetic diversity and genetic resources of date palm germplasm exist within the center of origin of date palm, a sizable part can be found in the wider center of diversity of the species (Jaradat 2013).

Long before modern transportation became available, oases in close proximity were more likely to have exchanged desirable cultivars than they would with more distant ones (Battesti 2013; Nabhan 2007; NENA-PGRN 2011). Consequently, we find that some elite cultivars of dates are associated with certain oases, whereas, most recently, we find some elite cultivars are widely distributed in several oases and modern plantations within and among countries (e.g., cv. Khalas in Saudi Arabia, Qatar, the UAE, and Oman and cv. Medjool in several countries of the Middle East and North Africa and in the USA). This indicates that some clonally propagated superior cultivars tend to be exchanged more often than inferior ones. The latter usually develop from seedlings that are left to grow and generally provide lower-quality dates and remain of local distribution and interest. The exchange of offshoots between farmers and between oases may have resulted in heterozygous and heterogeneous populations, whereas the introduction of seedlings, after fruit consumption, may have resulted in generating more diversity within and among oases (Munier 1981; Popenoe 1913; Rivas et al. 2012). However, consumer demands for fruit of specific elite cultivars led to their dominance through regional and international germplasm exchange and the loss of genetic diversity due to the disappearance of many cultivars with medium or low fruit quality (Allam and Cheloufi 2012; Hasnaoui et al. 2011; Jaradat and Zaid 2004).

With the advent of plant genetic resources collection, conservation, and utilization (IPGRI 2005), legal and regulatory phytosanitary systems have been developed in accordance with international directives and standards to regulate germplasm movement, exchange, and quarantine of plant material, especially of clonally propagated fruit trees such as date palm (Noriega et al. 2012). It is anticipated that germplasm movement and exchange will accelerate as countries in the southern hemisphere are establishing large-scale date palm plantations. However, some of the challenges that may limit germplasm exchange include the reluctance of some countries to make their date palm genetic resources available, unavailability of evaluation and characterization data, and inadequate information on genetic material conserved under traditional field genebank conditions or in vitro, the latter allows for the exchange of disease-free germplasm that can be conserved for a short or long period of time (Malaurie 2001).

In addition, the long and bureaucratic procedures involving international exchange of date palm genetic resources, even if governed by bilateral agreements, may hinder the process and especially where there is a multiplicity of institutional and administrative structures dealing with germplasm exchange (Noriega et al. 2012). The anticipated large volume of international date palm germplasm transfer, when coupled with the recent rapid advances in biotechnology, called for the development of date palm-specific regulations for its phytosanitary safety during transfer and movement across the globe. Therefore, the transfer of date palm germplasm should be carefully planned with quarantine authorities and should be accompanied with the necessary documentation (Frison et al. 1993; Noriega et al. 2012).

2.7 Threats to Biodiversity and Genetic Diversity

It was more difficult to document the process of genetic erosion and loss of biodiversity in date palm than initially expected because time-series data are not generally available for the species in its center of origin or center of diversity. Ecological and socioeconomic factors are affecting the delicate equilibrium of oasis agroecosystems; these include land degradation, genetic erosion, inappropriate agronomic practices, frequent droughts, aquifer depletion, desertification, sand encroachment (Jaradat 2011), and the introduction of exotic plant species into remote oases (Nabhan 2007). In addition, date palm biodiversity and production potential are threatened by a number of biotic and abiotic stresses. Date palm orchards in North Africa are aging; almost one-third of productive date palm trees in Algeria are beyond the limits of their productive years (Hamza et al. 2011; Hasnaoui et al. 2011), and almost half of the Tunisian productive date palms are more than 50 years old (Mohamed and Rachid 2006; Zehdi et al. 2004a, b).

For most *Phoenix* species, the status of genetic vulnerability is not well known or documented. Although many of the species are cultivated as ornamentals, there are probably few *pure Phoenix* species in ornamental plantings due to its readiness to hybridize with many genetically compatible species. However, due to its extensive cultivation (in oases and, most recently, in modern plantations), *P. dactylifera* as a species may not be considered threatened, whereas wild date palm germplasm may become extinct due to habitat destruction and climate change. Nevertheless, existing genetic diversity in cultivated date palm can be lost due to these factors if they result in the loss of local cultivars having specific genetic structures (Alhudaib et al. 2007).

Molecular evidence of hybridization between the endemic *Phoenix canariensis* and the widespread *P. dactylifera* L. in the Canary Islands was detected using random amplified polymorphic DNA (RAPD) markers. Interspecific hybridization between an endemic species (*P. canariensis*) and a common one (*P. dactylifera*) may result in putting the endemic species at risk if hybrid progeny and progeny from advanced hybridization are vigorous and fertile; or the common species may become at risk if the hybrid progeny is sterile or has reduced vigor (González-Perez et al. 2004). Another example is the relatively recent (~300 years) introduction of date palm to the oases of the Baja California peninsula. The exotic species became a keystone species and in many cases, growing along with or replacing the native fan palm (*Washingtonia filifera* and *W. robusta*) (de Grenade 2013). In these, and in similar situations (e.g., *P. atlantica* in Morocco and *P. sylvestris* in the Indian subcontinent), the introduction of *P. dactylifera* may pose serious threats to the genetic integrity and conservation of the endemic species; the latter may become at risk from genetic assimilation or from crossbreeding depression. In addition, extreme cases of invasiveness are increasingly becoming an issue of great global concern, especially in light of the ever-increasing scale of human movement and trade globalization of date palm genetic resources (Fiaboe et al. 2012).

In spite of individual cases where date palm cultivation has diminished or even vanished (e.g., wadi Alajal in Libya), overall date production worldwide has increased over time. However, threats to genetic diversity increased during the last ~30 years partly due to the introduction of improved mass propagation methods, such as tissue culture, of a limited number of date palm cultivars to the exclusion of many others (Racchi et al. 2013). Eventually, mass propagation of pure and a limited number of date palm genotypes would inevitably lead to genetic vulnerability (Diaz et al. 2003). The loss of biodiversity has serious implications for species, ecosystem services, and people who depend on environmental and natural resources for their livelihood.

On-farm conservation to promote genetic diversity of date palm was encouraged in several Middle Eastern and North African countries as a potential method of genetic resources conservation strategy. If farmers are motivated, they may be willing to participate in date palm genetic resource conservation. Having different cultivars on the farm can help lower the risks of total crop failure in the case of natural disasters; farmers may have a greater incentive to grow many and diverse cultivars. Farmers may be willing to give up extra income or yield to obtain a more stable and environmentally adaptable date palm cultivar (Ezebilo et al. 2013). Farmers who grow date palm for family consumption often have different tastes and are likely to grow different cultivars to meet their demand (Jaradat and Zaid 2004). Family size may dictate the number of date palm cultivars; the more the diversity of date palm cultivars in a family garden, the smaller the unit value.

2.7.1 Biotic and Abiotic Threats

A large number of published papers, report recommendations, and workshops (see Jaradat 2013 for a review) highlighted the threats of biotic and abiotic stresses to the millions of date palms in the Middle East and North Africa. Specifically, the red palm weevil (*Rhynchophorus ferrugineus* (Olivier)) (Alsaoud and Ajlan 2013; Fiaboe et al. 2012) and bayoud, caused by *Fusarium oxysporum* f.sp. *albedinis* (Sedra 2013), are threatening the region's multimillion dollar date industry and the very survival of the date palm trees. The red palm weevil is causing severe damage in date palm orchards in eastern Arabia, Iraq, and Egypt; the insect is considered as a major threat to *Phoenix theophrasti*, the native palm species in Crete, Greece. This pest was first known in India in the early 1900s; however, it was not until 1918 that the weevil presented a serious threat to the date palm industry in Iraq. Although integrated pest management has been employed to combat the insect, unfortunately and for all practical purposes, the most effective control measure is to cut down the infected trees and destroy them at an early stage to prevent the weevil from spreading over large areas. Recently, ecological niche modeling was suggested (Fiaboe et al. 2012) as a means of predicting the potential worldwide distribution of the red palm weevil for the integrated pest management to be effective in reducing its impact on the date palm genetic diversity.

Genetic diversity among isolates detected by RAPD analysis suggested that some local genetic differentiation has occurred since the Algerian populations of *Fusarium oxysporum* f.sp. *albedinis* were spread and established in new oases (Fernandez et al. 1997; Sedra 2013). The high incidence and severity of bayoud in North African date palm orchards could be attributed to the susceptibility of the wide spread of genetically uniform date palm cultivars (e.g., Medjool in Morocco and Algeria). It was estimated that bayoud already destroyed more than 13 million trees in these two countries during the twentieth century. Molecular markers have been recently identified in date palm as potential markers of resistance to this disease (Sedra 2013).

A number of abiotic stresses, such as soil and water salinity, caused by the presence of excessive amounts of soluble sodium chloride salts that hinder or affect the normal function of plant growth, are increasingly becoming serious threats to the expanding date palm industry in several Middle Eastern and North African countries (Eljuhani 2010; Jaradat 2011). More recently, drought, due to lengthy rainless periods and drying up of many desert water wells, resulted in increased water and soil salinity. Salinity problems develop in oasis agroecosystems due to mismanagement of water and soil resources under high evapotranspiration demand of hyperarid environments (Jaradat 2013; Wrigley 1995). Although some farmers are learning how to manage their date palm orchards under increasingly saline conditions, the need for a holistic solution to the salinity problem is greater than ever. Date palm is one of the most salt-tolerant fruit trees; however, large varietal differences have been documented in the species. Research results indicated that salt tolerance depends on growth stage and edaphic and management conditions. Recent results of RAPD analysis revealed that specific DNA fragments may characterize genes coding for tolerance to these abiotic stresses and are expected to serve as markers for early evaluation and screening for salinity and drought tolerance in date palm (Chao and Krueger 2007).

2.7.2 Anthropogenic Threats

Anthropogenic threats, especially those acting as drivers of climate change, often have multiple effects, including changes in biodiversity, species composition, and ecosystem functioning (Isbell et al. 2013); the long-term impacts of these drivers depend on how fast they can decrease biodiversity and restructure date palm populations. The within-species diversity of the date palm is increasingly being reduced by pressures for higher productivity and concentration of the market on a few high-quality cultivars, such as Medjool cv. in Morocco and Algeria and the Deglet Noor cv. in Tunisia, due to consumer demand; this consumer pressure is spreading to other countries such as Egypt and Jordan (Jaradat pers obs). Furthermore, and in response to biotic stresses (diseases and insects), selecting an even smaller number of resistant cultivars is a further threat to the diversity of the species (Alhudaib et al. 2007; Alsaoud and Ajlan 2013). This may be even more damaging if the resistance to a particular disease or insect proved to be short-lived because of changing climatic conditions or through a change in the virulence of the pest (Fernandez et al.

1997; Fiaboe et al. 2012). Economic and social factors also impact the diversity of date palm orchards; as a result, the composition of these orchards as to the number of cultivars witnessed a sharp decline in recent years (Eljuhani 2010).

Among the socioeconomic factors which influence peoples' livelihoods and the oasis agroecosystems are the marginalization of indigenous communities and the fast erosion of local cultures and indigenous knowledge. For example, Deglet Noor occupies ~60 % of date palm plantations in Tunisia and Algeria and continues to expand at the expense of other, less desirable, cultivars due to market forces (Dhieb et al. 2012). Furthermore, traditional propagation using offshoots of elite cultivars having desirable fruit quality traits resulted, even in the center of origin of date palm, in genetic erosion and in the confinement of cultivars with distinctive fruit types to certain oases. This process has been exacerbated by the massive propagation of elite date palm cultivars using tissue culture and potentially other mass propagation methods, at the expense of less popular but genetically valuable cultivars (Al-Bahrany and Al-Khayri 2012; Jain 2010; Zaid and de Wet 2005).

The oasis represents the climax of rigorous management of scarce water and land resources in alliance with the date palm. Human interactions that shaped oasis agroecosystems and enabled them to provide multiple ecological and socioeconomic services to meet the needs of local populations have always been a key to oases sustainability. Plant species and crop cultivars have been carefully selected from natural ecosystems or human-made introductions over centuries of experimentations (Nabhan 2007). This diversity and its associated indigenous knowledge are fundamental assets for the inhabitants of the oases and constitute a strategic portfolio of livelihood options. Unfortunately, the indigenous knowledge associated with the oasis agroecosystems, in general, and with date palm (bio)diversity and its management, in particular, is being gradually lost and needs to be maintained to ensure the sustainable life in the oases.

The traditional social water management system has been largely replaced by modern governing bodies with little or no coordination in some oases (e.g., Egypt, Morocco, Tunisia) (Battesti 2013), while it persisted for millennia in others (e.g., Oman) (Gebauer et al. 2007). In the latter case, indigenous knowledge in managing crop, water, and land resources in a *model* oasis agroecosystem is evident in maintaining high-quality irrigation water, the elaborately built soil structure of terraces, a system of water distribution designed to match crop needs during their different growth stages, adequate drainage, and the lack of salinization in ancient mountain oases. However, it remains to be seen if indigenous knowledge can help manage these *aflaj*-based systems to withstand the challenges posed by anthropogenic and climate change. As water becomes increasingly scarce, diversity is lost, cropping systems change, and social institutions are weakened; therefore, the need for documenting indigenous knowledge and values associated with life in the oases has become urgent (Jaradat 2011). The process should include systematic and comprehensive documentation of local and traditional knowledge on date palm bio(diversity) and proper functioning of the oasis agroecosystem, cultivar identification, water and soil resources, and land-use options. A fundamental question is how to gather, document, and use this knowledge for the improvement of living conditions in the oases.

2.7.3 Climate Change

The rate of climate change is projected to be so rapid during the twenty-first century that many crops and their wild relatives will not be able to adjust their distributions to new areas that would be suitable for their survival. Of particular concern is the impact of the general warming trend in tropical and subtropical ecosystems (Arpaia et al. 2012) where most palms including *Phoenix dactylifera* exist. According to IPCC (2007), a "significant extinction of plant species is expected" when global average temperature increases by more than 3.5 °C; therefore, the potential for loss of date palm biodiversity, termination of evolutionary potential, and disruption of its environmental services must be taken more seriously (Dawson et al. 2011). In view of climate change, the capacity to enrich genetic diversity of palms in general and date palm in particular becomes an essential component of their *ecological dynamics* in order to adapt them to future global climate change. Abundance and shifts in the species distribution induced by climate change may affect not only the survival of the species but also the biodiversity-related ecosystem services.

The climate is changing and, as a consequence, some areas that are climatically suitable for date palm will become unsuitable in 50–70 years, including major areas in its center of origin and diversity (i.e., parts of the Arabian Peninsula, southern Iraq, western Iran, and North Africa). Other parts of the world may become more suitable for date palm, including parts of North and South America and Australia (Shabani et al. 2012). Global circulation models (IPCC 2007) suggested that dry and hyperarid regions could become hotter and drier. If climate change increases the frequency and intensity of droughts, it would lead to more desertification and the loss of genetic diversity, genetic resources, sustainability, and productivity of vulnerable oases.

The impact of global climate change on genetic diversity within populations and species will be manifested through spatiotemporal changes in the distribution of genetic variants as the ranges of species and populations change; changes in levels of phenotypic plasticity of populations and individuals as they respond to new environmental conditions; and evolutionary adaptation to changing environmental conditions (Pauls et al. 2013). These changes may reduce genetic diversity in populations and species, including date palm. Simulation studies suggested that temperature and drought stresses will play important roles in date palm adaptation and distribution in the twenty-first century, and the future distribution of date palms will most probably be impacted by climate change (Shabani et al. 2012). Recent examples from the Sahara Desert highlight the adverse impact of climate change on fragile oasis agroecosystems. Ten oases in southern Morocco have lost ~40 % of their vegetation due to the combined effects of drought, depletion of groundwater, high temperature, and sand encroachment (Bodian et al. 2012), whereas date palm orchards, along with their genetic diversity, disappeared altogether from wadi Alajal in Libya due to the depletion of desert aquifers (Racchi et al. 2013).

Climate change and human activities (i.e., anthropogenic climate change) play different but equally important roles in oasis evolution on different temporal scales.

The impact of climate change on oasis evolution is continuous and is manifested over large areas, whereas the impact of human activities is local and disconnected (Shabani et al. 2012; Turner et al. 2010). The evolution or development of a more sustainable and productive oasis and desertification are largely related to the abundance or shortage of water resources in the oasis. Changes in spatiotemporal water resources in the oasis are key determinants of the process of oasis evolution, whereas the effect of human activities on oasis evolution is manifested through the direct and indirect impact on water resources (Battesti 2013; Popenoe 1913), all of which impact, directly and indirectly, the biodiversity and genetic diversity of date palm. Also, climate change will likely affect the geographic distribution of pests and diseases. For example, the future geographic distribution of the red palm weevil was explored using ecological niche modeling. Areas where the pest was not reported yet were found to be suitable for invasion by this insect in different parts of the world as a result of climate change (Fiaboe et al. 2012).

These and similar consequences of climate change may point to the inability to timely and adequately quantify and value oasis agroecosystem services before they reach a *tipping point*. Furthermore, the accelerated degradation of oases and the loss of their biodiversity pose the question: *to what extent does the inherent buffering capacity provided by oasis agroecosystems enhance societal ability to adapt to and mitigate anthropogenic climate change*? Therefore, it is necessary to ensure that the oasis agroecosystems are sustainable by being resilient to future changes in global climate, markets, and other social and economic pressures. The genetic diversity of the date palm (and of other crops in the oasis) is an important component of that resilience and needs to be enhanced by ensuring that the wide range of existing cultivars is not further reduced. Also, it is important not to allow the market forces to dominate or dictate the selection of the cultivars grown or favored in the future.

2.8 International Collaboration on Date Palm Genetic Resources

The future collaboration on biodiversity and genetic resources of date palm will depend largely on advanced knowledge and information about the dynamics, management, and sustainability of the oasis ecosystem and in-depth understanding of the genetic diversity of the species and its wild relatives using analytical and predictive power of modern genetic analyses procedures (Chao and Krueger 2007; Eljuhani 2010; Jaradat 2011). Two outputs are anticipated from this collaboration on date palm genetic resources, these are:

(a) Indicators, metrics, and tools for evaluating the status and trends of date palm genetic resources, genetic diversity, and agrobiodiversity in the oasis ecosystem. The aim is to identify and compile appropriate indicators, metrics, and tools for evaluating the current status and ongoing changes in date palm genetic diversity and biodiversity, especially in view of climate change and in response to anthropogenic factors. The development of these tools, metrics, and indicators will utilize and apply theoretical concepts and applied results from biological, environmental, ecological, and geographic information system (GIS) fields of research and development.

(b) Measurable indicators, metrics, and standardized methodologies for genetic diversity and agrobiodiversity developed by teams of international collaborators will be used by scientists, date palm/oasis researchers, and practitioners to evaluate, improve, and advance date palm biodiversity throughout the world. As a result of this scientific knowledge and awareness of spatiotemporal changes in genetic diversity and agrobiodiversity, stakeholders will be able to implement mitigation measures to significantly reduce genetic diversity and agrobiodiversity loss in the oasis agroecosystems. Furthermore, stakeholders will be able to recognize the value of, and implement, specific indicators and measures, for improved resilience, sustainable conservation, and productivity of the oasis agroecosystem.

(c) Valuation tools for genetic diversity and agrobiodiversity for researchers and policymakers will be developed, tested, and refined for a model oasis system and its biodiversity components. This output includes the identification and evaluation of economic valuation methodologies, decision-support systems (DSS), incentive tools, and, where applicable, policy interventions, for the sustainable conservation and use of date palm genetic resources, genetic diversity, and agrobiodiversity in the oasis agroecosystem. An additional suboutput will be capacity building of stakeholder and a wide-ranging dissemination of research and development findings and demonstration of these findings to farmers, entrepreneurs, and the scientific and research community around the world.

A number of critical disciplines are indispensible for the proper functioning of such an international collaborative project on date palm genetic resources; these include, but not limited to, expertise in systematics and biodiversity conservation, diversity and ecology of soil microbiology, diversity and ecology of date palm economic insects, diversity and ecology of date palm diseases, molecular biology, population and quantitative genetics and genetic diversity analysis, geographic information systems (GIS)/geographical positioning systems (GPS) technology and land use planning, and, most importantly, plant quarantine to prevent the spread of diseases (especially bayoud) and insects (especially the red palm weevil), to new or uninfested oases, plantations, or regions.

There is a need to improve communications and access to information and inform the public and policymakers of issues relevant to date palm and its current and future status. Knowledge sharing on date palm as a model using ontology was suggested (Sherimon et al. 2012) as a means of exchange of information and knowledge; it defines a common vocabulary for researchers and others who need to share information in a particular domain, such as date palm genetic resources. Moreover, this approach allows for common understanding of the structure of the information among different users. Information from different sources needs to be brought together with advanced querying functionality on regional and global levels (Nawar and Mackay 2010).

A web-based delivery of data and information to scientists, students, horticulturists, and other stakeholders will facilitate and accelerate global communication and collaboration on all aspects of date palm biodiversity, genetic diversity, management, genetic resources, and biotic and abiotic stresses. Although proprietary methods are available for interactively searching and updating databases through web interfaces, these methods generally require varying costs to maintain licensing agreements. Publicly available software that requires minimal or flexible licensing costs provides a cost-effective alternative to stakeholders (research laboratories, research stations and research centers, field genebanks, departments of agriculture, horticulturists, and the public) who consider access to databases via a web-accessible interface (Sherimon et al. 2012). Such a public domain web-based tool can be configured not only to support collaborative activities among researchers in various locations but also to provide relevant data to collaborators, clients, and other stakeholders. Data and information structures can be modeled according to the format in use by the Genetic Resources Information Network (GRIN, www.ars-grin.gov/npgs/) which handles all aspects of plant genetic resources at a large scale, including passport, characterization, and evaluation data. Data and information providers will need a method to communicate with and either download, upload, or query data and information.

2.9 The Future of the Date Palm

The future of date palm as a dioecious monocot fruit tree largely depends on two interrelated components of a long-term and global strategy; these are in-depth understanding of the genetic diversity of the species and its wild relatives using analytical and predictive powers of quantitative trait loci, somatic cell hybridization, and genomics to overcome some of the genetic research limitations (Alsaoud and Ajlan 2013; Jain 2010) and develop advanced knowledge and information systems about the dynamics, management, and sustainability of the oasis agroecosystem (Sherimon et al. 2012; Yin et al. 2012; Zhang et al. 2012). While it is not too difficult to define the conditions for achieving a sustainable and resilient oasis agroecosystem, managing such a system requires understanding of ecosystem functions in changing climatic, social, and economic conditions. Crucial aspects of those ecosystem functions may depend on certain components that have not previously been studied in sufficient detail in relation to the oasis agroecosystem, in general, and the date palm, in particular (Jaradat 2011). The sustainability and provision of multiple ecosystem services, including provisioning, supporting, regulating, and cultural services, will largely depend on a highly diverse genetic base of date palms.

In-depth assessment of the genetic vulnerability of date palm to climate change, desertification, and biotic and abiotic stresses requires knowledge of the extent and distribution of its genetic diversity, both of which depend on the species evolution and unique breeding system, past genetic bottlenecks, and ecological, geographical, and anthropogenic factors. Therefore, it is necessary to ensure that the oasis agroecosystems are sustainable by being resilient to future changes in global climate, markets, and other social and economic pressures. The genetic diversity of the date palm (and other perennial, annual, and forage crops and their wild and weedy relatives in the oasis) is an important component of that resilience and needs to be enhanced by ensuring that the wide range of existing species and cultivars is not further reduced, especially due to market pressure and climate change (IPCC 2007; Pauls et al. 2013; Shabani et al. 2012).

The impact of climate change may affect date palm in many ways, including biodiversity, genetic diversity, adaptation, and production potential. Therefore, the conservation, regeneration, distribution, and utilization of date palm genetic resources will be more urgent in the foreseeable future (Eljuhani 2010; Ezebilo et al. 2013). Date palm wild and feral genotypes that are geographically and ecologically widespread can be assumed to be under less threat of genetic erosion than those localized or restricted to a specific habitat; the latter should be given priority in conservation (Francisco-Ortega et al. 2000; Gaitto et al. 2003; Shapcott et al. 2009).

Although clonal propagation using offshoots and tissue culture maintains heterozygosity and genetic purity of female cultivars, it promotes genetic uniformity and may accelerate genetic erosion or enhance vulnerability of the date palm to environmental stresses. Therefore, the maintenance of genetic variation within and among oases remains a central question in the study of evolutionary biology and the production of genetically diverse populations of date palm. Strongly selected traits, especially through mass propagation using tissue culture, are expected to have low levels of genetic variance and lower heritability, whereas traits that are closely associated with fitness likely will have higher levels of genetic variance but lower heritability than weakly selected traits.

The future of date palm improvement will be centered on comparisons of its genome across species and cultivars, and some of the best opportunities may lie in using combinations of new genetic mapping strategies and evolutionary analyses to direct and optimize the discovery and use of genetic variation that can be used for yield and quality improvement and adaptation of the date palm to future climatic and edaphic conditions (Aldous et al. 2011; Dhanapal 2012; Fang et al. 2012; Yang et al. 2013). Future advances in developing elite date palm cultivars will depend on the identification or development of molecular and phenotypic markers that may assist in identifying economically and horticulturally important traits and cultivars. Detailed analyses of date palm populations originating from different geographic locations will help in understanding their genetic structures and will reveal the extent of geneflow through seed, seedlings, tissue culture, or offshoots, between populations and its impact on population structure and fruit quality (Bodian et al. 2012; Elhoumaizi et al. 2006; Gaitto et al. 2003).

2.10 Research Needs

It is pertinent to assume that the bulk of future research on date palm will be carried out in Middle Eastern and North African countries where dates are an important economic commodity and the date palm is a culturally significant fruit tree. However, recent research, development, and breakthroughs from outside this region are increasing and promise to revolutionize the future of the date palm (e.g., Ataga et al. 2012; de Grenade 2013; Diaz et al. 2003; Fang et al. 2012). The establishment of an International Consortium for Palm Research (ICPR) was proposed (Al-Khayri and Niblett 2012) to foster innovative international research collaboration for palm improvement, productivity, and utilization on a global basis. Research priorities on the extremely serious red palm weevil, *Rhynchophorus ferrugineus* (Olivier), vectoring the very serious red ring nematode (*Bursaphelenchus cocophilus* Baujard), and *Fusarium* diseases (bayoud, *Fusarium oxysporum* f.sp. *albedinis*), bud root disease (*Phytophthora palmivora*), and the phytoplasmas causing Alwijam disease have been recently emphasized and may constitute an initial set of collaborative regional or international research projects.

Research and development questions with significant importance for the future of date palm biodiversity and genetic diversity may include the following:

(a) What are the functions and services the date palm can provide within, and contribute to the survival of, oasis agroecosystems?
(b) How can the date palm impact and interact with other components of the biodiversity complex within oasis agroecosystems, and what are the practical implications of these interactions?
(c) How is the genetic diversity partitioned within and among populations and within and among traditional oasis and modern plantations?
(d) What are the scientific and practical implications for the conservation of this genetic diversity?
(e) What are the pros and cons of mass vegetative reproduction of date palm through tissue culture, and what are the consequences of this technology on total diversity and vulnerability of the species?
(f) Where are the *hot spots* for key tree traits for biotic and abiotic stress tolerance and for fruit quality traits, and how are these utilized efficiently?

More importantly, there is a need for capacity building and strengthening of existing research centers (Al-Khayri and Niblett 2012; Sherimon et al. 2012), development of a rapid, reliable and efficient means for information exchange, storage and retrieval of data to facilitate the utilization of germplasm (Gotor et al. 2008), and establishment of regional and national date palm field genebanks in the Middle East and North Africa (Alghamdi 2001; Bettencourt et al. 1992; Krueger 1998, 2011) and other parts of the world. Collectively, these would enhance research on current and future problems important to local production, both in traditional oases and large-scale plantations. Building relational databases on wild and domesticated species, tree and fruit phenotypic and biotechnological attributes of populations and cultivars, and the development of a *Digital Atlas* will help document and provide online information for research, conservation, and sustainable utilization of date palm genetic resources. Finally, the development of alternative markets for date palm by-products will create incentives to grow more and diverse date palm cultivars, encourage the development of a wide range of products based on phenotypic and fruit traits, and enhance the role of date palm as a functional genetic resource (Hasnaoui et al. 2011; Jaradat 2011).

Improving date quality is a major breeding and selection objective; it depends on access to improved cultivars adapted to local climate conditions and on the skill and dedication of scientists and date producers (Rivas et al. 2012). However, for a successful date palm industry, accurate estimates of genetic diversity and its partitioning, especially for fruit quality traits and tolerance to biotic and abiotic stresses within and among gene pools in its center of origin and center of diversity, are important considerations. Increased global trade and germplasm exchange present a number of challenges to date production whether in traditional oases or in large-scale plantations. Such challenges stem from increased pressure on water supplies, salinity, and biotic stresses caused by traditional, invasive, or emerging pests and diseases (Allam and Cheloufi 2012; Alsaoud and Ajlan 2013).

Genes or gene complexes of potential value and use in meeting these future challenges may well be present in nonelite date palm cultivars found in small gardens or traditional oases, but their presence and characteristics are largely unknown. Therefore, traditional farmers should be encouraged to replant orchards with locally produced and highly heterozygous and heterogeneous offshoots or seedlings. Replacement of old or dead date palm trees with only elite and foreign cultivars will diminish genetic diversity and hasten genetic erosion of locally adapted cultivars. Opportunities exist for date palm improvement through biotechnological research to identify and quantify genetic diversity components in the species, identify and clone genes and gene complexes for biotic and abiotic stresses, and utilize the generated information for future research and development.

The assessment of interspecific hybridization and introgression between species or subspecies is important for the implementation of appropriate genetic conservation strategies and for the assessment of overall biodiversity. Efficient management of wild relatives of date palm needs to identify and conserve the remaining unique populations and to evaluate the extent to which they are endangered by the introduction of the cultivated species. Efficient genetic transformation methods can be utilized to incorporate desired traits in newly developed cultivars. New technologies of genetic manipulation allow the transfer of selected gene(s) to a specific genotype in only a single generation that would not be possible by conventional breeding (Rivas et al. 2012). It is anticipated that access to seed of backcrosses that are identified as being female before germination and genotyped at trait loci to show similarity to the mother plant will soon become available to date palm geneticists and breeders (Aldous et al. 2011). These developments lay the foundation for future date palm research at the genomic level by providing the first genome-wide gene set, multi-variety polymorphism set, and sex-linked regions within relevant chromosomes for this species.

It is postulated that any molecular tool that provides a shortcut to breeding would be valuable for date palm improvement. Qualitative traits, such as disease and pest resistance, or sex of seedlings and immature plants, if determined at an early stage of growth and development using linked molecular markers without phenotyping, will allow breeders to select elite and novel genotypes with desirable traits at an early stage as candidates for breeding programs. The establishment of cultivar-specific molecular signature will make it possible to precisely identify those date palm cultivars which are being used in drug formulations (Alquraini et al. 2011). Emerging sequenced genomes of date palm may open the way for further detailed analysis of the genetic changes that have occurred during the history of date palm domestication and cultivation, including its productivity (Goldschmidt 2013).

An ambitious research and development proposal (Elshibli and Korpelainen 2008) was suggested to investigate the date palm genome at a worldwide scale and the identification of proper markers that may assist in identifying the economically and horticulturally important cultivars. Detailed analyses of date palm populations from different ecogeographical regions will promote the understanding of their genomes and will reveal the true extent of gene flow and germplasm exchange between populations. Moreover, species distribution models, bioclimatic models, and ecological niche models are becoming increasingly valuable tools in predicting date palm adaptation during the next ~100 years to climate change and can inform strategic planning to identify, possible at the cultivar level, new ecogeographical regions for date palm production and minimize economic impacts in areas that may be adversely impacted by climate change while preparing farmers and policymakers to take advantage of new opportunities in regions that may be positively influenced by the anticipated climate change (Shabani et al. 2012).

2.11 Conclusions and Recommendations

The domesticated date palm is closely related to a variable aggregate of wild and feral palms distributed over the southern, warm, and dry Middle East region as well as the northeastern Saharan and north Arabian deserts. The wild and feral palms show close morphological similarities and parallel climatic requirements with the domesticated species. In addition, these palms are interfertile with the cultivars and are interconnected with them through occasional hybridization. Date palm is composed of genetically discrete clones representing highly heterozygous cultivars without the benefits of a dynamic mutation-recombination system.

Domestication of the date palm has led to the increase in fruit size and pulp quality and to a shift to vegetative propagation, but not total elimination of sexual reproduction. Genetically, the vegetative mode of propagation resulted in the immediate fixation of desirable tree and fruit traits in highly heterozygous female cultivars. The strong artificial selection and vegetative propagation of date palms in oasis agroecosystems greatly altered their original genetic structure. The date palm is the dominant component upon which the oasis agroecosystem is based. The tremendous advantage of the date palm tree is its resilience, its long-term productivity, and its multipurpose attributes. However, some of its unique characteristics (i.e., slow growth, dioecy, the slow offshoot-based propagation system and the difficulty of predicting adult characteristics of the seedlings) have severely restricted its improvement.

Vegetative propagation by offshoots is the only method to maintain genetic integrity of date palm cultivars. Man-made selection resulted in increased yield and fruit size and palatability and reduced branching (offshoots) and facilitated its propagation. Traditional oases continue to play a vital role in the maintenance and enrichment of date palm genetic resources and their genetic diversity through multiple processes and dynamic conservation practices. Date palm (bio)diversity and its link to the properties of the oasis agroecosystem have cultural, intellectual, aesthetic, and spiritual values that are important to the general public.

A biodiversity-based model for sustainable agriculture in oasis agroecosystems is potentially the most cost-effective and durable solution for the problems associated with or emanating from anthropogenic, biotic, and abiotic stresses. Due to the unique, and often fragile, oasis agroecosystem, it is imperative to understand the combined social and ecological functions of its biodiversity, determine its contribution to ecosystem services and aesthetic value to society, and evaluate options for sustainable utilization of this biodiversity. The within-species diversity of the date palm is being reduced by pressures for higher productivity and concentration of the market on a few high-quality cultivars due to consumer demand. Furthermore, the accelerated degradation of oases and the loss of their biodiversity pose the question: to what extent does the inherent buffering capacity provided by oasis agroecosystems enhance societal ability to adapt to and mitigate anthropogenic climate change? Therefore, it is necessary to ensure that the oasis agroecosystems are sustainable by being resilient to future changes in global climate, markets, and other social and economic pressures.

The genetic diversity of the date palm and biodiversity in the oasis are important components of that resilience and need to be enhanced by ensuring that the wide range of existing cultivars is not further reduced. The future of date palm largely depends on developing advanced knowledge and information about the dynamics, management, and sustainability of the oasis agroecosystem and in-depth understanding of the genetic diversity of the species and its wild relatives. It is necessary to ensure that the oasis agroecosystems are sustainable by being resilient to future changes in global climate, markets, and other social and economic pressures. The genetic diversity of the date palm is an important component of that resilience and needs to be enhanced by ensuring that the wide range of existing species and cultivars is not further reduced, especially due to market pressure.

Combining comprehensive sequence information with in-depth knowledge of the morphological and physiological diversity of palms, in general, and the date palm, in particular, and well-understood phylogeny promises to answer many questions about genome evolution and function in this species. A small subset of carefully selected molecular markers can serve as a starting point for a set capable of distinguishing between the 2,000–5,000 date palm cultivars assumed to be cultivated throughout the world. This resource will allow future comparisons of traits, such as fruit quality and ripening time, and to dissect genomic differences between male and female date palms.

Public awareness may help create conducive environment for the conservation and sustainable utilization of date palm genetic diversity. In addition to the foregoing, this objective can be achieved by organizing more focused date fares, where the public, including policymakers and producers, can experience the diversity of dates and date products and begin to appreciate the importance of making use of more and diverse date palm cultivars. However, two research and development questions remain to be critically debated, researched, and answered in this regard: What are the short-term prospects of developing new fruit- or tree-based products to enhance the sustainable utilization of date palm genetic diversity, and how can sustainable utilization of traditional and new products through the standardization of fruit quality traits, opening new markets, and public engagement be promoted?

References

Ahmed M, Bouna Z, Lemine F et al (2011) Use of multivariate analysis to assess phenotypic diversity of date palm (*Phoenix dactylifera* L.) cultivars. Sci Hortic 127:367–371

Ahmed TA, Alhadidi SH, Alqaradawi AY, Radwan O (2013) Determination of inter- and intra-specific genetic variations among Qatari date palm cultivars using inter simple sequence repeat (ISSR) markers. Afr J Biotechnol 12:2540–2546

Akkak A, Scariot V, Marinoni DT et al (2009) Development and evaluation of microsatellite markers in *Phoenix dactylifera* L. and their transferability to other *Phoenix* species. Biol Plant 53:164–166

Al-Bahrany AM, Al-Khayri JM (2012) Optimizing *in vitro* cryopreservation of date palm (*Phoenix dactylifera* L.). Biotechnology 11:59–66

Aldous EK, George B, Al-Mahmoud ME et al (2011) *De novo* genome sequencing and comparative genomics of date palm (*Phoenix dactylifera* L.). Nat Biotechnol 29:521–528

Alghamdi AS (2001) Date palm (*Phoenix dactylifera* L.) germplasm bank in King Faisal University, Saudi Arabia. Survival and adaptability of tissue cultured plantlets. Acta Hortic 450:241–244

Alhudaib K, Arocha Y, Wilson M, Jones P (2007) Identification and molecular characterization of a phytoplasma associated with Al-Wajim disease of date palm in Saudi Arabia. Arab J Plant Prot 25:116–122

Alkhalifah NS, Askari E (2003) Molecular phylogeny of date palm (*Phoenix dactylifera* L.) cultivars from Saudi Arabia by DNA fingerprinting. Theor Appl Genet 107:1266–1270

Alkhalifah NS, Askari E, Khan AES (2012) Molecular and morphological identification of some elite varieties of date palms grown in Saudi Arabia. Emir J Food Agric 24:456–461

Al-Khayri JM, Niblett C (2012) Envision of an international consortium for palm research. Emir J Food Agric 24(5):470–479

Allam A, Cheloufi H (2012) Biodiversity of fruit species in the valley of Oued Righ: the case of the area of Touggourt (Algeria). Fruits 67:1–5

Alquraini F, Khan S, Alhemaid FM et al (2011) Assessing molecular signature for some potential date palm (*Phoenix dactylifera* L.) cultivars from Saudi Arabia, based on chloroplast DNA sequences *rpoB* and *psbA-trnH*. Int J Mol Sci 12:6871–6880

Alsaoud A, Ajlan A (2013) Effect of date fruits quantity on the numbers of red palm weevil, *Rhynchophorus ferrugineus* (Oliver), captured in aggregation pheromone traps. Agric Biol J North Am. doi:10.5251/abjna.2013.4.4.496.503

Alyahyai R, Alkhanjari S (2008) Biodiversity of date palm in the Sultanate of Oman. Afr J Agric Res 3:389–395

Arpaia S, Battafarano R, Chen L-Y et al (2012) Assessment of transgene flow in tomato and potential effects of genetically modified tomato expressing Cry3Bb1 toxins on bumblebee feeding behavior. Ann Appl Biol 161:151–160

Askari E, Al-Khalifah NS, Ohmura T et al (2003) Molecular phylogeny of seven date palm cultivars by DNA fingerprinting. Pak J Bot 35:323–330

Ataga CD, Mohammed AH, Yusuf AO (2012) Status of date palm (*Phoenix dactylifera* L.) genetic resources in Nigeria. Int J Life Sci Pharm Res 2:L46–L51

Barfod AS, Hagen M, Borchsenius F (2011) Twenty-five years of progress in understanding pollination mechanisms in palms (Arecaceae). Ann Bot 108:1503–1516
Barrow SC (1998) A monograph of *Phoenix* L. (Palmae: Coryphoideae). Kew Bull 53:513–575
Battesti V (2013) L'agrobiodiversité du dattier (*Phoenix dactylifera* L.) dans l'oasis de Siwa (Égypt): entre ce qui se dit, s'écrit et s'oublie. La Rev d'Ethnoécol, hal-00707908 version 3. http://hal.archives-ouveretes.fr/hal-00707908
Bekheet SA, Taha HS, Saker MM (2001) *In vitro* long-term storage of date palm. Biol Plant 45(1):121–124
Bekheet SA, Taha HS, Elbahr MK (2005) Preservation of date palm cultures using encapsulated somatic embryos. Arab J Biotechnol 8(2):319–328
Bekheet SA, Taha HS, Solliman ME, Hassan NA (2007) Cryopreservation of date palm (*Phoenix dactylifera* L.) culture *in vitro*. Acta Hortic 736:283–291
Bettencourt E, Hazekamp T, Perry MC (1992) Directory of germplasm collections. 6.1. Tropical and subtropical fruits and tree nuts. IBPGR, Rome
Bodian A, Elhoumaizi MA, Ndoye Ndir K et al (2012) Genetic diversity analysis of date palm cultivars from Figuig oasis (Morocco) using SSR markers. Int J Sci Adv Technol 2(3):96–104
Chao CCT, Krueger RR (2007) The date palm (*Phoenix dactylifera* L.): overview of biology, uses, and cultivation. Hortic Sci 42:1077–1082
Charlesworth D (2013) Plant sex chromosome evolution. J Exp Bot 64:405–420
Cherif E, Zehdi S, Castillo K et al (2012) Male-specific DNA markers provide genetic evidence of an XY chromosome system, a recombination arrest and allow the tracing of parental lineages in date palm. New Phytol 197:409–415
Chevalier A (1952) Recherches sur les *Phoenix africains*. Rev Int Bot Appl 32:205–236
Cruz-Cruz CA, González-Arnao MT, Engelmann F (2013) Biotechnology and conservation of plant biodiversity. Resources 2:73–95
Dawson TP, Jackson ST, House JI et al (2011) Beyond predictions: biodiversity conservation in a changing climate. Science 332:53–58
de Grenade R (2013) Date palm as a keystone species in Baja California peninsula, Mexico oases. J Arid Environ 94:59–67
Dhanapal AP (2012) Genomics of crop plant genetic resources. Adv Biosci Biotechnol 3:378–385
Dhieb A, Elleuch A, Kriaa W et al (2012) Molecular characterization and *in silico* analysis of RNA polymerase alpha subunit gene (*rpoA*) in date palm (*Phoenix dactylifera* L.) cv. Deglet Nour. Genet Genomics 34:599–608
Diaz S, Pire C, Ferrer J, Bonete MJ (2003) Identification of *Phoenix dactylifera* L. varieties based on amplified fragment length polymorphism (AFLP) markers. Cell Mol Biol Lett 8:891–899
Dransfield J, Uhl NW, Assmusen CB et al (2008) Genera Palmarum: the evolution and classification of palms. Kew Publishing, Royal Botanic Gardens, Kew
Eissa EA, Abdelrazak AB, Elsharabishi SF, Rizk RM (2009) Morphological and molecular genetic characterization of soft date palm (*Phoenix dactylifera* L.) cultivars in Egypt. Egypt J Genet Cytol 38:269–284
Elhoumaizi MA, Devanand PS, Fang J, Chao CCT (2006) Confirmation of "Medjool" date as a landrace variety through genetic analysis of "Medjool" accessions in Morocco. J Am Soc Hortic 13:403–407
Eljuhani LI (2010) Degradation of date palm trees and date production in Arab countries: causes and potential rehabilitation. Aust J Basic Appl Sci 4:3998–4010
Elkhishin DA, Adawy SS, Hussein EHA, Elitriby HA (2003) AFLP fingerprinting of some Egyptian date palm (*Phoenix dactylifera* L.) cultivars. Arab J Biotechnol 6:223–234
Elmeer K, Mattat I (2012) Marker-assisted sex differentiation in date palm using simple sequence repeats. 3 Biotech 2:241–247
Elmeer K, Sarwath H, Malek J et al (2011) New microsatellite markers for assessment of genetic diversity in date palm (*Phoenix dactylifera* L.). 3 Biotech 1:91–97
Elshibli S, Korpelainen H (2008) Microsatellite markers reveal high genetic diversity in date palm (*Phoenix dactylifera* L.) germplasm from Sudan. Genetica 134:251–260
Elshibli S, Korpelainen H (2009a) Excess heterozygosity and scarce genetic differentiation in the populations of *Phoenix dactylifera* L.: human impact or ecological determinants. Plant Genet Res 7:95–104

Elshibli S, Korpelainen H (2009b) Biodiversity of the date palm (*Phoenix dactylifera* L.) in Sudan: chemical, morphological and DNA polymorphisms of selected cultivars. Plant Genet Res 7:194–203

Eltarras A, Tawatti N, Almaliki F (2002) Genetic fingerprinting of some KSA date palm using modern molecular biotechnological techniques. Egypt J Hortic 29:217–228

Ezebilo EE, Elasafi M, Garkava-Gustavsson L (2013) On-farm diversity of date palm (*Phoenix dactylifera* L.) in Sudan: a potential genetic resources conservation strategy. Sustainability 5:338–356

Fang Y, Wu H, Zhang T et al (2012) A complete sequence and trascriptomic analyses of date palm (*Phoenix dactylifera* L.) mitochondrial genome. PLoS One 7:e37164. doi:10.1371/journal.pone.0037164

FAOSTAT (2012) http://faostat.fao.org/

Fernandez D, Ouinten M, Tantaoui A, Geiger J-P (1997) Molecular records of micro-evolution within the Algerian population of *Fusarium oxysporum* f. sp. *albedinis* during its spread to new oases. Eur J Plant Pathol 103:485–490

Fiaboe KKM, Peterson AT, Kairo MTK, Roda AL (2012) Predicting the potential worldwide distribution of the red palm weevil *Rhynchophorus ferrugineus* (Olivier) (Coleoptera: Curculionidae) using ecological niche modeling. Florida Entomol 95:659–673

Francisco-Ortega J, Santos-Guerra A, Kim S-C, Crawford DJ (2000) Plant genetic diversity in the Canary Islands: a conservation perspective. Am J Bot 87:909–919

Frison EA, Putter CAJ, Diekmann M (1993) FAO-IBPGR technical guidelines for safe movement of coconut germplasm. D/T0870E/1/1.94/4000. FAO-IBPGR, Rome

Gaitto FA, Grattapaglia D, Vencovsky R (2003) Genetic structure, mating system, and long-distance gene flow in heart of palm (*Euterpe edulis* Mart.). J Hered 94:399–406

Gao Q, Yue G, Li W et al (2012) Recent progress using high-throughput sequencing technologies in plant molecular biology. J Integr Plant Biol 54:215–227

Gebauer J, Luedeling E, Hammer K et al (2007) Mountain oases in northern Oman: an environment for evolution and in situ conservation of plant genetic resources. Genet Res Crop Evol 54:465–481

Goldschmidt EE (2013) The evolution of fruit productivity: a review. Econ Bot 67(1):51–62

González-Benito ME, Clavero-Ramírez I, López-Arnada JM (2004) The use of cryopreservation for germplasm conservation of vegetatively propagated crops. Span J Agric Res 2(3):341–351

González-Perez MA, Caujape-Castells J, Sosa PA (2004) Molecular evidence of hybridization between the endemic *Phoenix canariensis* and the widespread *Phoenix dactylifera* L. with random amplified polymorphic DNA (RAPD) markers. Plant Syst Evol 247:165–175

Gotor E, Alercia A, Rao VR et al (2008) The scientific information activity of Bioversity International: the descriptor lists. Genet Res Crop Evol 55:757–772

Gschwend AR, Weingratner LA, More RC, Ming R (2012) The sex-specific region of sex chromosome in animals and plants. Chromosome Res 20:57–69

Hammadi H, Mokhtar R, Mokhtar E, Ali AF (2009) New approach for morphological identification of date palm (*Phoenix dactylifera* L.) cultivars from Tunisia. Pak J Bot 41:2671–2681

Hammadi H, Vendramin GG, Ali AF (2011) Microsatellite diversity among Tunisian date palm (*Phoenix dactylifera* L.) subpopulations. Pak J Bot 43:1257–1264

Hamza H, Elbekkay M, Benabderrahim MA, Ferchichi Ali A (2011) Molecular and morphological analyses of date palm subpopulations in southern Tunisia. Span J Agric Res 9(2):484–493

Hasnaoui A, Elhoumaizi MA, Hakkou A et al (2011) Physico-chemical characterization, classification and quality evaluation of date palm fruits of some Moroccan cultivars. J Sci Res 3:139–149

Honnay O, Jacquemyn H (2008) A meta-analysis of the relation between mating system, growth form and genotypic diversity in clonal plant species. Evol Ecol 22:299–312

IPCC (2007) Fourth Assessment Report. In: Parry ML, Canziani OF, Palutikof JP et al (eds) Climate change 2007. Impacts, adaptation and vulnerability. Cambridge University Press, Cambridge/New York

IPGRI (2005) Descripteurs du palmier dattier (*Phoenix dactylifera* L.). Institut International des Resources Phytogenetiques, Rome. http://ipgri.cgiar.org.et; http://www.magrebdatepalm.org

Iqbal M, Munir M, Khan M (2009) Floral characteristics of the different male date palms and their response to fruit setting and yield of cv Dhakki. Pak J Agric Res 22:36–41

Iqbal M, Munir M, Niamatullah M (2011) Physico-chemical characteristics of date palm (*Phoenix dactylifera* L.) cultivars at various maturity stages under environmental conditions of Dera Ismail Khan. J Agric Res 49:249–259

Isbell F, Reich B, Tilman D et al (2013) Nutrient enrichment, biodiversity loss, and consequent declines in ecosystem productivity. Proc Natl Acad Sci 110:11911–11916, www.pnas.org/cgi/doi/10.1073/pnas.1310880110

Jain SM (2010) Date palm genetic diversity conservation for sustainable production. Acta Hortic 882:785–792

Jaradat AA (2011) Biodiversity of date palm. Land use, land cover and soil sciences. In: Encyclopedia of life support systems (EOLSS), developed under the auspices of the UNESCO. Eolss Publishers, Oxford. Available: http://www.eolss.net. 1 Apr 2013

Jaradat AA (2013) Multilevel analysis of date palm biodiversity and genetic diversity. Proceedings of the fifth date palm conference, Al Hassa, 3–5 Nov 2013

Jaradat AA, Zaid A (2004) Quality traits of date palm fruits in a center of origin and center of diversity. J Food Agric Environ 2:208–217

Jubrael JMS (2005) Assessment of AFLP-based genetic relationships among date palm (*Phoenix dactylifera* L.) varieties of Iraq. J Am Soc Hortic Sci 130:442–447

Karim K, Chokri B, Amel S et al (2010) Genetic diversity of Tunisian date palm germplasm using ISSR markers. Int J Bot 6:182–186

Khanam S, Sham A, Bennetzen JL, Aly MAM (2012) Analysis of molecular marker-based characterization and genetic variation in date palm (*Phoenix dactylifera* L.). Aust J Crop Sci 6:1236–1244

Khieralla HSM, Bader SM, Baum M, Hamwieh A (2011a) Assessment of genetic diversity of some Iraqi date palms (*Phoenix dactylifera* L.) using AFLP markers. Afr J Biotechnol 10:9570–9575

Khieralla HSM, Bader SM, Baum M, Hamwieh A (2011b) Genetic diversity of Iraqi date palms revealed by microsatellite polymorphism. J Am Soc Hortic Sci 136:282–287

Krueger RR (1998) Date palm germplasm: overview and utilization in USA. In: Proceedings first international conference on date palms. Al-Ain, March 1998, pp 2–37

Krueger RR (2011) Date palm germplasm. In: Jain SM, Al-Khayri JM, Johnson DV (eds) Date palm biotechnology. Springer, Dordrecht, pp 313–336

Malaurie B (2001) Medium- and long-term conservation and safe international exchange of germplasm from food and cash tropical crops. Acta Hortic 560:69–77

Masmoudi-Allouchi F, Châari-Rkhis A, Kriâa W et al (2009) In vitro hermaphrodism induction in date palm female flower. Plant Cell Rep 28:1–10

McKey D, Elias M, Pujol B, Duputie A (2010) The evolutionary ecology of clonally propagated domesticated plants. New Phytol 186:318–332

Meilleur BA, Hodgkin T (2004) In situ conservation of crop wild relatives: status and trends. Biodivers Conser 13:663–684

Meyer R, DuVal AE, Jensen HR (2012) Patterns and processes in crop domestication: an historical review and quantitative analysis of 203 global food crops. New Phytol 196:29–48

Mohamed BS, Rachid H (2006) Description phenopomologique de 15 cultivars tunisiens de palmier dattier (*Phoenix dactylifera* L.). Plant Genet Res News Lett 148:10–18

Munier P (1981) Origine de la culture du palmier-dattier et sa propagation en Afrique: notes historiques sur les principales palmeraies africaines. Fruits 36:689–706

Nabhan GP (2007) Agrobiodiversity change in a Saharan Desert Oasis, 1919–2005: historic shifts in Tasiwit (Berber) and Bedouin crop inventories of Siwa, Egypt. Econ Bot 61:31–43

Nawar F, Mackay M (2010) A global information portal to facilitate and promote accessibility and rational utilization of ex situ plant genetic resources for food and agriculture. IAALD XIIIth World Congress, Montpellier, 26–29 Apr 2010

Near East and North Africa Plant Genetic Resources Network (NENA-PGRN) (2011) A conservation and sustainable use of plant genetic resources: a strategy for the Near East and North Africa Region. Nov 2011

Nixon RW (1951) The date palm: "tree of life" in subtropical deserts. Econ Bot 5:274–301

Noriega IL, Halewood M, Lapeña I (2012) The multilateral system of access and benefit sharing: Case studies on implementation in Kenya, Morocco, Philippines and Peru. Bioversity International. http://biodiversityinternational.org/

Othmani A, Bayoudh C, Drira N et al (2009) Regeneration and molecular analysis of date palm (*Phoenix dactylifera* L.) plantlets using RAPD markers. Afr J Biotechnol 8(5):813–820

Ouarda HE, Walker DJ, Khoja ML (2012) Phenotypic and molecular DNA variation in Tunisian cultivars of date palm. Afr J Biotechnol 11(22):6034–6042

Ould Mohamed Salem A, Trifi M, Sakka H et al (2001) Genetic inheritance analysis of four enzymes in date palm (*Phoenix dactylifera* L.). Genet Res Crop Evol 48:361–368

Ould Mohamed Salem A, Rhouma S, Zehdi S et al (2008) Morphological variability of Mauritanian date palm (*Phoenix dactylifera* L.) cultivars as revealed by vegetative traits. Acta Bot Croat 67:81–90

Pauls SU, Nowak C, Balint M, Pfenninger M (2013) The impact of global climate change on genetic diversity within populations and species. Mol Ecol 22:925–946

Ponnuswami V (2010) Genetic diversity in palmyrah genotypes using morphological and molecular markers. Electr J Plant Breed 1(4):556–567

Popenoe PB (1913) Date growing in the Old World and the New. West India Gardens, Altadena

Racchi ML, Bove A, Turchi A et al (2013) Genetic characterization of Libyan date palm resources by microsatellite markers. 3 Biotech 4:21–32. doi:10.1007/s13205-013-0116-6

Rivas M, Barbieri RL, da Maia LC (2012) Plant breeding and in situ utilization of palm trees. Ciênc Rural, Santa Maria 42:261–269

Rivera D, Obón de Castro C, Carreno E et al (2008) Morphological systematic of date palm diversity (*Phoenix*, Arecaceae) in Western Europe and some preliminary molecular results. Acta Hortic 799:97–104

Sedra MH (2013) Genetic diversity analysis of Moroccan cultivar genotyping and rapid screening for bayoud disease resistance in date palm using molecular techniques. Acta Hortic 994:271–286

Sghaier-Hammami B, Drira N, Jorrin-Novo J (2009) Comparative 2-DE proteomic analysis of date palm somatic and zygotic embryos. J Proteomics 73:161–177

Shabani F, Kumar L, Taylor S (2012) Climate change impacts on the future distribution of date palms: a modeling exercise using CLIMAX. PLoS One 7:e48021. doi:10.1371/journal.pone.0048021

Shapcott A, Dowe JL, Ford H (2009) Low genetic diversity and recovery implications of the vulnerable Bankouale palm *Livistona carinensis* (Aracaceae), from north-eastern Africa and the southern Arabian Peninsula. Conserv Genet 10:317–327

Sherimon PC, Krishnan R, Vinu PV (2012) Knowledge sharing using ontology- a date palm model. Int J Adv Res Comput Sci Softw Eng 2(3):316–321

Siljak-Yakovlev S, Benmalek S, Cerbah M et al (1996) Chromosomal sex determination and heterochromatin structure in date palm. Sex Plant Reprod 9:127–132

Tengberg M (2012) Beginning and early history of date palm garden cultivation in the Middle East. J Arid Environ 86:139–147

Terral J-F, Newton C, Ivorra S et al (2012) Insights into the historical biogeography of the date palm (*Phoenix dactylifera* L.) using geometric morphometry of modern and ancient seed. J Biogeogr 39:929–941

Turner WR, Bradley BA, Estes LD et al (2010) Climate change: helping nature survive the human response. Conserv Lett. doi:10.1111/j.1755-263X.2010.00128.x

Walsborn R (2008) Date sector report and value chain development program. USAID/Iraq/Inma Agribusiness Program Contract No 267-C-00-07-00500-00. 59 pp. USAID. Washington, DC.

Weiblen GD, Oyama RK, Donoghue MJ (2000) Phylogenetic analysis of dioecy in monocotolydons. Am Nat 155:46–58

Wrigley G (1995) Date palm (*Phoenix dactylifera*). In: Smartt J, Simmonds NW (eds) The evolution of crop plants, 2nd edn. Longman, Essex

Yang M, Zhang X, Liu G et al (2013) The complete chloroplast genome sequence of date palm. PLoS One 5:e12762. doi:10.1371/journal.pone.0012762

Yin Y, Zhang X, Fang Y et al (2012) High-throughput sequencing-based gene profiling on multi-staged fruit development of date palm (*Phoenix dactylifera* L.). Plant Mol Biol 78:617–626

Younis RAA, Ismail OM, Soliman SS (2008) Identification of sex-specific DNA markers for date palm (*Phoenix dactylifera* L.) using RAPD and ISSR techniques. Res J Agric Biol Sci 4:278–284

Zaid A, de Wet PF (2005) Botanical and systematic description of the date palm. http://www.fao.oeg/DOCREP/006/Y3460E/y3460e05.htm#bm05.3/

Zehdi S, Trifi M, Billotti N et al (2004a) Genetic diversity of Tunisian date palm (*Phoenix dactylifera* L.) revealed by nuclear microsatellite polymorphism. Hereditas 141:278–287

Zehdi S, Sakka H, Rhouma A, Ould Mohammed Salem A et al (2004b) Analysis of Tunisian date palm germplasm using simple sequence repeat primers. Afr J Biotechnol 3:215–219

Zhang G, Pan L, Liu W et al (2012) Large-scale collection and annotation of gene models for date palm (*Phoenix dactylifera* L.). Plant Mol Biol 79:521–536

Zhao Y, Williams R, Prakash CS, He G (2013) Identification and characterization of gene-based SSR markers in date palm (*Phoenix dactylifera* L.). BMC Plant Biol 12:237. http://www.biomedcentral.com/1471-2229/12/237

Zohary D (2004) Unconscious selection and evolution of domesticated plants. Econ Bot 58:5–10

Zohary D, Hopf M (2000) Domestication of plants in the Old World, 3rd edn. Oxford University Press, Oxford

Part II
Africa

Chapter 3
Date Palm Status and Perspective in Egypt

Shawky A. Bekheet and Sherif F. El-Sharabasy

Abstract Cultivation of date palms in Egypt goes back thousands of years. The date palm tree has great socioeconomic importance and nutritional value in Egypt. Its traditional use as a primary source of food and by-products and its ecological benefits in oasis agriculture make it an important fruit tree and the best crop to be cultivated. Egypt is the most productive country of date palm fruit in the world. There is a high potential for increasing the production area of date palm to fulfill local consumption in the whole country and to produce date fruits for export purposes. Presently, the Egyptian Government and private sector are convinced of the potential of date production and are striving to establish commercial date plantations and promote viable date production. Despite its important roles in Egyptian agriculture and the efforts made, date palm cultivation, production, processing, and marketing are still beleaguered with several problems. The low quality of cultivars, poor farm management, pests and diseases, marketing obstacles, and insufficient applicable research are the major constraints of the date palm in Egypt. Moreover, inconsistent quality in combination with low-grade packaging provides a considerable challenge for Egyptian date producers to fill the unmet domestic and international demand. By contrast, an abundance of capable manpower and arable land, low labor costs as well as the wide range of date palm diversity are among the most important competitive factors. This chapter discusses the current status, constraints, and the available avenues that can be used to develop the date palm industry in Egypt.

Keywords Agricultural constraints • Date palm • Development • Egypt • Prospects

---

S.A. Bekheet (✉)
Plant Biotechnology Department, National Research Center,
El-Tahrir St, Dokki, Cairo 12622, Egypt
e-mail: shawky005@yahoo.com

S.F. El-Sharabasy
Central Laboratory for Date Palm Research and Development,
Agricultural Research Center, Gamaa St, Giza 12619, Egypt
e-mail: sharabasydates@yahoo.com

J.M. Al-Khayri et al. (eds.), *Date Palm Genetic Resources and Utilization: Volume 1: Africa and the Americas*, DOI 10.1007/978-94-017-9694-1_3

3.1 Introduction

Date palm (*Phoenix dactylifera* L.) is one of the oldest plants cultivated by man. From earliest records of predynastic Egypt, the date palm fruit was used as a beer sweetener (excavation of a vat in Hierakonpolis, Upper Egypt, 3,450 BC). However, the cultivation of date palm did not become important in Egypt until somewhat later than that of Iraq (about 3,000–2,000 BC) (Zohary and Hopf 2000). Date palm cultivation has had a very important influence on the history of Egypt. Without date palms, no large human population could have been supported in the desert regions, since it tolerates high temperatures, drought, and salinity more than other fruit crops. Currently, the date palm is playing an important role in Egyptian agriculture, and it represents a significant part in the reclamation program. The date palm is the source of a wide range of products and services, including many necessities of life. The primary product of the date palm is fruit, which is rich in protein, vitamins, and mineral salts. All secondary products of the palm resulting from annual pruning have essential uses for the cultivator. Moreover, modern technological developments have made it possible to look at the palm as a raw material source for industrial purposes. On the other hand, the date palm is a crop capable of establishing a sustainable system in subsistence agricultural areas and thus plays an important social role in reinforcing the subsistence base of a large population group by helping them to remain settled in rural areas versus migration to urban centers. In this respect, the date industry supports over one million Egyptian families, and it is estimated that a 40 ha commercial date plantation requires some 8,000 work days a year (Bazza 2007). Egypt has been the world's largest producer of dates since 1974 and reports very high average yields as compared to other countries (FAOSTAT 2010). There are 14 million trees, occupying 30,934 ha, which represents 6.32 % of the fruit cultivated area in Egypt (FAO 2002), and date production of Egypt represents about 20 % of the total world production (FAOSTAT 2009).

Despite the invaluable role of date palm in the agriculture sector, the industry is beleaguered with the following problems: (1) a dearth of information on the status of the Egyptian date palm industry; (2) fruits of most Egyptian date palms cultivars are very moist (moisture content over 50 %), and they cannot be kept at ambient conditions for more than a few days; (3) about one-half of Egyptian date palm cultivars are from seeds and represent an important source of genetic diversity, however with low agronomic qualities; (4) a large number of good quality date palm cultivars have been lost after building of the Aswan Dam; and (5) diseases and pathogen pests, particularly red palm weevil, are a serious threat to date palm plantations in Egypt. Red palm weevil (RPW) has destroyed thousands of palms all over the country, causing great economic loss to the growers. Therefore, much more attention should be given to improve the characteristics of local cultivars, expand date palm plantations and increase date palm production, and improve postharvest handling and marketing for the development of date palm industry.

3.2 Cultivation Practices

The development of Egyptian agriculture occurred within the context of agricultural operations which began about 2,500 BC, as evidenced by ancient texts. The basic tools of agriculture, the ax, the hoe, and the plow, are independent Egyptian inventions. Moreover, predation by insects, rats, birds, and hippopotami, in addition to tax collectors, is poignantly reported in ancient texts (Durant 1954). In addition, Egypt gave the world the first hydraulic engineering and systematic irrigation (Fig. 3.1).

3.2.1 Irrigation

Although Egyptian farmers have long experience in irrigation, date palms do not receive much attention regarding irrigation or fertilization. Egyptian date producers believe that date palms can grow and bear fruits under drought conditions and do not require much irrigation. However, all studies show that the water requirement of

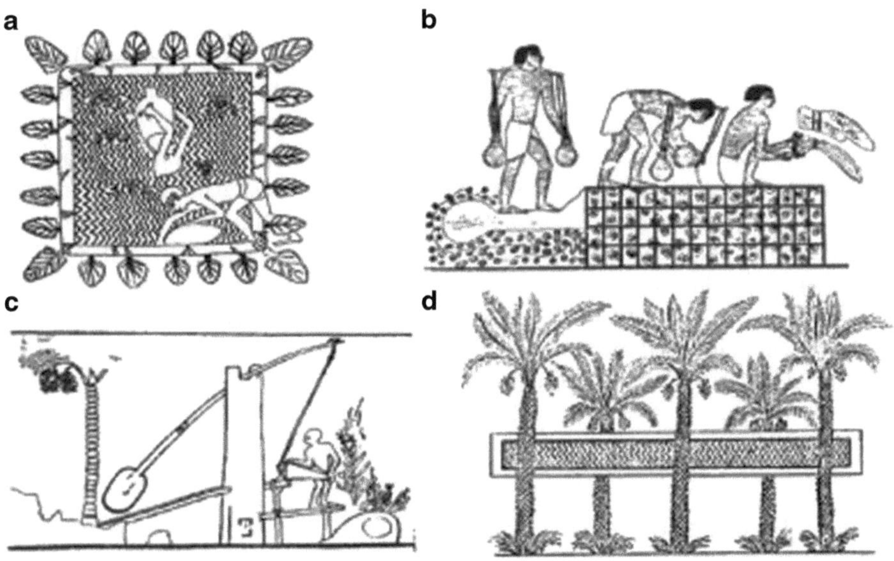

Fig. 3.1 Irrigation technology in ancient Egypt. (**a**) Drawing water in pots from a lily pond. From a tomb at Thebes, Egypt, ca 1,450 BCE (Singer et al. 1954, Fig. 343). (**b**) Irrigating and harvesting in a vegetable garden. Gardeners carry pots attached to a yoke and pour water into checkerboard furrows; another is tying onions into bundles. From a tomb at Beni Hasan, Egypt, ca. 1,900 BCE (Singer et al. 1954, Fig. 360). (**c**) Irrigation of a date palm orchard by a *shaduf*, a water-lifting device consisting of a beam holding a long pole in which a bucket is suspended at one end and a large lump of clay acts as a counterpoise. The water is funneled to a mud basin at the foot of the palm. From a tomb at Thebes, ca. 1,500 BCE (Singer et al. 1954, Fig. 344). (**d**) Date palm with water storage pond in a distorted perspective (Hyams 1971, p. 18)

date palms must be provided to produce high-quality fruit and high yield. There are many types of irrigation systems used in the date palm orchards in Egypt. The implementation of these systems depends on the water resources and quality, age of the palm trees, and soil composition. The irrigation methods currently in use are furrow and basin. Egyptian farmers have traditionally irrigated date palm fields by dividing them into small basins measuring 10 × 10 m. These basins provide farmers with sufficient water control that allows uniform irrigation even in somewhat unleveled fields. This irrigation method is used mainly in the old lands, especially in the northern part of the Delta area in Rasheed, Edco, and Domyate and also in Aswan. About 47 % of the total area of plantations and about 35 % of the total productive date palm trees use the basin irrigation method (Hussein and Hussein 1983). Recently, drip irrigation has been introduced to date palm plantations in Egypt. This method is being used routinely in the newly reclaimed land in Nobaria and the resort area in south Sinai and the Red Sea. On the other hand, the main source of irrigation water in oases is naturally flowing springs, or it is pumped from wells. This type of irrigation is carried out through a particular system of side channels especially designed to cope with an insufficient supply of water. In order to estimate water requirements of date palm in Egypt, a number of field trials were conducted by Allam et al. (1973). They reported that dates grown in Lower Egypt use an average of 10,280 m^3/ha, whereas in Upper Egypt average consumption rises to 14,880 m^3/ha. These results also showed that increasing irrigation frequency increased the size, weight, moisture content, and the total soluble solids of the fruits. The final recommendation was to irrigate 12 times a year. However, date palm trees in areas where the ground water table is high did not require any surface irrigation. Generally, to achieve the best response from irrigation of date palm, it is suggested to: (1) irrigate before pollination in order to stimulate the growth of the fruit bunches; (2) irrigate when the fruit bunches are developing, when the stalk is pendent, and when fruit is maturing; (3) reduce irrigation amounts and frequency near crop maturity to lower crop moisture content and avoid deterioration; (4) irrigate only in the early morning or late evening to avoid the heat; and (5) stop irrigation for about 40 days at the end of November, especially if the dates are being intercropped with clover or other annuals.

3.2.2 Fertilization

Adoption of a proper fertilization program of date palm, including adequate rates, appropriate sources, and efficient application methods and timing, is an important strategy for obtaining better fruit yield and quality. The amount of fertilizer needed by a single tree varies with soil type and depth, as well as by date cultivar and age. In this respect, fertilizer trials have been made on different Egyptian date palm cultivars in order to increase yield as well as to improve fruit quality (Attalla et al. 1988; Kassem et al. 1997; Soliman and Osman 2003; Soliman and Shaaban 2006). With respect to potassium fertilization, it was reported that applying potassium

Table 3.1 Fertilization recommendation for date palm in Egypt according to Central Administration for Agricultural Extension Services, Egypt

Nutrient	Type and amount of fertilizer/palm	Time of application
Organic fertilizer	100 kg	November (around the base)
Nitrogen	Ammonium sulfate (20.6 %) (6 kg)	Four equal quantities are applied in March, May, July, and September
Potassium	Potassium sulfate (1.5–2 kg)	Three equal quantities are applied in March, May, and July
Phosphorus	Calcium superphosphate (0.5–1 kg)	Applied with organic fertilizer
Magnesium	Magnesium sulfate (1–1.5 kg)	Three equal quantities are applied in March, May, and July

generally improves growth, yield, and fruit quality of some Egyptian date palm cultivars. El-Hammady et al. (1991) found that the highest yields and fruit quality of Siwy cv. dates were obtained by adding 2 kg potassium sulfate/palm yearly. However, Shawky et al. (1999) recommended 1.5 kg of potassium sulfate/date palm each year. Recently, Marzouk (2010) in a study on Zaghloul cv. stated that to produce higher yield and enhance overall quality of dates, the application of nitrogen fertilizer such as ammonium sulfate is advisable. Generally, the recommended fertilization program of mature palm in Egypt is summarized in Table 3.1.

As the majority of the reclaimed lands of Egypt have poor soil, organic wastes are applied as soil amendments for improving soil properties at the same time as slow release fertilizers. In this respect, investigation of physical and chemical properties of fruits of date palm planted in clay soil and amended with organic manures mixed with chemical fertilizers (NPK) was carried out by Osman (2009). It was found that pronounced beneficial effects of organic manures were reflected in the physical and chemical properties and some leaf mineral content. These results showed that the average yield (kg/tree), fruit and flesh weight, fruit dimensions, and fruit size were significantly higher as affected by organic manures mixed with chemical fertilizer. Also, Attalla et al. (2003) in their study on Zaghloul and Samany cvs. found that the trees fertilized with chicken manure alone or urea in the first and second seasons, respectively, had significantly higher pinnae potassium content. Recently, Kassem and Marzouk (2010) stated that applying organic manure either alone or combination with mineral fertilizers increased per palm yield and enhanced fruit color as compared with mineral fertilization alone. Chicken manure and cow dung resulted in the best fruit weight, fruit flesh weight, and fruit length.

3.2.3 Pollination

As the date palm is dioecious and naturally cross-pollinated, artificial pollination is required using a compatible pollen source for commercial cultivation. The average pollen-bearing capacity of a good male palm is sufficient to pollinate 50 female palms, determined by both the number of flowers and the pollen quantity per flower

(Zaid and de Wet 2002). Ben Salah and Hellali (1998) reported the direct effect of pollen type on some fruit characteristics outside the embryo and the endosperm. Moreover, the male parent used has great impact on date palm qualities, even on the flavor or aroma of the fruit. The direct influence of pollen on fruit physical and chemical characteristics is known as metaxenia, which affects time of fruit ripening and fruit color (Al-Delamiy and Ali 1970), weight of fruit and seed (El-Ghayaty 1983), and fruit size (Abdelal et al. 1983). So, the selection of suitable male parents as pollinizers is important for improving the quantity and quality of dates. Since ancient times, Egyptian date palm growers have been aware of the importance of artificial pollination when they noticed the effect of pollen grains of certain male cultivars on the size of fruits and seeds (xenia) as well as time of fruit ripening (metaxenia) of female cultivars. Male trees are selected and exchanged between farmers even over long distances of the country. It is notable that the male palms differ greatly by their production of pollen from one area to another. Farag et al. (2012) investigated the metaxenic effects of two date palm male pollens on physical and chemical properties of Zaghloul cv. The results proved that pollinizer A (Balteem 14) caused a significant increase in fruit weight, flesh weight, rag weight, fruit size, total soluble solids, reducing and nonreducing sugars, anthocyanin, vitamin C content, and crude fibers content and a significant decrease in chlorophyll a, b contents as compared with pollinizer B (Balteem 19).

In Egypt, male flower clusters are cut in the morning after the spathe splits open, to prevent wind or bees from causing a loss of pollen; clusters are dried in heated rooms before separation from the spadix. Shaheen (1986) found that delaying pollination after female spathe splitting reduced the percentage of fruit sets. He further reported that maximum set fruit with most Egyptian date palm cultivars was obtained from pollination within 3 or 4 days after the female spathe opening. Pollination is done in Egypt by hand with pollen selected from males blooming at the same time as the females to provide fresh pollen in order to get successful fruit setting. Strands of male flowers are cut from a freshly opened male spathe, and two or three of these strands are placed between the strands of the female inflorescence. Storage of pollens is not done in Egypt. In a commercial plantation, one male tree is enough to pollinate 25–30 female palms. Pollen grains of more than one male are sometimes mixed and then used for pollination. This may result in genetically mixed populations of date palm within orchards and oases.

3.2.4 Pest and Disease Control

The date palm tree and its fruits are subject to attacks by several pests that are well adapted to the Egyptian environment. Red palm weevil is considered the most destructive pest in date palm plantations in Egypt. Transfer of date palm offshoots as a planting material has played a major role in rapid proliferation of the pest. Moreover, farming practices such as over-irrigation exacerbate the problem and make the palm trunks ideal habitat for weevils. Various control measures have been

Table 3.2 Common pests of date palm in Egypt and the recommended control methods

Pest	Attacked portion	Management and control
Batrachedra amydraula	Fruits	Agricultural practices, i.e., covering date bunches with porous plastic cloth and thinning of bunches and/or diverging strands of bunches
		Chemical treatments using malathion (57 %) after 1 week of pollination and repeated three times every 2 weeks
Ephestia calidella	Fruits	Agricultural practices, i.e., covering date bunches and burning the falling fruits
		Chemical treatments using 1,140 bromomethyl
Oligonychus afrasiaticus	Leaves and fruits (hababouk stage)	Chemical treatments using mineral oils or sulfur
Phonapate frontalis	Leaflets and inflorescence	Agricultural practices, i.e., removing the infested leaflets
Parlatoria blanchardi	Petioles, leaflets, and fruits	Agricultural practices, i.e., removing and burning leaves
		Chemical treatments using mineral oils plus malathion (57 %)

Central Administration for Agricultural Extension Services, Egypt (2001)

considered in controlling the spread of RPW, including containment/destruction of infested plants and chemicals (diazinon, profenofos, and fipronil by regular spray, injection of infested trees) and biological control; palms in the late stage of attack have to be eradicated. In this connection, early detection of the weevil in a date-growing area is imperative. Field and cultural practices are important components to prevent weevil infestation. Moreover, integrated pest management based on pheromone traps and biological control should be developed. In addition, new breeds of date palm resistance to insects notably RPW are strongly demanded.

There are other serious pests which attack both date fruits and date palm trees in Egypt. In order to control these pests, several methods have been investigated, i.e., insecticides (Sayed and Ali 1995), mechanical control (Sayed and Temerak 1995), biocide (Sayed and Ali 1995), and natural products (Sayed et al. 2001). Moreover, different treatments of date storage facilities including pheromone-baited traps, spraying with malathion 57 % EC, and phosphine tablets (Phostoxin) were evaluated for controlling insect infestation and reducing date losses (Ali et al. 2007). Results suggest that applying phosphine tablets at a rate of three tablets per cubic meter of store house area was an effective treatment of insect control and limited date losses, followed by spraying the storeroom before storing dates with malathion 57 EC at a rate of 150 ml/100 l water. Pheromone-baited traps gave poor results for insect control. In this regard, the most common insect pests (excluded RPW) of Egyptian date palms and the recommended control methods are presented in Table 3.2.

Egyptian date palm cultivars are afflicted with several fungi resulting in decline of both growth and production. Several date palm fungi have been recorded such as *Graphiola phoenicis* (Briton-Jones 1925), *Diplodia* sp. (Fawcett 1930), *Thielaviopsis*

Fig. 3.2 Symptoms of heart rot in date palm cv. Zaghloul: (**a**) The folded leaves died, while the outer leaves appear healthy. (**b**) Advanced stage of infection

paradoxa (Fawcett 1931), *Phomopsis* sp. (Melchers 1931), *Mauginiella scaettae* (Sabet and Michael 1965), *Chaetosphaeropsis* sp. (Mostafa et al. 1971), *Botryodiplodia* sp., and *Chaetomium* sp. (Rashed 1991). Recently, 21 fungal species belonging to 15 genera were isolated from diseased date palm samples collected from different Egyptian localities. *Thielaviopsis paradoxa* was the most prevalent fungus followed by *G. phoenicis*, *Diplodia phoenicum*, *Botryodiplodia theobromae*, and *Fusarium oxysporum* (El-Deeb et al. 2007). During the last few years, date palm growers have complained about a destructive disease identified as heart rot infecting the vegetative system (Fig. 3.2). Infections are all characterized by partial to complete necrosis of the tissue. Decay is most serious when it attacks the terminal bud leading to the death of the tree, and no treatment is recommended. However, annual pruning of old infected tissue and leaves and their immediate burning is advised. In addition, root rot caused by several fungi, i.e., *Armillaria mellea*, *Diplodia* sp., *Macrophomina* sp., *Rhizoctonia* sp., and *Fusarium* sp., is considered one of the important diseases leading to death of high percentages of cultivated offshoots of date palm in Egypt. The recommended treatments are immersion of the offshoots in Benlit (3 g/l) + Rizulx (2 g/l) or Vitavax Thiram (3 g/l) + Rizulx (2 g/l) or Tobsin M.7 (3 g/l) + Rizulx (2 g/l).

3.2.5 Agroforestry Utilization

Date palm plantations once rehabilitated and/or established in various desert areas of Egypt improve the microenvironment and contribute to control desertification. Date palms can grow under unfavorable conditions where many of other fruit species may fail, and they provide protection to intercrops from heat, wind, and even cold. Moreover, date palm trees provide enough space for intercropping even if they are fully grown as they do not cover much area, being a very tall tree. Otherwise, the date palm is responsible for opening vast desert territories for human settlement and the development of oasis agroecosystems. Besides the fruit, the date palm over the centuries has also provided a large number of other products which have been extensively used in all aspects of daily life. The trees provide the raw materials for housing, furnishings, and many handcrafts. Furthermore, the trunk was used extensively in the supporting structures for water-lifting mechanisms such as in the Egyptian waterwheel, the Sakiyeh (Mason 1927). Besides woven products, leaflets are also used for making cord which is used to tie up bundles of nursery stock or other temporary purposes in horticulture. Cultivation of date palm constitutes one of the most successful agricultural activities in arid and semiarid regions and ranks as the second most important fruit crop after olive in the five Egyptian oases (e.g., Siwa, Bahariya, Farafra, Dakhla, and Kharga) which rely on the largest fossil water reserve (Nubian Sandstone aquifer) in the world (Monier and Fawzy 2006). Moreover, due to their resilience, long-term productivity, and multipurpose attributes, date palm trees are considered to be the most successful fruit crop cultivated in the new reclaimed land in the Toshki, El-Ewinates, and Sinai areas.

3.3 Genetic Resources and Conservation

Date palms have been cultivated and subjected to selection by man since ancient times. Exploitation of the date palm by humans probably began as simple gathering of the fruits, fronds, trunks, and other usable parts of the tree. Dispersal of date palm germplasm was probably originally by seed, which gave rise to the many local cultivars that are found in the Middle East (Krueger 2011). Local date palm cultivars with outstanding adaptation to climatic, edaphic, and management factors are the products of centuries of interaction between farmers, the genetic and breeding systems of the date palm, and the environment (Jaradat 2011). Thousands of cultivars of date palm exist in the different growing countries. These cultivars have been developed by continuous selection performed by date palm growers mainly to improve crop yield and quality. Currently, there are as many as 5,000 date cultivars around the world with differences in fruit color, flavor, shape, size, and ripening time (Jaradat and Zaid 2004).

3.3.1 Current Status and Prospect of Genetic Resources

Genetic diversity and structure of the gene pool complex of date palm have been shaped and greatly altered by natural and human selection, clonal propagation, and spatiotemporal exchange and movement of germplasm. Egyptian date cultivars have been developed over thousands of years by selection of seedlings possessing desirable characteristics, which have been propagated. Farmers also have used genetic variation to select and develop new cultivars from the date palm cultivars currently being grown. Date palm trees are dispersed in all regions of the country across nearly intolerable conditions of environmental adversity and varying climatic conditions. In Egypt, the total number of date palm trees is 14 million including 12 million at the bearing stage (FAOSTAT 2009). There are an estimated 6 million productive date palm trees planted separately all over the country (not in groves), bordering the farms or even in the middle of the farms, managed with the cropping pattern on the farm. On the other hand, the number of date palm cultivars in Egypt is variable due to different identification characters and the absence of suitable methods of description, which depend mainly on different references and local experiences. A total of 52 date palm cultivars have been identified although only 20 are widely distributed commercially (Rizk et al. 2004). There is a great number of seedling varieties resulted from sexual reproduction and some of them are highly desirable. These cultivars are distributed in the Nile Valley, oases, and desert districts. Although serious epidemics threaten cultivars and slow traditional multiplication methods are a hindrance, little effort has been made to develop new cultivars with good quality fruit and resistance to diseases. These efforts have slowly been abandoned in the face of the long life cycle of the plant that makes the establishment of segregating populations, backcrosses, and recurrent selection an enormous burden to the whole breeding program.

3.3.2 Threats and Degradation

There are many factors threatening the genetic diversity and cultivars of date palm, including natural disasters and environmental and climatic changes. Such factors could result in genetic erosion due to crop failure and loss of cultivars. Moreover, biotic factors such as pests and diseases can attack date palms resulting in negative impacts on the genetic variability within species (Bendiab et al. 1993). Red palm weevil insect currently is considered the most serious pest of date palm in Egypt. Since it was introduced accidentally to Egypt in 1992 (Cox 1993), thousands of healthy trees have been damaged or lost (El-Sebaey 2004). On the other hand, land use changes which include construction and building of roads, factories, canals, dams, and new residential areas are other threat factors. It is important to mention that date palm tree numbers decreased from 2.5 million to just over 1 million in the Aswan area due to the building of the High Dam (Hussein et al. 1993). Moreover, thousands of tons of commercial fertilizer are required each year to replace the fertility of flood-derived alluvial soils. Otherwise, desertification and soil salinization are other factors threatening the Egyptian natural habitat of date palms. In this respect, saline sea and lake water are the main reasons for the increase of soil salinity in which sodium chloride (NaCl) is the dominant salt (El-Assioti 1992). Furthermore, many existing date palms are aging and have stopped producing offshoots for propagation. Therefore, a number of neglected and underutilized cultivars of date palm are expected to be lost forever, and consequently this will reduce date palm diversity in the country. This situation makes conservation of genetic resources of date palms an imperative; there is a great need to put mechanisms in place for the protection of the important Egyptian date palm cultivars.

3.3.3 Conservation Efforts

Conservation of Egyptian date palm genetic resources has become an important issue in the development of date production and for food security in the country. Conservation is important not only to preserve the good quality cultivars but also to ensure future access to valuable genes for plant improvement programs. The Agricultural Research Center (ARC) is mainly responsible for the introduction, local collection, and the production and release of improved germplasm cultivars. In Egypt, over the last two decades, much more attention has been given to the evaluation of introduced date palm cultivars under local conditions, including foreign germplasm, e.g., Barhi and Medjool cvs. are used commercially after their involvement in crop improvement programs. Otherwise, improvement of agricultural practices, salinity, and disease control in addition to irrigation water are considered the most important issues for date palm cultivation, and conservation in the traditional growing regions has been given more attention. Moreover, farmers have been encouraged to replant date palms with locally produced offshoots or tissue culture-derived plants. By contrast, documentation of traditional cultivars, considered one of the important aspects of in situ conservation of date palms, is still elusive. In this connection, documentation for on-farm conservation may be particularly effective when it involves farming communities. On the other hand, a substantial effort was carried on by Egyptian researchers in the conservation of date palm germplasm using modern technology. In this respect, protocols for in vitro conservation of Egyptian date palm cultivars have been recognized (Bekheet et al. 2001b, 2007; El-Dawayati 2008; Hassan 2002). The two main types (slow growth and cryopreservation) of in vitro storage have been developed for preservation of date palm germplasm. With respect to date palm conservation, a complementary strategy for maintaining genetic diversity should employ a combination of methods including nature reserves, gene banks, and others, as no single method can conserve all the diversity. For such a strategy, three actions are necessary: (1) strengthen the national conservation program; (2) contribute to international collaboration; and (3) improve conservation technologies.

3.3.4 Germplasm Banks

Germplasm banks of date palm cultivars are considered one of the important aspects for the maintenance of genetic resources. In Egypt, the Ministry of Agriculture and Land Reclamation has activated the already existing Plant Genetic Resources

Program from 1994. According to the Ministerial Decree No. 1920 of 2003, the National Gene Bank (NGB) was established to be officially responsible for the conservation and maintenance of plant, animal, and microorganisms' genetic resources in Egypt's agricultural sector. The NGB consists of the following sections: (1) conservation (which includes true seed conservation, tissue culture conservation, conservation in liquid nitrogen, and field conservation), (2) evaluation and taxonomy, and (3) documentation and information. The cold storage facility of the NGB contains more than 30,000 accessions (NGB 2007). In this context, Egypt has already become a signatory to The International Treaty of Plant Genetic Resources for Food and Agriculture (ITPGRFA) and The Convention on Biological Diversity (CBD). Moreover, Egypt is a member of the FAO Global System. Through this system, Egypt has gained a number of benefits, such as technical cooperation, the approval of several projects, and the flow of technical information.

Date palm conservation in Egypt is mainly the responsibility of the Central Laboratory of Date Palm Research and Development gene bank (CLDPRD). In this respect, a total of 52 cultivars were collected, representing dry, semidry, and soft date palm fruits in Egypt. For characterization and management of date palm, a system of descriptors was recognized (Rizk and El-Sharabasy 2006). It contains standard passport data (17 traits), details of ecogeography (13 traits), ethnobotany (15 traits), management (18 traits), and characterization (58 vegetation traits, 27 fruit traits, 22 seed traits, and 41 inflorescence traits) and evaluation activities (113 traits). Moreover, a pilot field gene bank of date palm is managed by CLDPRD. Trait data were collected from trees growing in their natural habitats as well as from trees growing on CLDPRD farms. Otherwise, economic date palm cultivars, developed by CLDPRD, and some imported ones have been distributed to farmers all over Egypt, followed by extensions service delivered to the farmers. To overcome the constraints and to hinder the systematic loss of valuable germplasm, we recommend documenting all date palm germplasm in the country, to use micropropagation technologies to rapidly propagate the promising cultivars in order to increase the areas of date palm, to use in vitro techniques in complementarity with the other conservation methods for establishment of Egyptian date palm gene banks, and to import new cultivars with desired characteristics.

3.3.5 *Quarantine Regulations*

As in other countries, quarantine laws in Egypt are strict, and importation and exportation of plant genetic resources are possible only if regulations are followed. In this regard, the Central Administration of Plant Quarantine (CAPQ) supports both the governmental and the private sector producers by reducing the damage arising from the introduction and spread of harmful organisms. According to the agricultural quarantine laws, all imported shipments destined for disinfection must be submitted within 7 days of their initial inspection. Therefore, a request should be made for the shipment treatment before the expiry date. Any further delay treatment would threaten the country's plantations. Such shipments shall either be re-exported or destroyed at the title holder's expense without indemnity (compensation). On the other hand, the exported shipments shall be disinfected at the behest of its exporter or if the importing country's legislation so requires. A certificate to this effect shall be issued to the party concerned. This applies also to the passage of in vitro materials through quarantine. Planting out of imported vegetatively propagated material is only possible if the germplasm has been inspected and tested in isolated greenhouses according to existing rules and regulations. Egypt began to apply the concept of Plant Quarantine in 1904, with issuance of law No. 10, followed by law No. 21 of 1906 and law No. 1 of 1916. This legislation was updated in 2001 by ministerial decree No. 3007, with modification of scientific names for many pests taken into consideration; in addition pests and treatments not listed in previous legislation were added. Modernization of this law is to comply with international updates to regulate the importation of plants which are infected with foreign pests. According to Ministerial Decree No. 3007 of 2001 (Ministry of Agriculture and Land Reclamation), the unrecorded pests and diseases of date palms to be declined entry to Egypt are: bayoud caused by *Fusarium oxysporum f.* sp. *albedinis*, Omphalia root rot caused by *Omphalia* spp., Khamedj caused by *Mauginiella scattae*, and fruit rots caused by *Macrosporium* spp. and *Citryomyces remosus*. To further develop the Egyptian agricultural quarantine system, it is recommended to establish a department for quality control, to update the information on quarantine pests and their distribution, and to develop the diagnosis and test laboratories.

3.4 Plant Tissue Culture

3.4.1 Role and Importance

Date palm is propagated sexually through seeds and vegetatively by offshoots. Propagation through seeds has many limitations like a high percentage of male plants, slow growth, and progeny variation. The propagation by offshoots is also inefficient because the number produced by each palm tree is limited. The present need for planting material of date palm in Egypt is estimated at more than 20 million trees. To overcome these problems and meet the demand for planting material, it is necessary to develop a method of date palm propagation using plant tissue culture. Tissue culture has made it possible to rapidly produce large number of date palms from a single offshoot (Shaheen 1990). High-quality and uniform planting material can be multiplied on a year-round basis under disease-free conditions anywhere, irrespective of the season and weather. Moreover, date palm micropropagation can make field establishment success close to 100 %, and transportation of plants from the nursery to the field is simple and easy. In biotechnology, several techniques to conserve vegetatively propagated species have recently been developed, and some are undergoing rigorous testing. Preservation of plant cells, meristems, and somatic embryos has become an important tool for the mid- and long-term storage of germplasm using minimum space and maintenance (Shuji et al. 1992). In this respect, different in vitro methods are employed for preservation of date palm germplasm depending on the storage duration required. So, application of tissue culture for date palm has important roles in: (1) increasing the date palm cultivated areas in Egypt and subsequently date productivity, (2) replacing the old orchards of date palm with juvenile and high-quality date palms, and (3) conserving genetic diversity of the Egyptian date palm in particular the superior and rare cultivars such as those growing in Aswan and the oases.

3.4.2 Research and Development

During recent decades, extensive research efforts have been made to propagate Egyptian date palms through tissue culture. The first attempt was reported by El-Hennawy and Wally (1978). Later, research was carried out systematically in micropropagation of different cultivars aimed at developing methods applicable to large-scale propagation. In this context, Belal and El-Deeb (1997) succeeded in inducing rooting on in vitro proliferated shoots of date palm cvs. Zaghloul and Samany. Likewise, Bekheet and Saker (1998) studied the impact of plant growth regulators on direct and indirect shoot bud proliferation from shoot tips of Zaghloul cv. Also, El-Hammady et al. (1999) developed a reliable method for in vitro rooting on proliferated shoots of Siwy cv. Several reports appeared on micropropagation of different Egyptian date palm cultivars through somatic embryogenesis using shoot tip explants (Bekheet et al. 2001a; El-Sharabasy et al. 2001a; EL-Shiaty et al. 2004; Gabr and Abd-Alla 2010; Gadalla 2007; Gadalla et al. 2004). Organogenesis based on the direct regeneration of cultured shoot tips was achieved by Taha et al. (2001). An applicable protocol for mass propagation of seven cultivars (i.e., Zaghloul, Samany, Hayany, Amhat, Siwy, Selmi, and Malkaby) through direct organogenesis was developed (Hegazy 2008). In vitro multiplication of date palm through direct organogenesis is shown in Fig. 3.3. Another micropropagation method for date palm using immature inflorescences was reported by El-Kosary et al. (2007) and Abul-Soad (2011).

Research on date palm micropropagation proved that transfer of plantlets derived from in vitro conditions to the permanent field site is a critical factor limiting the success of tissue culture technique. In this regard, Shaheen (1990) reviewed the factors affecting the successful production of free-living date palm, including length of plantlets, strength of root system, humidity conditions, and number of leaves and composition of the soil. However, El-Sharabasy et al. (2001b) in their study on micropropagation of date palm reported high survival percentage (80 %) with a planting medium containing equal parts of peat, sand, and vermiculite. In parallel, molecular and biochemical techniques were successfully used to detect the somaclonal variations in tissue culture-derived plants. A wide range of variability has been reported using Random Amplified Polymorphic DNA (RAPD) and Amplified Fragment Length Polymorphism

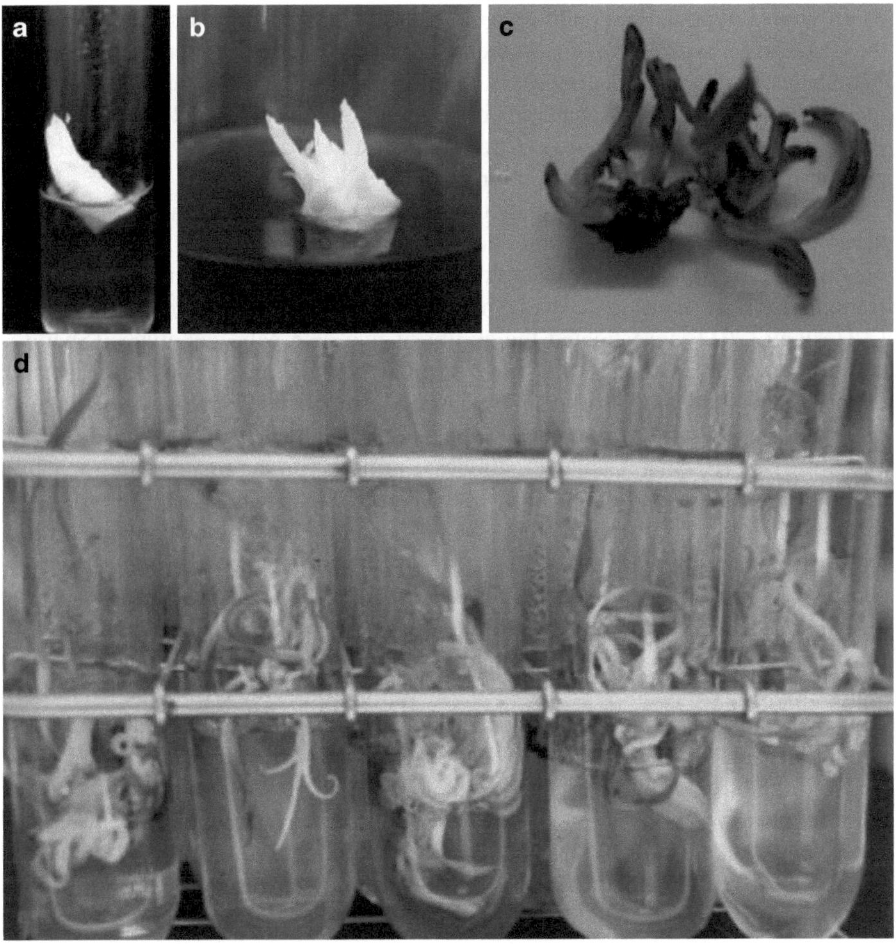

Fig. 3.3 Organogenesis of date palm cv. Zaghloul: (**a**) Shoot tip culture. (**b**) Initial subculturing. (**c**) Direct shoot proliferation. (**d**) Multiplication of propagules

(AFLP) markers (Moghaieb et al. 2011; Saker et al. 2000). Moreover, the inter-simple sequence repeat (ISSR) technique was utilized to assess the genetic stability of the micropropagated plantlets (Abd-Alla 2010).

3.4.3 Scaling-Up Production and Other Applications

The application of tissue culture propagation (micropropagation) of date palm enables large-scale mass multiplication and production of genetically uniform plants. Moreover, in vitro propagation makes available a large number of cloned plants in a short time. In this respect, tissue culture protocols for mass propagation of different Egyptian date palm cultivars were optimized for mass propagation by the CLDPRD. Furthermore, government-supported tissue culture facilities in Cairo, Ain Shams, and Assut universities, in addition of NRC, have developed systems for the production of planting materials of different plant species included date palm. Otherwise, several research projects to establish standard protocols for commercialization of date palm micropropagation have been financed.

On the other hand, biotechnology tools of tissue culture and genetic engineering have been employed to play a potential role in date palm breeding in Egypt. In this regard, salt tolerance of tissue culture of three cvs., i.e., Samany, Siwy, and Bartamoda, were investigated (El-Sharabasy et al. 2008a). Recently, Helaly and El-Hosieny (2011) investigated the ability of two date palm genotypes (Shamiya and Amry) to tolerate drought stress throughout their micropropagation period. Otherwise, isozyme polymorphism was used as a biochemical marker to distinguish drought tolerance of in vitro embryogenic callus and plantlets of Sakkoty, Zaghloul, and Siwy cvs. (El-Sharabasy et al. 2008b). Furthermore, the first research work on protoplast isolation and callus formation from the protoplasts in date palm was achieved in Egypt by Rizkalla et al. (2007). In addition, the genetic engineering of date palm as an alternative for characteristic improvement of this long life-cycle species was optimized. In this context, the particle bombardment transformation system for the Egyptian date palm cv. Siwy was reported (Saker et al. 2007). Furthermore, Saker et al. (2009) studied different factors which influenced transient expression of *Agrobacterium*-mediated transformation of the *GUS* gene in embryogenic callus of date palm. Recently, El-Rakashy et al. (2011) field evaluated transgenic date palm (TDP), expressing endotoxin *Cry3Aa* gene, against RPW. They demonstrated that the TDP retained stable production of *Cry3Aa* protein and were resistant to RPW. In this context, different biotechnological approaches developed for propagation, conservation, genetic improvement, and production of secondary metabolites of date palm cultivars in Egypt were summarized by Bekheet (2013).

3.4.4 Research and Commercial Labs

In vitro propagation of different cultivars of date palm using either somatic embryogenesis or organogenesis has been recognized in recent years as a routine procedure in several research and commercial laboratories to produce large numbers of plants at a competitive cost. For this purpose and others, government-supported plant tissue culture laboratories have been established at several locations in Egypt. The most up-to-date are at (1) the Egyptian Major Crop Improvement Program (EMCIP) situated at the Agriculture Research Center and (2) the Development of the Agricultural Systems Project Laboratory situated at the College of Agriculture, University of Cairo. One of the first laboratories in Egypt involved in selection for somaclonal variation is at the Agronomy Department, Al Azhar University. Furthermore, large-scale tissue culture labs for mass propagation and distribution of disease-free planting materials were established at the Ministry of Agriculture. In this regard, the Horticulture Research Institute, which belongs to the Ministry of Agriculture, has four central tissue culture laboratories. The major tissue culture and germplasm conservation research laboratory is located in Giza, to carry out departmental research programs on fruits, timber and forestry trees, and floriculture. The second major tissue culture laboratory to produce fruit seedlings at a commercial scale and the third tissue culture laboratory is located in Dokki, Giza, to serve the research programs of the departments of vegetable and medicinal and aromatic plants, and the last is located at Zohreia Garden, also in Giza. The role of the private sector in date palm micropropagation is negligible when compared with its role in mass propagation of banana and strawberry. Recently, Egyptian Hollandia Tissue Culture and Trade established a tissue culture laboratory for micropropagation of date palm, produces annually about 25,000 plants. At the present, Barhi and Bartamoda are the most micropropagated cultivars.

3.5 Cultivar Identification

Determination of genetic variability and proper cultivar identification in date palm would be of major significance in improvement programs and in germplasm characterization and conservation to control genetic erosion. Cultivars are commonly identified by a wide range of morphological features describing both trees and fruits. The morphology of leaves, spines, and fruits is the most common morphological or phenotypic characters used for the date palm (Nixon 1951). By contrast, molecular characters more quickly and accurately reveal genetic differences without the obscurance of environment and provide significant advantages in genetic analysis and germplasm characterization. In this respect, proteins and/or DNA attributes are used successfully for cultivar identification and for studying the genetic diversity of date palm cultivars. Currently, there are a number of molecular techniques available for characterization of the variation at the DNA level, e.g., RAPD, AFLP, and ISSR. Such techniques are able to reveal a virtually unlimited number of markers (Eissa et al. 2009).

3.5.1 Morphological Descriptors

From the point of morphological attributes and due to the large number of date palm cultivars in Egypt, there are no specific vegetative morphological criteria to distinguish closely related cultivars. Morphologically, the key distinctive attributes including fruit shape, dimension, and color as well as leaf attributes are commonly used for the identification of date palm cultivars in Egypt. Otherwise, some cultivars are distinguishable by their growth habits which include both the pattern and vigor of growth. In this respect, a total of 77 morphological attributes were applied for distinguishing the soft date palm cultivars in Egypt (Table 3.3) (Eissa et al. 2009). Moreover, systematical relationships of 21 dry date cultivars growing in Egypt were addressed based on 101 morphological

Table 3.3 Morphological attributes used in the study of soft date palm cultivated in Egypt (Eissa et al. 2009)

	Cultivars								
Morphological attributes	Samany	Zaghloul	Bent-Eisha	Hayany	Oraiby	Om-Elferakh	Amhat	Selmi	Barhi
Vegetative Attribute									
Trunk width (1. homogenous, 2. base broader than above)	2	1	1	1	1	2	2	1	2
Crown shape (1. dense, 2. moderate dense, 3. loose and open from the middle)	1	1	2	1	3	2	2	2	1
Leaf length (cm)	570	530	500	500	455	525	480	490	375
Leaf width (at the middle) (cm)	105	55	95	90	75	90	90	75	90
Colors of leaf (1. dark green, 2. green)	2	2	2	1	1	1	1	1	1
Midrib color (1. dark green, 2. glossy green, 3. light green)	3	2	3	2	3	1	3	2	2
Leaf curvature (1. high curved, 2. moderately curved, 3. slightly curved)	1	3	3	2	2	1	3	2	2
Leaf curvature point (l. at middle of blade, 2. at second half of blade, 3. at all blade)	2	2	1	1	3	3	3	2	1
Petiole length (cm)	50	110	50	65	60	45	35	80	45
Petiole to leaf ratio	0.09	0.21	0.10	0.13	0.13	0.09	0.08	0.16	0.12
Petiole thickness at the middle (cm)	5	3	3	3.5	3	4	4	4	4
Petiole width at the middle (cm)	10	5.5	7	8	7	8	6	6	8
Petiole shape (1. slender, 2. base stout than above)	2	1	1	1	1	2	1	2	1

Leaf base width at the attachment point (cm)	17	10	10	11	12	16	11	14	14
Color of leaf base abaxial surface (1. dark green, 2. light green)	2	1	2	1	1	2	1	1	2
Blotches on leaf base abaxial surface (1. small reddish brown blotches, 2. absent)	2	1	1	1	1	1	2	1	1
Blade length (cm)	420	405	390	400	340	400	310	300	250
Blade to leaf ratio	0.74	0.76	0.78	0.8	0.75	0.75	0.64	0.61	0.67
Number of pinnate	243	213	231	151	213	201	185	175	149
Pinnate density (1. dense, 2. lax, 3. very lax)	1	1	1	3	1	2	1	1	1
Length of blade/number of pinnate ratio	1.7	1.9	1.7	2.6	1.6	2	1.7	1.7	1.7
Pinnate length (cm)	65	65	67	60	45	60	65	55	45
Pinnate width (cm)	4.0	2.5	3.0	3.5	3.5	3.2	3.5	2.5	4.5
Pinnate shape (1. lanceolate, 2. linear)	2	2	2	1	1	2	1	2	1
Pinnate apex (1. acute, 2. soft end)	1	1	2	2	1	2	2	1	1
Pinnate nature (1. semi-drooping, 2. non-drooping, 3. semi-erect)	3	2	3	2	2	1	3	3	3
Pinnate-rachis angle (°)	57.5	40	45	60	70	50	57.5	55	60
Apical divergence (°)	67.5	50	70	90	85	70	80	75	95
Middle divergence of pinnate (°)	110	65	90	110	135	95	110	95	120
Valley angle (1. small, 2. large, 3. absent)	2	1	1	3	1	3	2	1	1
Spine area length (cm)	45	15	60	45	55	80	135	110	80
Spine area to leaf ratio	0.08	0.03	0.12	0.09	0.12	0.15	0.28	0.23	0.21
Spine shape (1. pyramids, 2. pyramids with longitudinal fold, 3. mixed)	2	1	1	1	2	2	2	3	2
Shorter spine length (cm)	9	4	5	6	5	7	3	4	6
Longer spine length (cm)	21	9	22	12	11	25	24	15	15
Spine base (l. not raised, 2. raised, 3. raised and pulvinous)	3	1	2	1	3	3	1	2	3

(continued)

Table 3.3 (continued)

Morphological attributes	Cultivars								
	Samany	Zaghloul	Bent-Eisha	Hayany	Oraiby	Om-Elferakh	Amhat	Selmi	Barhi
Spine type (1. single, 2. di, 3. mixed)	2	1	1	1	3	3	3	3	1
Spin color (1. dark green, 2. light green, 3. yellowish green)	3	1	1	2	1	2	2	2	1
Spine nature (1. flexible, 2. rigid)	2	1	1	1	2	2	2	1	2
Spine rachis angle (°)	55.0	40.0	62.5	40.0	55.0	50.0	35.0	45.0	55.0
Fruit Attributes									
Fruit length (cm.)	5.85	6.16	3.96	5.60	3.95	6.85	3.50	5.15	3.70
Fruit width (cm.)	3.35	2.85	2.27	2.72	2.82	2.98	2.15	3.15	2.82
Fruit weight (gm)	29.0	22.4	10.1	21.8	13.2	22.6	8.9	30.0	15.5
Fruit volume (cm^3)	30.0	26.0	8.30	20	11.5	15.0	8.30	27.5	15.0
Fruit density (W/V)	0.97	0.86	1.21	1.09	1.15	1.51	1.07	1.09	1.037
Seed length (cm)	3.85	3.55	2.95	3.35	3.25	4.30	2.25	2.65	2.50
Seed width (cm)	1.18	1.12	0.80	1.02	1.05	1.10	0.90	0.88	1.10
Seed weight (gm)	2.40	2.23	1.02	1.76	1.53	2.35	0.96	1.18	1.61
Seed volume (cm^3)	2.00	1.70	1.00	1.66	0.70	1.33	1.33	1.00	1.33
Weight of seed/weight of fruit	0.08	0.10	0.10	0.08	0.12	0.11	0.11	0.04	0.11
Volume of seed/volume of fruit	0.07	0.07	0.12	0.08	0.06	0.09	0.16	0.04	0.09
Fruit shape (1. cylindrical, 2. ovate-elongate, 3. obviate-elongate, 4. falcoid-elongate, 5. ovate, 6. obviate)	2	1	6	1	5	1	4	3	4
Fruit apex (1. obtuse, 2. blunt, 3. retuse)	2	1	3	1	1	1	1	2	1
Fruit base (1. obtuse, 2. retuse, 3. truncate, 4. truncate and emarginated)	3	3	1	3	2		3	4	2

Fruit color (khalal) (1. pale red, 2. shiny red, 3. pale yellow, 4. orange)	4	2	2	2	1	1	3	4	2
Fruit color at maturity (l. red, 2. dark red, 3. yellow-orange mottled pale red, 4. pale brown, 5. brownish black, 6. reddish black)	3	2	5	5	2	1	6	4	1
Fruit skin nature at maturity (1. smooth and united with flesh, 2. smooth and loose from flesh)	1	1	2	2	1	1	2	1	1
Fruit skin appearances (1. shiny, 2. not shiny)	1	1	1	1	1	2	1	2	1
Fruit skin thickness (cm.)	0.16	0.12	0.05	0.04	0.06	0.09	0.02	0.05	0.05
Fruit flesh thickness (cm.)	0.95	0.85	0.65	0.8	0.75	0.94	0.65	0.85	0.80
Fruit flesh color (1. white, 2. whitish yellow, 3. cream-brown)	1	1	1	3	2	2	3	1	1
Fruit flesh texture (1. soft, 2. firm, 3. fibrous, 4. dry)	2	2	1	1	2	2	1	2	3
Fruit flavor (1. poor, 2. good, 3. excellent)	1	2		2	1	1	2	2	2
Fruit taste (1. delicious, 2. delicious sweet)	2	2	1	2	1	1	1	2	1
Fruit pulp (1. stout, 2. less stout at apex)	1	2	1	2	2	2	2	2	2
Fully maturation (1. early, 2. internecine, 3. late)	2	2		1	3	3	1	3	1
Seed shape (1. cylindrical, 2. elliptical)	1	1	1	1	2	1	2	2	2
Seed color (l. cream, 2. pale brown)	1	2		1	2	2	2	1	2
Seed apex (1. obtuse, 2. blunt, 3. retuse)	2	3	1	2	1	3	2	1	2
Seed base (1. obtuse, 2. blunt, 3. caudate, 4. truncate)	3	4	1	4	1	4	1	2	4
Seed surface (1. smooth, 2. rough)	2	2	2	2	1	2	1	2	1
Seed transverse grooves (1. absent, 2. moderately)	2	2	2	2	1	2	1	2	1

(continued)

Table 3.3 (continued)

Morphological attributes	Cultivars								
	Samany	Zaghloul	Bent-Eisha	Hayany	Oraiby	Om-Elferakh	Amhat	Selmi	Barhi
Seed micropyle position (1. toward the apex, 2. at the middle, 3. toward the base)	2	1	3	3	2	1	1	2	3
Seed micropyle elevation (1. not sunken, 2. moderately sunken)	2	2	1	2	1	1	1	1	1
Ventral furrow shape (1. regular, 2. broadest at base, 3. broad at both ends)	1	1	3	2	3	1	1	3	1
Seed ventral furrow ends (1. Open at one end, 2. open at both ends)	2	1	2	2	2	2	1	2	2

characters of trunk, crown, leaves, fruits, and seeds (Rizk et al. 2004). Recently, a study was conducted by Abd El-Baky (2012) on differentiation between some soft and semidry Egyptian cultivars of date palm using morphological and anatomical features. Results obtained by numerical analysis technique represent the taxonomic similarity between cultivars, proved that the cultivars which grow under similar environmental conditions and had the same needs like temperature, humidity, and soil types were more close taxonomically to each other than to the other cultivars with different needs.

3.5.2 Molecular Descriptors

Molecular markers have been applied as a direct approach to detect genetic variation in date palm cultivars. In this context, RAPD and AFLP fingerprints have been used to identify date palm accessions in Egypt (Adawy et al. 2005; El-Khishin et al. 2003; Soliman et al. 2003). In the same way, El-Assar et al. (2005) used AFLP markers to study genetic diversity of 21 named accessions and 9 unknown pedigrees of date palm. Forty-seven samples collected from eight locations in Egypt were investigated using four sets of AFLP markers. Also, soft (Zaghloul, Samany), semidry (Amry, Aglany, Siwy), and dry cvs. (Sakkoty, Bartamoda, Gondaila, Malkaby, Shamiya) have been characterized using RAPD techniques (El-Tarras et al. 2002). Likewise, ten cultivars, which represent all semidry cultivars in Egypt, were subjected to identification, characterization, and documentation using RAPD analysis (Soliman et al. 2006). RAPD-PCR and ISSR-PCR analyses of nine cultivars of soft cvs. (Samany, Zaghloul, Bent-Eisha, Hayany, Oraiby, Om-Elferakh, Amhat, Selmi, Barhi) in Egypt were studied to identify and describe DNA markers by Eissa et al. (2009) (Fig. 3.4). Recently, Moghaieb et al. (2010) identified an unknown cultivar grown at Matrouh Governorate, comparing it with the other known cultivars grown in the same area, based on RAPD and ISSR analyses. The data indicate that the unknown cultivar was closely related to the cvs. Frehi and Oshkingbil. In addition, several research efforts were conducted using molecular markers in order to permit early identification of gender in date palm (Ahmed et al. 2006; Bekheet et al. 2008; Soliman et al. 2003).

3.5.3 Research and Commercial Labs

The morphogenic and genetic variation and genetic relationships among date palm cultivars are essential for establishing guidelines for conservation and utilization of date palm genetic resources in Egypt. Morphological characters have traditionally provided signatures of cultivar genotype and purity. Such morphological features are sensitive to environmental factors and can be observed only at the mature tree stage. However, using morphological traits alone, discrimination among closely related cultivars is often unreliable, because of the influence of environmental conditions. Clearly, an integrated approach is needed incorporating biochemical and molecular

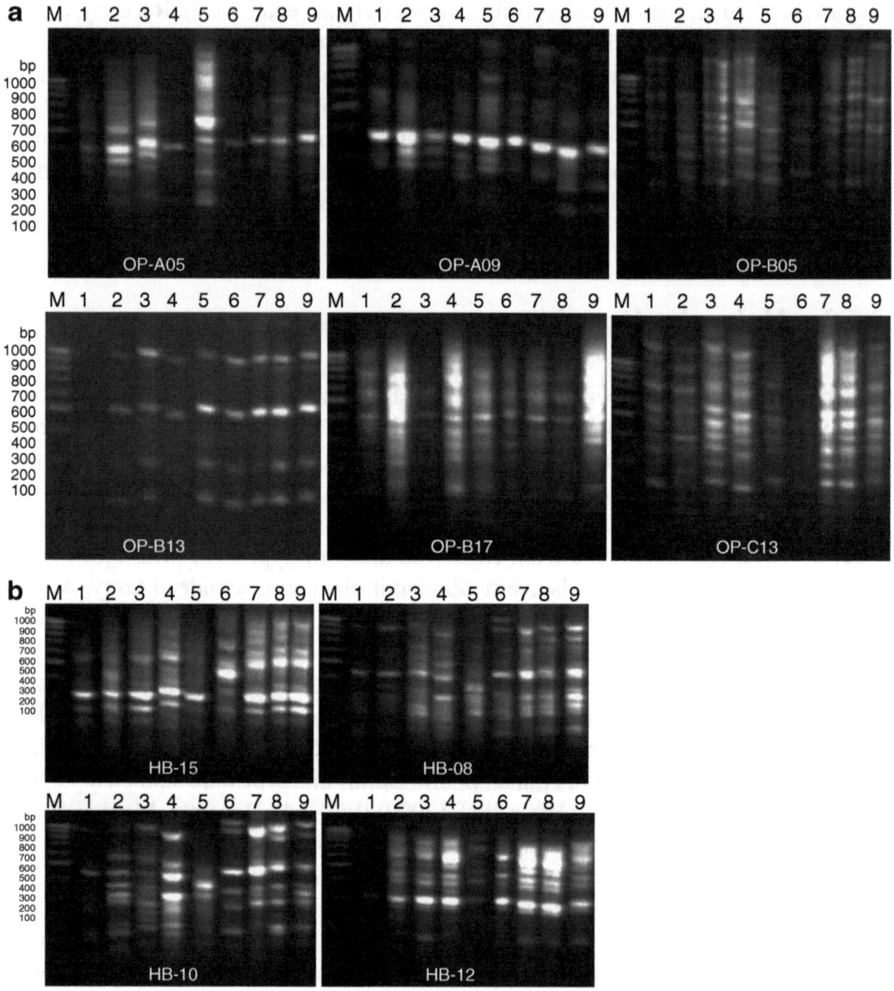

Fig. 3.4 Molecular profile of nine date palm cultivars grown in Egypt using (**a**) RAPD and (**b**) ISSR with different primers. M: 100 bp ladder marker. Lanes 1 through 9 refer to date palm cultivars (*1* Samany, *2* Zaghloul, *3* Bent-Eisha, *4* Hayany, *5* Oraiby, *6* Om-Elferakh, *7* Amhat, *8* Selmi, *9* Barhi)

markers to improve the knowledge of date palm taxonomy and diversity. Recently, molecular markers have provided complementary and accurate tools for genetic analysis and germplasm characterization. DNA typing has proven to be the most convenient method for accurately identifying date palm cultivars and for analyzing their genetic diversity and phylogenic relationships. In spite of effective research efforts exerted on both morphological and molecular identification of date palm cultivars in Egypt, commercial laboratories working in this field are almost nonexistent. Therefore, CLDPRD provides consultation and testing labs which help in the selection of cultivars compatible with environmental factors. However, the commercial molecular characterization is usually achieved in the Agricultural Genetic Engineering Research Institute (AGERI). Other organizations considering activities on molecular characterization are the Plant Biotechnology Department, National Research Center, as well as the Department of Genetics, Faculty of Agriculture, and Ain Shams University. Conversely, the role of the private sector is limited to the import and preparation of specialized chemicals used in molecular analysis, such as primers.

3.6 Cultivars Description

3.6.1 Growth Requirements

Date palm plantations are spread all over Egypt, wherever water is available, and the palm is considered the major component of the Egyptian oasis ecosystem. Egyptian date palm cultivars are classified into three types based on the fruit moisture content, namely, dry tamar, semidry *agwa*, and soft rutab (Amer 2000). Date fruit production in Egypt is dependent on the availability of certain heat requirements according to cultivar. The fruits of the soft cultivars are usually of an oblong shape, and they differ in color. The heat requirement of such cultivars is about 2,100 units during the growth season (May to October) with an average daily temperature of 25 °C. The most important cultivars of the soft group are: Zaghloul, Samany, Hayany, Bent-Eisha, Amhat, Om-Elferakh, and Bergy. These types of cultivars are principally located in the Nile Delta and along the Mediterranean coast. The heat requirements of the semidry cultivars are higher than those of the first group. These cultivars need to be planted in districts where the average daily temperature prevailing during the growth season is 27 °C. But this average can reach 29 °C in the oasis of New Valley where the cultivar Siwy is very well adapted. The main cultivars of semidry cultivars are: Siwy (mainly located in the New Valley, Fayoum, and Giza Governorates), Amry (principally located around Facous and Abou Kabir and Aglany in Sharkia Governorate. By contrast, the dry cultivars are located in Aswan and Qena Governorates where the heat requirements needed for these cultivars are available (Riad 1993). The most important cultivars of dry dates are Sakkoty, Bartamoda, Malkaby, Gondaila, El Gargoda, and El Shamiya. Date palm trees are successfully cultivated across a wide range of soil types in Egypt. Moreover, date palm can tolerate the two extremes, i.e., drought and water logging. Date palm tree can tolerate drought more than most of fruit trees, but if water quantities around the root system reach the wilting point, the growth of the leaves and the fruit will be affected.

3.6.2 Cultivar Distribution and Statistics

In Egypt, there are many cultivars that are significantly different in their vegetative growth and yield characteristics. Most of these cultivars are landraces that spread geographically within certain areas all over the Nile Valley and the Eastern Desert Oasis. The distribution of date palm culture in Egypt follows a geographic pattern, including locations for the successful production of soft, semidry, and dry types of dates (Fig. 3.5).

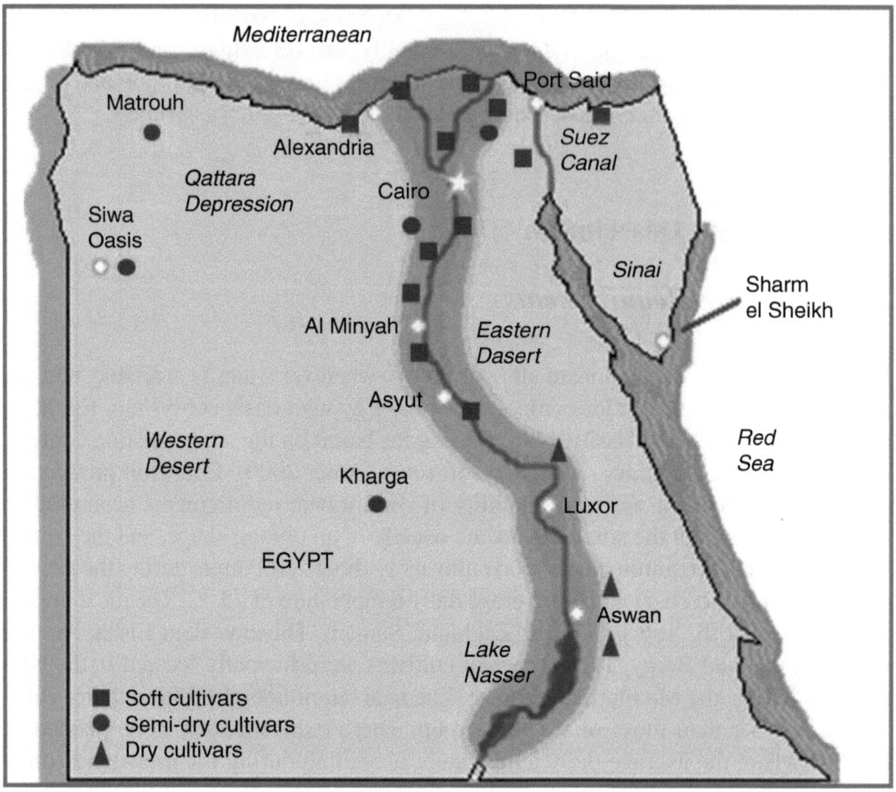

Fig. 3.5 Geographical distribution of date palm cultivars in Egypt

The Egyptian cultivars include the soft (43.5 %), semidry (19 %), and dry (2.5 %); productivity varies from 75 to 150 kg per palm, the highest of all Arab countries (Abd El-Baky 2012). Productive seedling trees account for 35 %. First of all, date plantations extend into the Nile Delta where there are about one-third of the productive date palms of Egypt. In this area, the particular climatic conditions lead to specific common traits of the date palm behavior and of its cultivation: because of the lack of heat and of the relatively high humidity, date fruit maturation is incomplete or very slow at the late stage. From above Cairo (south) to Aswan, in the short strip of cultivated land that extends along the Nile River and, also, in the Fayoum Depression, the number of productive date palms reaches one-half the country's total number. Because about two-thirds of these are seedling dates, they present very important genetic diversity, and their agronomic qualities differ greatly. The western part of Egypt, from the Nile to the Libyan frontier, is occupied by a vast desert plateau. It is cut, from the southeast to northwest, by a succession of depressions called the New Valley. In this valley, various oases are located that are, namely, from north to south: Siwa, Bahariya, Farafra, Dakla, Kharga, and Fayoum. The total number of productive date palms is estimated at about 700,000. These date palms are very heterogeneous and, on average, of low fruit quality. Date palms

Table 3.4 Number of the productive trees, yield, harvesting date, and distribution area of the most important date palm cultivars grown in Egypt

Cultivar type and name	No. productive female trees	Yield/tree (kg)	Harvesting date	Distribution area
Soft cultivars				
Zaghloul	1,043,061	116	Early October	Nile Delta and the Mediterranean Coast
Samany	648,411	130	Mid-October	Mediterranean Coast and Sharkia Governorate
Bent-Eisha	340,774	128	Late November	Nile Delta and Sharkia Governorate
Amhat	253,498	107	Early September	Giza and Fayoum Governorates
Oraiby	224,440	120	Late November	Nile Delta Governorates
Hayany	2,538,131	122	Mid-September	Sharqia, Domyat, Ismailia, and Behera Governorates
Others	4,107,019	80–100	From September to November	North and Upper Egypt, New Valley, and Sinai
Semidry cultivars				
Siwy	1,822,419	98	Late October	Oases of the West Desert and Giza and Fayoum Governorates
Amry	79,860	175	Late October	Sharkia and Ismailia Governorates
Aglany	167,730	163	Mid-October	Sharkia and Ismailia Governorates
Dry cultivars				
Sakkoty	99,016	55	Early November	Upper Egypt especially Aswan
Bartamoda	13,307	50	November	Upper Egypt especially Aswan
Gondaila	17,220	35	November	Upper Egypt especially Aswan
Malkaby	4,410	40	Late October	Upper Egypt especially Aswan
Shamiya	2,000	35	October to November	Upper Egypt especially Aswan
Gargoda	3,000	30	October to November	Upper Egypt especially Aswan

Central Administration for Agricultural Extension Services, Egypt (2001)

are also present but in smaller number in the South and North Sinai, along the Red Sea and in Matrouh Governorate. The total number of productive date palms in these areas is about 600,000 trees (Riad 1993). Based on horticultural descriptions, there are more than 27 economic cultivars produce different types of dates. The productivity of the most economic date palm cultivars grown in Egypt and their distribution area are illustrated in Table 3.4.

3.6.3 Nutritional Aspects

Date palm fruits provide unique functional and nutritional values. Recent research demonstrated numerous health benefits associated with consuming date palm fruits to enriched nutrition values of different kinds of food (El-Sohaimy and Hafez 2010). The date palm fruit is rich in many minerals such as potassium, calcium, magnesium, phosphorus, sulfur, and copper. Also, it is rich in some vitamins such as C and B2, thiamine, and nicotinic acid (El-Warraki et al. 1989). Moreover, the date fruit consists of 70 % carbohydrates (mostly sugars), making it one of the most nourishing natural foods available to humans. The glucose to fructose ratio in dates varies between 1 and 2 depending on the cultivar and ripening stage. A small amount of the carbohydrates found in dates is represented by polysaccharides such as cellulose and starch (Shinwari 1993). Dates were also reported to contain 0.5–3.9 % pectin, thought to possess health benefits (Al-Shahib and Marshall 2003). Yousif et al. (1996) showed that when dates are eaten alone or mixed with plain yoghurt, they have low glycemic indexes. In this context, date fruits constitute a substantial part of the diet of the Egyptian people, especially during the month Ramadan. According to Robinson (1972), 15 date fruits provide more than 80 % of daily body requirement of magnesium, 70 % of sulfur, 25 % of potassium, 20 % of calcium, and a substantial amount of the requirements of iron, manganese, copper, and zinc.

The classification of Egyptian date palm fruits into dry and soft types, mainly based on the texture of the ripe fruit, is related to the moisture content and different forms of sugar as well as sugar acidity. Sucrose is the main sugar in most of the semidry and dry cultivars, while reducing sugars (fructose and glucose) are predominant in the soft cultivars. The flesh of the fruit of the high moisture content cultivars at the stage at which they are eaten is over 50 %. However, the fruits of semidry date cultivars have moderate moisture content and a high percentage of inverted sugar along with a low percentage of sucrose. On the other hand, mature fruits of dry date cultivars (tamar) contain a low moisture percentage (15–20 %) and high percentage of sugar (65–70 %) in which sucrose represents a significant part (Hussein et al. 1979). The fruits of Egyptian date palm cultivars were investigated for their phytochemical characteristics. Mousa (1981) observed that total sugar content ranged between 75 and 85 % in the fruits of six seedlings of date palm trees grown at Ismailia Governorate. Likewise, Youssef et al. (1999) evaluated the chemical composition of eight date cultivars from different areas of Upper Egypt. Total sugar contents ranged from 73.65 to 81.77 % for dry cultivars and from 75.10 to 87.27 % for semidry cultivars. Nonreducing sugars (41.85–46.52 %) were the dominant sugars of dry cultivars, while reducing sugars (71.83–79.08 %) were present in high amounts in the semidry cultivars.

3.6.4 Morphological and Fruit Characteristics

Date palm cultivars are commonly identified by a wide range of morphological features describing both the trees and the fruits. The morphological properties include: diameter of trunk, length of leaf, width of leaf base, length of blade, length of spine area and number of the spines, number of leaflets, and length and width of leaflets. In addition, fruit characteristics were found to be one of the most important features that differentiate cultivars into distinct groups. The most important fruit characters are shape and dimension, color, apex, and pulp. The fruit morphology of Egyptian date palm cultivars has been studied by several researchers. Hussein and Hussein (1983) reported that the average weight, length, and diameter of Sakkoty (dry cultivar) ranged from 6.11 to 8.84 g, 27.75 to 29.45 mm, and 12.08 to 13.18 mm, respectively. Mansour (2005) indicated that the fruit length ranged from a minimum of 2.80 cm in Aglany cv. to a maximum of 5.92 cm in Zaghloul cv. Moreover, the diameter of the fruit apex of Zaghloul (0.70 cm) was significantly more than the other studied cultivars. Otherwise, the investigator observed five colors in fruits of the date palm cultivars: bright yellow in Aglany, yellow with red spots in Samany, orange in Amry, bright red in Zaghloul, and scarlet red in both Hayany and Bent-Eisha. Some physical and chemical fruit characteristics of date palm cultivars grown in Siwa Oasis, Egypt, were studied by El-Wakil and Harahash (1998). In this connection, trunk, leaf, and physical fruit characteristics of the most important Egyptian date palm cultivars are shown in Table 3.5.

Moreover, variability in fruit morphology of the best known soft, semidry, and dry cultivars is illustrated in Figs. 3.6 and 3.7. On the other hand, anatomical fruit characteristics of eight soft cultivars of Egyptian date palm were investigated by Sakr et al. (2010). Such studies add precise evidence to the taxonomic identity of eight soft cultivar of date palm, which includes the most important Egyptian date palm cultivars.

3.7 Dates Production and Marketing

3.7.1 Practical Approaches

To ensure good fruit yield and overall production, a thorough understanding is necessary of the horticultural practices that affect tree growth and productivity. In this respect, fruit thinning is the most critical cultural practice in the date palm production chain that affects fruit development, quality, and yield. Studies by Egyptian researchers have found that flower thinning enhanced fruit quality and regulated the yield of Zaghloul, Hayany, Halawy, Siwy, and Amry cvs. (Marzouk et al. 2007). Likewise, Zaghloul fruit weight, dimensions, total soluble sugar (TSS), and sugar contents were increased due to strand shortening or reduction in their numbers (El-Kassas 1983). Moreover, optimum yield with good fruit quality was obtained with 15–30 % of fruits thinned as early as 2–4 weeks after pollination (Abdel-Hamid 2000; Al-Wasfy and Mostafa 2008; Hammam et al. 2002; Khalifa et al. 1987). Recently, Mostafa and El-Akkad (2011) studied the effect of bunch and strand thinning on yield and fruit quality of Zaghloul and Hayany cvs. A remarkable improvement in fruit quality was noted, after the fruit thinning in comparison to the no thinning. Thinning either bunches or strands leads to a significant increase in the fruit weight, size, and flesh percentage as compared to no thinning. Furthermore, fruit thinning also positively

Table 3.5 Trunk, leaves, and physical fruit characteristics of most important Egyptian date palm cultivars

Cultivar type and name	Vegetative characteristics		Fruit characteristics			
	Trunk	Leaves	Length, mm	Diameter, mm	Color	Consume stage
Soft cultivars						
Zaghloul	Cylindrical, moderate stout with sparse crown	Leaflets not splitting, inconspicuous spines	60	30	Red	Khalal and Rutab
Samany	Most vigorous and strongly built cv. in Egypt, crown moderate dense	Leaflets 6 cm board, terminal leaflets split, spines in pairs except five lower solitary	50–60	25–30	Yellow mottled red	Khalal and Rutab
Bent-Eisha	Cylindrical with moderate dense crown	Leaves not drooping, apical leaflets splitting, short conspicuous spines in pairs become solitary at base	35–40	25–30	Dark red	Khalal and Rutab
Amhat	Moderately stout with sparse crown	Leaves not drooping, leaflets not splitting, numerous moderate-length spines solitary all over rachis	35	25	Pale yellow	Rutab
Oraiby	Cylindrical with moderate dense crown	Leaflets not splitting into halves, short conspicuous spines arranged in pairs become solitary	40–50	25–30	Dark red	Khalal and Rutab
Om-Elferakh	Moderately stout with moderate dense crown	Leaves not drooping, spines are conspicuous, short	60–70	30	Dark red	Khalal and Rutab
Hegazy-Abyad	Moderately stout with sparse crown	Leaves not drooping and leaflets splitting	40–48	22–25	Dark yellow	Khalal and Rutab

Table 3.5 (continued)

Cultivar type and name	Vegetative characteristics: Trunk	Leaves	Fruit characteristics: Length, mm	Diameter, mm	Color	Consume stage
Hayany	Cylindrical in all its parts with moderate dense crown	Leaves not drooping, leaflets splitting into two halves. Solitary spines on rachis	40–50	25–30	Dark red	Khalal and Rutab
Bergy	Moderately stout with moderate dense crown	Leaves not drooping and leaflets not splitting into halves	40	30–35	Bright Yellow	Khalal and Rutab
Semidry cultivars						
Siwy	Stout with sparse crown	Leaflets board not splitting into halves	35–40	20–25	Yellow	Khalal and Tamar
Amry	This cultivar has a sparse crown with drooping leaves	Leaflets not, spines in pairs become close solitary at lower part	50	35	Red	Khalal and Tamar
Aglany	Dense crown with moderate drooping leaves	Leaflets sometimes split into halves, spines in pairs, and five lower pairs closed together	35–40	22–25	Light yellow	Tamar
Dry cultivars						
Sakkoty	Cylindrical with sparse crown	Leaves drooping and leaflets not split	40–60	18	Yellow	Tamar
Bartamoda	Cylindrical trunk has moderately dense crown	Leaves not drooping, leaflets not splitting, spines in pairs become sparse solitary at base	50–60	20	Brown	Tamar
Gondaila	Cylindrical in all its parts, with sparse crown	Leaves slightly drooping, leaflets slender not splitting, all spines solitary	40–45	25	Yellow	Tamar

(continued)

Table 3.5 (continued)

Cultivar type and name	Vegetative characteristics		Fruit characteristics			
	Trunk	Leaves	Length, mm	Diameter, mm	Color	Consume stage
Malkaby	Stout with very sparse crown	Leaves are splitting into halves. Spines in pairs become solitary and sparse at the base	55–65	25	Red	Tamar
Shamiya	Cylindrical in all parts with dense crown	Short slender leaflets not splitting, spines in pairs	70–80	30	Light brown	Tamar
Gargoda	Stout with moderate length	Short cylindrical leaflets, spines in pairs become sparse solitary	40–45	15–20	Yellow turning brown	Tamar

Central Administration for Agricultural Extension Services, Egypt (2001)

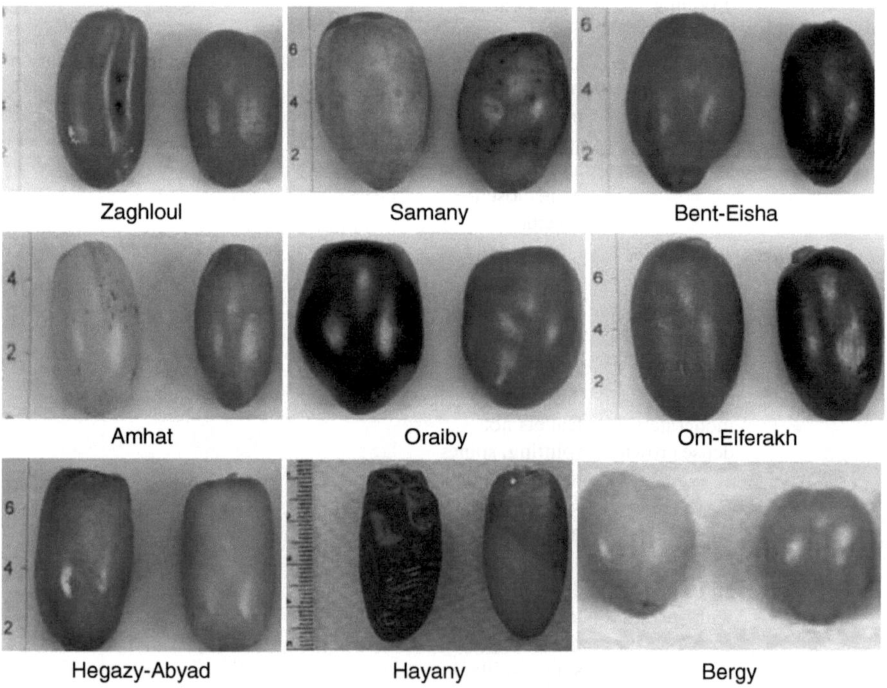

Fig. 3.6 Variability in fruit morphology of the best known soft cultivars (Zaghloul, Samany, Bent-Eisha, Amhat, Oraiby, Om-Elferakh, Hegazy-Abyad, Hayany, and Bergy) of date palm in Egypt

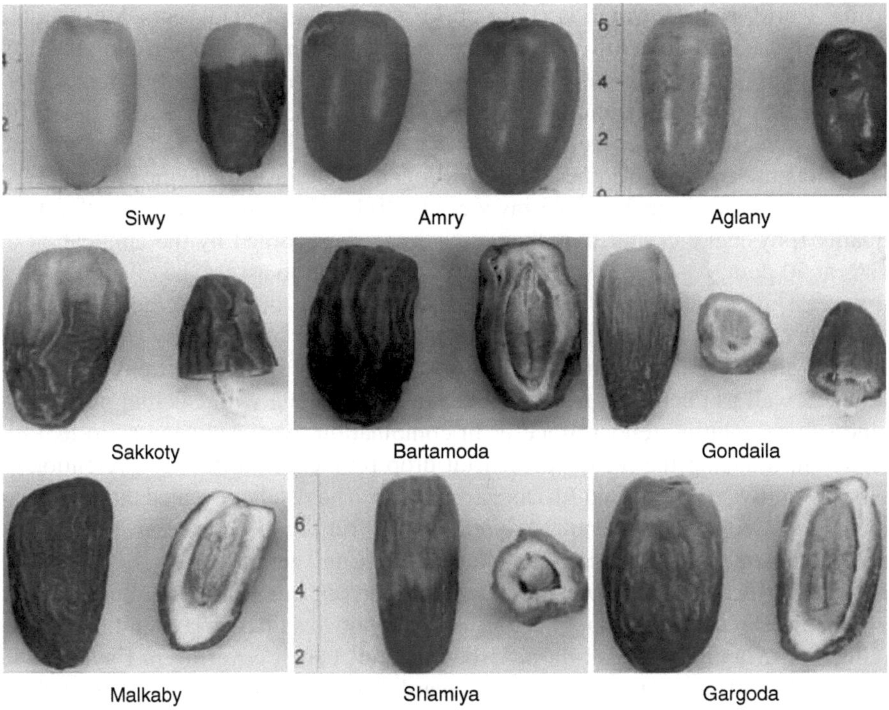

Fig. 3.7 Variability in fruit morphology of semidry cultivars (Siwy, Amry and Aglany) and best known dry date palm cultivars (Sakkoty, Bartamoda, Gondaila, Malkaby, Shamiya, and Gargoda) in Egypt

improved the date palm fruit chemical properties. Manual thinning as well as pollination processes are generally practiced in Egypt. With regard to thinning, there is much concern about the use of chemicals and their affects on environmental pollution and human health. Therefore, development of a more safe and economic thinning agent for the date palm is critically required, especially under harsh conditions. In the context of practical aspects of date palm production in Egypt, we recommend establishment of specific centers including pilot farms aimed at investigating irrigation and fertilization methods, studying the control of date palm diseases, selecting superior local male and female clones, introducing new date palm cultivars, conserving the local clones, and propagating date palm offshoots.

3.7.2 *Optimization of Yield*

At present, Egypt is the leading country in terms of average yield as well as the total production of dates. Average yield of the date palm in Egypt is about 102 kg per tree, calculated on the base of the bearing palms. This figure is very high compared to the world average which is about 50 kg per tree. In recent years considerable attention has been given by date producers to the possibility of controlling the fruit quality of Egyptian date palm by means of chemical application. Soliman (2007) reported that the application of 100 ppm GA_3 to pollinated fruit, 50 days after full bloom, increased average fruit weight, flesh weight, fruit length, fruit diameter, and fruit moisture content. Another study was conducted to investigate the effect of GA_3 application on fruit quality (Ghazawy et al. 2011). The results indicated that high quality (physical and nutrition characteristics) was recorded by the application of GA_3 at 50 ppm. Also, an investigation was carried out to assess the effect of active dry yeast as a foliar spray on yield and fruit quality of Hayany and Siwy cvs. (Gadalla et al. 2011). Results revealed that the spraying of active dry yeast at 75 g/l induced a striking improvement in fruit set, fruit retention, and fruit physical properties as well as decreased fruit drop. Likewise, spraying with calcium nitrate and zinc sulfate solution, either alone or in combination, was evaluated as to its efficiency in reducing the incidence of fruit drop by increasing the fruit retention of Hayany cultivar (El-Baz and El-Dengawy 2001). The results revealed that the treatment increased the force required to remove the fruit, reduced excessive fruit drop, and increased yield. Spraying ethephon at 500 ppm within 2 weeks after spathe emergence increased fruit weight and enhanced fruit quality (Bassal and El-Deeb 2003). By contrast, early application of ethephon resulted in a significant yield reduction (Maximos et al. 1980). Also, date fruit bunch covering in Egypt led to increased yield and fruit quality. In this regard, Kassem et al. (2011) investigated the response of Zaghloul cv. productivity, ripening, and quality in response to different polyethylene bagging. Spathe bagging was carried out at pollination using polyethylene bags at a thickness of 30 μm. It was found that polyethylene bagging, with removal at later growth stages, caused a significant early fruit ripening; increased fruit weight, length, diameter, and yield; as well as decreased the percentage of tip cracked fruits at harvest comparing with no bagging.

3.7.3 Harvest and Postharvest Operations

The stage of maturity at which date palm fruit is picked varies considerably depending on weather conditions, consumer preference, cultivar, processing technique, and picking method. Moreover, as not all female flowers are produced at the same time, the stage of maturity of the dates is also staggered for the different bunches. In Egypt khalal, rutab, and tamar dates are harvested as whole bunches when the majority of dates are ripe. Gathering may continue in the same cultivar from 3 to 4 weeks. In most cases workers have to climb the date palm in order to reach the fruit brunches using several techniques to ascend the palm. The climber is usually barefoot and uses a rope that is connected to a wide harness. There are no mechanical date harvesters in Egypt but just aids such as wooden ladders that allow the workers to reach the fruits more quickly, reducing fatigue and the risk of injury. Laborers cut a bunch of dates and place them on mats on the ground, and the fruits are then removed from their bunch and carefully packed and then transported to the packing house. Egyptian date fruits are consumed at the fresh stage (khalal and rutab) only during the harvesting season, while dry dates are stored in traditional packing boxes to be consumed for the whole year or to export.

In Egypt, fresh dates are usually washed and then separated into several categories according to texture, moisture content, and appearance. Overripe fruits are removed, as well as damaged or small mature fruits. Then dates are pitted prior to packing depending on the intended market. According to research results, postharvest treatments have a great role in the improvement of fruit quality during cold storage as well as during the marketing period. The optimal temperature to store the fruits is from 0 to 4 °C and at a relative humidity of 85–90 % to be kept for a month or more depending on the cultivar (Al-Redhaiman 2005). A recent study was carried out on Zaghloul, Samany, Amhat, and Siwy cvs. to investigate the efficiency of using hot water and sodium carbonate as alternatives to the use of chemicals to control postharvest damage during cold storage (Hafez et al. 2012). Results obtained revealed that cold storage treatment prevents pathogenic molds for up to 40 days in both treated and untreated fruits. Hot water and sodium carbonate treatments are more effective to reduce weight loss (%) and fruit decay (%) compared with untreated fruits, in all the investigated date palm cultivars. Also, these treatments delay fruit decay for up to 20 days with cold storage in Amhat and Siwy cvs. Otherwise, in a study on the storage of Siwy cv., Nazam El-Din and Abd El-Hameed (2001) reported that treating dates with sulfate dioxide led to the best color and that unacceptable color appeared in dates packed in plastic containers. With regard to packaging, different types are used, including plastic, cardboard boxes, and wooden crates. Usually, date are packed in open-top cartons 1, 3, or 5 kg according to buyers' requirement (fresh date), 5 or 10 kg carton (semidry dates), and 10 or 25 kg carton (dry dates). In spite of the importance and broad culture areas of conventional date palm cultivation, field postharvest losses are high, and methods for measuring product quality and the use of date products and by-products need improvement.

3.7.4 Commercial Producers and Major Farms

The Ministry of Agriculture and the private sector are convinced of the potential of date fruit production and are striving to establish plantations and to promote a modern date industry in Egypt. In this respect, the Siwa Oasis producers offer both fresh dates and dried dates for commercial use. The three most important local cultivars are Siwy, Frehi, and Azzawi. Three other cultivars, i.e., Ghazaal, Takdat, and Amnzou, are highly valued for their fruit qualities, but are produced in such small quantities that they are at risk of extinction. Production and commercialization of the dates in the Siwa Oasis are mainly controlled by SCDEC (Siwa Community Development Environmental Conservation), established by the Italian Development Cooperation's Siwa Environmental Amelioration Project. Zaghloul date is the most important commercial soft cultivar in Egypt and in high demand in the Arab

Table 3.6 Main date palm growing areas of date palm in Egypt

Production area	Governorate	Type of fruit
Al-Salhiya	Sharqia	Soft and semidry
El-Hosaynia	Sharqia	Soft and semidry
North Domyat	Domyat	Soft
Edco	Beheira	Soft
Rashid	Beheira	Soft
Nubaria	Alexandria	Soft
Siwa Oasis	Matrouh	Semidry and soft
Dakhla Oasis	New Valley	Semidry
Al-Arish	North Sinai	Soft
Nasr El-Noba	Aswan	Dry
Aswan	Aswan	Dry
Kom Ombo	Aswan	Dry
Edfu	Aswan	Dry

markets. Thus far, Egyptian date palm orchards consist primarily of smallholder farmers. Because there is no comprehensive information available about the date palm producers in Egypt, only the most important production or commercialized areas can be included in Table 3.6.

3.7.5 Marketing Status and Research

Date fruits have a unique marketing nature which is more specialized than other horticultural crops. These characterizations are: (1) storage in normal environments at low cost is possible; (2) dates do not spoil quickly; and (3) date marketing and service is limited in scope and not complicated or costly. In fact, Egyptian date production is characterized by inconsistent quality and quantity and poor packing, with the majority of sales occurring in informal markets at low prices. The internal trade of dates in Egypt, which represents 1 million mt (about 99 % of the total production), is a key limited element with regard to foreign trade (El-Anany 2007). In spite of the low quality, the conventional marketing of dates is very lucrative in Egypt. The date marketing season begins from September to December based on cultivars and areas of production. Dates in the khalal stage of ripeness appear first in the harvesting season and therefore find a ready market. Date marketing in Egypt is unsophisticated; there is a series of intermediaries between producers and consumers. Three types of markets exist: (a) farmers' markets, (b) assembly markets, and (c) wholesale markets. All soft dates are marketed by wholesalers and intermediaries, while cooperative societies share in marketing of dry dates. Commercial dates arrive at the market from three different sources: farmers, merchants, and processors. In this respect, Mohamed et al. (1999) studied the nature of date marketing in both Aswan and New Valley provinces and its related problems. Their results indicated that the wholesalers sell the crop to the retail merchants, while some of them use marketing channels of commissioners to sell the crop to the capital (Cairo) date marketers or sell to consumer directly or export the crop. Also, the research found that the price of dates is limited by demand and the total amount produced, as well as tourist activities in the Aswan area; if Ramadan occurs in a summer month, there is increased demand. The long marketing chain along with lack of storage facilities results in spoilage of about 20 % of fresh produce before reaching the consumer. Like other horticulture crops, date commerce is in the hands of the private sector. However, the government assists the system by offering physical infrastructure particularly wholesale markets and communication, market promotion, market intelligence, and regulatory measures to remove difficulties in business operations. All traditional date palm farmers (small in scale) have operated outside of formal organizations, institutions, and government policies. We, the authors, faced difficulties in obtaining data or statistics on the quantity of dates marketed within the governorates due to the multiplicity of marketing streams. Despite achievements, date quality remains a major concern for Egypt. Appearance, composition, and packaging have not always been up to standards and have negative effects on local consumer purchases and the country's competitiveness in foreign markets. Investments in date postharvest handling and packaging operations are recommended to meet the increasing Egyptian demand for high-quality dates.

3.7.6 Current Import and Export

Date palm fruits are considered one of the most suitable crops for exportation, when they are harvested and marketed at three stages of their development (khalal, rutab, and tamar). The world date fruit export market is about 420,000 mt per annum (2005), and the ten leading exporting countries are: Iran, Pakistan, Saudi Arabia, Tunisia, United Arab Emirates, Iraq, Algeria, Israel, France, and Egypt. However, the world date import market is about 630,000 mt per annum (2005), and the ten leading importing countries are: India, France, UK, Malaysia, China, Indonesia, Germany, Italy, Canada, and Spain (FAOSTAT 2005). Although the cultivation of date palms has developed considerably and great attention has been given to date production in Egypt, the level of date exportation is considered low compared with other date producing countries. Only 10,000 mt are exported per year, accounting for less than 1 % of the total volume of production (Saleh 1999). Egyptian dates exported range in price from USD 500 to 1,500 per mt, depending on quality, and 60 % are on the lower end of that scale, unable to compete with dates grown in Saudi Arabia, Tunisia, and Morocco (FAO 2004). The most attractive Egyptian cultivar for export is Hayany followed by Siwy. In this respect, the amount of the exported dates in 2002 was about 5,006 mt of fresh dates, stuffed dates, and unstuffed dates. The stuffed dates are more desirable in terms of total quantity exported of 4,117 mt (82 % of the total). On the other hand, despite the large amount of the production, Egypt is importing small quantities of dates. The amount imported in 2002 was about 325 mt (USD 500/mt) (El-Anany 2007).

The date palm sector in Egypt does not have a national marketing authority to provide a clear strategic policy for exporting dates and classifying them for export or grading them in terms of quality and pricing. There is a growing consensus that smallholders should become involved in the production of high-value crops for export; this would be the most effective way to stimulate the rural nonagricultural economy toward positive growth. In this context, the Horticultural Export Improvement Association is working to improve the capacity of Egypt to export high-quality horticultural products including dates. Moreover, the Union of Producers and Exporters of Horticultural Crops is aiming to develop the cultivation of horticultural crops in a scientific manner, increase the area under production, and develop and increase exports of horticultural crops including dates. Hindering date exports are the lack of knowledge, pre-harvest and postharvest technology, market information, and economic analysis regarding the returns to date palm investments. Therefore, export orientation should go along with specialized farms, packing stations, and warehouses in order to comply with international quality requirements. Moreover, new institutional arrangements should be developed for mediating disputes between farmer associations and exporters of date palm.

3.8 Processing and Novel Products

3.8.1 Industrial Processing Activities

Date palm fruits are mostly eaten fresh but also dried, pounded into paste, or fermented to produce alcohol or vinegar. Semidry cultivars are processed into paste and date syrup called "dibs" which is used in some recipes. Date syrup is probably the most common derived date product. Since ancient times, date syrup, also known as *date molasses*, was produced as an incidental by-product in the storage of bagged, humid dates. The surplus date fruits of inferior grades, which are unacceptable for the packing industry, are sold at low prices. These fruits differ in their composition but still a good source of sugars, minerals, and other substances (Ramadan 1995). Date palm syrup was evaluated for its physicochemical characteristics and compared with sugarcane syrup (molasses) (Ramadan 1998). Results of the organoleptic evaluation proved that date palm syrup is the more highly desirable.

Date products may also be in the form of stuffed and coated dates and date cake with walnuts. Also, dates are used as a component of food preparations like sweets, confectionary, breakfast foods, baking products as well in dried fruit and nut snack mixtures. Recent innovations include chocolate-covered dates and products such as sparkling date juice, used in some Islamic countries as a nonalcoholic version of champagne, for special occasions and during the month of Ramadan (Mahmoudi et al. 2008). Besides the use of fruits for human consumption, a number of by-products derived from dates also have various uses. In this regard, a number of date constituents are extracted from date fruits and used as natural additives and as functional foods. In addition, pharmaceutical, cosmetic, and medicinal products are derived from dates since an aqueous extract was found to significantly inhibit lipid

Fig. 3.8 Home furniture (**a**) and crate (**b**) made from leaf midrib of date palm tree

peroxidation and protein oxidation in a dose-dependent manner (Allaith 2008). Furthermore, date wine was used by ancient Egyptians as an alcoholic beverage for pleasure, nutrition, medicine, ritual, remuneration, and funerary (embalming) purposes (Cherrington 1925). Ancient Egyptians had at least 17 types of beer and 24 kinds of wine from dates, some of which were used as ingredients of medicines. Among of these products was a special type of distilled spirit known as *arrack*, which is still manufactured in rural areas of Egypt (Nazir 1970).

The date palm over the centuries has also provided a large number of other products which have been extensively used in all aspects of daily life. Ancient Egyptians used palm trunks for roofing and leaves for basket making (Darby et al. 1977). Otherwise, leaves were used to manufacture of sandals especially for priests and temple workers for whom the use of animal substances was not allowed (Nazir 1970). The technological developments have made it possible to look at the palm as a raw material source for more industrial purposes. In this respect, the midrib is used in Mashrabia handicrafts (e.g., window latticework) as a substitute for imported beech wood. Another interesting application was the use of midrib in the core layer of blackboards as a substitute for the imported spruce wood without sacrifice of the utilization property of the product. In addition, triple-layered particleboard was successfully made of palm midrib as a substitute for casuarina wood (El-Mously 2001). At present, date palm residues serve as a raw material for a wide spectrum of micro-, small-, and medium-scale industries in Egypt that could be a very efficient vehicle for the indigenous development of local communities in the region. In this connection, the Egyptian Government (through the Authority of Building and Development of Village) has provided loans to farmers for small projects using secondary products of the date palm tree. Currently, there are several successful small projects utilizing date tree fronds to make home furniture in rural areas of Egypt (Fig. 3.8).

3.8.2 Commercial Date Processers

Recently, Egypt has experienced a remarkable development in date fruit processing. Date-based industries in Egypt are: agwa, stuffed dates, unstuffed dates, dibs, jams,

Table 3.7 Dates processing factories belonging to Egyptian General Authority for Manufacturing in Egyptian pounds (EGP)

Governorate	Number of factories	Production value (1,000 EGP)	Investment costs (1,000 EGP)	Employment
Cairo	1	11	57	6
Domyat	1	113	93	6
Beheira	2	424	5,672	90
Ismailia	1	0	3,725	57
Giza	4	967	15,435	292
Fayoum	1	1,300	1,105	65
New Valley	2	4,358	8,700	567
Matrouh	4	694	887	78
North Sinai	1	250	166	58
Total	17	8,117	35,840	1,219

molasses, and date baking products. Also, the residue and poor quality fruits are used for alcohol, vinegar, and citric acid production. There are many date factories, some belonging to the Egyptian General Authority for Manufacturing and others in the private sector. Currently, date-processing factories are of concern to private sector. In 2005 there were 17 date factories scattered in 9 Governorates of Egypt (Table 3.7). Although considerable in terms of development, there are some districts with date-processing and packing operations but without local store houses for cool storage that would allow for long-term storage. It is realistic to admit that date-processing infrastructure in Egypt to the present is far from satisfying the needs.

3.8.3 *Secondary Metabolites*

In recent years, secondary metabolites of date fruits have received special attention given their health benefit claims and potential use in the booming industries of functional foods and pharmaceuticals (Biglari et al. 2008). Because of its tannin content, the fruit is used medicinally as a detersive and astringent for intestinal troubles. Moreover, it is administered as a treatment for sore throat, colds, and bronchial coughs. In addition, dates have demulcent, expectorant, and laxative effect. Pollen is reportedly ingested to enhance fertility (Darby et al. 1977). In this context, date palm pollen grains have been used historically in Egypt to enhance/restore fertility, especially in women. Furthermore, date palm kernels, with their natural fats, are also included in many concoctions used to reduce skin wrinkles due to its antiaging properties (Bauza 2002) and to prevent irritant contact dermatitis. Important research work has been accomplished on extraction of essential oils and polyphenols from date seeds. Three main phenolics (hydroxycinnamates; flavonols, flavan-3-ols, flavan-3,4-diols, and proanthocyanidins) have been detected in dates (El-Hassni et al. 2004). In this respect, the nonpolar and polar successive extracts of

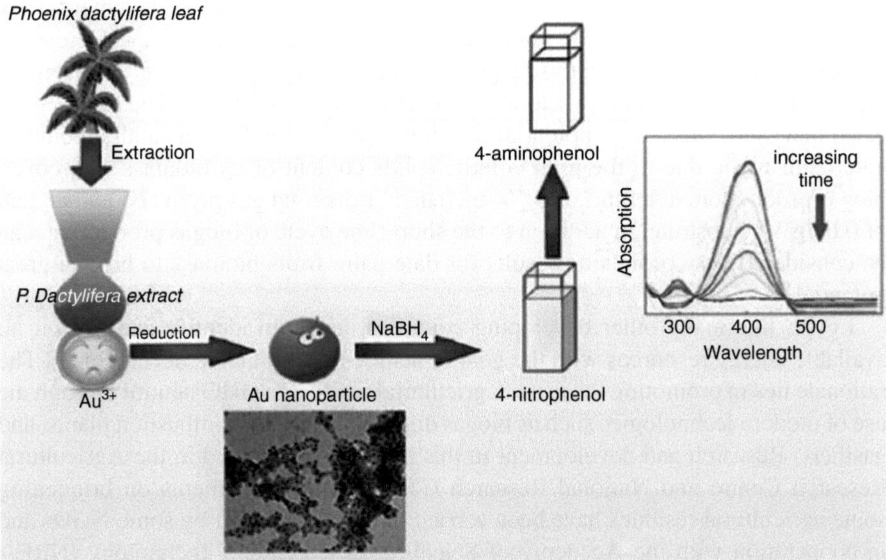

Fig. 3.9 Date palm extract-assisted phytosynthesis of gold nanoparticles. Source: Zayed and Eisa 2014

the seeds of date palm, which is widely distributed in El-Dakhla Oases, were tested for the antioxidant activity by an in vitro bioassay technique according to the b-carotene bleaching method, as well as for the estrogen-like activity (Ammar et al. 2009). The results showed that both extracts have antioxidant and estrogen-like activity with different degrees. The nonpolar and polar extracts showed complete safety. Moreover, enhancement of phenolics and peroxidase production in suspension cultures of date palm was recently reported (Taha et al. 2010). On the other hand, leaf extract of date palm was used as reductant and capping agent for the synthesis of gold nanoparticles (Fig. 3.9) (Zayed and Eisa 2014). Fourier transform infrared (FTIR) analysis suggests that the synthesized gold nanoparticles might be stabilized through the interactions of hydroxyl and carbonyl groups in the carbohydrates, flavonoids, tannins, and phenolic acids present in extract of date palm.

3.8.4 Bioenergy

The most common biofuels are ethanol and biodiesel, which are generally used for, or added to, gasoline and diesel, and also methane which can be used directly. Raw materials for ethanol production by fermentation are fruits and cellulose biomass crops which need atmospheric carbon dioxide for photosynthesis. In this context, date palm wastes have proved to be excellent biofuel sources (Hamada et al. 2002). Large quantities of date palm by-products are generated during harvesting, such as leaves, fruit stalk prunings, fruit wastes, and date seeds. However, a high percentage of this waste is currently disposed of in landfill without any further treatment. Very few studies have evaluated the use of date palm by-products as a renewable energy source (Abed et al. 2012). In this respect, a study of biogas production by anaerobic digestion of date palm pulp waste was conducted by Jaafar (2010). The results indicated that there was a high potential for biogas production from a small amount of date palm waste due to the high volatile solids content of its biomass. Moreover, biogas production, at a ratio of 67 % methane of the total gas produced and a yield of 0.6 l/g VS substrate, in addition to the short time cycle of biogas production, can be considered very promising results for date palm fruits biomass to be of a great potential.

Egypt, like many other developing countries, needs to identify and exploit all available energy resources with the goal of achieving sustainable development. The rationale lies in promoting the use of agricultural waste as a GHG neutral through the use of modern technologies such as biogas digesters, biomass combustion plants, and gasifiers. Research and development in this field was conducted in the Agricultural Research Centre and National Research Centre. Pilot experiments on briquetting some agricultural residues have been carried out by NREA and by some NGOs and in cooperation with the Academy of Scientific Research and Technology (NREA 2006). In Egypt, the total amount of biomass is on the order of 60 million mt/year; their gross calorific value is about 855 million GJ, which is equivalent to about 20 million TOE/year (Mubarak et al. 2006). The bulk of the potential material for bioenergy can be classified as agricultural residue, animal by-products, municipal solid waste, and sewage sludge. The estimated amount of agricultural waste in Egypt ranges from 30 to 35 million mt (AWRU 2005). Rice, wheat, and sugarcane are the three crops generating the greatest amount of waste. Also, Egypt is characterized by the presence of a great number of date palms abundantly distributed throughout the country. Considering the large number of date palms grown in Egypt, the date biomass is an ideal candidate as a feed stock for bioenergy production. Egyptian date production of 300,000 mt/year is an excellent potential source of cheap raw material for the production of approximately 150,000 m^3/year of pure ethanol.

3.9 Conclusions and Recommendations

Date palm cultivation has had great influence on the history of Egypt. Currently, the date palm is playing an important role in the Egyptian agriculture, and it represents a significant part in the national reclamation program. Besides its nutritional and calorific values, the date palm and its by-products are used daily by Egyptians. At the present, date palm constitutes the principal source of income and the basis of the economy for large numbers of Egyptians, inasmuch as one million families are supported by the date palm industry. The date palm tree is cultivated all over Egypt, from Alexandria in the north to Aswan in the south and from the Red Sea in the east to the New Valley and the oases in the west. Although there are many cultivars of date palm in Egypt, only 20 are widely distributed commercially. Besides, there are a great number of seedling date palms, as a result of sexual reproduction; some of them are highly desirable for fruit quality. Although Egypt is the world's leading date producer, it does not play an influential role in world trade of the fruit. This is because the date palm sector in Egypt has faced several problems, which can be summarized as: low fruit quality, diseases and pathogen pests particularly red palm weevil, the weakness of marketing services, and neglect of farmers in achieving efficient postharvest processes. In addition, Egypt at present does not process date fruits using modern techniques. In order to develop the date palm sector in Egypt, a complementary strategy for propagation, conservation, genetic improvement, and disease and pest control of the most economic cultivars, using both conventional and advanced methods, should be implemented. Moreover, modern technology of the harvesting, postharvest, and processing of dates should be introduced into the Egyptian date palm industry. In this respect, farmers and extension staff should be trained on date palm production and processing using workshops, seminars, and tours. The Ministry of Agriculture should promote community ownership of date palms through awareness and strengthening of planning and management committees by providing leadership skills training. Likewise, an education center should be established to train the beneficiaries who obtain loans and produce at home, especially women. Otherwise, grading dates should be taken into consideration according to international standards, and much care must be directed to modernization of date-processing equipment and facilities. Moreover, the government needs to establish a committee consisting of members from traditional producers and date companies. This committee could study the problems the date palm sector is facing and recommend solutions.

References

Abd-Alla MM (2010) Genetic stability on *Phoenix dactylifera* var. Karama produced *in vitro*. N Y Sci J 3:70–75

Abdelal AF, Mahmoud HM, El-Agamy SZ (1983) The effect of pollen source on fruit characteristics of Zaghloul dates (*Phoenix dactylifera*, L.). Assuit J Agr Sci 14:347–355

Abd El-Baky MA (2012) Using morphological and anatomical features as taxonomical evidences to differentiate between some soft and semi-dry Egyptian cultivars of date palm. J Hort Sci Ornam Plants 4:195–200

Abdel-Hamid N (2000) Effect of time, rate and patterns of thinning, leaf bunch ratio and male type on "Zaghloul" date yield and quality. Arab J Agr Sci Ain Shams Univ Cairo 8:305–317

Abed I, Paraschiv M, Loubar K et al (2012) Palm waste pyrolysis analysis. BioResources 7:1200–1220

Abul-Soad AA (2011) Micropropagation of date palm using inflorescence explants. In: Jain SM, Al-Khayri JM, Johnson DV (eds) Date palm biotechnology. Springer, Dordrecht, pp 91–117

Adawy SS, Hussein EHA, Ismail SEM, El-Itriby HA (2005) Genomic diversity in date palm (*Phoenix dactylifera L.*) as revealed by AFLPs in comparison to RAPDs and ISSRs. Arab J Biotech 8:99–114

Ahmed M, Soliman S, Elsayed E (2006) Molecular identification of some Egyptian date palm males by females varieties (*Phoenix dactylifera* L.) using DNA markers. J Appl Sci Res 2:270–275

Al-Delamiy NS, Ali SH (1970) The effect of different date palm pollen on maturation and quality of Zahidi date fruits. J Am Soc Hort Sci 94:638–639

Ali MA, Metwally MM, Hussein AE (2007) Possibility of minimizing insect infestation and losses of store semi-dry and dry dates. 1- Treatment of store house. Fourth symposium of date palm in Saudi Arabia, King Faisal University, Al-Hassa, 5–8 May. Abstracts p 272

Allaith AA (2008) Antioxidant activity of Bahraini date palm (*Phoenix dactylifera* L.) fruits of various cultivars. Int Food Sci Tech 43:1033–1040

Allam HM, Hussein F, Abdalla KM (1973) Seasonal requirements of water by Sakkoti dates grown at Asswan. Moshtohor Fac Res Bull 39:1–10

Al-Redhaiman KN (2005) Modified atmosphere extends storage period and maintains quality of (Barhi) date fruits. Acta Hort 682:979–986

Al-Shahib W, Marshall RJ (2003) The fruit of the date palm: its possible use as the best food for the future? Int J Food Sci Nutr 54:247–259

Al-Wasfy MM, Mostafa RAA (2008) Effect of different methods of fruit thinning on Zaghloul date palm production and fruit quality. Assiut J Agr Sci 39:97–106

Amer WM (2000) Date palm, (*Phoenix dactylifera* L) cultivars in Egypt. El Minia Sci Bull 13:1–15

Ammar NM, Al-Okbi SY, Mohamed DA, Abou El-Kassem LT (2009) Antioxidant and estrogen like activity of the seed of *Phoenix dactylifera* L. palm growing in Egyptian Oases. Report Opin 1(3):1–8. http://www.sciencepub.net

Attalla AM, Haggag MN, Attia MM, Ibrahim AMF (1988) Fruit and leaf mineral content of four date palm varieties grown in Alexandria. J Agr Res Tanta Univ 14:319–331

Attalla AM, Shahein AH, Kassem HA, Aly HSH (2003) Effect of applying different organic and inorganic nitrogen sources to Zaghloul and Samany date palm cultivars on: I. leaf and fruit mineral content. International conference on date palm in Kingdom of Saudi Arabia. Qasseem Branch, King Saudi Univ, Riyadh, pp 209–221

AWRU (2005) Agricultural Waste in Egypt, Agricultural Waste Recycling Unit (AWRU), Ministry of State for Environmental Affairs (MSEA), Egyptian Environmental Affairs Agency (EEAA). Cairo, Egypt

Bassal MA, El-Deeb MD (2003) Effect of thinning and growth regulator on yield and fruit quality of Zaghloul date palm. Zagazig J Agr Res 29:1815–1837

Bauza E (2002) Date palm kernel extract exhibits antiaging properties and significantly reduces skin wrinkles. Int J Tiss React 24:131–136

Bazza M (2007) Irrigated date palm production in the Near East. Workshop on irrigation of date palm and associated crops, in collaboration with Faculty of Agriculture, Damascus University Damascus, Syria, 27–30 May

Bekheet SA (2013) Date palm biotechnology in Egypt. App Sci Rep 3:144–152

Bekheet SA, Saker MM (1998) *In vitro* propagation of Egyptian date palm: II- direct and indirect shoot proliferation from shoot tip explants of *Phoenix dactylifera* L. cv Zaghlool. In: Proceedings first international conference on date palm, Al Ain, 8–10 Mar, pp 150–157

Bekheet SA, Saker MM, Taha HS, Moursy HA (2001a) Plant regeneration *via* somatic embryogenesis of date palm (*Phoenix dactylifera* L.). Arab J Biotech 4:111–118

Bekheet SA, Taha HS, Saker MM (2001b) *In vitro* long-term storage of date palm. Biol Plant 45:121–124

Bekheet SA, Taha HS, Solliman ME, Hassan NA (2007) Cryopreservation of date palm (*Phoenix dactylifera* L.) cultured *in vitro*. Acta Hort 736:283–291

Bekheet SA, Taha HS, Hanafy M, Solliman ME (2008) Morphogenesis of sexual embryos of date palm cultured *in vitro* and early identification of sex type. J Appl Sci Res 4:345–352

Belal AH, El-Deeb MD (1997) Direct organogenesis of date palm (*Phoenix dactylifera* L.) *in vitro*. Assiut J Agr Sci 28:67–77

Ben Salah M, Hellali R (1998) Metaxenic effects of nine pollinators on three palm date varieties (*Phoenix dactylifera*, L.) growing in Tunisia coastal oasis. First international conference on date palms. United Arab Emirates Univ, Al Ain, 8–10 Mar. Abstract p 61

Bendiab K, Baaziz M, Brakez Z, Sedra H (1993) Correlation of isoenzyme polymorphism and Bayoud disease resistance in date palm cultivars and progeny. Euphytica 65:23–32

Biglari F, AlKarkhi AFM, Easa AM (2008) Antioxidant activity and phenolic content of various date palm (*Phoenix dactylifera* L.) fruits from Iran. Food Chem 107:1636–1641

Briton-Jones HR (1925) Mycological work in Egypt during the period 1920–1922. Tech Bull Agr 49:129

Central Administration for Agricultural Extension, Egypt (2001) Date palm. Bulletin No. 710, 2001, Ministry of Agriculture and Land Reclamation (in Arabic). Cairo, Egypt

Cherrington EH (1925) Standard encyclopedia of the alcohol problem, vol 6. American Issue Publishing, Westerville

Cox ML (1993) Red palm weevil, *Rhynchophorus ferrugineus* in Egypt. FAO Plant Prot Bull 41:30–31

Darby JW, Ghalioungi P, Louis G (1977) Food: the gift of Osiris, vol 2. Academic, London

Durant W (1954) The story of civilization. Part I. Our oriental heritage. Simon and Schuster, New York

Eissa EA, Abd El-Razek AB, El-Sharabasy SF, Rizk RM (2009) Morphological and molecular genetic characterization of soft date palm (*Phoenix dactylifera* L.) cultivars in Egypt. Egypt J Genet Cytol 38:269–284

El-Anany SG (2007) Date palm in Egypt- study in economic geography. MSc Thesis, Faculty of College of Education, Ain Shams Univ, Cairo

El-Assar AM, Krueger RR, Devanad PS, Chao CT (2005) Genetic analysis of Egyptian date (*Phoenix dactylifera* L.) accessions using AFLP markers. Gen Res Crop Evol 52:601–607

El-Assioti I (1992) Environmental profile of the Nile irrigation system in Egypt. Egyptian Dutch Panel Meeting 1, Wageningen

El-Baz ET, El-Dengawy FA (2001) Effect of calcium and zing sprays on fruit dropping nature of Hayany cultivar. 1. Yield and fruit quality. Second international conference on date palm, Al-Ain, United Arab Emirates, 25–27 Mar. Abstracts p 177

El-Dawayati MM (2008) Using tissue culture technology to storage some plant tissues of date palm. PhD Thesis, Department of Pomology, Faculty of Agriculture, Cairo University, Giza

El-Deeb HM, Lashin SM, Arab YA (2007) Distribution and pathogenesis of date palm fungi in Egypt. Acta Hort 736:421–429

El-Ghayaty SH (1983) Effect of different pollinators on fruit setting and some fruit properties of the Siwi and Ahmadi dates. In: Proceedings 1st symposium on date palm. King Faisal Univ, Al-Hassa, pp 72–82

El-Hammady AM, Khalifa AS, Montasser AS (1991) Effect of potash fertilization on Sewy date palms. II. Effect on yield and fruit quality. Egypt J Hort 18:199–210

El-Hammady AM, Wanas WH, Abo-Rawash M, Awad AA (1999) *In vitro* propagation of date palm. 2. Factors affecting rooting and acclimatization of *in vitro* proliferated shoots of date palm "Sewy" cv. In: Proceedings international conference of date palm, Assiut Univ, Asyut. 9–11 Nov, pp 25–46

El-Hassni M, J'Aiti F, Dihazi A et al (2004) Enhancement of defense responses against Bayoud disease by treatment of date palm seedlings with a hypoaggressive *Fusarium oxysporum* isolate. J Phytopathol 152:1–8

El-Hennawy H, Wally Y (1978) Date palm (*Phoenix dactylifera* L.), bud differentiation *in vitro*. Egypt J Hort 5:81–82

El-Kassas SE (1983) Manual and chemical thinning of Zaghloul dates. Assiut J Agr Sci 14:221–233

El-Khishin DA, Adawy SS, Ebtissam HA et al (2003) AFLP fingerprinting of some Egyptian date palm (*Phoenix dactylifera* L.) cultivars. Arab J Biotech 6:223–234

El-Kosary S, Bakr EI, Hassan GM, Mtwaly AA (2007) Micropropagation of selected date palm trees using inflorescence explants. Fourth symposium of date palm in Saudi Arabia, 5–8 May. Abstracts p 141

El-Mously H (2001) The industrial use of the date palm: an eloquent example of sustainable development. Second international conference on date palm, 25–27 Mar. Al-Ain. Abstracts p 146

El-Rakashy S, Martin D, Johanson C (2011) Field evaluation of transgenic date palm that were genetically engineered to resist red palm weevil through expressing of the *Bacillus thuringiensis*

Coleopteran- specific endotoxin *Cry3Aa* transgene. First scientific conference for the development of the date palm and date sector in The Arab World, Riyadh, 4–7 Dec. Abstracts p 55

EL-Sebaey Y (2004) Field evaluation of certain insecticides against red palm weevil *Rhynchophorus ferrugineus* Oliv. (*Coleoptera, Curculionidae*) in Egypt. Egypt J Agr Res 82:1591–1599

El-Sharabasy SF, Bosila HA, Mohamed SM et al (2001a) Micropropagation studies on Zaghloul and Sewi cultivars of date palm (*Phoenix dactylifera* L.) 2. Shoot and root formation. In: Proceedings second international conference on date palm, Al-Ain, 25–27 Mar, pp 540–546

El-Sharabasy SF, Bosila HA, Ibrahim IA et al (2001b) Micropropagation studies on Zaghlool and Sewi cvs. of date palm (*Phoenix dactylifera L.*): III. Plantlet acclimatization. In: Proceedings second international conference on date palm, Al Ain, 25–27 Mar, pp 523–530

El-Sharabasy SF, Wanas WH, Al-Kerdany AY (2008a) Effect of salinity stress on some date palm cultivars during proliferation stage *in vitro*. Arab J Biotech 11:273–280

El-Sharabasy SF, Wanas WH, Al-Kerdany AY (2008b) Date palm cultivars *in vitro* screening to drought tolerance using isozymes. Arab J Biotech 11:263–272

EL-Shiaty OH, El-Sharabasy SF, Abd El-Kareim AH (2004) Effect of some amino acids and biotin on callus and proliferation of date palm (*Phoenix dactylifera* L.) Sewy cultivar. Arab J Biotech 7:265–272

El-Sohaimy SA, Hafez EE (2010) Biochemical and nutritional characterizations of date palm fruits (*Phoenix dactylifera L.*). J Appl Sci Res 6:1060–1067

El-Tarras A, Hossni Y, lbanna EA, Sef Eldin A (2002) Genetic diversity within Egyptian date palm cultivars based on RAPD analysis. Egypt J Hort 29:205–216

El-Wakil HE, Harahash MM (1998) Evaluation of some date palm cultivars grown in Siwa Oasis. In: Proceedings first international conference of date palm, Al-Ain, 8–10 Mar, pp 583–601

El-Warraki AG, Oweda EG, Youssef M (1989) Studies on Al Madinah Al-Monawara dates, 1-some physical and quality characteristics. Ann Agr Sci Fac Agr Ain Shams Univ 34:519–643

FAO (2002) Date palm cultivation. Rev. 1. FAO, Rome

FAO (2004) The marketing potential of date palm fruits in the European market. Commodities and Trade Division of the Food and Agriculture Organization of the United Nations (http://www.fao.org/es/ESC/)

FAOSTAT (2005) Crop Production, Statistics Division, Food and Agriculture Organization of the United Nations. Rome http://faostat.fao.org

FAOSTAT (2009) Crop Production, Statistics Division, Food and Agriculture Organization of the United Nations. Rome http://faostat.fao.org

FAOSTAT (2010) Crop Production, Statistics Division, Food and Agriculture Organization of the United Nations. Rome http://faostat.fao.org

Farag KM, Elsabagh AS, ElAshry HA (2012) Fruit characteristics of "Zaghloul" date palm in relation to metaxenic influences of used pollinator. Am Euras J Agr Environ Sci 12:842–855

Fawcett HS (1930) An off-shoot and leaf stalk disease of date palms due to *Diplodia*. Phytopathology 20:339–344

Fawcett HS (1931) Observations on the culture and diseases of date palms in North Africa. Date Grow Inst Ann Rep 8:4

Gabr MF, Abd-Alla MM (2010) Micropropagation of *Phoenix dactylifera* L. var karama. N Y Sci J 3:64–69

Gadalla EG (2007) High frequency somatic embryo production and maturation plantlets in date palm (*Phoenix dactylifera* L.) through suspension culture. Egypt J Agr Res 85:349–365

Gadalla EG, Hassan MM, Hamed AM (2004) Effect of cytokinins and auxins on growth and development of Egyptian semi-dry date palm (cv. Amry) somatic embryos. The second international conference on date palm El-Arish, pp 124–138

Gadalla EG, Abo Rekab ZA, Nady NH (2011) Role of yeast as a new biofertilizer for improving yield and fruit quality of both Hayany and Sewy date palm cultivars. First scientific conference for the development of date palm and date sector in the Arab World. 4–7 Dec, Riyadh. Abstracts p 111

Ghazawy HS, Bakr EI, El-Kosary S, El-Bana A (2011) Effect of Gibberellic acid on fruit quality of Samani and Zaghloul date palm cultivars. First scientific conference for the development of the date palm and date sector in the Arab World, Riyadh, 4–7 Dec. Abstracts p 25

Hafez OM, Saleh MA, Naguib MM (2012) Quality improvement and storability of some date palm cultivars by safe postharvest treatments. Austral J Basic Appl Sci 6:542–550

Hamada JS, Hashim IB, Dharif FA (2002) Preliminary analysis and potential uses of date pits in food. Food Chem 76:135–137

Hammam MS, Sabour A, Ebeed S (2002) Effect of some fruit thinning treatments on yield and fruit quality of Zaghloul date palm. Arab Univ J Agric Sci Ain Shams Univ Cairo 10:| 261–271

Hassan MM (2002) In vitro studies on somatic embryogenesis conservation of date palm. Ph. D. Thesis, Department of Pomology, Faculty of Agriculture, Cairo University, Giza

Hegazy AE (2008) Micropropagation of Egyptian date palm c.v Selmy through floral buds culture. J Agr Sci Mansoura Univ 33:2803–2815

Helaly NM, El-Hosieny HAM (2011) *In vitro* selection and photosynthetic characterization of date palm regenerated seedling as affected by water stress. Am J Plant Phys 6:126–143

Hussein F, El-Khatny S, Wally YA (1979) Date palm growing and date production in the Arab and Islamic World. Ain Shams Ain Shams University Press, Egypt (in Arabic)

Hussein F, El-Kholy MH, Abou-Sayed Ahmed TA (1993) Organic-chemical constituents of some Egyptian dry-date cultivars grown at Aswan. Zagazig J Agr Res 20:1313–1321

Hussein F, Hussein MA (1983) Effect of irrigation on growth, yield and fruit quality of dry dates grown at Asswan. In: Proceedings first symposium on the date palm, Al-Hassa, pp 168–172

Hyams E (1971) A history of gardens and gardening. Praeger, New York

Jaafar KA (2010) Biogas production by anaerobic digestion of date palm pulp waste. Al-Khwarizmi Eng J 6:14–20

Jaradat AA (2011) Biodiversity of date palm. In: Encyclopedia of life support systems: land use, land cover and soil sciences. Eolss Publishers, Oxford, p 31

Jaradat AA, Zaid A (2004) Quality traits of date palm fruits in a center of origin and center of diversity. Food Agr Environ 2:208–217

Kassem HA, El-Shabrout MB, Attia MM (1997) Effect of nitrogen and potassium fertilization on yield, fruit quality and leaf mineral content in some Egyptian soft date varieties. Alex J Agr Res 42:137–157

Kassem HA, Marzouk HA (2010) Effect of different organic fertilizer sources on improving fruit nutritional value and yield of Zaghloul dates. Fourth international date palm conference, Abu Dhabi. 15–17 Mar. Abstracts p 25

Kassem HA, Omar AKH, Ahmed MA (2011) Response of Zaghloul date palm productivity, ripening and quality to different polyethylene bagging treatments. Am Euras J Agr Env Sci 11: 616–621

Khalifa AS, El-Kady AI, Abdalla KM, El-Hamady AM (1987) Influence of thinning patterns and leaf/bunch ratio on "Zaghloul" dates. Ann Agr Sci Fac Agri Ain Shams Univ Cairo Egypt 32:637–647

Krueger RR (2011) Date palm germplasm. In: Jain SM, Al-Khayri JM, Johnson DV (eds) Date palm biotechnology. Springer, Dordrecht, pp 313–335

Mahmoudi H, Hosseininia G, Azadi H, Fatemi M (2008) Enhancing date palm processing, marketing and pest control through organic culture. J Organ Syst 2:29–39

Mansour HM (2005) Morphological and genetic characterization of some common *Phoenix dactylifera* L. cultivars in Ismailia region. MSc Thesis, Botany Department, Faculty of Science, Suez Canal University

Marzouk HA (2010) Soil fertilization study on Zaghloul date palm grown in calcareous soil and irrigated with drainage water. Am Euras J Agr Envir Sci 10:728–736

Marzouk HM, El-Salhy AM, Abdel-Galil HA, Mahmoud AE (2007) Yield and fruit quality of some date palm cultivars in response to some flower thinning rates. Fourth symposium on date palm in Saudi Arabia. King Faisal Univ, Al-Hassa, 5–7 May. Abstracts p 110

Mason SC (1927) Date culture in Egypt and the Sudan. Bull 1457, US Dept. of Agriculture, Washington DC

Maximos SE, Aziz ABA, Desouky IM (1980) Effect of GA_3 and ethephon on the yield and quality of Sewy date fruits (*Phoenix dactylifera* L.). Ann Agr Sci Moshtohor Egypt 12:251–262

Melchers LS (1931) A check list of plant disease and fungi occurring in Egypt. Tran Kansas Acad Sci 34

Moghaieb REA, Abdel-Hadi AA, Ahmed MRA, Hassan AGM (2010) Genetic diversity and sex determination in date palms (*Phoenix dactylifera* L.) based on DNA markers. Arab J Biotech 13:143–156

Moghaieb REA, Abdel-Hadi AA, Ahmed MRA (2011) Genetic stability among date palm plantlets regenerated from petiole explants. Afr J Biotech 10:14311–14318

Mohamed AI, Ramadan MH, Mohamed MA (1999) Marketing farm of dates crop in Aswan and New Valley Governorates. International conference on date palm, Assiut Univ, Asyut, 9–11 Nov, pp 313–334

Monier MAE, Fawzy AM (2006) Plant diversity around springs and wells in five oases of the western desert, Egypt. Int J Agr Biol 8:249–255

Mostafa RAA, El-Akkad MM (2011) Effect of fruit thinning rate on yield and fruit quality of Zaghloul and Haiany date palms. Austral J Basic Appl Sci 5:3233–3239

Mostafa AF, Michael SH, El-Arosi H (1971) Black leaf spot of date palm. Phytopathol Medit 10:128–130

Mousa IA (1981) Evaluation and studying of some seedling date palms grown in Ismailia province. MSc Thesis, Fac Agr Zagazig Univ, Ash Sharqiyah

Mubarak A, Khalil A, Mostert W, Kaseb S (2006) Renewable energy sector in Egypt, bioenergy industry, a study carried out for Industrial Modernization Program (IMP) by Energy Research Center (ERC) Faculty of Engineering, Cairo University, Project: IMC/PS-217, Giza

National Gene Bank (2007) Country report on the state of plant genetic resources for food and agriculture. Arab Republic of Egypt. Cairo, Egypt, p 13

Nazam El-Din AMM, Abd El-Hameed AKE (2001) Study on the storage of Egyptian Siwi date variety (semi-dry date). The second international conference on date palm, 25–27 Mar. Al-Ain. Abstracts p 121

Nazir W (1970) Plants in ancient Egypt. Egyptian Co, Cairo (in Arabic)

Nixon RW (1951) The date palm -"Tree of Life" in the subtropical deserts. Econ Bot 5:274–301

NREA (2006) New and Renewable Energy Authority (NREA), Biomass energy, http://www.nrea.gov.eg/biomass_energy.htm. Accessed: Feb 2006

Osman SM (2009) Response of Sakkoty date palm cultivar propagated by tissue culture-derived to different sources of fertilization. World J Agr Sci 5:631–638

Ramadan BR (1995) Biochemical, nutritional and technical studies on dates. PhD Thesis, Food Sci Tech Dept Fac Agr Assiut Univ, Asyut

Ramadan BR (1998) Preparation and evaluation of Egyptian date syrup. First international conference on date palm. Al-Ain, 8–10 Mar, pp 86–99

Rashed MF (1991) Pathological studies on fungi causing deterioration of date palm offshoots in Egypt. MSc, Fac Agr, Cairo Univ, Giza

Riad M (1993) The date palm sector in Egypt. CIHEAM – Options Méditerranéennes: Série A. Séminaires Méditerranéens 28:45–53. Available at: http://om.ciheam.org/om/pdf/a28/96605879.pdf

Rizk RM, El Sharabasy SF, El Banaa A (2004) Morphological diversity of date palm (*Phoenix dactylifera* L.) in Egypt. I. Dry date palm cultivars. Egypt J Biotech 16:482–500

Rizk RM, El-Sharabasy SF (2006) A descriptor for date palm (*Phoenix dactylifera* L.) characterization and evaluation in gene banks. Am Euras J Agr Environ Sci 1:133–145

Rizkalla AA, Badr-Elden AM, Nower AA (2007) Protoplast isolation, salt stress and callus formation of two date palm genotypes. J Appl Sci Res 3:1186–1194

Robinson J (1972) Normal and therapeutic nutrition, 13th edn. MacMillan Co, New York

Sabet Y, Michael IF (1965) Occurrence of Khamedj disease (inflorescence rot) of date palm in A.R.E. J Microbiol A R E 1:207–212

Saker MM, Bekheet SA, Taha HS et al (2000) Detection of somaclonal variation in tissue culture-derived date palm plants using isozyme analysis and RAPD fingerprints. Biol Plant 43:347–351

Saker MM, Allam MA, Goma AH, Abd El-Zaher MH (2007) Optimization of some factors affecting genetic transformation of semi-dry Egyptian date palm cultivar (Sewi) using particle bombardment. J Genet Eng Biotech 5:57–62

Saker M, Ghareeb H, Kumlehn J (2009) Factors influencing transient expression of *Agrobacterium-mediated* transformation of Gus gene in embryogenic callus of date palm. Adv Hort Sci 23:150–157

Sakr MM, Abu Zeid IM, Hassan AE et al (2010) Identification of some date palm (*Phoenix dactylifera*) cultivars by fruit characters. Indian J Sci Tech 3:338–343

Saleh SA (1999) The Egyptian date palm and its future under the recent changes. The international conference on date palm, Assiut Univ, Asyut, 9–11 Nov, pp 247–265 (in Arabic)

Sayed AA, Ali AG (1995) Timing of application of certain organophosphates versus of biocide to control the cosmopterigid, *Batrachedera amydraula* Meyr infesting date palm fruits in New Valley. Assiut J Agr Sci 26:253–259

Sayed AA, Temerak SA (1995) Mechanical, chemical and biological control of *Cadra* spp in date palm trees at Kharga Oasis, New Valley Governorate. Assuit J Agr Sci 25:51–58

Sayed AA, Temerak SA, Vergoulas P (2001) Comparative performance of *Bacillus thuringiensis* sups Kurataki and the natural product, Spinosad for the control of the lesser date moth *Batrachedra amydraula* Meyr infesting date palm trees in New Valley, Egypt. Assuit J Agri Sci 32:184–189

Shaheen MA (1990) Propagation of date palm through tissue culture, a review and an interpretation. Ann Agr Sci Ain Shams Univ Cairo Egypt 53:885–909

Shaheen MR (1986) Pistil receptivity in three cultivar of date palm (*Phoenix dactylifera* L.). Proc 1st Hort Sci, conference, Tanta Univ, Tanta, pp 489–499

Shawky L, Yousif M, El-Gazzar A (1999) Effect of potassium fertilization on "Seewy" date palm. Ann Agr Sci Ain Shams Univ Cairo 44:727–735

Shinwari MA (1993) Date palm. In: Macrae R, Robinson RK, Sadler MJ (eds) Encyclopaedia of food science, food technology & nutrition, vol 2. Academic, London, pp 1300–1305

Shuji N, Akira S, Yoshihiko A, Tsunetomo M (1992) Cryopreservation of asparagus (*Asparagus officinlis* L.) embryogenic cells and subsequent plant regeneration by a simple freezing method. Cryo Lett 13:379–388

Singer C, Holmyard EJ, Hall AR (1954) A history of technology, vol 1, Fall of ancient empires. Oxford Univ Press, London

Soliman KH, Rizk RM, El-Sharabasy SF (2006) Genetic polymorphism of semi-dry date palm (*Phoenix dactylifera* L.) cultivars in Egypt. J Biotech 22:261–273

Soliman SS (2007) Effect of GA_3 on yield and fruit characteristics of Sakkoty date palm under Aswan conditions. The fourth symposium of date palm in Saudi Arabia, 5–8 May. Abstracts p 111

Soliman SS, Ali BA, Ahmed MM (2003) Genetic comparisons of Egyptian date palm cultivars (*Phoenix dactylifera* L.) by RAPD-PCR. Afr J Biotech 2:86–87

Soliman SS, Osman SM (2003) Effect of nitrogen and potassium fertilization on yield, fruit quality and some nutrients content of Samany date palm. Ann Agr Sci Ain Shams Univ Cairo 48: 283–296

Soliman SS, Shaaban SHA (2006) Macronutrient changes of Samany date palm leaflets and fruits as affected by nitrogen and potassium application. Egypt J Appl Sci 21:641–660

Taha HS, Abdel-Kawy AM, Fathalla MA, El-Shabrawi HM (2010) Implement of DMSO for enhancement and production of phenolic and peroxides compounds in suspension cultures of Egyptian date palm (Zaghlool and Samany) cultivars. J Biotech Biochem 1:1–10

Taha HS, Bekheet SA, Saker MM (2001) Factors affecting the multiplication of *in vitro* date palm bud cultures. Biol Plant 44:431–433

Yousif AK, Alghamdi AS, Ahmed A, Moustfa AI (1996) Processing and evaluation of date juice milk drink. Egypt J Dairy Sci 24:277–288

Youssef MKE, El-Rify MN, El-Geddawy MAH, Ramadan BR (1999) Nutrient elements and vitamins content of some New Valley dates and certain date products. The international conference on date palm, Assiut Univ, Asyut, 9–11 Nov, pp 201–210

Zaid A, de Wet PF (2002) Pollination and bunch management. In: Zaid A (ed) Date palm cultivation. FAO, Rome, pp 145–175

Zayed MF, Eisa WH (2014) *Phoenix dactylifera* L. leaf extract phytosynthesized gold nanoparticles; controlled synthesis and catalytic activity. Spectrochim Acta A 121:238–244

Zohary D, Hopf M (2000) Domestication of plants in the old world: the origin and spread of cultivated plants in West Asia, Europe, and the Nile Valley. Oxford University Press, Oxon

Chapter 4
Date Palm Status and Perspective in Algeria

Nadia Bouguedoura, Malika Bennaceur, Souad Babahani, and Salah Eddine Benziouche

Abstract Date palm is the axis of Algerian Saharan oasis agriculture creating a microclimate suitable for the cultivation of fruit trees, cereal crops, and vegetables. Date palm cultivation is subject to abiotic and biotic constraints including diseases like bayoud which destroyed millions of palm trees in southwestern Algeria and continues to expand despite prophylactic measures taken by the Plant Protection Services. Traditional and modern techniques are utilized equally in the operations of small and large farms. Various problems related to agricultural practices keep the yield per tree low in comparison to the surrounding regions. Approximately 18 million date palms are cultivated on an area of 169,380 ha; out of these, ten million trees are producing an annual yield of 500,000 mt of dates. Exports of Algerian dates are small because of weak marketing strategies. A program for the development and expansion of date palm agriculture was implemented by the Ministry of Agriculture and Rural Development (MADR) in the recent years. Moreover, research on propagation, improvement, and evaluation of Algerian date palm cultivars is receiving attention by researchers in various universities and research

N. Bouguedoura (✉)
Faculty of Biological Sciences, Laboratory of Research in Arid Areas,
University of Science and Technology Houari Boumediene,
El Alia Bab-Ezzouar, 32, Algiers 16111, Algeria
e-mail: nadiaboug@gmail.com

M. Bennaceur
Department of Biology, Faculty of Nature Sciences and Life,
University of Oran, El M'Naouer, 524, Oran 31000, Algeria

Faculty of Biological Sciences, Laboratory of Research in Arid Areas,
University of Science and Technology Houari Boumediene,
El Alia Bab-Ezzouar, 32, Algiers 16111, Algeria
e-mail: mabennaceur@yahoo.fr

S. Babahani
Laboratory of Saharan Bio-Resources, Kasdi Merbah University,
511, Ouargla 30000, Algeria
e-mail: bbhsouad@yahoo.fr

S.E. Benziouche
Agronomy Department, Faculty of Exact Sciences and Natural and Life Sciences, Mohamed Khider University, 145, RP, Biskra 07000, Algeria
e-mail: sbenziouche@yahoo.fr

J.M. Al-Khayri et al. (eds.), *Date Palm Genetic Resources and Utilization: Volume 1: Africa and the Americas*, DOI 10.1007/978-94-017-9694-1_4

institutes. This study describes the research development and the knowledge gained during the last four decades in Algeria to enhance the date palm cultivation which constitutes the pillar of the agroecosystems of Algerian Saharan oases.

Keywords Axis of Saharan agriculture • Bayoud • Biotechnology • Constraints • Deglet Noor • *Phoenix dactylifera*

4.1 Introduction

Date palm (*Phoenix dactylifera* L.) is the main crop of both traditional and modern Algerian Saharan agriculture. The economy of the southern provinces (*wilayates*) is based primarily on date palm cultivation and utilization of its fruit by-products such as paste, flour, syrup, vinegar, alcohol, yeast, and confectionery. This provides a major source of income for oasis inhabitants. All parts of the date palm are used, including the leaves and trunks which are used for basketry and house construction. The fruit is consumed in fresh and dry forms, processed to produce syrup (Mimouni and Siboukeur 2011), or fermented to produce wine and vinegar (Ould El Hadj et al. 2012). Leaflets and seeds are used in animal feeding.

Following an increase in living standards, local people have diversified their diet (Brac de la Perrière and Dubost 1987), and consequently, dates have lost their position as a main food as compared to the previous years. Nonetheless, it is still a highly appreciated fruit in some countries for cultural reasons as well as in the Western industrialized countries where the date is considered a health food. Algeria has the world's best plantations of dates cultivar Deglet Noor, which are in high demand in global markets, and the country aims to increase its production to satisfy local consumption and to expand exports.

This chapter is a summary of research advances made in relation to date palm propagation, utilization, and conservation, conducted in Algeria since the 1970s, following the escalating problem of the deadly bayoud disease of date palm. It also describes cultivation and management of the crop and its socioeconomic importance to Saharan agriculture.

4.1.1 Historical and Current Agricultural Aspects

According to Bouguedoura et al. (2010), in the early twentieth century, date palm was cultivated as a subsistence crop but diversified and based on the local economy and water control through a system of *foggaras* (underground conduits), groundwater, stream water, and other sources. At that time, 4.5 million date palms were being exploited.

During the colonial period, the number of palms increased to 6.7 million, cultivation techniques improved, and the understory cultures in particular fruit trees were introduced. The Deglet Noor cultivar became the most desired dessert fruit and the fruit of export.

Reorganization of date palm cultivation began after the country's independence with actions undertaken, supported by FAO, leading to the creation of research stations in southern Algeria. It was also the period of rural exodus which led to the loss of knowledge and know-how. Farmers currently prefer more profitable crops such as cereals and vegetables. Therefore, traditional date palm plantations have declined, and maintenance of the *foggaras* is neglected.

During the 1980s, new areas of Saharan agriculture were created, especially in Adrar, El Oued, Biskra, Ouargla, and Ghardaïa. In the same period, in those regions, an electric power network was developing, and new water resources were mobilized. The number of date palms rose from 8 to 9 million in 1990 by the creation of large areas in Biskra, El Oued, El Guerrara, El Meniaa, Adrar, and In Salah. Since 2000, Algerian date palm groves have witnessed a further expansion that reached 13.5 million trees occupying 120,830 ha in 2002 and at present 18 million trees on 169,380 ha.

Currently, oasis date palm cultivation occupies the regions situated south of the Saharan Atlas Mountains. It begins at the Moroccan border, in the west, and ends at the east Tunisian-Libyan border. From north to south in Algeria, it extends from the southern Saharan Atlas Mountain foothills at Reggane in the west, Tamanrasset in the center, and Djanet in the east (Fig. 4.1).

4.1.2 Importance to Country Agriculture

Date palm is grown in numerous oases spread over the southern part of the country, where the climate is hot and dry. The oases are living spaces which have been artificially established in the midst of a large arid area where water is present. In these locations, a *ksar* (a village made out of clay) was built and date palms were planted around it. These oases systems of complex intensive production are maintained with a very fragile balance. Given the geography of Algeria, it is possible to describe several regions of date palm cultivation (Fig. 4.1):

(a) In the Atlas Mountains foothills (Ksour Ouled Naïl, Zibans, and Aures), there is an oasis chain that marks the gateway of the Sahara.
(b) In the east, Zibans (Biskra), Oued Ghir, Oued Souf (El Oued), and the basin of Ouargla especially with the Deglet Noor cultivar of high commercial value.
(c) In the west, Saoura (Beni Abbes), the Touat (Adrar), the Gourara (Timimoun), and the Tidikelt (Reggane) where palm groves include cultivars of relatively low commercial quality. It is in this area where the only truly bayoud-resistant cultivar, Taqerbucht, exists.
(d) At the center. El Golea, the M'zab (Ghardaïa), and Laghouat.

There are different types of oases depending on the nature and operation of water resources, the type of the soil, and topography. Four types have been distinguished (Zella and Smadhi 2006):

(a) Oases in erg (dune field) depressions, where irrigation water is sourced from groundwater by wells and drilling (Ouargla oasis)
(b) Oases in Ghouts where irrigation water is drawn up by capillary action (Souf oasis).

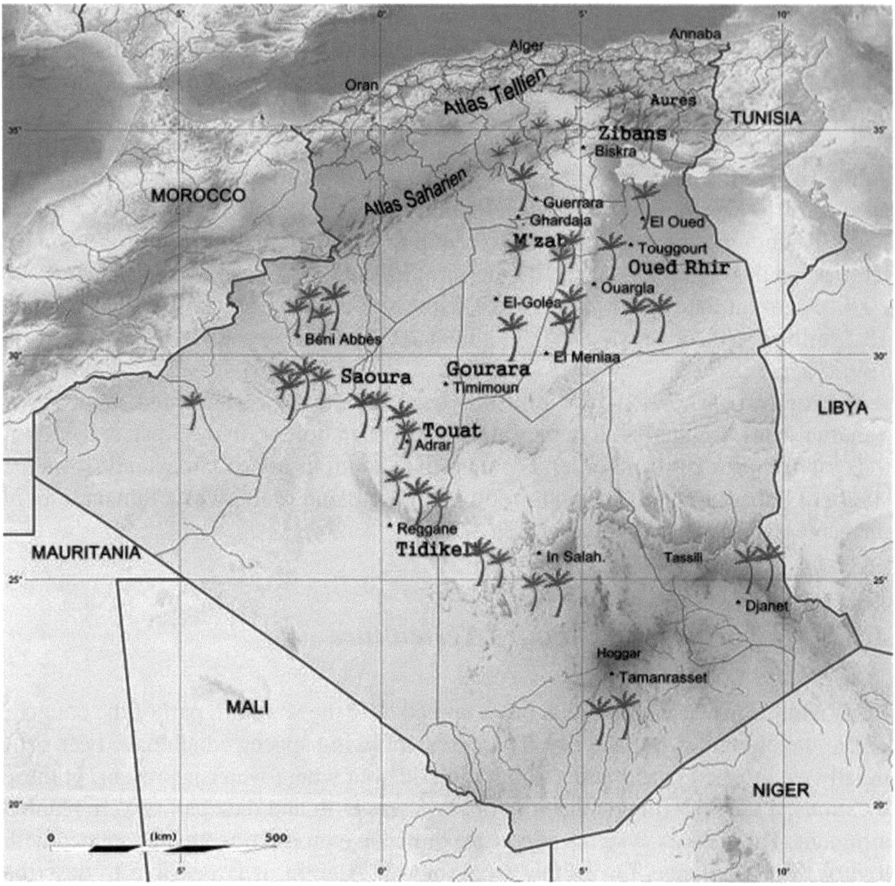

Fig. 4.1 Map of Algeria indicating the different areas with date palms; those in *red* are bayoud infested, those in *green* not infested (Source: N. Bouguedoura Research Laboratory of Arid Areas (LRZA))

(c) River oases, supplied with water from rivers (Oued of Ghoufi, Oued M'zab, Oued Saoura).
(d) Oases of depressions, supplied with water by *foggaras* (Touat, Gourara, and Tidikelt).

4.1.3 Production Statistics and Economics

Date production in Algeria varies annually which correlated with the alternate bearing of the date palm, cultural practices, climatic hazards, and region of cultivation.

This production has surged from 205,907 mt in 1990 to 755,000 mt in 2011, an increase of 266 %. The average production during the period of analysis is estimated at 420,290 mt (Fig. 4.2).

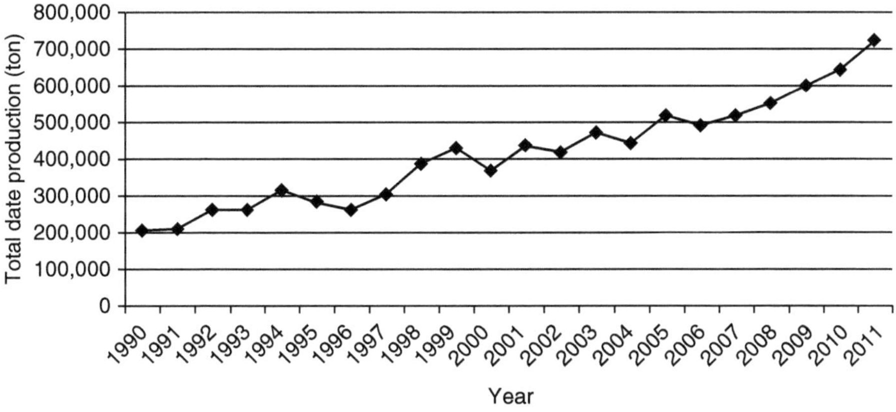

Fig. 4.2 Date production in Algeria 1990–2011 (Source: Benziouche (2012))

Over 92 % of the increase in date production is due to new plantations established within the framework of the Accession to the Agricultural Land Ownership (APFA) and the National Program of Agriculture Development (PNDA) (Benziouche 2010), as well as to the strong recent interest given to this crop.

Statistics show that date production in Algeria is mainly concentrated in the southeastern part of the country, which is responsible of 76 % of national production. The province of Biskra ranks first with nearly 31 % followed by El Oued (27 %) and Ouargla with 18 %. The edaphic and pedoclimatic specificities, as well as the crop management and the market value of cultivars, justify the importance of the production in these regions. In date groves elsewhere, production is less important, contributing 24 % of total national date production allocated as follows: southwest (15 %) and south center (9 %).

4.1.4 Current Agricultural Problems

Date palm is the keystone species of the Algerian oasis ecosystem which is affected by multiple factors: genetic erosion from bayoud disease caused by a soil fungus, *Fusarium oxysporum* f. sp. *albedinis* (Foa), and dominance of the cultivar Deglet Noor to increase exports. Besides, the human population of the Algerian Saharan region has quadrupled during the past 30 years which has created demographic pressures. Consequently, urbanization increases at the expense of date palm groves. The progression of desertification and sand dune encroachment threatening oases are becoming increasingly important. Salinization of soil and water, due to poor drainage management, causes reduction in the number of palms and fruit yield. Finally, the aging of many date palms as well as diseases causing degeneration of date palms have led to adverse changes in fruit quality and a significant drop in yield. For many years, farmers have been abandoning dates in favor of more profitable crops such as vegetables.

4.2 Cultivation Practices

Date palm is a species demanding considerable maintenance and specialized cultivation practices. Algerian date palm farmers often are engaged in other farm activities and do not pay adequate attention to those aspects that affect the date crop quantity and quality. Nevertheless, many efforts are being made within the context of expansion of date palm areas through various development projects. Unfortunately, these efforts and their recommendations have not been followed to improve cultivation practices. The rarity of applied research on date palm cultivation practices impedes advances in farmers' knowledge in this area.

4.2.1 Research and Development

Research on date palm gained significance beginning in the 1970s related to bayoud disease control issues, supported by FAO for the three Maghreb countries of Algeria, Morocco, and Tunisia. As a result, research about resistant cultivars and biotechnological approaches prevailed.

However, this approach required basic studies to better understand date palm biology (Bouguedoura 1979, 1980, 1982, 1991, 2012), identification of the *Fusarium* responsible for bayoud disease, host-parasite relationships (Bounaga 1985), and morphological characterization of cultivars which led to them to be inventoried (Belguedj 2002; Hannachi et al. 1998) and the formulation of standard date palm descriptors (IPGRI et al. 2005).

The few cultivars supposedly resistant to bayoud were found to bear fruit of low commercial quality, but they are important in the context of their utilization in the fight against the disease. For this reason, their multiplication by biotechnological processes continues, mainly through the use of somatic embryogenesis.

Another research approach developed is the formation of cultivars combining the resistance and quality of dates, either by mutagenesis or protoplast fusion (Chabane et al. 2007, 2010). In fact, this task is daunting using classic guided crossing because the date palm is dioecious and heterozygous. This technology approach was attempted in 1984 by the United Nations Development Program and FAO (PNUD/FAO/RAB/84/018; PNUD/FAO/RAB/88/024) with the aim of obtaining new clones resistant to bayoud and producing good-quality fruit. For this purpose, 19 American male date palms were selected by this regional project, while females have been chosen from among the best cultivars of North Africa: Deglet Noor, Medjool, Taqerbucht, and Feggous.

The first plants obtained by mutagenesis, developed by the National Institute of Agronomic Research of Algeria (INRAA) team, were putatively resistant based on in vitro selection of calli, cultivated on a medium containing fungal culture extract; plants are currently being tested in infected fields in Ghardaïa (RAF/5/049 and ALG/5/023 Projects).

The first plantlets obtained by protoplast fusion from the resistant cultivar Taqerbucht and the susceptible cultivar Deglet Noor showed a double level of ploidy compared to the original plants (Chabane 2007). The fusion phase and characterization of somatic hybrids remain to be explored. Characterization of cultivars by biochemical or molecular approaches remains a fundamental goal.

4.2.2 Description of Current Cultivation Practices

The rehabilitation of date gardens is done by planting offshoots (*hachan* or *djebbars*) which are separated and weaned with great care by taking multiple steps. It is necessary to carefully clear the base of the offshoot in order to identify the point of attachment with the mother palm, which is cut and detached with a chisel. After cleaning to remove the injured roots and *cornafs* (petioles), the offshoot is usually planted in spring or fall. The planting hole of 1.3 m is filled up to 30–40 cm of its depth with a mixture of manure (3–5 kg per tree) and loose soil. Offshoots should be planted angled towards the north and surrounded by 3 or 4 date leaves to protect them from the frequent heat and winds.

On traditional farms, regular plant spacing is not followed, but on new plantations being developed, spacing is 8–12 m, depending on the adopted culture system. The most common spacing used is 9×9 m (Benziouche and Chehat 2010; Bouammar 2009).

In Algeria, soil tillage is not a very common practice for the date palm. This operation is only carried out to establish the irrigation *seguias* (canals) or the placement of boards to divert water for understory crops. The work is often manual using simple equipment along with some mechanized operations.

Studies of mineral fertilization of the date palm in El Arfiane and Ain Benoui (Algerian Southeast) helped to determine the following formula in terms of inputs of fertilizer for date palms per hectare: 36,900 kg of nitrogen, 36,900 kg of phosphoric acid, and 73,800 kg of potassium hydroxide (Munier 1973). The planting density is often standard at 120 palms/ha. Fertial (2010) proposed 600 U of N per ha, applied at three stages: flowering, fruit set, and fruiting.

In most date palm groves, farmers generally apply organic fertilizer. Applications are irregular because of the high price of the fertilizer; they are carried out every 2 years, or even 4–6 years. In the Ziban oases, the recommended standards of the Institute of Technical Development of Saharan Agriculture (ITDAS) of 100 kg of manure/year/tree are not followed (Benziouche and Chehat 2010).

Regarding irrigation, farmers use wells with pendulum water lifts (Touat, Ouargla), *noria* or *saniya* water wheels (M'zab Oued Ghir), *foggaras* or *qanate* (Adrar), and the *ghouts* system of capillary water (Oued Souf) (Daoud and Halitim 1994). Currently, well drilling is dominant especially in potential new areas. Monciero (1950) in El Arfiane gave an average coefficient/ha of 50 l/min, which corresponds to an annual volume of 26,383 m^3 per hectare distributed as follows: (a) cool season, October to March, 40 l/min/ha, irrigation per week and (b) hot period, April to September, 60 l /min/ha or even two irrigations per week.

The irrigation system most widely used consists of basins or ponds and diverting water from them using boards. In large farms, irrigation water is obtained by drilling to the Miopliocen and Albian aquifers. Localized irrigation can be expanded in these farms through water conservation. However, it should be noted that most Algerian date palm groves suffer from insufficient water due to poor management practices (Benziouche and Chehat 2010; Bouammar 2009).

In Algeria, field drainage is generally simple, using open ditches despite their disadvantages from silting, landslides, and the need of frequent dredging. Given the ineffectiveness of most drainage throughout the oases, the salinity of irrigation water as well as the soil is often high.

The culling of senescent palms is the main care given to date palm groves. This eliminates an average of 8–12 leaves/palm/year (Hussein et al. 1979). All active leaves should be maintained as the date production relies on them for photosynthesis. Other tree care practices such as pruning dead leaves and petioles are rarely done. Thorn cutting is not practiced in Algeria. Generally, farmers do not manage the number of leaves the tree carries, failing to take note of and removing leaves which are no longer active. Lack of proper date palm care is related to financial constraints and the shortage of labor (Fig. 4.3).

4.2.3 Pollination, Fruit Quality, and Metaxenia

In Algerian oases, farmers use any available pollen source for artificial pollination, because it appears there is no incompatibility between male and female date palms (Pereau-Leroy 1958). In rare cases, some female cultivars provide better production when pollinated by certain *dokkars* (male palms). This seems to be related to compatibility between the male and female cultivars.

During *dokkar* flowering, the farmers cut the male spathes to obtain fresh flower spikelets for pollination. In the case of excess pollen or when the male trees bloom earlier than females, the spikelets are placed on paper or fabric to dry. After drying, the spikelets are collected and stored in cardboard boxes or cloth bags and kept in ventilated enclosures away from heat and moisture. Previously, farmers kept dried spikelets between the leaves of date palms in a shaded area at their farms. This practice has declined as it does not preserve pollen viability. Refrigeration (spikelets or pollen powder) and freezing (pollen powder) are rarely used for pollen storage in Algeria (Boughediri et al. 1995). The traditional and most commonly used pollination practice in Algeria is to place a few male spikelets into the female inflorescences and to tie the spathe back together with a green leaflet.

Mechanical pollination is simple, mixing pollen with a thinner (e.g., talcum, wheat flour, ash) and using a hand or backpack duster, but this is practiced only on a few experimental stations or pilot farms. Tests have shown that it is possible to use only 9 % pollen in the mixture with a thinner (Babahani et al. 1997).

The date palm, like all fruit trees, is alternate bearing. Indeed, every 2 years, without human intervention, the date palm bears many fruits. The following year, the harvest will be less, so it is necessary to do fruit thinning to provide more consistency in the production and enhance the size and improve the quality of dates.

Fig. 4.3 Date palm groves. (**a**) Modern garden. (**b**) Traditional garden (Source: LRZA)

In Algeria, two methods of fruit production management are used: reducing the number of inflorescences on a tree and branch thinning of the fruit stalks. The reduction regime is usually done between May and June by reducing the number of inflorescences and is based on eliminating old and weak stalks. Fruit thinning is not widely practiced in Algeria, the exception being with the Deglet Noor cv. in the Ziban oases (Benziouche and Chehat 2010). There are different methods of thinning: cutting out the center branch of the fruit, reducing the number of fruit branch on the periphery, and a combination of the two. The combined method helps overcome the disadvantages of each while combining their advantages. Indeed, thinning at the periphery eliminates sterile flowers that may exist on the ends of the spikelets.

Thinning of the center fruit strands keeps the individual fruit surfaces dry, especially if the air is moist and warm (Babahani 2011). Some farmers practice branch thinning, reducing the number of fruits on each branch.

4.2.4 Pest and Disease Control

Certain pests such as the moth *Myelois* sp. and the spider mite *boufaroua* (*Oligomycin afrasiaticus*) attack date palms causing serious damage to date production and can lead to deadly diseases. These latter have been described by Bliss (1936), Brun and Laville (1965), and Carpenter and Klotz (1966). However, some of them remain poorly understood.

The main incurable disease and the one causing the most damage is bayoud. Other diseases attack the roots and cause trunk-base rot, terminal bud disease, and leaf disease; fruit and inflorescence diseases are not insignificant since most of them cause severe fruit loss. Recently, a viral disease called brittle leaf disease is worrying Algerian date growers after its appearance in an important region (Fig. 4.4).

Bayoud

Bayoud disease is caused by a fungal soil pathogen *Fusarium oxysporum* f. sp. *albedinis* (Foa) Malençon Snyd and Hans (Malençon 1934), abbreviated as *Foa*. The disease probably originated in the Draa Valley in Morocco, where it was first

Fig. 4.4 Cultural techniques. (**a**, **b**) Pollination, (**c**) thinning, (**d**) pruning (Source: Babahani (2011))

observed in 1890. From there, it spread eastward more than 2,000 km and in 50 years destroying two-thirds of Moroccan palms, ten million trees, as cited in Pereau-Leroy (1958), and palm groves of western Algeria numbering three million trees.

Foa is an imperfect ascomycete present in the soil which becomes virulent in contact with the roots of date palm. Various studies (Matheron and Benbadis 1985; Rahmania 2000) suggest that the penetration of the fungus is via the roots. It then invades the vascular system of the palms, causing gradual leaf drying and bleaching, hence the name (bayoud means *white*). When the terminal bud is reached, the tree dies within a few months or a few years after the onset of the disease.

All attempts at chemical control against this scourge have been unsuccessful (Bulit et al. 1967; Louvet and Toutain 1973; Saaidi 1979). Various control measures have been used to counteract the general effects of bayoud, such as improving farming practices, biological and chemical applications, as well as genetic control techniques. Furthermore, genetic control, the use of resistant cultivars, remains the most promising and least toxic to the environment.

The generalized resistance of cultivar Taqerbucht to bayoud is remarkable in the oases of Tidikelt and In Salah. It is a cultivar that is endemic to the western regions; it requires development and proliferation along with other resistant cultivars or the creation of new cultivars combining good fruit quality and strong resistance to the disease. Selecting resistant cultivars or *khalts* (seedlings) is a long-term undertaking, which has been going on since 1983 in Algeria.

Research on microorganism antagonists has resulted in the characterization and identification of new Saharan species of promise to combat bayoud (Sabaou and Bounaga 1987). The chemical industry continues to develop new control substances to control bayoud (Boulenouar et al. 2009).

Boufaroua

Boufaroua is an insect, an Acarid spider mite (*Oligonychus afrasiaticus* McGr), which feeds on dates, causing them to become dry. The fruits become hard and the epicarp turns a brown color and becomes dusty. To combat this pest, preventive measures are maintenance and cleaning of gardens and chemical treatment by dusting with sulfur and lime.

Djerb or Sem

Djerb or *Sem* insect attack is caused by white cochineal (*Parlatoria blanchardii* Targ.) which parasitizes young stems and can invade inflorescences. These date palm scale insects feed on sap and inject a toxin that affects the chlorophyll production. The covering of the leaf surface by the pest impedes respiration and photosynthesis. The physical result is buckling of the leaf surface; chemical treatment using petroleum oil during the winter period (Nadji 2003) is harmful. Preferable is biological control by the introduction of ladybirds which are predatory to cochineal, a method that is less used.

Date Moth

Larvae of the date moth, *Myelois ceratoniae* Zell. (Ectomyelois), damage date fruit making them unfit for human consumption; other species infest storage facilities. Control is by chemical treatment with malathion + parathion 2 + 1.25 % for each palm. Biological control using ladybirds gives good results, as well as sterilizing male worms by radiation and releasing them into the plantations.

Brittle Leaf Disease

This disease appears to be a physiological disorder although its etiology remains unknown. It affects the palm at the cellular level, mainly the chloroplasts and cell walls. Structural modifications, probably related to those physiological and molecular changes, cause leaf dryness. Brittle leaf disease has been reported in Algeria since 1976.

4.2.5 Agroforestry Utilization and Potential

In the Algerian oasis system, cereal crops underwent an expansion between 1987 and 1994. After 1994, there has been a decline due to lower yields, and this led to a decrease in investment in dates. Other crops such as vegetables, fodder, and other trees have undergone development. In particular, olive cultivation has expanded with very large new plantings in the southeast (Biskra, Ghardaïa). Animal rearing in oases is characterized by the dominance of goat and sheep; the products are almost exclusively produced for sale. Oases are supported by the sale of farm products. The state subsidies contribute to the financing of oasis operations. The support of oasis ecosystems requires coordination between entities of the agricultural sector, farmers, and investors to ensure sustainability.

4.2.6 Limitations and Prospects

A lack of means to counteract abandonment of some maintenance work in most date groves and the insufficiency of prophylactic treatments (maintaining orchards and collecting fallen fruit) are the main causes of unsatisfactory yield returns. Other factors, such as the lack of remedial and chemical control of date pests, which are rarely performed or done improperly without compliance with the technical instructions (dose, scheduling), contribute to this situation. Moreover, a lack of extension services, the low availability of recommended pesticides, and their high cost are also negative factors. None of the date farm operations apply the appropriate technical practices fully and properly such as artificial pollination, pruning, irrigation regimes, and harvest. In other words, some very important technical and economical operations are not realized at all or are done ineffectively. Some farmers rarely apply mineral fertilizers and manure, or they are applied at amounts far below the recommended standards.

The fight against diseases caused by pests requires a considerable investment. Many studies have been devoted to bayoud; chemical eradication was performed in the infested location of El Meniaa, and since then, this oasis remains unscathed. Attempted fumigations were also tested in the oases of M'zab. However, this approach is expensive and may cause environmental problems affecting human and animal health. On date palm plantations, efforts by agricultural services have encouraged farmers to remove diseased palms. In situations where bayoud has caused considerable damage, some farmers conduct various empirical practices. Where outreach efforts have had an impact, conscientious farmers apply preventive measures recommended by experts or plant protection agents. The spread of bayoud in the newly infested areas and in isolated oases is furthered by the lack of knowledge of the population regarding the mechanisms of transmission.

Actions carried out to combat bayoud on the ground are diverse. The main recommendation of distribution of the cultivar Taqerbucht involved establishment, in the area of M'Guiden, Timimoun, and El Meniaa, of a date grove exclusively of Taqerbucht as a field of mother palms. The introduction and gradual spread of this cultivar is remarkable in the oasis of Tidikelt. In the settlement of In Salah, Taqerbucht became among the most abundant cultivars. Elsewhere, the farmers have continued the distribution of this cultivar even if sometimes it does not appear well adapted.

The approach of utilizing genetic resources has been promoted to show the importance of each cultivar. These works helped to identify a dozen cultivars recognized as potentially resistant to bayoud. The need to confirm the resistance of promising cultivars has led researchers to develop new approaches to field and laboratory experimentation. The path of genetic control being both practical and promising, researchers have opted for the screening of natural seedling populations (*khalts*) followed by studies of their molecular markers.

4.3 Genetic Resources and Conservation

The oasis system plays ecological, social, economic, and cultural roles with the date palm as the main element. According to Benkhalifa (2007), citing Jain (1997), date palm genetic resources are classified into several categories:

(a) Traditional cultivars from an intuitive selection and used in traditional palm plantations. Those having Arabic or Berber vernacular names may be specific to a given region.
(b) Natural populations of male and female palm trees from seed (*khalts*). In general, they are not subject to scientific selection, but in some countries, the spread of date palms is only by seed. In Algeria, the population of *khalts* represents

nearly 10 % in some oases and can be a reservoir of new clones selected for some interesting characteristics such as fruit quality, productivity, resistance, and early bearing.

(c) The advanced cultivars or modern cultivars from controlled crosses. In Algeria, breeding programs in El Arfiane launched in 1940 and in Adrar in 1970 were ultimately unsuccessful.

(d) Controlled hybrids which include new genetic material obtained by biotechnological means.

In this classification, traditional cultivars and *khalts* are found in traditional oasis and are estimated to number nearly 1,000 cultivars.

4.3.1 Research in Genetics, Breeding, and Conservation

Because date palm is heterozygous and dioecious, it is very difficult to study, especially from the genetic and improvement perspective. Therefore, these domains have resulted in few scientific works. Initially, there was the need to understand the biology of the development of new combinations, which was very poorly known. Studies by Bouguedoura (1979, 1980) helped to understand the structure and evolution of different axillary productions of date palm and undertook multiplication and improvement using biotechnology, namely, tissue culture and haplomethods (Bouguedoura 1989). In Algeria, this is the issue of bayoud that has dominated the orientation of research. It also is why research into the recognition of *Fusarium*, its culture, and its genetics has developed along with investigations of soil resistance to this fungus (Amir et al. 1996).

Recently, significant results have been achieved in the production of date palm protoplasts; however, plant regeneration from protoplast-induced callus remains limited and requires further research (Assani et al. 2011; Chabane et al. 2007, 2010). Protoplast fusion made between the sensitive variety Deglet Noor and resistant variety Taqerbucht has not yet resulted in mass production of somatic hybrids. Some plantlets obtained shown 4 n level of ploidy (Chabane 2007). Moreover, induced mutagenesis is another potentially powerful biotechnological tool for date palm improvement (Jain 2012). Currently, selected mutants induced from cvs. Deglet Noor and Teggaza are being evaluated for disease resistance in bayoud-infested fields as a part of a joint project (No. AlG/5/023) realized by INRAA and sponsored by the International Atomic Energy Agency (IAEA) (unpublished results).

4.3.2 Current Status and Prospect of Genetic Resources

A census by researchers from Research Laboratory of Arid Lands (LRZA), INRAA, and the Commission for the Agricultural Development of Saharan Regions (CDARS) shows the rich genetic heritage of Algerian date palms (Acourene et al.

2007; Belguedj 2002; Belguedj and Tirichine 2011; Brac de la Perrière and Benkhalifa 1989; Brac de la Perrière and Bounaga 1990; Hannachi et al. 1998). However, Algeria's date palm heritage is yet to be fully recorded.

4.3.3 Threats and Degradation

Date palms represented by landraces suffer from strong genetic erosion over many years. Date palm diversity decreases in the so-called modern palm plantations where only a small number of cultivars are represented. Such is the case in Zibans and Oued Righ. Date palm is mainly grown under monoculture focused on the Deglet Noor cv. along with cvs. Degla Beida, Ghars, and Mech Degla, at the expense of the ancestral gene pool. This monoculture model has serious consequences, the most important being genetic erosion, which increases the fragility of the oasis ecosystem, increases dependence on foreign markets sensitive to geopolitical uncertainties, and reduces the diversification of food and income.

Each region of Algerian oases can have cultivars the equivalent of or superior to Deglet Noor. A broad range of cultivars represent an inexhaustible source of wealth because they offer a choice of various tastes and meet the needs of local populations. In addition, production from a larger number of cultivars extends the date harvest for up to 6 months.

Other factors behind genetic erosion in addition to bayoud disease are a lack of water for irrigation, soil salinization, and silting. Urbanization and the scarcity of manpower to care for date palms also accentuate this decline. The spread of *Fusarium* soil fungus, very visible in the oases of the southwest and in the center of Algeria, has caused the disappearance of a large number of cultivars. Only a dozen cultivars are resistant to or tolerant of bayoud (Brac de la Perrière and Benkhalifa 1991). More attention needs to be paid to prevent loss of the many date palm cultivars, because most of them are rare and ancient, such as M'charet and Tiwraghin.

4.3.4 Conservation Efforts

Some date palm regions are experiencing a significant decline in date palm diversity as observed by both scientists and farmers. This regression is evident by the disappearance or shortage of certain cultivars specific to some oasis soils. It was in this context that a United Nations Development Program, PNUD (RAB98/G31), entitled *Participatory management of date palm genetic resources in the oases of the Maghreb*, was implemented in Algeria, Tunisia, and Morocco in 2001–2005. The project's main objective was the preservation and sustainability of oasis ecosystem production by maintaining date palm genetic diversity.

In Algeria, the project was launched in the M'zab (Ghardaïa) oasis region where efforts are focused on two factors: first, the replacement of traditional cultivars with a small number of cultivars as favored by the national development programs and, second, the fight against market forces promoting certain cultivars of major national and international commercial value to the detriment of local cultivars. The project has made a significant contribution to the conservation of date palm biodiversity and the sustainable development of oasis systems.

The highlights and key findings from the comprehensive IPGRI et al. (2008) report are as follows:

(e) Implementation of a participatory methodology diagnostic for inventory and screening of cultivars among the dates of M'zab.
(f) In situ conservation of endangered cultivars (30 %) by planting offshoots from these cultivars or in vitro propagation in the laboratory of INRA of 24 cultivars screened and selected by the farmers themselves.
(g) Enhancement of the capacity of management to maintain genetic resources in situ of date palm has been strengthened by the integration of environment and development associations. Training in the areas of participatory diagnoses was given to associations, technicians and farmers to better manage and conserve in situ date palm genetic resources. Thus, farmers have fostered new selections from *khalts*.
(h) Research and development of alternative markets to promote the so-called cultivars, *common varieties*, by incentivizing part of the cost of transformation of the date fruit or the use of by-products of the date palm. Thus, the knowledge and the know-how (woodcrafts, woodworking, vinegar, syrup and jams, and other date products) from these cultivars were enhanced through date fairs initiated by the RAB project. These alternative markets are likely to counteract genetic erosion and make possible reintroduction of these cultivars in situ. The participatory approach of the project has brought various actors together (researchers, farmers, associations) for research and development.

4.3.5 Germplasm Banks of Genetic Resources

The conservation and production of date palm are facing problems related to biology. There are two strategies to conserve plant genetic diversity: in situ and ex situ conservation (UNCED 1992).

In Situ Conservation

Farmers have always protected some biodiversity in their gardens, preserving cultivars of traditional date palm groves. However, the number of cultivars is very small because the multiplication of selected ones is limited by the low number of offshoots produced during the life of the date palm or the advanced age of palms, when offshoot growth ceases.

Living date palm collections are located at stations of the Technical Institute of Saharan Agricultural Development (ITDAS) Ain Ben Noui and Feliache in Biskra and El Arfiane in Oued Righ. There is a field collection of 49 local date palm cultivars held by INRAA, but unfortunately, they do not have computerized or documented data.

Ex Situ Conservation

Date palm seeds have a long viability. However, seeds are heterozygous and therefore do not allow for the conservation of specific genotypes. It has therefore not been adopted in Algeria. Tests of pollen conservation were also made by Boughediri et al. (1995). Results showed that the pollen viability of date palm after 230 days of storage was higher when held at 4 °C in the presence of $CaCl_2$, compared to storage at 4 to −20 °C or after lyophilization. The use of biotechnology overcomes these difficulties. Tissue culture by somatic embryogenesis and organogenesis allows mass production of clones from selected individuals. Tissue culture is supported by INRAA and LRZA.

According to Engelmann (2010), cryopreservation (liquid nitrogen, −196 °C) is the only available technique at this time for the safe conservation of plant material grown in vitro with reduced input and costs.

Studies are being conducted to understand the biological and biophysical mechanisms involved in a cryopreservation protocol. An inventory of the conservation of date palm genetic resources revealed a lack of financial resources and organization to provide adequate support.

4.3.6 *Quarantine Regulations*

A series of regulations relative to plant protection have been issued by the Department of Legal Affairs and Regulations (DAJR) of the MADR in January 2013 (http://www.joradp.dz/). Application of the regulations to the date palm quarantine is designed to prevent the introduction and spread of harmful pests, just as it determines and organizes control measures to counteract them. The regulations governing this operation focus primarily on the list of plant pests and measures to monitor and control them.

Decree No. 93-286 of 11/23/93 regulates phytosanitary border control. It mandates inspection of all plants and plant products for import and export to prevent introduction of quarantined pests (insects, mites, nematodes, phytopathogenic fungi, bacteria, viruses, mycoplasma, and all other similar harmful pathogens) into the country.

Besides the objective of developing exports and standardizing the quality of exported products, control measures can also reduce the level of pest control necessary inside the country. Enforcement is carried out by sworn officers appointed by the phytosanitary authority. These enforcement agents may use the Institute of Plant Protection (INPV) or any other scientific body for any diagnosis or expertise needed for border control of plants. For the date palm, the list of harmful organisms whose introduction is prohibited includes *Fusarium oxysporum* f. sp. *albedinis* and red palm weevil, *Rhynchophorus ferrugineus*.

Decree No. 95-387 of 11/28/95 establishes the list of plant pests and measures to monitor and control them. Phytosanitary directives on specific measures are applicable to Foa. Other ministerial decrees have been promulgated: (a) Order No. 489/SM of 03/26/70 relating to the fight against bayoud date palm disease and (b) Order No. 01295 of 11/17/77 amending and supplementing Decree No. 489/SM of 03/26/70.

There are also orders regulating the prevention and/or the fight against bayoud in the provinces of Adrar, Ghardaïa, Ouargla, El Oued, and Biskra. The decree aims to limit the spread of pests from areas already contaminated to uninfected areas in the country. This charge is not only to inspect nurseries, fields, and date palms but also to monitor all established plantations.

4.4 Plant Tissue Culture

Traditional date palm breeding techniques have limitations. The rapid expansion of date palms in the general program of development of Saharan lands, and the need for rapid multiplication of large numbers of selected cultivars, required the establishment of tissue culture laboratories. This began in Algeria in the 1970s with the FAO program in the Maghreb countries.

4.4.1 Role and Importance

In Algeria, the date palm propagation program by tissue culture was implemented at a Maghreb countries' meeting in Algiers in 1972, organized by UNDP/FAO, following the loss of ten million palm trees in Morocco and three million in Algeria, due to bayoud. In fact, only these actions could respond to the need for quick replacement of the millions of lost date palms and to create new date areas. Furthermore, these techniques would fairly and quickly multiply resistant and rare endangered cultivars which could be multiplied by tissue culture.

4.4.2 Research and Development

Despite the wealth of accumulated knowledge about Foa and the host plant, and notwithstanding all the control methods applied, over a century after its appearance, bayoud has not been controlled or eradicated. It continues to expand its range and is now reported in Mauritania, with sporadic outbreaks occurring in Algeria. Researchers agree that the only way to combat the disease is to develop resistant cultivars, but very few have been found in Algeria. In general, those cultivars that do not bear commercial quality fruit can only be developed in the region where they originate. Vegetative propagation by traditional means is inadequate given the limited number of offshoots produced per individual palm.

Given the scarcity of resistant cultivars and their average fruit quality, another approach is to create new resistant cultivars through biotechnology, because conventional means by controlled crosses requires a long time period and the outcome is uncertain. For these reasons, propagation techniques and breeding by in vitro culture hold the most promise: somatic embryogenesis for the multiplication of resistant cultivars and protoplast fusion using resistant cultivars and cultivars bearing good-quality dates.

Somatic embryogenesis is a technique which was adopted by LRZA at the University of Science and Technology Houari Boumediene (USTHB) and INRAA and is focusing on several cultivars. However, improvements are needed to remove the time and synchronization constraints of germination of somatic embryos.

In vitro culture began in Algeria at INRAA and the Department of Botany at the University of Algiers in 1973. In 1979, the first results on a laboratory scale were reported (Bouguedoura 1979). This work was enhanced by the researchers of LRZA (Arban 2009; Bouguedoura 1991; Bouguedoura et al. 1990; Fergani 1998; Si-Dehbi 2009) using vegetative organs and young floral explants.

These various studies have shown that callus is a mandatory step for any multiplication process by in vitro culture in the date palm. The multiplication step of these calli using a medium enriched with casein allows production of large amounts of embryogenic callus which develops into plantlets where root development is enhanced by adding an auxin. In this context, INRAA introduced 13 rare in vitro cultivars selected by farmers in the Ghardaïa region; the rate of adoption has been significant.

4.4.3 *Scale-Up Production and Other Tissue Culture Applications*

Preparation of Cell Suspensions and Plant Regeneration

Callogenic strains are either directly oriented to a later embryogenic expression on agar and shoot formation or used for the initiation of embryogenic cell suspensions in liquid medium. During the initiation of cell suspension, the application of pectinase at 1 % in nodular callus for 3 h is sufficient to cause dissociation of the cells. These cells are the starting point for a viable embryogenic cell suspension. The growth of cell suspension culture results in the formation of cell aggregates in the maintaining medium, then in the wash medium without growth regulators. During the 5 weeks of culture in the maintenance medium, cells multiply rapidly to reach an exponential phase. After this phase, the rate of multiplication of the cell suspension declines. In liquid medium under continuous stirring at 90–100 rpm, the cell suspensions are organized into small embryogenic cell aggregates. These cells aggregates are structured over time to give proembryos able to regenerate somatic embryos and viable seedlings (Fig. 4.5).

Protoplasts, Regeneration, and Fusion

Protoplasts can also be isolated directly from callus clumps harvested from cell suspension cultures. Estimates show that the yield of protoplasts depends on the basic plant material. Thus, the initial material which revealed a high yield of protoplasts from the Deglet Noor cv. is calli from vegetative explants and floral explants with $12.01 \times 100{,}000$ and $3.88 \times 100{,}000$ protoplasts per gram of cells, respectively. Similarly, in cv. Taqerbucht, the highest rate of protoplasts was obtained from callus of vegetative explants and floral explants as $9.66 \times 100{,}000$ and $1.32 \times 100{,}000$ cells per gram, respectively. The protoplasts were then cultured at a density of 106 ml^{-1}. An early regeneration of the cell wall is visible under a microscope with the aid of calcofluor from the third day of culture. Cell divisions settled after about 2 weeks to form microcalli. The microcalli formed on the feeder layer or in the liquid medium and then subcultured on the callus medium become calli which are then transferred to a solid medium multiplication. Calli from microcalli formed from protoplasts plated on the regeneration medium; the formation of somatic embryos occurred easily after this step. They are transferred individually onto the germination medium for the development of plantlets (Fig. 4.5).

Protoplasts are also the material of choice for the fusion of genetic material from the two cultivars: one resistant to bayoud is Taqerbucht and another sensitive but with very good-quality fruit is Deglet Noor. The LRZA team focused on this technique for 8 years. After being able to produce protoplasts in large quantities from the two cultivars, it was possible to fuse them and obtain somatic hybrids with a 4n ploidy level, twice the parents' ploidy, as revealed by flow cytometry (Chabane 2007); it is still necessary to determine the genetics of the new production for a valid comparison.

4.4.4 Commercial Production

Research on somatic embryogenesis from callus or cell suspensions regenerated vitro plants, which were planted in Adrar, Touggourt, and El Meniaa. Currently, they are in the public domain for agriculture, university, and scientific research institutes. A private production laboratory exists in Ghardaïa, but their results are not yet known.

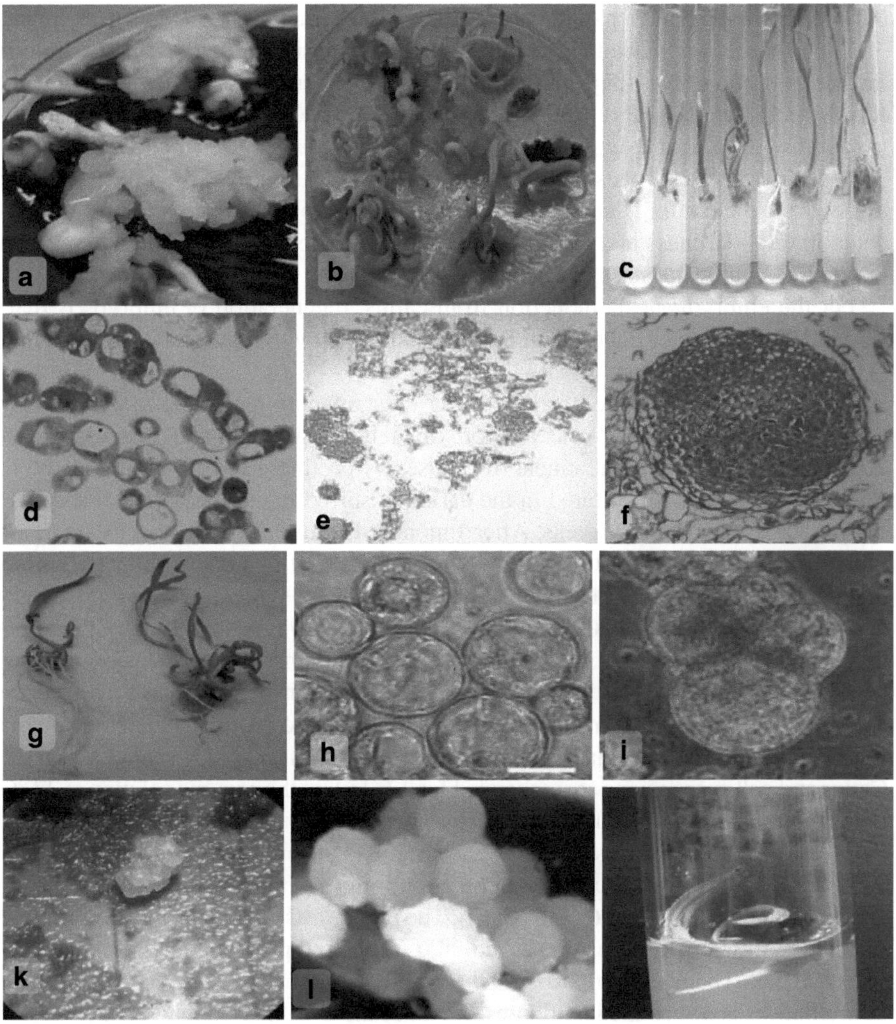

Fig. 4.5 Date palm tissue culture. (**a**) Friable callus from female flowers, (**b**) germination of somatic embryos regenerated from callus of apical shoot, (**c**) development of rooted plantlets from callus, (**d**) cytological aspect of cell suspension, (**e**, **f**) detailed histological cell aggregates with proembryos in cell suspension established from flowers calli, (**g**) development of rooted plantlets from cell suspension, (**h**) protoplasts isolated from embryogenic callus, (**i**) protoplast in division, (**j**) formation of microcallus, (**k**) embryos from microcallus, (**l**) evolution of microcallus to plantlets (Source: Chabane et al. (2007); Si-Dehbi et al. (2013a))

4.4.5 In Vitro Protocols

To initiate in vitro cultures, the plant material commonly used is the *heart* of the date palm offshoot or very young flowers from reproductive date palms. After sterilization in an antifungal solution, the heart is cut into different fragments of

0.5–1 cm, while the flower spikes are fragmented into one to two flowers per explant.

The explants are placed aseptically on a callus initiation medium consisting of the basic solution Murashige and Skoog (1962) (MS) supplemented with 2,4-D and IPA, Morel vitamins (Morel and Wetmore 1951), and 30 g L^{-1} sucrose. The medium is solidified with agar (Sigma) to 0.7 %. The pH is adjusted to 5.7, and the medium is dispensed into test tubes and then autoclaved for vegetative explants or in Petri dishes whose medium is autoclaved prior to floral explants. The cultures are incubated in total darkness at 27 ± 1 °C. Subcultures at intervals of 1 month are carried out until a significant mass of callogenesis is obtained.

For the embryogenesis induction from callus, regeneration of embryos is realized in MS medium enriched with BAP or without growth regulators. Their rooting is favored by the addition of ANA.

The cell suspensions are obtained either after mincing the callus and cultivating them in liquid medium in flasks at approximately 0.5 g of callus in dissociation on 20 ml of medium or after treatment of calluses by pectinase at 1 % for 1, 3, and 6 h. The recovered cells are cultured in the dark in a stirred liquid medium at a speed of 50 rpm for a period of 2–4 weeks. After 1 month of culture, clusters of cells are transferred to a medium free of IPA at a rate of about 0.5 g of plant material per 20 ml of medium. The cultures are then on continuous stirring at 90 rev/min with diffuse light at a temperature of 25–27 °C and a photoperiod of 3,000 Lux 12 h/12 h. Subcultures are made every 15–20 days. After a period of 2–3 months, the cell suspension, after histological analysis, is used as a source of production of protoplasts or somatic embryos. The isolation of protoplasts is done according to the technique based on the use of cellulases and pectinases adopted for bananas, *Musa* sp. (Assani et al. 2002). Callus and cell suspensions are macerated for 12–15 h in the enzyme solution at 27 °C in total darkness. The product of maceration is filtered using two sieves (100/25 μm). Purification of protoplasts is carried out according to the protocol described by Assani et al. (2006) by two centrifugations at 65 g for 5 min each in a washing solution consisting of 204 mM KCl and 67 mM $CaCl_2$. The protoplasts are then cultured in a liquid medium of the same composition as the callus. The protoplast yield was estimated by Nageotte cell. The results are expressed as yield per gram of callus. And the viability of protoplasts is determined by fluorescein diacetate (FDA) (Widholm 1972).

4.5 Cultivar Identification

The identification of selected cultivars over time is made by farmers themselves using their ancestral knowledge. But this knowledge is lost unless identification can be substantiated and documented by researchers.

Genetic resources of the date palm are from selections by farmers over thousands of years. This selection process is difficult because of the dioecious nature and long development cycle of date palm. Another relevant factor is that oasis agroecosystems are very fragile and unstable. The knowledge of the farmers of the identification as well as the classification and selection of the genetic diversity of date palm has revealed the presence of a large number of cultivars which have allowed the establishment of oases with date palms very specific to a region.

In Algeria, the selection is made to a small extent from seedling dates (*khalts*), but mainly from offshoots of cultivars from the same garden or of other areas. The protection and conservation of date palm genetic resources require a deeper understanding of biodiversity and the different mechanisms of selection by the farmers.

4.5.1 Morphological Characterization

Identification and naming of cultivars is based mainly on the morphological characteristics of the fruit and less importantly on the morphology of the tree itself. This description is very difficult as it is possible for the same cultivar to exhibit morphological differences from one palm to another. Only experienced farmers can distinguish among cultivars in their own gardens, identifying cultivars from visual observations and fruit taste.

The scientific use of morphological markers has been standardized across the Maghreb as a result of RAB/98/G31/A/1G/71 Project and IPGRI et al. (2005). Cultivar identification is based not only on qualitative and quantitative fruit and seed characteristics but also on certain other vegetative characteristics.

As to the progress of surveys, new data are added to the identification records, specifically resistance to bayoud and other diseases, habits and uses of dates, and various by-products. The characteristics considered are so numerous and varying with the age of the trees that it seems impossible to distinguish cultivars by this phenological method.

The work of Brac de la Perrière and Benkhalifa (1989) on the variability of landraces in southwestern Algeria, based on the morphological characteristics of the fruit and seed, showed distinctions between some cultivars as well as intra-cultivar variability, which seems genetic. There is in fact an intra-cultivar variability due to environmental influences that affect date fruit morphology. Thus, fruits of cultivar Deglet Noor from Tolga or Biskra are very good, while those of the same cultivar grown in M'zab are generally drier and smaller and therefore of a much lower quality.

The simplest morphological characterization is still based on the fruit characteristics, but those mentioned above are produced after a vegetative phase of several years. Thus, new accurate and faster methods of identification are imposed based on biochemical and molecular characterization.

4.5.2 Biochemical Characterization

Two main types of compounds have been analyzed to study the genetic diversity of date palm in Algeria: flavonoids and isoenzymes. Phenolic compounds are considered very useful biochemical markers in the qualitative and quantitative comparison among different cultivars. They open the way for intraspecific patterning and identify several cultivars.

Regarding the use of polyphenols, work began in the 1990s with the objective to make an inventory of the cultivars of the *Phoenix dactylifera* L. by searching biochemical markers. A total of 50 individuals belonging to nine cultivated dates were analyzed for their flavonoid content. The description of HPLC profiles of glycosyl flavones and glycosyl flavonols initially enabled, for the first time, the identification of 15 compounds (Ouafi 2007).

Structural analysis showed that flavonic aglycones are specific markers to the date palm. A second study showed that the diversity of the flavonic glycosides, considered as markers, permits distinguishing two sets of cultivated date palms: some are quite homogeneous (Ahartane, Aghamu) and others are not (Deglet Noor, Taqerbucht). The second set corresponds to cultivars submitted to intensive cultivation and phenotypical selection by date palm farmers. A negative correlation is shown between the content of flavones and the degree of lignification in the date palm which suggests that the process of lignification is quicker in resistant cultivars; thereby, the resistance to bayoud will be constitutive (Ouafi and Bounaga 2012).

Moreover, a test of conformity was made between the adults and vitro plants of cv. Feggus using the flavonic content. A second test was conducted between male and female plants of cvs. Deglet Noor and Degla Beida, using the same compounds. It revealed that flavonoids cannot constitute chemical markers related to the gender; there is conformity between the male and female palm trees of the same cultivar (Ouafi and Bounaga 2008, 2010).

Bennaceur et al. (2010) characterized phytoconstituents of secondary metabolism of leaflets from 20 date palm cultivars. A phytochemical screening was established and used to highlight phenolic acids, flavonoids, and tannins. However, these metabolites have not been identified and did not allow the identification of cultivars tested in a single way.

Isoenzymes which are multiple molecular forms of enzyme have similar catalytic activities in an organism. The existence of isoenzymes in a species allows a better adaptation in metabolism to respond to the needs of a tissue or a particular stage of development. A study by Bennaceur et al. (1991) of the leaflets of 186 individuals belonging to 31 different Algerian date cultivars from different oases found that of the seven enzyme systems tested, only five systems were selected (alcohol dehydrogenase, ADH; diaphorase, DIA; glutamic-oxaloacetic, GOT; acid phosphatase, ACP; endopeptidase, leucine aminopeptidase, LAP; phosphoglucomutase, PGM) which allowed the identification of 20 cultivars. An intra-cultivar variability was detected among cvs. Hartane, Taqerbucht, Tazerzayt, and Ghars. Individuals of cv. Deglet Noor were homogeneous based on the isoenzyme profile.

All biochemical compounds (flavonoids and isoenzymes) were not totally discriminatory for differentiating between Algerian cultivars but are considered useful markers in other date palm cultivars.

4.5.3 Molecular Characterization

Studies of RAPD (random amplified polymorphic DNA) markers were conducted by Bouchireb and Clark (1997) using 6 arbitrary primers of 39 individuals from 8 cultivars. Benkhalifa (personal communication, 1990) studied 385 individuals representing male and female date palms and other species of *Phoenix* with 9 arbitrary primers. RAPD data easily isolate the date palms in relation to other species of the genus *Phoenix*. The results showed that intra-cultivar variability does not exist for some popular cultivars such as Deglet Noor and confirm the synonymy for cases like Tilemsu/Hmira; similarity to the cultivar Feggus was modified by the dialects of different regions and the presence of an intra-cultivar variability in the case of cv. Taqerbucht.

Clusters obtained in the date palm cultivars do not show a genetic structuration depending on the geographical origin (cultivars introduced from Morocco, Iraq, and Egypt and undifferentiated Algerian cultivars), resistance or susceptibility to Foa, morphological or taste characteristics of dates, or sex differentiation.

Before the development of date palm microsatellites, 24 microsatellite primers of coconut (*Cocos nucifera*) and 18 microsatellite primers of oil palm (*Elaeis guineensis*) were tested on 50 individuals of 10 Algerian cultivars of date palm. The results showed polymorphism and transferability of these markers to the date palm (Bennaceur et al. 2000).

The use of 12 pairs of universal chloroplast primers of tobacco, *Nicotiana tabacum*, on 60 cultivars of Algerian date palms using the SSCP and CAPS generated a low polymorphism in these cultivars (Bennaceur et al. 2002).

Billotte et al. (2004) developed the first microsatellite of date palm. Their use by many researchers allowed a breakthrough in the study of the genetic diversity of the date palm and the establishment of key cultivar identification in several countries (Zehdi et al. 2004).

In Algeria, microsatellites were used by Moussouni (work to be published). Analysis of genetic diversity and the establishment of the key of the molecular determination of Algerian cultivars using 18 microsatellites and a chloroplast genome minisatellite allowed molecular characterization of Algerian date palm cultivation heritage.

A new approach to the study of the genetic diversity of the date palm is the analysis of chloroplast polymorphism. It is the study of genotyping psb ZtrnfM-locus and 20-locus sequencing chloroplast. This study showed the presence of two major haplotypes distributed in all Algerian oases (Moussouni et al. 2011). These results were used to ascertain the date palm's center of domestication, by comparison with those obtained on cultivars from the Middle East, North Africa, Europe, and the Americas.

To determine the genetic relationship among 17 Algerian date palm cultivars, Conserved DNA-Derived Polymorphism (CDDP) markers are used. Seven informative primers were selected from 11 CDDP primers based on their ability to produce clear and repeatable polymorphic and unambiguous bands among the cultivars. A total of 43 bands were produced; 31 of them were polymorphic (72.1 %). This method allowed to distinguish clearly between two clusters with the first group including 15 of the 17 cultivars tested and the second included only 2 cvs., Deglet Noor and Feggus, which have different origin as compared to the other 15 cvs. (Si-Dehbi et al. 2013b).

4.6 Cultivar Description

In Algeria, nearly 1,000 cultivars have been inventoried, and their distribution shows a very marked breakdown into eastern, center, and western portions of the country. Fifty cultivars are found in two or three regions, but most cultivars are endemic to the region and their area of origin. Brac de la Perriere and Benkhalifa (1989) found a very high rate of endemism of 70 % for the dates of the southwestern and more than 60 % on average in those of the southeastern parts of Algeria.

4.6.1 Growth Requirements

Among the nearly 1,000 cultivars surveyed within a total population of 18 million date palms, there also exists an impressive number of *khalts* or *dgouls*; these are *francs* from seeds that grow randomly in the oases and palm groves, especially those ravaged by bayoud or abandoned by their owners.

Khalts represent up to 10 % of the population of date palms, and they are a valuable resource for new selections by farmers. Depending on fruit quality, productivity, and more importantly the resistance to diseases, a *khalt* can be selected to be cloned and become a cultivar (the term *variety* is inappropriate in the case of date palm). It will receive a designation (e.g., Ghars) and therefore can preserve genetic diversity multiplied vegetatively by offshoots.

In a natural seedling date grove, approximately equal percentages of male and female palms will be present. However, in a cultivated date palm grove, male palm trees (*dokkars*) are generally reduced to 1–2 % of the total population to conserve water and allow more space for female palms. A male palm tree which produces good-quality pollen can pollinate a hundred female palms (*nakhla* in Arabic).

Pollination is achieved manually and males are not given their own appellations. In some cases of vegetative similarity, a male palm is identified with respect to the female which it resembles.

4.6.2 Cultivar Distribution

Date palm genetic resources are mainly represented by traditional cultivars which are female individuals selected by farmers according to their bearing (early July for early- and December for late-bearing cultivars), productivity, suitability for storage (long or moderate duration or unsuited for storage), market value, nutritional value, taste qualities, and their resistance to drought and diseases, in particular bayoud (Ben Saadoun and Boulahouat 2010). Cultivars receive local vernacular names; very often, a Berber name for the fruit may indicate a geographic location, the name of the village, or even an individual farm owner.

Random selection by humans over millennia has led to the evolution of almost 1,000 Algerian cultivars, each adapted to slightly different types of soil, temperature, and humidity. Cultivars are not evenly distributed across the different oases.

Through the action plan of the MADR, which aims to promote agricultural products, the Algerian date palm cultivation sector has registered and recorded 994 cultivars thus far. Among these, 7.7 million palms are of cv. Deglet Noor, three million of soft date cvs., and eight million of dry date cvs. (Table 4.1 and Fig. 4.6).

4.6.3 Cultivar Production Statistics and Economics

Algeria has an agricultural resource of 18 million date palms occupying an area of 169,380 ha, with ten million palms currently in production. Domestic production of date fruits is now estimated at more than 500,000 mt per year. It should be noted that Algeria produced 700,000 mt of dates in the 2011/2012 season, including 30,000 mt only for export. The level of export is expected to increase to 60,000 mt by the 2014/2015 season. According to experts, the potential of the sector can exceed this volume. Agricultural services have 40 processing and packaging units scattered across the Saharan region. Over a period of 6 years, 83,000 date farmers have been identified.

In eastern Algeria, cv. Deglet Noor fruits intended for export to the north continue to increase and now represent close to 50 % of the planted population of dates. Fruits of dry date cvs. Degla Beida and Tinnaser are exported to countries in sub-Saharan Africa. Sometimes, dates such as the cv. Hmira are exported to Russia and China. Among the emerging cultivars, Tafezwin is exported to South American countries. Bentqbala cv. fruit, in a frozen state, is renowned in the local market in Ghardaïa (East). Agaz, an early-bearing cv., grown in Tidikelt (West), is frequently marketed in Ouargla and Ghardaïa.

4.6.4 Nutritional Aspects

Quantitative and qualitative analyses were performed on the seeds of 9 Algerian cultivars, 8 from Ghardaïa and 1 from Adrar (Khiat et al. 2011). The study allowed an appreciable quantitative and semi-qualitative estimation of flavonoids from date seeds, which is variable from one cultivar to another. Similarly, Mansouri et al. (2005) showed that the content of phenolic compounds is highly variable in the fruits of different cultivars.

Table 4.1 Cultivar of the three Algerian date palm areas

Area	Region	Location	Number of cultivars identified/reported	Names of cultivars identified
East	Zibans	Biskra, Tolga	9/140	Arechti, Degla Beida, Deglet Noor, Ghars, Ghazi, Mech Degla, Tantboucht, Tinicine, Zoggar Moggar
	Oued Souf	El Oued, El Meghaier, Djamaa	37/70	Arechti, Degla Beida, Deglet Noor, Ghars, Ghazi, Mech Degla, Tantboucht, Tinicine, Zoggar Moggar, Halimi-Halwa (Halwaya), Kesba, Khodri, Loulou, Masri-okrya, Tachelilt, Tacherwint, Tachlikt, Takermust, Takhedrayt, Tantbucht, Taoudent, Tarmount, Zaghraya, Zehdi, Deglet Noor, Ghars, Takermoust, Tanslit
	Oued Righ	El Arfiane, Ouargla, Touggourt	22/200	Aliyane, Beidh H'mam, Bentqbala, Bouldjib, Degla Beida, Deglet G'rara, Deglet Mechta, Dguel El Hadj, El Caber, El Kid, Ghars-Halwa, Hamraya, Tafezwin, Akermoust, Tanetboucht,, Tanslit, Taoudanet, Tawragha, Tazegakht, Tinicine
	Aures	Khenchela	3/220	Buzrur, 'Alig, Buhles, Mech Degla, Tanghimen, Tabanist, Khadaji
	Tassili	Batna	3/180	
Center	M'zab	Ghardaïa, Berriane, Guerrara, Zelfana	26/140	Tamezouaret, Tanaguarout, Tanetboucht, Tawragha, Tazerzayt, Tazizawt, Timdjouhart, Timedwel, Tinnaser, Tissibi, Adham Bent Q'bala, Ajujil, Baydir, Bent Q'bala, Bouarous, Chikh, Degla Beida, Deglet Noor, Gachouch, Ghars, Naser Ou Salah, Oucht, Sab'a Bedraa, Taddela (El Dala), Tademamt, Tafezwin, Taqerbucht (Akerbouch)
West	Touat	Adrar, Timimoun	8/190	Bamekhlouf, Feggus, Hmira, Ouarglia, Taqerbucht, Takerbucht Beida, Takerbucht Hamra, Taqerbucht Safra
	Saoura	Bechar, Béni Abbes	14/80	Adham Boula, Adham Tirnou, Adhamet El Rob, Cherka, Deglet Talmine, Feggus, Hmira, Hartan, Kenta, Khomira, M'charet, Taqerbucht, Timliha, Tinnaser
	Tidikelt		4/60	Tgazza, Taqerbucht, Cheddakh, Agaz

Source: Benkhalifa (1998), Selmani (2012, unpublished)

Fig. 4.6 Fruits of some Algerian date palm cultivars (Source: Bouguedoura – LRZA)

Concerning carbohydrate compounds, studies were conducted on three date cvs. from Ouargla, at tamar stage, representing the three classes of dates (soft, semidry, dry) (Sayah and Ould Hadj 2010). Results showed a difference in the sugar content from one cultivar to another. The semidry cultivars contained various sugars (reducing and nonreducing) with the dominance of reducing sugars. Soft cultivars are rich in total reducing sugars (glucose and fructose) the result of the transversion of sucrose by invertase during maturation. Dry dates present a great amount of sucrose which is responsible for their hard consistency after sugar crystallization.

4.7 Dates Production and Marketing

Many studies have shown that the date sector faces great difficulties in its operations and has not yet achieved its upstream and downstream objectives, although various support policies have been implemented to benefit the industry.

4.7.1 Practical Approaches

Studies of some Algerian date groves reveal that the date cultivation practices are good in some regions such as the Zibans (Benziouche and Chehat 2010). However, practices occasionally fall below the standards recommended with regard several operations. Indeed, the degree of compliance with standards varies from one region to another, from one operation to another, from one plot to another within the same property, and even from one palm to another in the same plot.

No particular solution applies fully and appropriately to this situation. Despite the fact that certain operations are economically and technically important, they are not performed or, if they are, performed very poorly in an unskilled manner without regard for the ideal suggested time frame for the operations (Benziouche 2006). As a result, they have no positive effects on the technical and economic performance in terms of the quantity and the quality of date production.

4.7.2 Optimization of Yield

Analysis of the production structure by cultivar shows that the lion's share of production is dominated by cv. Deglet Noor (nearly 49 %), followed by Degla Beida with 32.50 %, while Ghars and similar cultivars accounted in 2011 for 18.5 % of the total production.

Higher yields recorded during the period of analysis (1990–2011) are significant, from 33 kg/palm increasing to over 50 kg/date palm in 2011 (Fig. 4.7). However, these average yields remain very low and less than the expected 70 q/ha, compared to standard levels of 95 q/ha registered in the USA and 60 q/ha in Tunisia.

At the national level, the best average yields per palm are recorded in the oases of El Oued and Biskra with 61 kg/tree, but they do not exceed 29 kg/tree in the oases of southwest and 37 kg/tree in the south-center. However, the best producers realize up to 150 kg/tree in Tolga with cv. Deglet Noor.

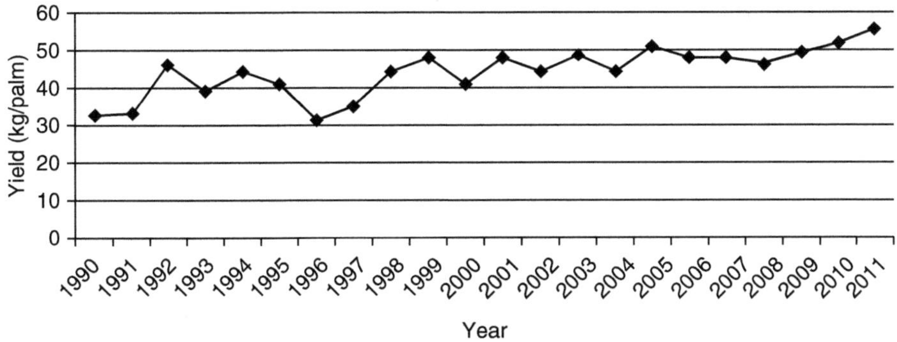

Fig. 4.7 Evolution of the average yield of date palms in Algeria during the period 1990–2011 (Source: Benziouche (2012))

The analysis of per cultivar yield shows that the highest average value is recorded by the Deglet Noor cv. with 60 kg/date palm. Other common cultivars including Ghars do not exceed 39–50 kg/date palm. Crop management, agroclimatic conditions, planting density, age of the palms, and other constraints are the major reasons for these differences (Benziouche and Chehat 2010). Nearly 78 % of increased national yield of date palms results from the actions of PNDA to improve the knowledge and technical aspects of cultivation.

4.7.3 Harvest Mechanization

As with the other transactions that fall within the technical process of date palm cultivation, date harvest throughout Algeria is strictly traditional, and no mechanization is reported except for a few attempts made in the palm groves of D'haouia in Oued Souf between 1998 and 2000. Unfortunately, this attempt to modernize the harvest failed due to mechanical constraints of two sophisticated machines and not taking into account the local topography. Thus, the methods of picking are homogeneous for all farmers, except for some differences which depend on various factors related to the cultivar, climate, and business requirements (Benziouche 2006). Despite the fact that mechanical harvesting is done in other countries, Algerian farmers are skeptical and suspicious of it, and hence, they do not view it as a credible alternative.

Benziouche and Chehat (2010) consider that farm unit fragmentation also impedes the possibility of mechanization because it becomes impractical on microplots making any development, intensification, and investment inefficient and difficult to undertake.

All these obstacles to mechanization leave it to manual operations to complete the harvest on time, and especially to avoid any risks related to climate conditions during harvest, disease outbreaks, and price fluctuations (Benziouche 2012). These traditional techniques have many adverse effects, especially the shortfall due to the high rate of fruit loss.

4.7.4 Postharvest Operations

In the Saharan regions of Algeria, date fruit sorting is homogeneous among all producers with the operation accomplished through traditional means. No modernization of the sorting is practiced. After harvesting and transporting the fruits to sheds in plastic crates, dates are sorted into different categories of maturity to create homogeneous lots. Sorters usually stand or squat around a large screen (1 or 2 m) fixed on iron legs 1 m in height. These screens are used to remove all the objects that are smaller than the fruit. Every worker has before him or her empty cases equal in number to categories being sorted and into which the fruits are placed (Benziouche 2012). Sorting follows commercial practices and standards imposed by the purchasers. It is almost always done under very rustic and primitive conditions. Therefore, it is common to find on the market poorly sorted nonhomogeneous dates.

However, date sorting is more or less well done depending on the cultivar, capacity, and type of producers. Indeed, for the cv. Deglet Noor, some farmers are able to distinguish 10 categories. In the region of Zibans, major producers distinguish 7 categories while small producers differentiate 4–5 categories. Previously, farmers of El Oued region separated 16 different categories (Emil 1934). Sorting Deglet Noor fruits, the dominant cultivar in the majority of date palm regions of Algeria (Zibans, Oued Souf, Oued Righ, M'zab, Ouargla), is in descending order of scale according to their degree of maturity: *martoba* (very soft Deglet Noor, good quality), *fraza* (dry Deglet Noor, dates not marketed), *bser* (dates not mature enough), and *hchef* (dry dates, without seeds due to abnormal fertilization) (Belguedj 2004).

4.7.5 Marketing Status and Research

Marketing Status

Date marketing in Algeria is not very successful and fails to meet the preferences and expectations with regard to the existing means and reforms of the institutions responsible for the promotion of this sector. The date market in Algeria is clearly segmented, and each segment is justifiable for a particular approach in terms of marketing mix. Indeed, analysis shows that the Algerian dates are known and do not need improvement, but merely effective treatment against pests and diseases and reliable classification to satisfy world consumers.

In Algeria, date fruit is perceived as a *generic* product although each cultivar has specific characteristics making it suitable for a particular purpose or market. This is not the case for most of the dates to be found on the various national markets.

For packaging, one of the most important elements of marketing, at the national level, there are two types of presentation: bulk or packaged dates. Nearly 90 % of the dates are mostly sold loose. In this market, there are many types of cartons of different shapes, qualities, and weight, from 250 g to 10 kg. Marketing to gain a niche with European consumers, where date consumption has a hedonistic dimension, some processing units use fancy packaging to differentiate the requirements of importers.

For advertising and promotion, and with a view to develop the export of dates, Algeria has established several approaches. However, they are not used effectively and are not well attuned to the international economic date commerce. For example, there is not an Algerian airline magazine in which to promote dates, although such magazines have proven their value in countries like Tunisia. Also beneficial are any advertising spots on ethnic radio programs for the Muslim communities in Europe. The lone technique that is experiencing a boom in recent years is Internet advertising which has become a more and more essential means of communication. Some exporters and packaging facilities have begun to use the Internet for advertising and e-commerce of Algerian dates.

The other method used to promote Algerian dates is through participation in national and international fairs and exhibitions. These events represent a real opportunity to make contacts with customers. Thanks to subsidies from the National Agency for the Promotion of Foreign Trade (ALGEX), exporters have participated in these fairs with many beneficial results. Increased exports are due to recent participation in trade fairs.

Research Status

As a result of Algerian reforms, new institutions have been established and others restructured. Research on Saharan agriculture is currently the responsibility of several research institutions of different disciplines, namely, LRZA, research stations of INRAA in Touggourt and Biskra, and regional stations of the National Institute of Plant Protection (INPV) in Biskra and Ghardaïa. Much of this research is supported by the Technical Institute of Saharan Agricultural Development (ITDAS), headquartered in Biskra, and two regional stations in El Arfiane and Ouargla. In 1994, a new research center was established, the Centre of Scientific and Technical Research in the Dry Areas (CRSTRA), with headquarters in Biskra.

However, despite these encouraging investments, there is currently a lack of research on practical issues such as technical requirements of date palm irrigation amounts specific to local climatic and soil conditions. Several objectives were assigned to different research institutions in Saharan agronomy in order to promote the effectiveness of the research and development of applied research. This research is mainly focused on improving the technical and economic performance of the date sector, identification of water-conserving irrigation systems, protection against pests and diseases, improvement of crop management, and mechanization of date palm farming operations. However, according to Ababsa (2001), it seems that the reforms and policies implemented so far have not led to major improvements. Indeed, little attention is paid to other equally important disciplines for this undertaking, including socioeconomics.

The present situation is characterized by the absence of a coherent research program and the inability and inefficiency of some researchers in solving practical field problems experienced by farmers. This situation is explained to a large extent by the technical, economic, political, and social constraints. In addition, research results are often valued by neither policy makers nor farmers, due to the inefficiency of the current extension service system and the disconnection which exists among the key trilogy of training, research, and extension.

4.7.6 Current Date Imports and Exports

Despite the increase in production, the growth of date exports from 1990 to 2011 (Fig. 4.8) shows that Algeria exported only 12,743 mt per year, on average, during that time period, with a maximum of 26,000 mt in 2011 and a minimum of 3,763 mt in 1994. This data series is characterized by fluctuations, but the quantity of exports increased during the most recent season, 2012/2013.

The average value of Algerian date exports was USD 27,660 million in the same period. This value varies between a minimum of USD 10,440 million in 2001 and a maximum of USD 79,120 million in 1995. Statistics confirm a general upward trend of value due in part to rising prices.

The integration rate (expressed as the ratio between the quantity of dates exported and the quantity produced in the same period) has a clear trend during the period 1990–2011, from 4.19 % in 1990 to 6 % in 2011. The average of this ratio throughout the period is estimated at 3.52 %.

Algeria exports mainly the Deglet Noor cv. and small amounts of other fresh dates such as cv. Tafezwin. Deglet Noor cultivar continues to dominate with 86 % of the average quantities of dates exported between 1996 and 2011 and accounting for

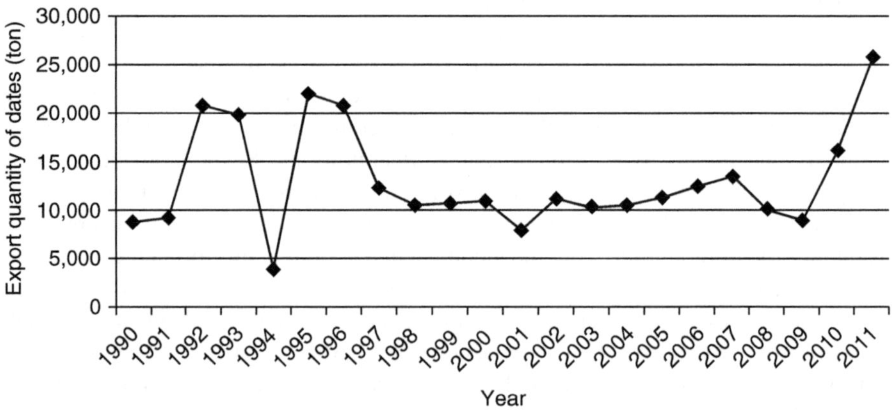

Fig. 4.8 Evolution of the quantities of dates exported during the period 1990–2011 (mt) (Source: Benziouche (2012))

nearly 94 % in terms of value. Regarding soft dates, two categories dominate: Ghars cv. fruit and date paste with a total of 12 % of exports in volume during this period.

The remainder consists of dry and similar cvs., e.g., Degla Beida and Mech Degla, which represent only 2 % of the export volume. Most of these exports are principally intended for the European Union accounting for 94.82 % of Algeria's exports of dates in value and 95.46 % in quantity exported in 2011. In the EU, France is the largest importer with 77 % of the quantity and 80 % of the value of date exports between 2000 and 2011, due to the lack of efforts to diversify exports to other markets. Algeria occupied seventh place with 3.12 % of the average global exports of dates during the period 1990–2010.

4.8 Processing and Novel Products

4.8.1 Industrial Processing Activities

In Algeria, there is little processing at the industrial or semi-industrial scale for secondary or by-products from the date palm. There is almost a total lack of production of finished products based on date fruits, such as jam, juice, and vinegar; existing technology allows for the production of Ghars date paste, a secondary activity of processing dates. There are ten processing units of this product line. New units have recently begun the manufacture of pulp as exclusive activity.

Fruits are pitted, crushed, grounded, and pressed into 1 kg lots that are packed in cellophane and placed in a cardboard box. The product is intended for the local market or for export to France and Canada, where the Muslim community is the main consumer, especially for religious occasions and for the period of Ramadan (Belguedj 2004). Some processors use a certain type of packaging to export pitted and stuffed Deglet Noor dates.

In addition, some farmers extract certain products by traditional means from fruit and by-products, such as honey or jam; these derivatives are generally intended for household consumption. However, there are intentions expressed by operators in new products, namely, the production of ethanol, flour, dates, and vinegar. Insufficient knowledge of appropriate technologies, lack of a potential market, high cost of this operation, lack of coordination between research and industry, satisfaction of producers who benefit from the trade of this untreated fruit, and financial question are the main impeding factors. Indeed, by-products are neglected and of low commercial value while they could be an additional source of income for both the date palm farmers and processors.

To improve its export qualitatively and quantitatively, Algeria was among the first date-producing countries in the world to be interested in packaging techniques and to develop an important infrastructure for packaging. Historically, Algerian date packaging was done in Marseille, France, and tradesmen in that city opened offices in Algeria in the early years of independence, i.e., since the 1960s. Currently, Algeria has decided to take advantage of technological advances in processing and packing with the launch of a major investment program.

There were seven facilities which were created during the period of agrarian reform located in the key date production areas of the country, namely Biskra, Mghaier, Jamaa, El Oued, Tolga, Touggourt, and Ouargla. The theoretical capacity of these seven processing facilities is 28,000 mt per year. Each facility had at its inception fumigation chambers, sorting lines, ovens for heat treatment, packaging lines, and refrigerated warehouses.

The management of these facilities was initially the responsibility of the Office of Fruits and Vegetables (OFLA) and the National Office of Dates (OND) established in 1971. Management is now in the hands of private companies in the form of Companies for Action (SPA).

Following economic reforms, many private factories were constructed for date packaging, with different capacities and technical levels, such as the Date Packaging Company of Biskra (SOCODAT). But, especially since 2000, the industrial sector has undergone profound changes under the National Agricultural Development Plan (PNDA). This program also has encouraged the maintenance and modernization of old processing facilities, increased the storage capacity, and created new private factories (Benziouche 2012).

Since the restructuring of date processing, the packaging sector now also operates an industrial factory serving about 35 private processing factories with different capacities. However, the majority of these facilities are poorly designed and do not meet industry standards. Although the capacity of processing and packaging of dates has more than doubled over the last decade, it is still insufficient and characterized by an uneven geographic distribution. There is a high concentration of facilities in southeastern Algeria such as Biskra, which has 29 private factories, while in the southwest region, there is no modern packaging factory (Benziouche 2012).

4.8.2 Survey of Commercial Date Processers

The demand for dates in Algeria varies from one region to another depending on the cultivar, how the fruits are sorted, and the type of packaging used. In this regard, Deglet Noor fruit left on the strand is very popular with residents of Algeria's central coast, i.e., Algiers, Boumerdes, and Tipaza, and for export. While the category *martoba* (very soft Deglet Noor) is more appreciated in the west of the country (Oran), the *fraza* (very dry Deglet Noor) is in demand by traders from the highlands of eastern region especially during winter.

Biskra traders are interested in any category of Deglet Noor requested by exporters and by many packaging factories in that province. In addition, Ghars cv. fruit is usually packaged in goatskin or textile bags (Benziouche 2007) or processed into a pulp in some packaging factories and is requested by all regions of the country because of its wide use in confectionery (Benziouche 2012). Degla Beida cv. is in high demand by traders in Touggourt, Tamanrasset, Ouargla, Adrar, and Illizi, and it is exported to sub-Saharan African countries as part of a barter economy (Benziouche 2012). For date by-products such as honey, vinegar, flour, and jam, small amounts are to be found in local markets in areas of date palm growing (Benziouche 2012).

4.8.3 Secondary Metabolites

Research has been done on the secondary metabolites in some date cultivars grown in Algeria, but the results are at the laboratory-research stage. One of the first studies by Gaceb and Rahmania (2013) investigated the bioactive natural compounds from date palm leaves, for the purpose of enhancement of the species and the subsequent industrial use of the molecules identified. They have identified new natural substances in the date palm including two steroidal saponins, diosgenin and yamogenin acetate, and saponins with a Spirostane core. In addition, the research showed, for the first time, the presence of a tetracyclic triterpene saponin. Other research focused on date fruit fiber and its phenolic compounds (Benchabane et al. 2013). The role of the fraction *fibers* focused on their effects in human disease and diabetes control. Some results show that the content of phenolic compounds differs from one cultivar to another.

4.8.4 Bioenergy

The goal of bioenergy technology is to recycle waste, utilizing it as fuel to produce energy and provide a partial alternative to nonrenewable fossil fuels. This is a relatively new area of research at the global level; in Algeria, it is under development on an experimental basis (Aziza et al. 2008). Several pilot projects are planned in Algeria, such as the Center for the Development of Renewable Energies (CDER) project studying methane recovery from cull dates and fruit processing of by-products. The proposed project involves the development of a method of treatment and extraction of bioethanol or fuel after fermenting date juice. This process creates a considerable amount of waste of a different nature, lignocellulosic and oilseed saccharase (Houari 2012). Trials are being conducted within the framework of National Research Projects (PNR) for the production of organic, renewable, and clean energy while promoting some cultivars of low market value, so as to avoid genetic erosion of Algeria's heritage date palm germplasm pool (Houari 2012). Thus, a project is being implemented in Biskra as the first facility to manufacture Nakhoil, a nonpolluting date-based ethanol. According to Bousdira (2010), the biomass from date palm cultivation reveals a very interesting energy potential on the national scale; these numbers would reach, respectively, for wood and dates, 658,108 and 1,176,285 MWh.

4.9 Conclusions and Recommendations

In Algeria, date palm cultivation has existed for many centuries (the *foggaras* irrigation system since before the sixth century). Date palm growing has advanced since the colonial period with a significant increase in the number of date palms and gardens. In recent years, the date palm sector has been marked by strong momentum and a significant rise in production. This increase is due to various agricultural development programs launched by the government, as well as monitoring of date palm grove health and preventive treatment for several diseases that affect the palms. However, participation of the date sector in terms of Algerian exports remains meager and benign. Concerning the Deglet Noor cultivar, which is very popular in global markets, efforts should be made to establish a better international market position.

It is clear that cultivation techniques have not improved because the benefits of adopting industry's best practices have not been publicized and agricultural products are hard to obtain and costly. New farms are not properly applying appropriate technical practices in irrigation, pruning, pollination, thinning, and insect/disease control. The main socioeconomic constraint hindering the development of the sector is the social phenomenon of date palm grove abandonment because of aging farmers and lack of interest by their heirs to continue date production. This trend is particularly serious in the case of small farms where there are several sons, indivisible land ownership, and disagreement among heirs. Absentee landlords of low productivity date groves are an added negative factor. These circumstances make it difficult to increase production and investment on small landholdings. Another impediment is the advanced age of a large proportion of the producing date palms; 13 % of them are above 80 years and another 19.6 % are close to that age (Benziouche and Cheriet 2012).

Algerian date palm cultivation is constrained by the substandard level of crop management, lack of know-how of some farmers, low mechanization, water shortages, mismanagement in some groves, as well as a high number of users depending upon a single borehole water source. In addition, lack of effort to rehabilitate traditional irrigation systems, occasional failure to adopt modern irrigation techniques, and falling water tables due to overuse have forced some date producers to abandon their investments.

The consequences of persistent drainage problems in some oases of the country lead to declining economic returns and to a significant shortfall. If these obstacles can be overcome, it will be possible to improve the yield per date palm, currently well below the potential for several Algerian cultivars.

Research stimulated by the outbreak of the deadly bayoud disease in the late nineteenth century, and its spread in the twentieth and present centuries, has focused on the knowledge of the date palm and on the organisms responsible for bayoud and other diseases. This research has opened the way to the modern techniques of propagation through tissue culture and its different aspects with respect to the fast and high-volume propagation of interesting cultivars and also to creating new ones. The objective of producing quality date palm plantlets has not yet achieved large-scale application in Algeria, although there are foreign companies which produce them. However, the availability of offshoots in Algeria is not a problem, and thus, tissue culture remains mainly a research tool in university laboratories. It must be recognized that these studies have yet not found wide application because of the organizational structure of biotechnology research in Algeria.

Molecular characterization of the main Algerian cultivars, of the approximately 1,000 known, is very good, thanks to the progress on date palm genetics by researchers in Qatar (Elmeer et al. 2011).

Since the appearance of the red palm weevil in neighboring countries, the import of palms of any species is prohibited in Algeria. This action was taken to prevent any penetration of this insect into the date gardens.

Final hindrances to developing the date palm section are the inadequate extension system, a lack of effective programs to improve and promote date growing, and an absence of coordination on relevant subjects relative to training, research, and agricultural recommendations.

In terms of recommendations, they can be addressed by a program that addresses the production of all date cultivars, rehabilitates old and creates new plantations, expands exports of Deglet Noor and promotes other cultivars with high market value such as Degla Beida and Tafezwin, establishes processing facilities near Ksour and other oases for local processing and packaging, modernizes oasis farming techniques, and preserves and enhances date palm and other biodiversity.

Based on the subjects discussed, the following recommendations can be made:

(a) Adopt a comprehensive approach to integrated development, including rehabilitation of oases including those located some distance apart.
(b) Coordinate the actions of various actors in the date sector through a network that will include date palm farmers, researchers, investors, and traders.
(c) Induce farmers to adopt good farming techniques for better yield and performance and find ways for farmers to implement the results of applicable research.
(d) Develop research in mechanization and its integration into the sector.

To promote and develop the date industry, special efforts should be made to:

(a) Encourage investors to establish factories by eliminating obstacles to investment in this area and solving marketing problems and productivity.
(b) Support research and innovation having an important role must encourage applied research on the possibilities of transforming different parts of the date palm with a view to establishing a transformation industry of by-products in potential regions of date palm.
(c) Implement research results concerning the use of various tissue parts that make the trunk and leaves which can be used in the manufacture of paper and compressed wood from fibers and petioles.
(d) Further expand efforts to promote the export of quality dates packaged under Algerian brands. The economic organization of the sector is a prerequisite for improving the market products.
(e) Continue efforts in the reform of institutions to promote exports to strengthen partnership agreements.
(f) Intensify advertising campaigns.
(g) Create a database for the date sector.

Acknowledgments Our gratitude goes to Mrs. Samira Bouguedoura and Mr. Bouandel Aymen for assisting in the English translation of the original French text and for their moral support. The Association of Environmental Protection Beni Izguen (APEB) has made available several date cultivars of the Ghardaïa region. The pictures that we took were used to prepare a portion of Fig. 4.5. We express our gratitude for the assistance given.

References

Ababsa SF (2001) Le palmier dattier dans la recherche scientifique; consolidation et élargissement des approches. Rev Prosp Agr INRAA 1:93–103

Acourene S, Allam A, Taleb B, Tama M (2007) Inventaire des différents cultivars de palmier dattier (*Phoenix dactylifera* L.) des régions de Oued-Righ et de Oued-Souf (Algérie). Séch 18:135–142

Amir H, Amir A, Riba A (1996) Rôle de la microflore dans la résistance à la fusariose vasculaire induite par la salinité dans un sol de palmeraie. Soil Biol Biochem 28:113–122

Arban A (2009) Contribution à l'amélioration de l'embryogénèse somatique du palmier dattier (*Phoenix dacylifera* L.) cultivar Deglet Nur par l'utilisation d'explants floraux. Mémoire de Magister. USTHB, Alger

Assani A, Haicour R, Wenzel G et al (2002) Influence of donor material and genotype on protoplast regeneration in banana and plantain cultivars (*Musa* spp.). Plant Sci 162:355–362

Assani A, Chabane D, Haïcour R et al (2006) An improved protocol for microcallus production and whole plant regeneration from recalcitrant banana protoplasts (*Musa* spp.). Plant Cell Tiss Org Cult 85:257–264

Assani A, Chabane D, Shittu H, Bouguedoura N (2011) Date palm cell and protoplast culture. In: Jain SM, Al-Khayri JM, Johnson DV (eds) Date palm biotechnology. Springer, Dordrecht, pp 605–629

Aziza M, Bessah R, Kaidi F (2008) Bio alcool en Algérie:Etat actuel de la filière bioéthanol et perspectives de l'avenir. Premier Atelier National sur les biocarburants, Maroc

Babahani S (2011) Analyses biologique et agronomique de palmiers mâles "Dokkars" et conduite de l'éclaircissage des fruits chez les cultivars Deglet Nour et Ghars. Thèse de Doctorat. Ecole Nationale Supérieure d'Agronomie El Harrach, Alger

Babahani S, Allam A, Djabourbi N (1997) Utilisation de la farine du blé comme support pour le pollen du palmier dattier (*Phoenix dactylifera* L.). Rev INRAA 19:44–47

Belguedj M (2002) Les ressources génétiques du palmier dattier: caractéristiques des cultivars de dattier dans les palmeraies du Sud-est Algérien. Rev Ann INRAA N°01, p 130

Belguedj M (2004) Analyse diagnostic du secteur du palmier dattier en Algérie. Etude des marchés des produits du palmier dattier au Maghreb. PNUD, rapport d'étude

Belguedj M, Tirichine A (2011) Ressources génétiques du palmier dattier. Caractéristiques des cultivars de Ghardaia. Rev Ann INRAA N°02, p 139

Ben Saadoun N, Boulahouat N (2010) Le palmier dattier, raconté par un cultivateur. Ed BEDE

Benchabane A, Boutali A, Kechida F (2013) Extraction, fractionation and characterization of the fibers of some dates cultivars of Algeria. Acta Hortic 994:223–229

Benkhalifa A (2007) Les ressources génétiques du palmier dattier. Agriculture et développement rural durable, édité par l'Institut National de la Vulgarisation Agricole 4:25–27

Bennaceur M, Lanaud C, Chevalier MH, Bounaga N (1991) Genetic diversity of the date palm (*Phoenix dactylifera* L.) from Algeria revealed by enzyme markers. Plant Breed 107:56–57

Bennaceur M, Lebrun P, Billote N et al (2000) Polymorphisme du palmier dattier (*Phoenix dactylifera* L.) à l'aide de marqueurs microsatellites du cocotier (*Cocos nucifera* L.) ou du palmier à huile (*Elaeis guineensis* Jacq.). VIIes journées scientifiques: Des modèles biologiques à l'amélioration des plantes. Ed Hamon S, pp 599–600

Bennaceur M, Chevallier MH, Bouguedoura N (2002) Apport des marqueurs chloroplastiques à l'étude de la diversité génétique du palmier dattier. VIIIème journées scientifiques: Biotechnologies végétales: valorisations pour une agriculture durable. Ed El Hadrami I

Bennaceur M, Bengag A, Marouf A, Bouguedoura N (2010) Phytochemical profile and antioxidant activity of *Phoenix dactylifera* L., *Phoenix canariensis* L. and *Chamaerops humilis* L. Acta Hortic 882:1099–1108

Benziouche SE (2006) L'agriculture dans la vallée d'Oued Righ; Quelques éléments d'analyse. Université de Biskra, Rev Sci Hum 10:100–112

Benziouche SE (2007) Structure and operation of the phoenicicole small-scale farmings in the south area of Algeria. In: Langeveld H (ed) Changing European farming systems for a better future: new visions for rural areas. University Wageningen, Wageningen

Benziouche SE (2010) L'analyse de l'impact des politiques agricoles sur la production des dattes en Algérie par la MAP. Int Res J Finan Econ UK 44:68–77

Benziouche SE (2012) Analyse de la filière dattes en Algérie, constats et perspectives de développement. Etude du cas de la Daïra de Tolga. Thèse Doc en agronomie, ENSA El-Harrach

Benziouche SE, Chehat F (2010) La conduite du palmier dattier dans les palmeraies des Zibans (Algérie) Quelques éléments d'analyse. Eur J Sci Res 42:644–660

Benziouche SE, Cheriet F (2012) Structures et contraintes de la filière dattes en Algérie. Rev New Medit 4:49–57

Billotte N, Marseillac N, Brottier P et al (2004) Nuclear microsatellite markers for the date palm (*Phoenix dactylifera* L.): characterization and utility across the genus *Phoenix* and in other palm genera. Mol Ecol Notes 4:256–258

Bliss DE (1936) Rhizosis, a recently discovered disease of date palms. Date Grow Ann Rep 13(4):6

Bouammar B (2009) Le développement agricole dans les régions sahariennes Etude de cas de la région d'Ouargla et de la région de Biskra. Thèse de doctorat en sciences économiques, Université Ouargla

Bouchireb N, Clark MS (1997) The application of biotechnology to date palm culture. In: Watanabe KN, Pehu E (eds) Plant biotechnology and plant: genetic resources for sustainability and productivity. Academic, San Diego, pp 183–195

Boughediri L, Cerceau-Larrival MT, Dore JC (1995) Significance of freeze-drying in long term storage of date palm pollen. Grana 34:408–412

Bouguedoura N (1979) Contribution à la connaissance du palmier dattier *Phoenix dactylifera* L: étude des productions axillaires. Thèse de Doctorat Troisième Cycle, USTHB, Alger

Bouguedoura N (1980) Morphologie et ontogenèse des productions axillaires du palmier dattier. Comp Rend Acad Sci Sér D 291 N°10:857–860

Bouguedoura N (1982) Development and distribution of axillary buds in *Phoenix dactylifera* L. Proceeding of the first symposium of the date palm, pp 40–45

Bouguedoura N (1989) Résultats préliminaires de l'Androgenèse et de la gynogenèse chez le palmier dattier (*Phoenix dactylifera* L.). Proceeding du colloque Maghrébin sur la culture des tissus du palmier dattier à Marrakech PNUD/FAO/RAB/88/024, pp 105–111

Bouguedoura N (1991) Connaissance de la morphogenèse du palmier dattier (*Phoenix dactylifera* L.). Etude in *situ* et *in vitro* du développement morphogénétique des appareils végétatif et Reproducteur. Thèse Doctorat d'Etat, USTHB, Alger

Bouguedoura N (2012) Le palmier dattier: développement morphogénétique des appareils végétatif et reproducteur. Book Editions Universitaires et Internationales

Bouguedoura N, Michaux - Ferriere N, Bompar JL (1990) Comportement in vitro de bourgeon axillaire de type indéterminé du palmier dattier (*Phoenix dactylifera* L.). Can J Bot 68:2004–2009

Bouguedoura N, Bennaceur M, Benkhalifa A (2010) Le palmier dattier en Algérie: Situation, contraintes et apports de la recherche. In: Biotechnologie du palmier dattier. Ed IRD, Montpellier, France, pp 15–22

Boulenouar N, Cheriti A, Marouf A (2009) Effect of some poisonous plants extracts on *Fusarium oxysporum* f. sp. *Albedinis*. J Biol Sci 9:594–600

Bounaga N (1985) Contribution à l'étude de *Fusarium oxysporum* f. sp. *albedinis* (Killian et maire) Gordon, agent causal de la fusariose du palmier dattier. Thèse Doctorat d'état, USTHB, Alger
Bousdira K (2010) Potentiel énergétique de la biomasse phoenicicole en Algérie: évaluation et recommandations. CDER Ghardaïa, Algérie
Brac de la Perrière RA, Benkhalifa A (1989)1 Identification de cultivars de dattiers (*Phœnix dactylifera* L.) du sud-ouest algérien. Plant Gen Res Newsl FAO/IBPGR.78/79:13–20
Brac de la Perrière RA, Benkhalifa A (1991) Progression de la fusariose du palmier dattier en Algérie. Séch 2:119–128
Brac de la Perrière RA, Bounaga N (1990) Etude du verger phoenicicole d'une palmeraie traditionnelle (Beni-Abbès, sud-ouest algérien) Inventaire exhaustif et composition variétale. Revue du réseau pour l'amélioration de la productivité agricole en milieu aride 2:9–18
Brac de la Perrière RA, Dubost D (1987) La culture du palmier dattier en Algérie: un programme de recherche au service du développement. P H M Rev Hortic 273:49–52
Brun J, Laville E (1965) Observations sur un dépérissement de la couronne foliaire et du bourgeon terminal du palmier dattier. R I M Fruits 20:391–397
Bulit J, Louvet J, Bouhot D, Toutain G (1967) Recherches sur les fusarioses. I. Travaux sur le bayoud, fusariose vasculaire du palmier dattier en Afrique du nord. Ann Epiphyt 18:231–239
Carpenter JB, Klotz LJ (1966) Diseases of the date palm. Date Grow Inst Rep 43:15–21
Chabane D (2007) Amélioration du palmier dattier *Phoenix dactylifera* L. par fusion de protoplastes de deux cultivars Deglet Nour sensible et Takerbucht résistant au bayoud. Thèse Doctorat d'état, USTHB
Chabane D, Assani A, Bouguedoura N et al (2007) Induction of callus formation from difficile date palm protoplasts by means of nurse culture. C R Biol 330:392–401
Chabane D, Bouguedoura N, Assani A (2010) Importance of protoplast culture in the genetic improvement of date palm (*Phoenix dactylifera* L). Acta Hortic 882:185–192
Daoud Y, Halitim A (1994) Irrigation et salinisation au Sahara algérien. Sécher 5:151–160
Elmeer K, Sarwath H, Malek J et al (2011) New microsatellite markers for assessment of genetic diversity in date palm (*Phoenix dactylifera* L.). Biotech 1:91–97
Emil P (1934) Le dattier dans les oasis des Zibans et d'Oued Righ. Rapport de mission de l'OFALAC
Engelmann F (2010) Conservation des ressources génétiques du palmier dattier. In: Biotechnologie du palmier dattier. Editions IRD, Montpellier, France, pp 161–174
Fergani K (1998) Embryogenèse somatique du palmier dattier (*Phoenix dactylifera* L.) cultivar Deglet Nur. Thèse de Magister, USTHB, Alger
Fertial SPA (2010) Manuel: utilisation des engrais. Société de Fertilisants d'Algérie, Annaba
Gaceb R, Rahmania F (2013) Saponines stéroïdes: new molécules in the date palm (*Phoenix dactylifera* L.). Acta Hortic 994:215–221
Hannachi S, Benkhalifa A, Khitri D, Brac de La Perrière RA (1998) Inventaire variétal de la palmeraie Algérienne. Sélection et impression, Anep Rouiba Algérie
Houari B (2012) La valorisation du méthane à partir des décharges en débat. J Maghreb
Hussein F, El Khahtani S, Wali Y (1979) La production dattière dans les mondes arabe et islamique. Imprimerie Ain Schamss. (in Arabic)
IPGRI, PNUD, FEM/GEF, INRAA (2008) Gestion participative des ressources génétiques du palmier dattier dans les oasis du Maghreb Evaluation finale du projet, rapport national de l'Algérie. Ed. INRAA
IPGRI, INRAA, INRAM et al (2005) Descripteurs du palmier dattier (*Phœnix dactylifera* L.). Institut international des ressources phytogénétiques, Rome, Italie; Fonds pour l'Environnement Mondial, Washington, Etats-Unis, Programme des Nations Unies pour le Développement, New York, Etats Unis; Institut National de la Recherche Agronomique d'Algérie, du Maroc et de Tunisie
Jain SK (1997) Les ressources génétiques des plantes. La Rech 84:1067–1077
Jain SM (2012) *In vitro* mutagenesis for improving date palm (*Phoenix dactylifera* L.). Emir J Food Agr 24(5):400–407

Khiat M, Karrouche A, Rahmania F, Gaceb-Terrak R (2011) Flavonoïdes des graines de neuf cultivars algériens de dattier (*Phoenix dactylifera* L.). 1st international date palm symposium in Algiers, 13–14 Nov. Abstract, p 117

Louvet J, Toutain G (1973) Recherches sur les fusarioses. VII. Nouvelles observations sur la fusariose du palmier dattier et précisions concernant la lutte. Ann Phytopath 55:35–52

Malençon G (1934) La question du bayoud. Ann Crypt Exot 7:1–41

Mansouri A, Guendez E, Kokkalou E, Panagiotis K (2005) Phenolic profile and antioxidant activity of the Algerian ripe date palm fruit (*Phoenix dactylifera* L.). Food Chem 89:411–420

Matheron B, Benbadis A (1985) Etude comparée des premières phases de l'infection du Palmier dattier (*Phoenix dactylifera* L.) par deux formes spéciales de *Fusarium oxysporum* la f.sp. *albedinis* (agent du Bayoud) et la f. sp. *melonis*. Bull Soc Bot Fr Lett Bot 132:203–212

Mimouni Y, Siboukeur O (2011) Etude des propriétés nutritives et diététiques de sirops de dattes extraites par diffusion en comparaison avec des sirops à haute teneur en fructose (isoglucose) issus de l'industrie de l'amidon. Ann Sci Tech 3:1–11

Monciero A (1950) La fécondation mécanique du palmier dattier. Bull d'information. Office tunisien de standardisation, Tunis

Morel G, Wetmore RH (1951) Fern callus tissue culture. Am J Bot 38:141–143

Moussouni S, Pintaud JC, Bouguedoura N (2011) Etude du polymorphisme chloroplastique du palmier dattier (*Phoenix dactylifera* L.) en Algérie. 1st international date palm symposium in Algiers, Algeria,13–14 Nov. Abstract, p 61

Munier P (ed) (1973) Le palmier dattier. Maisonneuve et Larose, Paris

Murashige T, Skoog F (1962) A revised medium for rapid growth and bio assays with tobacco tissue cultures. Physiol Plant 15:473–497

Nadji S (2003) Atelier sur la protection intégrée du palmier dattier dans les pays de l'Afrique du Nord, Tunisie. pp 24–32

Ouafi S (2007) Contribution à l'étude biochimique du Palmier dattier, analyse des polyphénols des folioles de palme de cultivars du *Phoenix dactylifera* L. dans un but chimiotaxonomique. Thèse Doctorat d'Etat Univ. Sci Tech Houari Boumediene, Alger

Ouafi S, Bounaga N (2008) Les glycosides flavoniques marqueurs de cultivars algériens de palmier dattier (*Phoenix dactylifera* L.). Acta Bot Gallica 155:307–315

Ouafi S, Bounaga N (2010) Phenolic compounds as markers for Algerian cultivars of the date palm *Phoenix dactylifera* L. Acta Hortic 882:165–172

Ouafi S, Bounaga N (2012) Effect of Fusariosis on the metabolism of phenolic compound on date palm leaflets resistant or sensitive to the disease. J Food Sci Eng N°02, 213–217

Ould El Hadj MD, Cheick M, Hamdi W et al (2012) Etude comparative de la production d'éthanol brut à partir de trois variétés de dattes communes (Degla Beida, Tachrewit et Hamraya) reparties dans les différentes classes de dattes (molle, demi molle et sèche) de la cuvette de Ouargla (Sahara septentrional est algérien). Algérien. J Arid Envir 2:78–87

Pereau-Leroy P (1958) Le palmier dattier au Maroc. IFAC, Rabat

Rahmania F (2000) Contribution à la connaissance des relations histo-cytophysiologiques entre le palmier dattier, *Phoenix dactylifera* L. et l'agent du bayoud, *Fusarium oxysporum* f. sp. *albedinis* (Killian et Maire) Gordon. Thèse Doctorat d'état USTHB, Alger

Saaidi M (1979) Contribution à la lutte contre le Bayoud, fusariose vasculaire du palmier dattier. Thèse d'université, Dijon

Sabaou N, Bounaga N (1987) Actinomycètes parasites de champignons: étude des espèces, spécificité de l'action parasitaire au *Fusarium oxysporum* f. sp. *abedinis*. Can J Microbiol 33:445–451

Sayah Z, Ould Hadj MD (2010) Etude comparative des caractéristiques physico-chimiques et biochimiques des dattes de la cuvette de Ouargla. Ann Sci Tech 2:87–92

Si-Dehbi F (2009) Contribution à la production de suspensions cellulaires à partir d'explants végétatifs et floraux de palmier dattier (*Phoenix dactylifera* L.), Mémoire de Magister. USTHB, Alger

Si-Dehbi F, Chabane D, Bouguedoura N (2013a) Improvement of date palm (*Phoenix dactylifera* L.) by using cell suspensions. Acta Hortic 994:303–312

Si-Dehbi F, Zehdi S, Cherif E et al (2013b) CDDP as new molecular marker in detection of genetic variation among Algerian date palm (*Phoenix dactylifera* L.) varieties. Premier meeting international sur le palmier dattier: Ressources phytogénétiques du palmier dattier: Etat, caractérisation et défis de gestion. Djerba, Abstract, p 84
UNCED (1992) Convention on biological diversity. United Nations Conference on Environment and Development, Genève
Widholm JM (1972) The use of fluorescein diacetate and phenosafranine for determining viability of cultured plant cells. Stain Technol 47:189–194
Zehdi S, Sakka H, Rhouma A et al (2004) Analysis of Tunisian date palm germplasm using simple sequence repeat primers. Afr J Biotech 3:215–219
Zella L, Smadhi D (2006) Histoire de la gouvernance de l'eau. Larhyss J 5:7–18

Chapter 5
Date Palm Status and Perspective in Sudan

Mohamed M.A. Khairi

Abstract Date palm (*Phoenix dactylifera* L.) in Sudan is an economic and food security crop. Estimated annual date production is about 431,000 mt (FAOSTAT 2010) which is far below the country's potential. Sudan has been famous in the world for the production of dry dates. Its climatic transition from the extremely dry to humid conditions facilitates a parallel change in date cultivar distribution from dry dates in the north to semidry and soft dates southward. Palm and bunch management techniques, safer means of ascending date palms, and harvest, packing, and handling techniques are being improved. Six good local commercial cultivars are available, and research is coming up with better composition of cultivars by local selection and foreign introductions from tissue laboratories. Seven tissue culture laboratories are being developed for plant propagation, but these laboratories are not yet engaged in date palm propagation. No named local male date palms are known thus far, and males used for commercial pollination are random. Improvement of males by local selection and foreign introduction is in process. Sudan is still free from the devastating red palm weevil, but termites, scale insects, spider mites, storage pests, birds, and rodents are among the serious pests. Sudan is also free from the destructive *bayoud* disease caused by *Fusarium oxysporum* f. sp. *albedinis*. Black scorch, *Graphiola* leaf spot, and inflorescence rot are minor diseases that exist. Sanitation is the current means for controlling these diseases. A line of government-owned Kareema factory is designed for packing dates, and new modern factories are to be erected. Date processing by-products are limited to vinegar and medicinal alcohol. Utilization of date palm parts has been known for a long time and is being developed. Date fruits are marketed locally.

Keywords Cultivars • Cultivation • Diseases • Genetic resources • Harvest • Historical • Manufacturing • Marketing • Pests • Processing • Propagation

---

M.M.A. Khairi
Faculty of Agriculture, Dongola University, Dongola, Sudan
e-mail: khairy5@hotmail.com

J.M. Al-Khayri et al. (eds.), *Date Palm Genetic Resources and Utilization: Volume 1: Africa and the Americas*, DOI 10.1007/978-94-017-9694-1_5

5.1 Introduction

5.1.1 Historical and Current Agricultural Aspects

The Arabian Gulf is considered to be the probable origin of date palms (*Phoenix dactylifera* L.). However, there is strong evidence showing that dates were known in Sudan since an early period, based on the petroglyphs found on the tomb walls and roofs of palm logs 3,000 BC, linked to the ancient civilizations in Upper Egypt. Date palm parts were components of water wheels for lifting water from the Nile, roofing of houses, fences, ropes, and the popular edible fruit presented in various ways as food and drink. Because of suitable environment and growers who have known date palms for so long, an extensive stretch of green date palm plantations decorate the two sides of the River Nile banks north of latitude 18 °N. Date palms have also been grown in oases of Qaab west of Dongola, Salima west of Wadi Halfa, the Bara area of Kordofan, and the Kutum area of Darfur for a very long time. Recognized cultivars appeared in the northern fringe and spread southward replacing seedling date palms which then dominated the southern parts of date palm growing areas. Some of the superior cultivars are said to have originated in Algeria and imported to northern Sudan via Egypt. Indeed, Barakawi cultivar which dominates northern Sudan is grown in Upper Egypt under the names Sikkoati and Ibreemi. Some other cultivars like Medeena and Mishrig are believed to have been introduced from Saudi Arabia centuries back (Tothill 1948).

Date palm in Sudan is an economic and food security crop that is grown in arid regions of the country. Its role in food production, animal feed, income and employment generation, foreign exchange earnings, and raw material for industries makes it an important asset for national development. Statistics are lacking, but estimated annual production of dates in Sudan is about 431,000 mt (FAOSTAT 2010) which makes it about seventh in the list of top date producing countries of the world. But Sudan deserves a much higher rank among date producing countries if it could exploit its huge potential for date production, including the large stretch of area suitable for production of this crop which extends from latitude 21 °N at the Egyptian border south to about 12 °N across the entire width of the country, availability of irrigation water from the River Nile and underground, and a suitable climate for production of a wide range of dry to semidry and soft cultivars. The distribution of cultivars in Sudan is a smooth transition from dry dates in the north, at Wadi Halfa, to a mix of cultivars in the middle, at Abu Hamad, ending with semidry and soft dates at the southern end. This transition is an outcome of climatic conditions which are extremely dry at the northern border and a gradual increase in relative humidity southward which limits date production. A map of Sudan showing the various districts is shown in Fig. 5.1. A national program for promoting the date palm sector in Sudan is being implemented targeting the following objectives:

(a) Increasing date production with diversion toward semidry and soft dates without neglecting the wealth of dry dates
(b) Upgrading the quality and postharvest techniques of Sudan dates to meet international standards

Fig. 5.1 A map of Sudan showing various districts, UN Map No 4458 Rev2, March 2012

(c) Building the human and institutional capacities, with focus on cell culture and packaging facilities
(d) Improving the welfare of date palm growers
(e) Increasing the contribution of dates to the gross domestic product and foreign exchange earnings

Fig 5.2 Old multiple-stemmed date palm

5.2 Cultivation Practices

5.2.1 Cultural Operations

Date palm growers in Sudan have been accustomed to enjoy a decent date harvest by merely pollinating and harvesting date palms grown along the fertile River Nile banks. Groundwater level was within easy reach of date palm roots enabling the date palms to flourish without irrigation. Date palm culture in Sudan, therefore, remained traditional. Date palms of seed origin are quite common. Normal practice allows offshoots to develop on the parent, the result being after several years, a clump of trees competing with one another for light, water, and nutrients, often providing a thicket that harbors rats and other pests. Irregular layout of plantations that impede mechanization and service operations is common. Old dry leaves are seldom removed, hanging down impeding the operations of ascending the palms (Fig. 5.2). Cutting leaf bases and removing all the fibers, leaving a naked trunk that facilitates climbing trees by ropes, removal of thorns to facilitate operations of pollination, bunch management, harvest, and tree service were seldom practiced. To cope with changes of a sharp drop in the groundwater table, expansion in date palm growing areas beyond the fertile Nile banks, higher consumer demand, and threats of pests and diseases, date palm growers started to adjust their tree services.

Fig. 5.3 Date palm propagation by rooting in plastic containers

The value of single stem culture is being advocated. In recent plantations, date palms are laid out in regular lines that facilitate service and harvest operations. Spacing varies according to soil type, but usually 8 × 8 m. Bunch management techniques of fruit thinning, bending, and covering are being advocated. Irrigation practices are mostly basin, but drip and bubbler water saving systems are being introduced. Drip irrigation systems are being designed to emit 160 l per day for adult trees. Animal manure at the rate of 60 kg per palm annually has been recommended (Bashab et al. 2006) and is being applied. Some chemical fertilizers, mainly NPK, are being applied by some growers depending on soil conditions.

Local propagation of date palm is mostly by offshoots. While semidry cultivars can be propagated easily by offshoot separation from the mother tree and planting, dry date palm cultivars are too difficult to propagate that way. Dry cultivars are therefore propagated by rooting offshoots in plastic containers, then separation from the mother trees for planting out in the field. These containers are placed at the bases of cleaned offshoots, and a planting mix is added to cover the base of the offshoot. The mix is kept moist until the offshoot forms new roots inside the container (Fig. 5.3). This process usually takes about 9–12 months. The offshoot is then separated from the parent tree for propagation.

5.2.2 *Pollination*

The practice of pollination is being well performed in Sudan and has been since ancient times. But, while variation in the potency of pollen has been recognized a long time ago, implementation of these important observations was neglected. Growers uproot most males without realizing the need to keep and maintain the best. Vegetative propagation of males from good parents is not known in Sudan. So, named male date palms propagated vegetatively do not exist in Sudan. Fairly

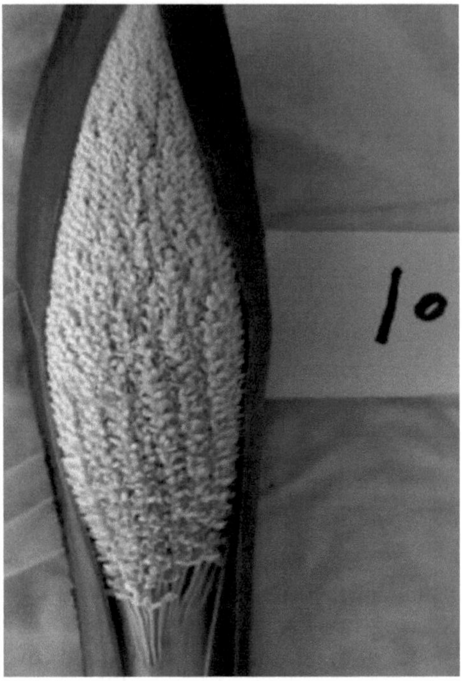

Fig. 5.4 Male no. 10 compatible with Gondaila cv

recently, Halfa 3 and Halfa 6 males have been recommended for Mishrig cultivars (Dawood 1997). Recent research work on some numbered male evaluations recommended male 3 for Barakawi cultivar and male 10 for Gondaila (Fig. 5.4), while Mishrig cultivar was found to be compatible with all males or even found to set fruit without pollination (Bashab and Khairi 2012). Ghannamy and Fahl Alain males were introduced recently from the United Arab Emirates. Ghannamy introductions have not flowered yet, and the Fahl Alain male did not perform like a regular male, sometimes setting false fruits.

Pollination is done manually by tying 3 or 4 strands of a male spathe and inserting them among the strands of a female spadix as soon as it bursts open. Several female spathes open at a time necessitating four or five visits from a pollinator whose payment used to be one bunch per tree for pollinating and subsequently cutting down the bunches when ripe. A pollinator today is paid two bunches per palm. Pollen storage and pollination by dusting is practiced in a very limited scale by pioneer date palm growers under the guidance of the Date Palm Society. This is a voluntary society that is leading the modernization process of date palm operations in the country. Its members are some well-experienced scientists in date palm culture, date palm growers, and individuals who are interested in date palm culture and literature. It also organizes date palm festivals during *rutab* fruit stage and at harvest times, in addition to participation in date shows and festivals in the region. The society pollinates date palms that are grown in public streets, provides pollen to date palm growers in the form of strands or powder, and organizes training courses for different levels of trainees in all aspects of date palm culture.

Fig. 5.5 Climbing date palm in Sudan

5.2.3 Harvest Operation

Dates in Sudan are harvested by workers climbing to the tops of date palms by stepping on the stubs of leaf bases. This is not a safe way of getting to the tops of date palms and incidences of accidental falling do occur. Safer means of getting to the tops of date palms such as using ropes are being advocated (Fig. 5.5). Ladders are seldom used. The use of machines may be coming soon with modernization of operation techniques, but the high cost of machinery and irregular layout of date palm gardens encumber the use of machinery.

5.2.4 Pest and Disease Control

Insects

Red Palm Weevil Sudan is still free from the devastating red palm weevil (*Rhynchophorus ferrugineus* Oliv.). Strict quarantine measures are being taken to keep the country free from this destructive pest. Tissue culture–propagated date palm seedling introductions are restricted to torpedo stage seedlings with no more than 4–6 first leaves to keep the weevil out of the country.

Termites *Microcerotermes diversus and Odontotermis classic* Sjosted are very serious pests that attack date palms from the roots to close to the tops of the date

Fig 5.6 Serious date pests of Sudan. (**a**) Termites, (**b**) green scale, (**c**) dust mite

palms. Furadan is applied at the bases of offshoots during planting and in basins around adult date palms for control. But infestation could be very severe (Fig. 5.6a).

Scale Insects Three types of scale insects attack date palms in Sudan. White scale (*Parlatoria blanchardi*) is an endemic damaging insect. Green scale (*Palmaspis phoenicis*) (Fig. 5.6b) is an exotic insect which was accidentally brought into the country in about the early 1980s. Although this pest does not inflict a serious threat in countries where it is endemic, its damage in Sudan is catastrophic, especially on the commercial dry cultivars Abattamoda, Barakawi, and Gondaila. The destructive nature of this pest in Sudan could be environmental, lack of predators, vulnerability of local cultivars to the pest attack, and failure to apply quarantine measures to eradicate the pest in its area of origin.

When the incidence of green scale attack was encountered in the USA with introduction of offshoots from the old world, all efforts of control by chemical means on adult date palms failed. Chemical control was only effective on young date palms. Accidental flames that broke out in 1906 with an earthquake in California revealed that ornamental *Phoenix canariensis* palms recovered from the flames. This observation led to an action of gas torching of scale-infested date palms in California and Arizona (Shamblin 1924). Scale insects were then eradicated from the USA and the country is still free. Similar measures of scale insect control by burning infested trees are being adopted by some date palm growers in Sudan. *Phenicoccus marlate* is a scale insect which exists on a minor scale.

Fig 5.7 Birds known to inflict damage to date palm in Sudan. (**a**) House sparrow, (**b**) white-vented bulbul, (**c**) bird nests hanging at the leaf tip

Spider Mites Dust mites (*Oligonychus afrasiaticus*, McGregor) and (*O. pratensis*) are very damaging to dates in the early stages of fruit development. Unless treated by a protective spray before date palm flowering and as soon as the symptoms of attack are detected, the damage could be extremely severe (Fig. 5.6c). Spraying of dust mites by Vertimec insecticide is a very effective means of control in Sudan.

Vertebrates

Birds The house sparrow, *Passer domesticus* (Linnaeus) (Fig. 5.7a), and white-vented bulbul, *Pycnonotus barbatus* Arsinoe (Fig. 5.7b), are destructive pests attacking dates on the palm in the sweet *khalal* and *rutab* stages of fruit development. Covering bunches with muslin or cloth bags reduces bird damage to some extent, but the best type of bunch covers is still to be found. Birds may also be driven away by loud sounding devices. Reduction of bird population by destruction of breeding nests (Fig. 5.7c) and traps is also a means to reduce bird damage, but they continue to be a major cause for reducing yield and lowering fruit quality.

Rats Rats may attack any part of the date palm. They also dig underground galleries damaging the palm roots and irrigation systems. Nile rat (*Arvicanthis niloticus*), roof (black) rat (*Rattus rattus* Rodentia), and white-bellied house mouse (*Mus musculus* L.) are known to cause considerable damage to dates in storage. These rats are chemically controlled by the use of anticoagulant rodenticides, such as Killrat, Storm, and Racomin. Multimammate rat (*Mastomys natalensis*) and gerbils (*Tatera robusta*) together with Nile rat are found in the fields and orchards scurrying on date palms. Control of these rats is carried out by baiting. The bait consists of crushed sorghum as a carrier to which a rat poison is added with adhesive. The common rodenticide used is zinc phosphide. Bait is prepared at the rate of 1 % zinc phosphate (1 kg zinc phosphide + 1 kg food oil + 98 kg crushed sorghum).

Storage Pests Some serious pests are known to attack dates on the palm as well as in storage. Such pests are controlled by dusting bunches with 5 % malathion 3 weeks before harvest. These dates are then collected in clean containers and taken to storage without mixing them with dates that fell on the ground and could have become contaminated. Stores are fumigated with methyl bromide to keep the dates clean. Three destructive pests in Sudan's date stores are: greater date moth *Arenipsis sabella hampsim*, raisin moth *Ephestia* sp., and grain saw beetle *Oryzaephilus surinamensis*.

Diseases

To date no thorough survey of date palm diseases has been conducted in Sudan, and some diseases are known only by local names. Sudan is still free from the devastating *bayoud* disease caused by *Fusarium oxysporum* f. sp. *albedinis*. But pathologists have identified other organisms that attack date palms (Fig. 5.8a–d).

Black Scorch *Thielaviopsis paradoxa* is one of the most common and easily identified diseases. It shows several morphological symptoms hence it has the different names of black scorch, bud rot and bending head. The main method of control is sanitation by collecting and burning of diseased plant parts. Severely infected palms are destroyed. Chemically, surrounding tissues are sprayed with a copper-based fungicide.

Graphiola Leaf Spot *Graphiola phoenicis* J. This fungus occurs only on a very limited scale in humid areas, and it is not being controlled chemically. Infected parts are pruned off as a sanitation measure.

Other Disease Agents *Mauginella scaettae*, *Aspergillus* sp., *Fusarium moniliforme*, *F. oxysporum*, and *Helminthosporium* sp. Like most fungi these diseases can be controlled by copper-based fungicides, but none of these fungi currently are controlled chemically. Sanitation is the means of control.

5.3 Genetic Resources and Conservation

The amount of biodiversity in Sudan's dates makes them ideal for local selection and cultivar improvement. Morphologically, diversity is exemplified by fruit color, size, shape and time of maturity, seed, fruit stalks, fronds, leaflets, spines, fiber, tree canopy, and other features. Names of some date palm cultivars are designations of

Fig 5.8 Date palm diseases of Sudan. (**a**) *Graphiola* leaf spot, (**b**) black scorch, (**c**, **d**) unknown

fruit shape like Bride Fingers (*Asabi Elarous*) or fruit color at *khalal* stage like Ashger and Ahmer (Albakr 1972). In Sudan, Nubian-named cultivars Galad (red) and Gol Bogo (wide top) refer to morphological plant parts. Chemical characters like tannin content and sugar concentration in various stages of fruit development are also factors of contribution to biodiversity of date palm cultivars (Elshibli and Korpelainen 2009, 2010). Natural selection by adaptation to environment and human selection based on morphological and chemical characters led to the current date palm cultivars of Sudan.

Molecular markers, being a nondestructive process utilizing small plant parts, have been advocated for identifying date palm polymorphism and sex determination in early stages of plant development. Random amplified polymorphic DNA (RAPD) fingerprints, amplified fragment length polymorphism (AFLP), and microsatellite markers have been developed to identify date palm polymorphism. Sakina Elshily reported photos of the detection of polymorphism at two microsatellite loci developed by Billotte et al. (2004) in 7 date palm cultivars from Sudan (Fig. 5.9).

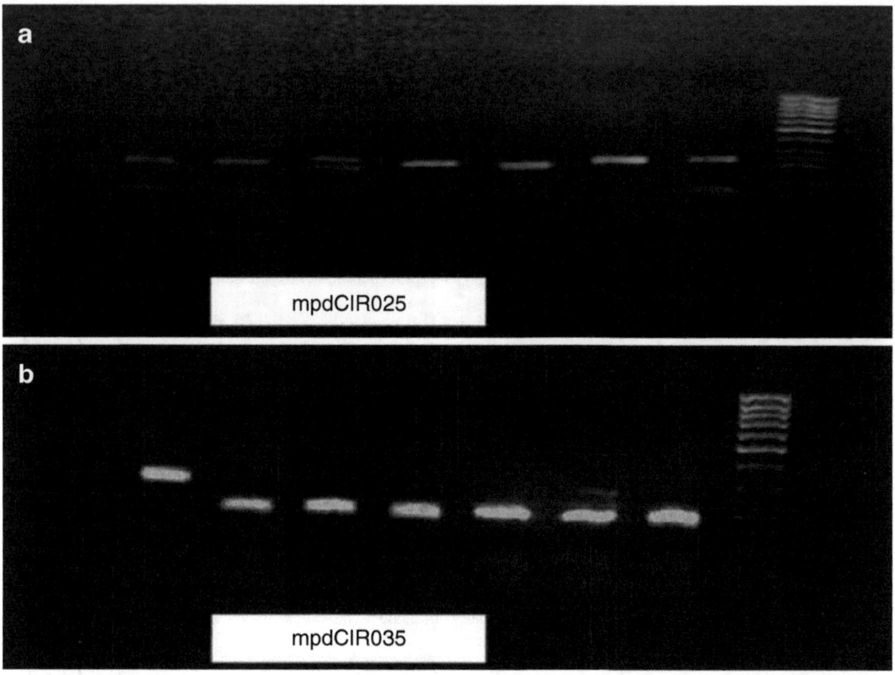

Fig. 5.9 Polymorphism detected at two microsatellite loci developed by Billotte et al. (2004) in seven date palm cultivars from Sudan; cvs. from left to right: Tonisi, Asada, Sultani, Galisoog, Siringai, Hamra, and Safra. (**a**) Shows the co-dominance of microsatellite markers, heterozygosity present at locus mpdCIR025 in the cv. Sultani. (**b**) Shows differences in the product sizes of the same locus mpdCIR035. Photos provided by Sakina Elshibli

Elshibli and Korpelainen (2008) reported mean expected heterozygosis (Hexp) of 0.84 in Sudanese cultivars and 0.80 in Sudanese males in a study of genetic diversity of 45 cultivars and 23 males from Sudan and Morocco. Moroccan cultivars were reported to have shown significant differentiation in relation to the Sudanese group.

There is no breeding program for date palms in Sudan, but the trend to maintain outstanding natural selections has continuously been going on. Date palms with lower height, stout stem, fewer spines, higher yield, and better fruit quality are desirable, and some local cultivars with such merits are known. In every date-growing area, there are few palms that are famous for their novel merits. Fruits of such date palms are kept for use by owners and their close friends only. Sultanat Badeen (Fig. 5.10a) is an example of a favorite novel date with extra-large fruit size and relatively low-sugar date. Validation of these unverified novel dates is underway. Two PhD theses submitted in December 2012 evaluated and recommended the follow-up of some local selections which were recommended for multiplication and observation (Bashab and Khairi 2012; Ibnouf and Khairi 2012). Research work by Ibnouf and Khairi (2012) at Merowe has initiated mapping processes for 29 date palm selections in the area revealing the opportunities to introduce new lines of superior cultivars from local selection to widen the base for germplasm of commercial cultivars. Novel selected lines are to be multiplied with tissue culture propaga-

Fig 5.10 (**a**) Sultanat Badeen novel date, (**b**) Zadna Company germplasm collection, Kadaro, Sudan

tion techniques. Zadna International Company has established a germplasm collection of tissue culture introductions at Kadaro in Khartoum North (Fig. 5.10b). A recent germplasm collection of tissue culture–propagated date palm introductions from Saudi Arabia and United Arab Emirates comprising 3,600 seedlings was established by Haj Beshier Mohamed Eid at Silait scheme in Khartoum North.

5.4 Plant Tissue Culture

The role of tissue culture as a quick means of propagation for production of true-to-type, disease-free date palms is well understood in Sudan, and the following tissue culture laboratories have been established:

(a) Department of Horticulture, Ministry of Agriculture, Khartoum
(b) Environment Research, Khartoum
(c) Faculty of Agriculture, University of Khartoum
(d) Lina, Kadaro, Khartoum North
(e) Sudan University of Science and Technology, Shambat
(f) Technology, Shambat, shared by Khartoum State and Ministry of Science and Technology
(g) Zadna, Kadaro, Khartoum North (under construction)

These laboratories are propagating mainly potato and banana. Propagation of date palms has not started yet, until the procedure is worked out. Collaboration between Zadna Company and Alrajihi laboratories in Saudi Arabia has been established to promote the processes of procurement of equipment, training of nationals, and getting the propagation process underway.

An influx of tissue culture–propagated date palm imports from Saudi Arabia, United Arab Emirates, and Iran began two decades ago. Torpedo date palm seedlings are imported, grown in nurseries (Fig. 5.11a), and distributed to various date palm growing areas. Starting mainly with Barhi and Khalas, about half a million date palm

Fig 5.11 (**a**) Zadna nursery at Kadaro, (**b**) Barhi at Gebel Awlia Alikhwa farm, (**c**) Medjool. Author's nursery at Silait, (**d**) Anbara at West Omdurman (Source: Khairi et al. (2010))

seedlings of the cultivars were introduced. Most of these introductions are just starting to set fruit. In summer rain and humidity damage-prone areas of Khartoum, Barhi, Khadrawy, and Khinaizy showed great success by being edible ahead of summer rains. Medjool and Sukkary (consumed usually at *tamr* stage) are early maturing, but success may only be achieved in years of low rainfall. Anbara cv. showed some rain damage at *rutab* stage in Khartoum, but some samples from West Omdurman showed Anbara approaching *tamar* stage. As more introductions come into bearing, success of more introductions may be achieved (Fig. 5.11b–d) Khairi et al. (2010).

5.5 Cultivar Identification

Date palm cultivars are identified using morphological and genetic characters. Cultivars vary in canopy structure, fruit shape, size, and chemical composition at various stages of fruit development.

Currently in local laboratories, fruits and seeds are weighed using sensitive balances. Verniers are used for measuring length and width of fruits and seeds. Chemical characteristics are determined by the method described by Lane and

Table 5.1 Total sugars, reducing sugars and sucrose % in twelve date cultivars grown in Sudan based on dry weight (Khalifa and Osman 1988)

	Total sugars %		Reducing sugars %		Sucrose %	
Cultivar	Rutab	Tamar	Rutab	Tamar	Rutab	Tamar
Abattamoda	75.23	79.19	35.97	44.17	39.26	35.00
Assada	65.87	69.79	41.50	62.92	24.37	06.88
Barakawi	71.07	71.88	48.69	56.67	22.37	15.21
Bureir	85.97	88.54	58.88	86.67	27.09	01.88
Deglet Noor	56.72	78.75	14.94	40.00	41.78	38.75
Gondaila	91.87	94.79	31.82	36.67	60.05	58.13
Madina	67.95	69.79	44.27	66.67	23.68	03.13
Mishrig Wad Khateeb	71.07	75.00	49.27	72.50	21.82	02.50
Mishrig Wad Laggai	67.95	70.84	45.83	69.58	22.02	01.25
Sultani	88.77	91.67	58.88	83.75	29.86	07.92
Umhat	67.95	72.92	45.93	67.92	22.02	05.10
Zaglouli	67.95	72.92	45.93	66.67	22.02	05.84

Eynon (1923). Genotyping of biodiversity to characterize variability at DNA level using molecular marker techniques is available in the laboratories of the Botany Department, Faculty of Agriculture, University of Khartoum, and the Agriculture Research Corporation laboratory in Khartoum. There are numerous laboratories using Lane and Eynon methods, for examples, (a) Faculty of Agriculture, University of Khartoum; (b) Sudan University of Science and Technology, Shambat; (c) the Central Laboratory at Soba, Khartoum; (d) University of Jazeera Laboratory; (e) Agriculture Research Corporation laboratory, Wad Medani; (f) Food Processing Research Center laboratory at Shambat, Khartoum North; and (g) Environment Research laboratory, Khartoum.

The genetic resources of Sudan's date cultivars are hard to trace as date culture in the country is prehistoric and there is so much variation within commercial cultivars. Some cultivars are believed to have been brought in from North Africa. Others are believed to have been brought from Arabia. Cultivars of indigenous seed origin are uncountable and locally known as *jaw* which is equivalent to *digil* in some countries. To characterize cultivars chemically, fruit samples are examined at progressive levels of maturity (*khalal*, *rutab*, and *tamar* stages). Khalifa and Osman (1988) made an assessment of sugar content in Sudan's commercial date cultivars at *khalal* and *rutab* stages using the Lane and Eynon method. In both *rutab* and *tamar* stages, the popular cultivar Gondaila showed the highest sugar content (Table 5.1).

5.6 Cultivar Description

Sudan has long been famous for the production of dry dates due to the sunny hot low relative humidity climate. Until recently the country has been relying on six indigenous commercial cultivars. Minor cultivars also exist (Osman 1984). Detailed data about morphological characters of these cultivars cannot be found, but basic information and photographs can be given. While these cultivars have been maintained by offshoot propagation, there is so much variation within each cultivar as no research for selection has yet been conducted. Facilitated by climatic variation, a number of dry and semidry cultivars are grown.

5.6.1 Dry Cultivars

These cultivars are grown in the northern end of the date palm belt between latitudes 22–18 °N and require 2,093–2,316 C total heat units and very dry climate. The textures of dry dates are so hard that their fruit moisture content does not exceed 20 %. The main commercial dry cultivars are described below:

Abattamoda This cultivar comes under several names but the correct name is Abattamoda. Abatta is the Nubian name for Barakawi, and this is an advanced style of Barakawi (Abattamoda). In his report about the possibilities for improving date culture in Sudan, Nixon (1967) stated that most of Abattamoda date palms are present in the region between Dongola and Wadi Halfa and, in quality, it is probably the best cultivar in Sudan (Fig. 5.12). It looks similar to cv. Deglet Noor and develops to a dry date if kept on the tree for long. However, if harvested at early maturity and well kept, it is a fancy soft date. Fruits are 5–6 cm long and 1.5–2 cm in diameter. Limited in number, this cultivar represents only about 2 % of the date palm population in Sudan (Al-Baker 1972). Abattamoda cv. fetches very high price, but there is not sufficient quantity of it in the market to make an impact.

Fig 5.12 Abattamoda, the top-quality date of Sudan

Fig 5.13 Barakawi Khalal (**a**) and Tamar (**b**), Gondaila Rutab (**c**), and Tamar (**d**)

Barakawi It is the most dominant and important dry date cultivar grown in North Sudan and forms about 85 % of the date palm population in Merowe Dongola area. Mason (1927) stated that "Sudan has in Barakawi one of the few first class dry dates in the world." It is a very hard date if left on the tree for long and therefore not very vulnerable to storage pest attack. Fruits are elongated, 4–5 cm long, 1.5 cm in diameter, and edible only in late mature *tamar* stage due to high tannin content in early fruit development stages (Fig. 5.13a, b).

Gondaila This is the second most important date palm cultivar in Sudan after Barakawi, representing about 5 % of the date palm population. The stem is stouter than Barakawi and the fibers are tougher. It is commonly treated as a dry date, but it does not have the hard texture of Barakawi cv. and tends to be more susceptible to storage pest attack. If it is harvested at early *tamar* stage and stored properly, Gondaila makes a decent-quality date. Fruits are about 4–4.5 cm long and 2–2.5 cm in diameter. A lack of uniformity in fruit color may be a disadvantage (Fig. 5.13c, d), but Gondaila has an excellent flavor and sells at one and a half times the price of Barakawi. It is a very popular cultivar.

Kulma According to Nixon (1967), this Kulma is the largest of the date fruits in Sudan. But Nixon did not like the fruits of this cultivar as he thought that the fiber content of the fruits is high. There are only a few hundred palms of this cultivar, most of them in Merowe province.

5.6.2 Semidry Cultivars

These cultivars require 1,400–1,500 C total heat units and moderately dry climate. The climatic zone of these cultivars starts at the Merowe area at about 18 °N latitude and increases southward dominating the Nile banks at Abu Hamad and southward until high humidity limits date production at about 12 °N latitude. The two commercial semidry date cultivars grown in Sudan are Mishrig Wad Laggai (Fig. 5.14a, b) and Mishrig Wad Khateeb (Fig. 5.14c).

The textures and quality of these dry cultivars are comparable to the best of the soft dates of Iraq (Nixon 1967). Mishrig Wad Khateeb cv. is rated below Mishrig Wad Laggai in quality because the flesh is less firm and not quite smooth in texture, the skin separates, and the seed is larger. Commercial growers prefer Khateeb which dominates the date-growing areas south of Abu Hamad because it is hardy and better

Fig 5.14 (**a**) Mishrig Wad Laggai Kimri and Khalal, (**b**) Mishrig Wad Laggai Tamar, (**c**) Mishrig Wad Khateeb Khalal, (**d**) Madina Khalal and Rutab

adapted to climate. But Mishrig Wad Laggai is much fancier. Fruits of both cultivars are small and require thinning to get a decent size. Semidry cultivars bear much heavier crop compared to dry cultivars, and a yield of 200 kg per palm is normal.

5.6.3 *Soft Cultivars*

These cultivars require 1,100–1,150 °C total heat units, and Madina is the only commercial soft date cultivar in Sudan; it is also the earliest ripening commercial date palm cultivar. Madina is handled and consumed almost entirely as a fresh date and may be in local markets by July (Fig. 5.14d). Fruits are large and appealing. The stem is stout compared to most Sudan's cultivars. If this cultivar is not pollinated as soon as the spathe splits open, the trees set false fruits. There are only a few thousand of cv. Madina palms in Sudan, and the recently constructed Marwa dam flooded the area of its main production causing great Madina palm losses.

5.7 Date Production and Marketing

Date production in Sudan is based on dry dates for which Sudan is very famous. Dry dates facilitate easy harvest, handling, storage, and transport in simple containers like sacks, bags, and skins. Being so firm, harvested dry date bunches are dropped from palm tops onto mats spread under the trees to minimize dust contamination. Soon after harvest, the dates are spread out in the sun to dry and stored in sacks and earthenware containers, ready for marketing. In local markets, dry dates are exhibited for sale in open containers (Fig. 5.15a). Recent trends, however, tend to direct

Fig 5.15 (**a**) Dry dates with measures in Khartoum, (**b**) soft Madina cv. sold in cartons

date production toward soft and semidry dates, which are more palatable, but require more sophisticated techniques for harvest, handling, packaging, and cold storage. *Khalal* and *rutab* dates are harvested by lowering bunches tied to ropes down to containers. Strands are cut out and lined in 1–3 kg or larger, in paper or plastic boxes for sale. In farmers' markets *khalal* dates are marketed in cartons (Fig. 5.15b).

Most gardens of date cultivars are small in size, not exceeding 2 ac (0.8 ha) in most cases. Some recent gardens, mostly tissue culture–based cultivars, however, are relatively large in area. Gandale garden in Makarab area, close to Damer city in River Nile State, comprises 43,000 date palms of mixed local and Iraqi cultivars. With the influx of tissue culture–propagated cultivars, multithousand date palm gardens were established, mostly in Khartoum State, during the past 6 years. Zadna gardens in Kadaro, Virgin Mary, and Goosy in Omdurman and date gardens in Gibait, Red Sea State, are each 3,000–4,000 palms in size.

5.8 Date Processing and Novel Products

Traditional date processing in Sudan started a long time ago. The most famous products are date semi bread and date paste. Date bread is usually prepared by mixing date paste with wheat flour and baked with the addition of some spices and ghee. This product is consumed during travel. Date paste is widely consumed as a well-known product throughout the country with the addition of some spices and ghee.

In 1958, a date-processing pilot plant was erected at Kareema in Northern State with designed capacity of 1,700 mt per annum, as a joint venture between the Ministry of Agriculture and FAO. Later, in 1965, it was handed over to the Ministry of Industry. This factory has storage, cleaning, grading, and packing facilities. It is currently a fairly old facility and rehabilitation needs to be considered. Currently, it is being operated by a private sector producing packed dry and semidry dates, soft dates packed in smaller consumer packs, date paste which is used in confectionary products, processing of date sweets, and production of alcohol from inferior quality dates.

Manufacturing date palm parts to make mats, containers, furniture, and other novel items has long been known in Sudan. Mats are made from leaflets. Containers are made from bunch stalks stitched with leaflets. Beds, furniture, and baskets are made from leaf rachis (Fig. 5.16a–d). Improvement of this industry is being encouraged.

5.9 Conclusions and Recommendations

Date palms in Sudan have long been recognized as a source of food security, economic, and shelter crop. The potential for date production of an assortment of dates is enormous, yet production in the past has focused on dry dates for which Sudan is

Fig 5.16 Manufacturing date palm parts. (**a**) Bed from leaf fronds, (**b**) mats and containers from leaflets, (**c**) furniture from leaf fronds, (**d**) basket and hats from palm leaflets

famous, together with some semidry dates. Traditionally, production has relied on six cultivars, but recently a policy of diversification by introduction of tissue culture cultivars and local selection from indigenous seedlings has been adopted. The country is still free from the devastating red palm weevil pest and bayoud disease, but yield is impaired from attack by various insect, mite, and vertebrate pests, as well as fungal, mycoplasma, and viral diseases, many of which are awaiting diagnosis and identification. Cultural operations are mainly traditional, but modern techniques are being introduced in field operations, palm and bunch management techniques, handling, and packaging to facilitate marketing and to meet growing consumer demands. Based on the present situation, implementing the following recommendations would enhance date palm production in Sudan:

(a) Establishment of a national palm and date center as a lead institution to promote the palm and date sector in the country. National institutions can interact with relevant regional and international institutions to exchange information and coordinate collaborative efforts to handle issues encountering the development of palm and dates at large.

(b) Creation of a cooperative institute to coordinate and back up national efforts for date production, packaging, and marketing. Postharvest handling and marketing operations are usually beyond the capacity of most individual growers. Pooling the national resources is always a lucrative and enumerative way to organize these processes. The Date Growers' Institute in the USA is an example of success of cooperatives in promoting agricultural products.
(c) Implementation of advanced techniques in field operations, palm and date management techniques, and subsequent processes.
(d) Strengthening infrastructure: well-equipped tissue culture and research laboratories, modern packing houses, packaging facilities, and training and extension facilities are all premises that are required for advancement of operations toward a quality product.
(e) Capacity building: a strong core of capable national staff to handle research, extension, marketing techniques, advertisement, and other operations needs to be built.
(f) Current programs for cultivar improvement targeting the best adapted and highest yielding cultivars for each climatic zone need to be pursued.
(g) Strengthening plant protection and quarantine facilities to assure protection of the palm and date sector, prevention of introduction of devastating insects and diseases into the country, and deployment of scientific and survey teams for identification of unknown diseases and advice on control measures.

References

Al-Baker A (1972) The date palm. A review of its past and present status; and the recent advances in its culture, industry and trade. Al-Ani Press, Baghdad

Bashab F, Khairi M (2012) Evaluation and selection of some promising males for pollination of 3 commercial date palm cultivars in Northern state of Sudan. PhD thesis Sudan Academy of Sciences, Agriculture Research Corporation Library, Wad Madani, Sudan

Bashab F, Ibnoaf M, Salih A (2006) Effect of organic manure on growth, yield and fruit quality of Barakawi date palms on high terrace soils of Northern Sudan. Agriculture Research Corporation, Ministry of Agriculture, Sudan

Billotte N, Marseilla CN, Brottier P et al (2004) Nuclear microsatellite markers for the date palm (*Phoenix dactylifera* L.): characterisation, utility across the genus *Phoenix* and in other palm genera. Mol Ecol Notes 4:256–258

Dawood H (1997) Studies to evaluate and characterize seedling male date palms for New Halfa Area. PhD thesis, University of Khartoum, Agriculture Research Corporation Library, Wad Madani, Sudan

Elshibli S, Korpelainen H (2008) Microsatellite markers reveal high genetic diversity in date palm (*Phoenix dactylifera* L.) germplasm from Sudan. Genetica 134:251–260

Elshibli S, Korpelainen H (2009) Biodiversity of date palms (*Phoenix dactylifera* L.) in Sudan: chemical, morphological and DNA polymorphisms of selected cultivars. Plant Genet Res Charac Util 7:194–203

Elshibli S, Korpelainen H (2010) Identity of date palm (*Phoenix dactylifera* L.) germplasm in Sudan: from the morphology and chemical characters to molecular markers. Acta Hort 859:143–153

FAOSTAT (2010) Agro-statistics data base. FAO, Rome
Ibnouf M, Khairi M (2012) Evaluation of some outstanding seedling date palm selections as potential cultivars at Merowe area of northern Sudan. PhD thesis. Sudan Academy of Sciences, Agriculture Research Corporation Library, Wad Madani, Sudan
Khairi M, Elhassan M, Bashab F (2010) The status of date palm cultivation and date production in Sudan. Acta Hort 822:37–42
Khalifa T, Osman A (1988) Physiochemical evaluation of some Sudanese date cultivars. Date Palm J 6(1):230–239
Lane J, Eynon L (1923) Determination of reducing sugars by means of Fehlings solutions with methylene blue as internal indicator. J Soc Chem Ind 42:32–36
Mason S (1927) Date culture in Egypt and the Sudan, vol 1457, USDA bulletin. USDA, Washington
Nixon R (1967) Possibilities for improving date culture in the Sudan. Report. Department of Agriculture, Ministry of Agriculture, Sudan
Osman AMA (1984) The performance of date palms in the Sudan. Acta Hort 143:231–237
Shamblin A (1924) Eradication and control of date scale. Date Grow Inst Ann Rep 1:13–14
Tothill JD (ed) (1948) Agriculture in The Sudan. Oxford University Press, London

Chapter 6
Date Palm Status and Perspective in Tunisia

Hammadi Hamza, Monia Jemni, Mohamed Ali Benabderrahim, Abdesselem Mrabet, Sana Touil, Ahmed Othmani, and Mohamed Ben Salah

Abstract Date palm is one of the most important fruit trees grown in southern Tunisia and represents a good cash crop for many farmers. Here, we provide an overview of the Tunisian date palm status. In fact, this important subtropical fruit crop is currently in danger due to several constraints such as anthropogenic spread of disease, water shortages, salinization, and irregular climatic conditions. In addition, Tunisian date palm is threatened by genetic erosion as a consequence of the predominance of the elite cultivar Deglet Noor in modern plantations and the disappearance of many cultivars with medium and low fruit qualities. A series of successful experiments were widely applied for micropropagation of endangered cultivars. Many exhaustive resource inventory programs of date palms have been done in Tunisia. Tunisian genetic diversity was studied by using morphological and molecular markers; some are related to agronomic traits. Tunisian date production increased during recent decades and represents more than 18 % of the national tree production. Tons of cull dates are rejected by processing industries, and attempts to develop new products are supported by research programs. However, the major products are currently destined for exportation because of limited local marketing and the dietary habits of the Tunisian people.

H. Hamza (✉) • M.A. Benabderrahim • A. Mrabet
Arid and Oases Cropping Laboratory, Arid Area Institute, Elfjé, Medenine 4119, Tunisia
e-mail: hamzapalmier@yahoo.fr; medali_abderrahim@hotmail.com; abdessalem_mr@yahoo.fr

M. Jemni
Département de Technologies de Dattes , Centre Régional de Recherche en Agriculture Oasienne de Degueche, Tozeur, Degueche 2260, Tunisia
e-mail: jemnimonia@yahoo.fr

A. Othmani
Laboratoire de Culture In Vitro, Centre Régional des Recherches en Agriculture Oasienne de Degueche, Tozeur, Tunisia
e-mail: degletbey@yahoo.fr

S. Touil
Department of Plant Protection, Centre Technique des Dattes, Kebilli 4200, Tunisia
e-mail: touilsana3@yahoo.fr

M.B. Salah
Research Regional Center on Oases Agriculture, Degache 6021, Tunisia
e-mail: bensalah_mohamed@yahoo.com

J.M. Al-Khayri et al. (eds.), *Date Palm Genetic Resources and Utilization: Volume 1: Africa and the Americas*, DOI 10.1007/978-94-017-9694-1_6

Keywords Date palm • Conservation • Diversity • Genetic erosion • Historical • Novel products • Production • Tunisia

6.1 Introduction

The date palm (*Phoenix dactylifera* L.) was introduced in Tunisia by the Phoenicians before the Roman occupation (Hodgson 1932). However, the Romans did not give much importance to date palm cultivation in comparison with other crops such as olives and vegetables. The development of date palm cultivation occurred after the settlement of southern Tunisia by Arabs, and this was observed especially in the region of Jerid. The date palm has played an important role in establishing oases and caravan trade routes. Gradually, the oases became more organized, which is reflected in the thirteenth century, when Ibn Chabbat (1221–1285) conceived of a judicious system of sharing water from sources in the oasis of Tozeur (Jerid region). In addition, in the fourteenth century, Tozeur's agriculture became sophisticated and this stimulated a demographic explosion (Rouissi 1969). In fact, many date palm cultivars grown in this area were of seedling origin or brought by travelers and pilgrims especially from the East. However, Deglet Noor cv. was introduced into Tunisia four centuries ago from Oued Souf, in western Algeria (Kearney 1906), and became the most appreciated cultivar of the farmers. After this period, high taxes imposed by the government and external attacks caused serious problems in these oases (Rouissi 1969). In consequence, many oases, especially Nafta, were destroyed and it took up to a century to rebuild them. In the nineteenth century, the oasis economy changed again after the decline of the caravan trade. Indeed, the French occupation of Algeria disrupted the human migrations between Algeria and the south of Tunisia. Consequently, the oasis plantations disappeared gradually.

At the beginning of the twentieth century, many water sources were developed by the indigenous people and nomads who became founders of oases. The French protectorate, with security objectives, contributed by creating small areas scattered around water sources especially in the Nefzaoua. Kearney (1906) reported that there were around 1.3 million date palms in Tunisia with 30,000–50,000 of Deglet Noor cv. Thereafter, the number increased and the oasis area was 7,300 ha with 2.65 million date palm trees (Hodgson 1932), and the most productive harvest was in 1920–1930 with 41,500 mt of dates.

After independence, many strategies have been tried, aimed at modernizing oases and especially to reduce water stress. Indeed, the state sold the nationalized ex-settlers' land to the famous STIL society (Société Tunisienne des Industries Laitières) which provided enormous potential for the export development of Deglet Noor dates (Hajji 1997). This economic progress prompted the authorities to create new modern oases at the end of the 1980s such as in the region of Rjim Maatoug (Nefzaoua) where more than 2,000 ha was established by the national army. To the present, 6,000 people have been settled in this desert region. In 2011, there were more than five million date palms growing in southern Tunisia on more than 40,000 ha of oases (Fig. 6.1). Tunisia's contribution to world date production has become important; it has increased significantly in recent decades. In fact, 190,000 mt were produced in 2011, 76 % represented by cv. Deglet Noor (GIF 2011).

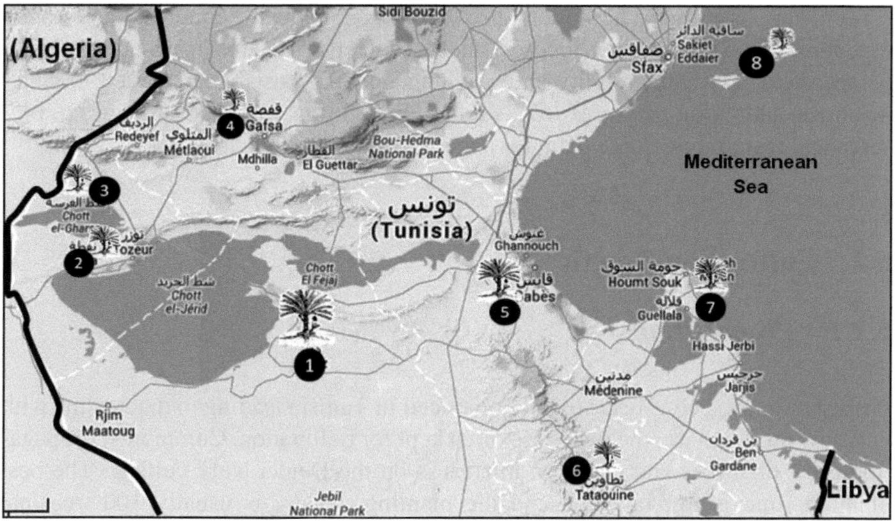

Fig. 6.1 Sites of Tunisian oases. *1* Nefzaoua, *2* Jerid, *3* Chebika and Tamerza, *4* Gafsa, *5* Gabès, *6* Tataouine, *7* Jerba, and *8* Kerkena

The date palm now is an important income source for farmers. In many cases, and principally in Nefzaoua, a date palm can provide a total of USD 100 per year worth of dates. This has increased farm income and encouraged the creation of new plantings. In addition many farmers have introduced intercropping with olives, pomegranates, grape vines, cucurbits, and forage crops which are the most adapted to Tunisian oases. In addition, since 2011 many agricultural cooperatives have been established, and they have mounted programs of oasis development like the introduction of new crops to optimize water use. Despite the fact that these attempts are only beginning, they can produce high-quality potatoes and farmers are becoming more convinced of the value of these changes.

Besides the operations under state control, a large expansion was carried out by private enterprise mainly in Nefzaoua where many date groves were created with unauthorized well drilling and based on the monoculture of Deglet Noor cv. Today, these *illegal* areas represent one-half of the date palm groves in this region. Consequently, the availability of water has become a serious threat to the sustainability of some oasis zones. This situation is aggravated by the threat of destructive diseases such as bayoud, caused by the fungus *Fusarium oxysporum* f. sp. *albedinis* (Baaziz et al. 2000; Louvet and Toutain 1973). It is estimated that bayoud disease killed more than 13 million susceptible trees in Morocco and Algeria in less than a century (Oihabi 2000a, b). Deglet Noor is very sensitive to this fungus which exacerbates the overall threat to Tunisian date palms by genetic erosion. In Tunisia, bayoud disease has not been reported, but other diseases are known, including unidentified diseases and those associated with the presence of certain fungi and insects.

Tunisian oases have important productive, ecological, economic, and social potentialities. This chapter focuses on the date palm status in Tunisia and its perspectives. The vulnerability and challenges facing oasis areas related to water resources and biotic pressures were studied. In addition, Tunisian strategies to preserve and restore oasis/date palm heritage are described and analyzed.

6.2 Cultivation Practices

6.2.1 Propagation

Propagation by seed is occasionally practiced in Tunisia leading to date palms with very low productivity; the males are often kept for pollination. Commonly, propagation is by offshoots and the main interest is on the Deglet Noor cultivar. The best planting time is in May/June and the planting density is usually 100 trees/ha. Offshoots are planted in a hole of 1 m^3 filled with pure soil.

Serious problems should be solved for date palm propagation, expansion of oases, and improvement of disseminated plant materials. The main problem is the smuggling of offshoots from neighboring oases or countries which may be infested with diseases or insects. A source of healthy offshoots with high growth vigor and fruit production is necessary.

6.2.2 Irrigation

The southern Tunisian oases are supplied with water from two principal aquifers, the Continental Intercalary (CI) and Complex Terminal (CT), which extend over 80,000 km^2 and are exploited with more than 1,200 well drillings (OSS 2009). The CT aquifer depth range is 30–500 m while the CI varies from 60 to 2,800 m (Omrani and Burger 2010). A recent inventory revealed 3,069 illegal drilling sites from the CT in the Nefzaoua region alone (MEDD 2010). Groundwater in Nefzaoua is increasingly overexploited, essentially during the summer period, and the area is threatened by salinization. The situation became aggravated in the Nefzaoua oases where the exploitation of the CT in 2011 exceeded by 166 % the allowable quantity (ODS 2011). Flood irrigation is the method most widely used in Tunisian date palm culture. This traditional irrigation system wastes large quantities of water through evaporation loses, and farmers are being required to adopt modern irrigation methods.

Ghazouani et al. (2009) summarized the irrigation constraints cited by many farmers: (a) waterlogging, (b) hot irrigation water, and (c) low frequency of delivery water. That is why the knowledge of experts and farmers should be combined to allow a critical evaluation to resolve these problems. In this regard, a comparative study was done to determine the optimum water amount to be applied according to plantation density (Omrani and Burger 2010). Three palm tree densities were tested: 64, 100, and 156 tree/ha. The best water efficiency in 2009 was obtained with water application of 8,500 m^3/ha/year for all date palm tree densities (Omrani and Burger 2010). However, when the experiment was repeated for 3 years (2009, 2010, and 2011), the results were different. Results showed that the most productive water volume was 17,000 m^3/ha/year at a density of 100 tree/ha and 20,000 m^3/ha/year for the other two densities. In the past 2 years, many farmers have begun to practice drip irrigation. This irrigation method was introduced and developed by the farmers for more efficient use of water; only the amount of water needed by the palm is applied. However, this method is very expensive for farmers when no grants are available from the government.

6.2.3 *Fertilization*

Soil fertility decline can be corrected by adequate application of mineral fertilizers, especially manure. In 2009, 660 mt of phosphate fertilizers, mainly diammonium phosphate (DAP), 613 mt of ammonium nitrate, and 116,000 mt of manure were used by farmers (MEDD 2010). For both mineral fertilizers, the average amount applied per hectare was around 16 kg, less than 150 g per date palm. Concerning manure, the average is about 20 kg per date palm tree. For the oases which suffer from soil fatigue, farmers bring in sand to renovate the field.

6.2.4 *Pollination*

March, April, and May is the normal pollination period in Tunisia. Farmers collect male spathes after they burst and insert, by hand, male inflorescence flowers into the female spathes. The number of flowers inserted depends on the female cultivar and the pollen quality. The selection of pollen is done by the farmers who prefer male inflorescences with a high pollen density. Pollination is usually done on a good sunny day as soon as the first break in the female spathe is observed.

Date palm pollination is manual and no mechanical method is employed. Short-term pollen storage or the practice of storing pollen from 1 year to another is practiced to pollinate early cultivars in which the female spathes open before the male spathes.

Bchini (2006) selected traits of male date palm trees which have a pollen effect (metaxenia) on maturation, early flowering, and fruit size in Deglet Noor cv. He concluded that pollen derived from early-bearing cultivars is favorable for earliness and pollen from a male seedling of a cultivar bearing large fruits is favorable for fruit size.

6.2.5 Bunch and Fruit Management

Fruit and bunch thinning is mainly practiced in Deglet Noor cultivar, but rarely in others. The aim of these practices is to enhance fruit size and consequently satisfy market preference, reduce damage due to humidity by a greater air circulation around the fruits while on the bunch, and reduce incidence of blacknose (darkening and shriveling). Tunisian farmers realize date-fruit thinning by reducing:

(a) The number of bunches per palm to keep a proper balance between the number of leaves and fruit bunches; a ratio of eight to ten leaves per bunch should be maintained, which guarantees good production the following year.
(b) The number of fruits per strand (at pollination): the lower third of the female inflorescence is removed.
(c) The number of strands per bunch (12–14 weeks after pollination): the lower third or slightly more of the bunch is removed mostly from the center.

In Tunisia, the Deglet Noor ripening season coincides with the rainy season and the rains can cause severe loss of fruit. Farmers use plastic bags to protect bunches against rain and humidity. Protection is applied at the early kimri stage. Over the last 5 years, farmers have used a new bag promoted by CRRAO (Center for Date Palm Research) Degache to cover the fruit bunches. This bag protects the fruits against both rain and insects; it has two parts: the upper part protects from rain and moisture and the lower is an insect net especially against the carob moth. Synchronized with bunch bagging, farmers usually carry out leaf pruning. Leaves are removed when they became dry. In the coastal oases, farmers practice thorn removal to facilitate pollination and handling of fruit bunches.

6.2.6 Date Palm Diseases and Pests in Tunisia

In southern Tunisia, particularly in the Jerid and Nefzaoua zones of high date production, the date palm is often prone to various insect attacks. The major pests found in Tunisia and their methods of control are described below.

6.2.6.1 *Ectomyelois ceratoniae* Zell.

The larva of the carob moth, *Ectomyelois ceratoniae*, infests the dates internally not only in the field before harvest but also during storage (Triki et al. 2003) (Fig. 6.2). It is a major constraint for date exports, especially for Deglet Noor fruit. To control this pest, farmers usually protect the fruit bunches and try to keep the plantation clean. Rarely the date traders fumigate harvested and stored dates. Many procedures should be taken to decrease the incidence of carob moth attack by the use of pheromone traps or by spraying the infested fruits with the biological pesticide *Bacillus thuringiensis* (Djerbi 1995). A recent study by Jemni et al. (2014) tested a treatment

Fig. 6.2 Major pest in Tunisian oases. (**a**) *Ectomyelois ceratoniae* Zell. (**b**) Date palm fruit attacked by *Oligonychus afrasiaticus* McGr. (**c**) *Parlatoria blanchardii*, white scale (photo by S. Touil). (**d**) Brittle leaf disease (photo by A. Namsi). (**e**) *Phoenix canariensis* attacked by the RPW in the city of Carthage, Tunis (photo by S. Touil). (**f**) *Phoenix dactylifera* attacked by *Phytophthora* sp. (photo by H. Hamza)

of Deglet Noor dates with NaClO, UV-C, O_3, and EW and showed a positive effect for lowering their natural infestation by carob moth as well as microbial growth after 30 days of storage at 20 °C. In particular, UV-C and EW were the most effective against moth without adverse effects.

6.2.6.2 *Oligonychus afrasiaticus* McGr.

The date palm spider mite, *Oligonychus afrasiaticus*, causes serious damage to fruits (Dhouibi 1991; Khoualdia et al. 1997) (Fig. 6.2). In Tunisia, the infestation time varies from year to year, ranging from the first to the third week of July (Ben Chaaban et al. 2011). To decrease pest damage, Palevsky et al. (2004) proposed several control techniques that should be practiced: (a) reduction of the overwintering populations, (b) application of physical barriers to prevent airborne mites from reaching fruit bunches, (c) choice of a properly timed seasonal acaricide application, and (d) postharvest rehydration treatment.

6.2.6.3 *Parlatoria blanchardii* Targ.

Four generations of white scale insect, *Parlatoria blanchardii*, per year have been detected on date palms in southwestern Tunisia (Khoualdia et al. 1993). The most injurious is the spring generation. All green parts of the trees are damaged, with infestation higher on the palms at the base of the crown than in the center (Fig. 6.2). Variability in cultivar resistance to the white scale is underscored. Indeed, Kentichi cv. is significantly more resistant than Deglet Noor, Aligue, or Khouat allig (Khoualdia et al. 1993). In Tunisia, biological control against white scale was tried by the introduction of an exotic predator in the Segdoud region (oases of Gafsa). The exotic ladybird beetle, *Chilocorus bipustulatus* L. var. *iranensis*, is giving promising results (Khoualdia et al. 1997).

6.2.6.4 *Oryctes agamemnon* Burm.

Oryctes agamemnon, the Arabian rhinoceros beetle, was accidently introduced in the late 1970s into the date palm plantations of Mrah Lahouar in the Jerid region by offshoots brought in from the Persian Gulf region, where the pest originates (Baraud 1985). Ten years later, the beetle infected Rjim Maatoug oases in the Nefzaoua region. Gradually, this insect affects more date palms especially Deglet Noor cv. The symptoms appear only at very advanced stages and threaten sudden death of the tree (Soltani 2004). The only practical preventive method to reduce insect populations is their collection at the adult or larval stages. Because of the large infested area, the collection takes a long time and requires a sophisticated light trap. In addition, it should be done before the reproductive period of the beetle, a practice which has not been followed (Soltani 2010).

6.2.6.5 Brittle Leaf Disease

Brittle leaf disease (BLD) has been reported from Tunisia since the 1980s. Namsi et al. (2007) reported that there are 40,000 trees affected by this disease. It has been observed on most Tunisian cultivars including Deglet Noor, Tozerzaid, Akhouat Alig, Ammy, and Kinta, as well as on seedling trees and pollinator trees (Fig. 6.2). Kentichi cv. seems to be relatively tolerant. No data are available on other possible hosts, for example, on ornamental palms. BLD does not involve a pathogen but it is related to manganese (Mn) deficiency in the palm trees (Namsi et al. 2007). Currently, a tentative treatment is being done on affected date palm by adjusting the Mn quantities by trunk injection (Namsi Ahmed, personal communication, 2013).

6.2.6.6 *Rhynchophorus ferrugineus* Oliv.

Red palm weevil (RPW), *Rhynchophorus ferrugineus*, originates from Asia, where it is a serious pest of coconuts (Rahalkar et al. 1985). RPW is a polyphagous insect (Murphy and Briscoe 1999) which can infect many species of fruit and ornamental palm trees.

Following a notification in December 2011 from the city of Carthage (Fig. 6.2), Tunis, by the General Directorate of Protection and Quality Control of Agricultural Products, quarantine specialists visited the site where RPW was observed in order to establish a diagnostic (Chebbi 2011). The first observation noted tens of ornamental Canary Islands palms (*Phoenix canariensis*) exhibiting dried leaves. The preventative measures recommended were not followed quickly enough and the attack expanded progressively in area. Programs should be installed as soon as possible before this problem affects date palms.

6.2.6.7 *Phytophthora* sp.

Infection with water mold, *Phytophthora* sp., causes *belâat* disease and the death of young fronds along with an odor of acetic and butyric fermentation (Zaid et al. 2002). Usually, this disease is rare in Tunisia, but in the spring of 2012 and 2013, it was noted in many oases especially those in the Nefzaoua area (Fig. 6.2). Keeping date plantations clean is recommended to prevent attacks of water mold. Spraying with maneb or Bordeaux mixture at the rate of 8 l/palm can control the disease during early stages (Zaid et al. 2002).

6.2.6.8 *Ommatissus binotatus* var. *lybicus*, De Berg.

Ommatissus binotatus var. *lybicus*, the dubas bug, is a newly detected insect pest found for the first time in the Tamerza oasis in 2010. Thereafter, it was recorded in different areas of the Jerid oasis such as Chebika, Tozeur, Hezwa, and Nafta (Zouba and Raeesi 2010a). Biological and chemical controls are effective (Aldryhim 2008).

6.2.6.9 *Raoiella indica* Hirst

Red palm mite (*Raoiella indica*) is another newly discovered insect pest detected for the first time in Tunisia in July 2010 on Deglet Noor cv. It was observed initially in the old oasis of Nefzaoua and subsequently in different areas of the Jerid oasis such as Tozeur, Degache, and Nafta (Zouba and Raeesi 2010b).

6.3 Genetic Resources and Conservation

The date palm is widespread in the southern Tunisia, primarily in two main regions: continental and littoral oases (Fig. 6.1). Dominant are the continental oasis with 33,723 ha or 82 % of the total oases (Table 6.1). These oases contain more than 89 % of the total number of date palms in the country and contribute up to 85 % of the national date production (Sghaier 2010).

The continental oases are divided into Saharan and mountain oases. The Saharan oases are located in Jerid and Nefzaoua. The mountain oases are situated in Tamerza, Chebika, and Gafsa regions and are characterized by the predominance of common Tunisian cultivars.

In Nefzaoua, certain cultivars are ubiquitous in every oasis such as Deglet Noor, Alig, Bisr Helou, Horra, and Kinta (Fig. 6.3 and Table 6.2). But, other cultivars are rarely present (Hamza et al. 2006). The distribution of cultivars shows the dominance of the Deglet Noor cv. with 82 %. In second place is Alig cv. with 6.35 %. The latter has a fairly significant value in the local market and it is well known for its stress tolerance. The cultivation of male date palms, *dhokkars*, depends on the number of female palm of the plot. In the Jerid oases, the cultivar structure is almost the same. Deglet Noor cv. is about 60 % (MEDD 2010) and the Kentichi cv. is specific to the region, and it is more ubiquitous than the other common cultivars with 2.66 %.

Littoral oases are mainly in the region of Gabès (Fig. 6.1) and occupy 7,080 ha or 17 % of the total (Table 6.1) and contain 10 % of the total number of palms in Tunisia. The date palm cultivation patterns are different; there are other cultivars like Bouhattem, Garn Ghazel, Eguiwa, Lemsi, and Arichti (Ben Salah 2012) (Table 6.2).

Tunisian oases are also classified into two types: traditional and modern. Traditional oases occupy an area of 15,051 ha, or 37 % of the total. They contain 46 % of the palm trees; the majority are made up of common cultivars planted at a relatively high density of 166 trees per ha. These oases are characterized by small

Table 6.1 Types of Tunisian oases regarding their geographical origin

Oases	Location	Area (ha)	Percentage
Continental Saharan oases	Nefzaoua	22,980	56
	Jerid	8,363	20
Continental mountain oases	Gafsa	2,380	6
Littoral oases	Gabès	7,080	17
Total		40,803	100

Source: Sghaier (2010)

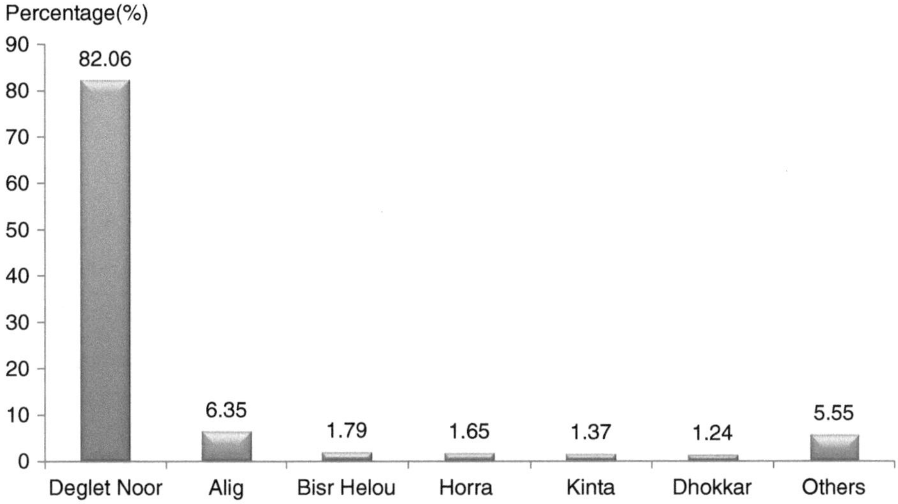

Fig. 6.3 Structure of date palm cultivars in Nefzaoua (Source: Hamza et al. 2006)

Table 6.2 Some characteristics of the main varieties in Tunisian oases

Variety	Distribution	Precocity	Fruit color at tamar stage	Fruit consistency	Seed weight/ fruit (%)
Deglet Noor	Continental oases	Late	Amber	Semisoft	9.7
Alig	All oases	Late	Dark brown	Semidry	9.9
Bisr Helou	All oases	Season	Light brown	Dry	17
Kinta	All oases	Season	Amber	Dry	13.6
Kentichi	Jerid	Late	Reddish	Dry	16.33
Bouhattem	Littoral oases	Season	Dark brown	Soft	14
Lemsi	Littoral oases	Season	Dark brown	Soft	14
Arichti	All oases	Late	Dark brown	Dry	12

Source: Ferchichi and Hamza (2008), Rhouma (2005)

farm size, resulting from property subdivision through inheritance. Modern oases occupy 25,752 ha (63 %) of the total area of the oases. They are characterized by larger farms and a lower planting density, fewer than 115 palm trees/ha with a high proportion of Deglet Noor cv.

In Nefzaoua, where many modern oases have been established, a correlation can be noted between the age of the oasis and the percentage of Deglet Noor (Fig. 6.4), the modern extensions showing a trend toward monoculture (Hamza et al. 2006). However, the hectarage of other date farms is decreasing and some disappearing in spite of their important nutritional and economic value. It is necessary to elaborate a strategy of preservation to protect production, adaptation, and resistance of the genetic inheritance. For this purpose, conservation plots were created in order to preserve the threatened cultivars. Four stations, located in Degache, Ibn Chabbat I, Ibn Chabbat II, and Atilet, are maintaining live specimens of 100 endangered cultivars.

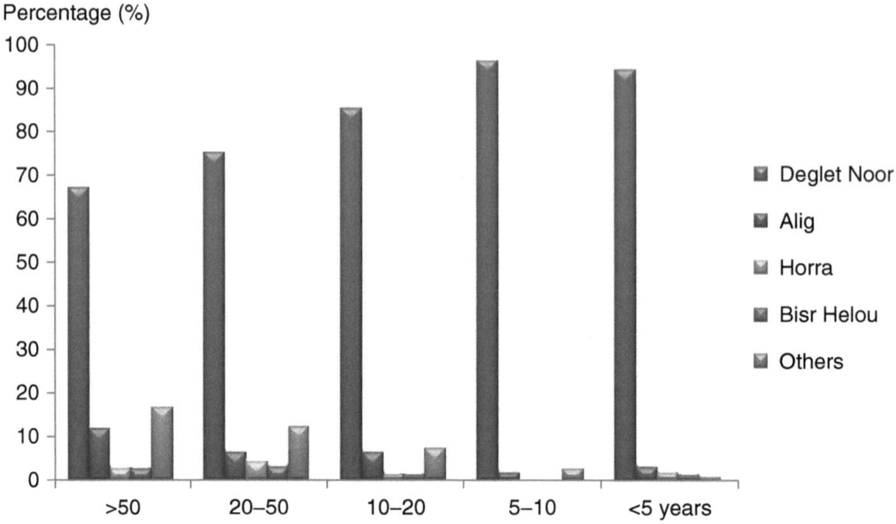

Fig. 6.4 Percentage of variation of date palm cultivars related to the oasis age (Source: Hamza et al. 2006)

6.4 Plant Tissue Culture

Many date palm cultivars are characterized by their good fruit quality but also their rarity. In vitro culture techniques contribute to the preservation of such endangered genotypes. In Tunisia, in vitro culture investigations are very highly developed and perfected. Two Tunisian laboratories are leading this effort: the Laboratory of Plant Biotechnology, Faculty of Sfax, and the Laboratory of In Vitro Culture, CRRAO Degueche.

Several plant tissue culture methods have been successfully adapted to a number of Tunisian cultivars such as Deglet Bey, Boufeggous, Gondi, and Cheddakh. To the present, Tunisian laboratories have produced up to 1,600 of plantlets by tissue culture. Different approaches have been established for the micropropagation but the most used is somatic embryogenesis. It offers a higher potential for mass propagation, allows for rapid propagation, and preserves date palm genetic resources (Fki et al. 2011a; Othmani et al. 2010). An improvement in somatic embryogenesis production of Deglet Noor cv. was made by Sghaier et al. (2008). A detailed discussion of the date palm somatic embryogenesis process and the influencing factors was recently provided by Fki et al. (2011b). Tunisian investigations by Drira and Benbadis (1985) and Fki et al. (2003) showed that callogenic capacity of inflorescences is higher than that of leaves. Furthermore, Fki et al. (2003) reported that the 2,4-D auxin is the most popular for callogenesis. The same authors in 2011 used reduced amounts of 2,4-D (0.2 mg l^{-1}) to establish adventitious bud growth of the introduced cultivar Barhi, from juvenile leaves, and this contributed to limiting the risk of somaclonal variation.

In addition, Othmani et al. (2009) successfully obtained in vitro regeneration of the elite date palm cultivar Deglet Bey (Mnakher) through both somatic embryogenesis (Fig. 6.5) and direct shoot formation (Fig. 6.6) from young leaf explants cultured on MS agar-solidified medium supplemented with 10 mg l^{-1} 2,4-dichlorophenoxyacetic

Fig. 6.5 Induction of somatic embryogenesis and plant regeneration from leaf explants of date palm cv. Deglet Noor. (**a**) Primary callus formation after 6 months of culture. (**b**) Induction of an embryogenic callus from primary callus after 10 months of culture. (**c**) Differentiation of somatic embryos from embryogenic callus. (**d**) Maturation of somatic embryos after 1 month of transfer on MS medium deprived of plant growth regulators. (**e**) Plantlet derived from the conversion of a matured somatic embryo. (**f**) Potted plant after 1 year of transfer to the greenhouse. (**g**) Plant after 2 years of transfer to the greenhouse (Photos by A. Othmani)

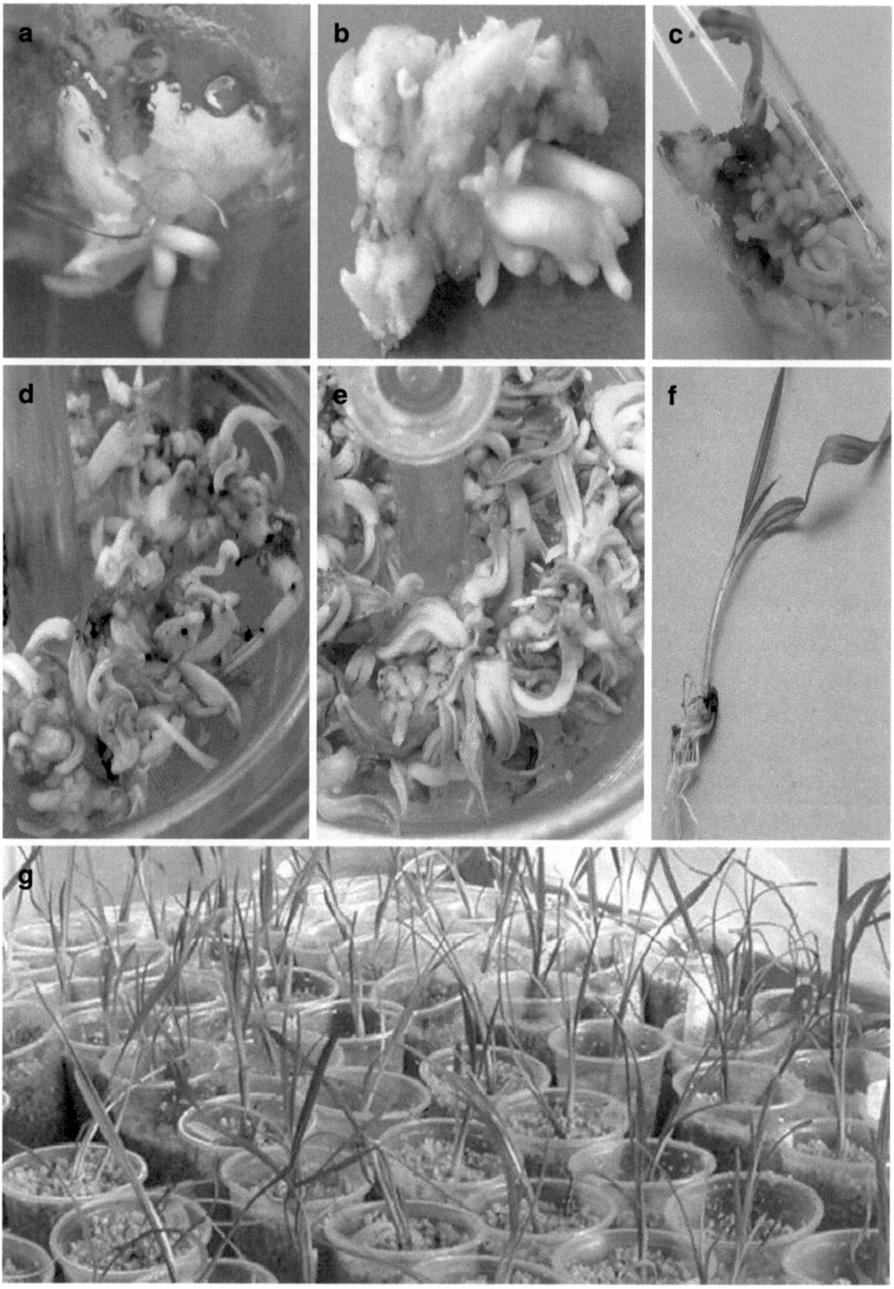

Fig. 6.6 Induction of shoot organogenesis and plant regeneration from leaf explants of date palm cv. Deglet Noor. (**a**) Induction of primary shoots from a leaf explant after 10 months of culture. (**b**) Initiation of shoot multiplication after transfer of explants on MS medium supplemented with NAA and BAP. (**c**) Shoot cluster in active multiplication phase. (**d**) Initiation of elongation of shoots. (**e**) Induction of roots from elongated shoots. (**f**) Plantlet obtained from the conversion of an elongated shoot after 1 month of culture on MS medium without plant growth regulators. (**g**) Potted plants after 3 months of transfer into the greenhouse (Photos by A. Othmani)

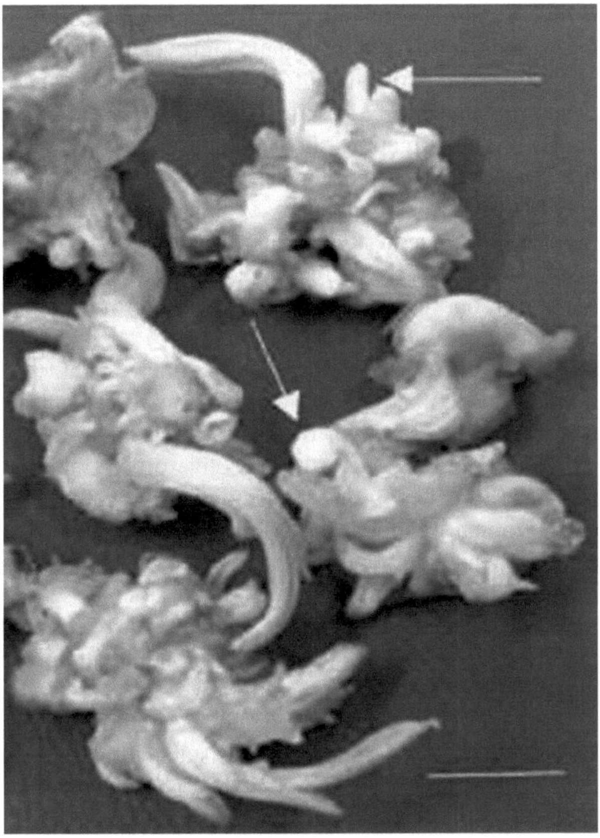

Fig. 6.7 Multiple-shoot induction (*arrows*) from shoots obtained on agar-solidified medium that were cultured for 2 weeks in the TIB system containing proliferation medium. The immersion frequency was 5 min every 8 h. *Scale bar*: 0.5 mm (Othmani et al. 2009)

acid for 8 months. They have developed somatic embryogenic suspensions to improve differentiation of embryogenic callus and culture of shoots in a temporary immersion bioreactor (TIB) to enhance proliferation of regenerated shoots. The principles, advantages, and disadvantages of TIB were recently described in detail by Othmani et al. (2011). The immersion frequency tested so far appears to be suboptimal for date palm callus growth and could be optimized by testing different durations. These authors reported that the embryogenic calli of date palm cv. Deglet Bey turned brown and died using a RITA® bioreactor with immersion frequency of 5 min every 8 h. They found that the temporary immersion system was better than the solid medium only for shoot proliferation (Fig. 6.7). Despite many promising results, culturing in bioreactors proves to be more complicated than culturing on agar-solidified media in terms of hyperhydricity and contamination risks. As for the expression of exogenous contamination, often it can be controlled by good sterile techniques; however, endogenous contamination cannot be easily controlled in repeated subcultures. The contamination problem can be controlled using embryogenic callus and shoot clusters pretested to be free from endophytic bacteria. Nevertheless, subculture of affected plant material in antibiotic-amended liquid medium did not avoid this problem. The antibiotics tested were cefotaxime (250 mg l^{-1}) and streptomycin (500 mg l^{-1}).

To preserve contamination-free material and prevent somaclonal variation, cryostorage of date palm somatic embryos was done in Tunisia. For the purpose of creating a cryobank of proliferating tissues, Fki et al. (2013) tested the possibility of generating and cryopreserving highly proliferating meristems of the Kheneizi cultivar. These investigations showed that regeneration rates using standard vitrification, droplet-vitrification, and encapsulation-vitrification protocols reached 26.7, 60.0, and 40.0 %, respectively. Only explants smaller than 3 mm in diameter were found to survive cryogenic treatments. In addition, sucrose pre-culture, cold hardening, and loading solution pretreatments showed significant effects on regeneration rates. Moreover, our results indicate that both sucrose pre-culture and cold acclimation of explants increased proline content.

On the other hand, Masmoudi-Allouche et al. (2009, 2010) achieved in vitro hermaphrodism induction in date palm female flowers which opens opportunities for the investigation of an in vitro self-fertilization process. In fact, several date palm cultivars were the objective of sex modification under these particular tissue culture conditions. Masmoudi-Allouche et al. (2009) used an appropriate hormonal treatment based on the addition of IBA (indole-3-butyric acid) and BAP (6-benzylaminopurine) at different concentrations. Further investigations by Masmoudi-Allouche et al. (2011) were done to search for putative variations that may have occurred on the initial genome due to the application of plant growth regulators. The results revealed that hormonal treatment entailed no detectable genetic variation in the treated date palm flowers.

Date palm tissue culture techniques should be adopted as a central part of innovation. The creation of new pest-resistant cultivars is very important. In fact, several experiments have been started by the Laboratory of Plant Biotechnology, Faculty of Sfax, in collaboration with the International Atomic Energy Agency (IAEA) to select cultivars resistant to bayoud disease (FAO/IAEA 2001). Selection was based on the resistance against the active toxins of the pathogen applied to irradiated in vitro cultures. The isolation of bayoud from leaves and the production of toxin were achieved in Morocco.

6.5 Cultivar Identification

An exhaustive resource inventory program of date palms has been done in Tunisia. A number of methods have been used to analyze genetic diversity in germplasm accessions, breeding lines, and populations. These methods have relied mainly on two genetic markers: morphological and molecular.

6.5.1 Morphological Characterization

As to morphological markers, the most common characters used to identify different cultivars in date palm are the phenotypic expression of leaves, spines, and fruit characters. Several studies have described the importance of morphological traits in identifying Tunisian date palm cultivars (Ben Salah 1993; Ben Salah and Hellali 2004; Rhouma 1994, 2005). Despite these descriptions, it remains very difficult to identify cultivars, especially outside the fruiting period. In fact, owing to the great adaptive flexibility of this species, many farmers cannot recognize cultivars outside their oasis farm. Indeed, the majority of morphological characters are an adaptive response to the environment. Hamza et al. (2009) sifted through morphological characters that are not controlled by edaphic or climatic factors. These characters were selected on the principle of their low environmental plasticity and strong genetic control, important in cultivar identification, and also they have an intra-genotype measure reproducibility (Hamza et al. 2009, 2011). Nine characters were selected. Six characters are vegetative: spiny midrib part length (%), apical divergence angle (°), maximal pinna width at the top leaf (cm), solitary spine number (%), spine length in the middle (cm), and maximal spine angle (°). Three reproductive characters describe the inflorescences and fruit: bunch length without spikelet (%), spikelet length without fruits (%), and fruit internal cavity ratio.

6.5.2 *Molecular Characterization*

Sakka et al. (2004) used chloroplast DNA to identify Tunisian cultivars following the method of polymerase chain reaction-restriction fragment length polymorphism (PCR-RFLP). These markers have the advantage of being codominant and can scan all the DNA extract. The results are highly reproducible and great diversity was observed as a function of the enzyme and probe used. However, the RFLP technique requires a large amount of DNA and markers used for nuclear DNA are limited (Dowling et al. 1996). To detect variability, it is necessary to use more enzymes which make analysis expensive. Some enzymes have more restriction sites which makes the results dependent on gene activation. The DNA fragment migration on gel is logarithmic, making it difficult to detect fragments of larger sizes.

Random amplification polymorphism DNA (RAPD) is widely used in the identification of date palm cultivars in Tunisia and to study their phylogenetic relationships (Ben Abdallah et al. 2000; Trifi et al. 2000). This technique is important in identifying the genetic fingerprints of date palm cultivars and in the early detection of genotypes. The advantage of RAPD is its simplicity as it does not require a large amount of DNA. However, Benkhalifa (1999) reported that RAPD shows a low rate of polymorphism. Moreover, it is widely criticized for the reproducibility, the structure of the primers, the dominance of markers, and the independence of loci.

Inter-simple sequence repeat (ISSR) markers were used to assess the polymorphism in Tunisian cultivars (Zehdi et al. 2002; Zehdi-Azouzi et al. 2011). Among 12 ISSR primers used in the study, only 7 generated polymorphic fragments. In addition, using other ISSR primers (Fig. 6.8), Hamza et al. (2012) detected genetic differentiation between Tunisian subpopulations in terms of fruit consistency, and soft fruit cultivars were significantly differentiated from dry fruit cultivars. The ISSR markers have the advantage of being highly polymorphic. Thus, it has been reported that these markers are more variable than RFLP and RAPD techniques (Nagaoka and Ogihara 1997). Moreover, they are more reproducible than those

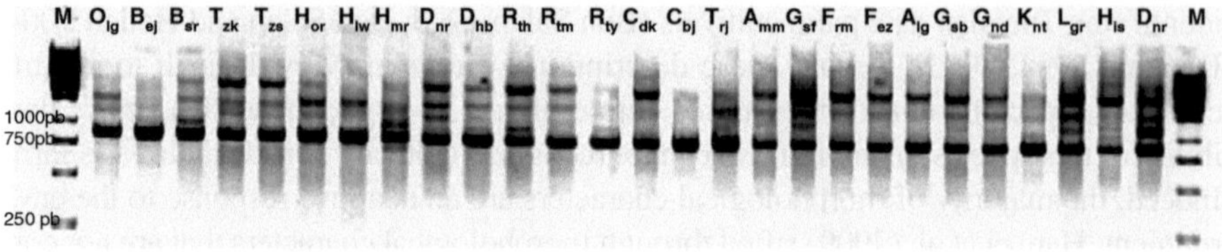

Fig. 6.8 Banding patterns in different date palm cultivars by ISSR primer D12. *M* 1 Kb ladder, O_{lg} Om Leghlez, B_{ej} Bejjou, B_{sr} Bisr Helou, T_{zk} Tezerzayet Kahla, T_{zs} Tezerzayet Safra, H_{or} Horra, H_{lw} Halwa, H_{mr} Hamra, D_{nr} Deglet Noor, D_{hb} Dhahbi, R_{th} Rtob Houdh, R_{tm} Rotbayet Elmansoura, R_{ty} Rotbayet Yagouta, C_{dk} Cheddakh, C_{bj} Choddekh Ben Jbir, T_{rj} Tronja, A_{mm} Ammary, G_{sf} Ghars Souf, F_{rm} Fermla, F_{ez} Fezzani, A_{lg} Alig, G_{sb} Gosbi, G_{nd} Gondi, K_{nt} Kentichi, L_{gr} Loghrabi, H_{is} Hissa (Hamza et al. 2012)

generated by the RAPD technique (Nagaoka and Ogihara 1997). However, ISSR markers have drawbacks such as sensitivity to the variation of the amount of template DNA as well as the annealing temperature, which must be carefully developed (Bornet and Branchard 2001).

The random amplified microsatellite polymorphism (RAMPO) technique is a modification of the RAPD to avoid the lack of reproducibility and dominance. Rhouma (2008) found a high genetic diversity in Tunisian cultivars compared with the results of RAPD and ISSR.

Rhouma (2008) used six amplified fragment length polymorphism (AFLP) primer combinations to characterize Tunisian cultivars, and a total of 428 polymorphic bands were generated. AFLP markers have the disadvantage of the need for high technical expertise and high-quality DNA.

The specific simple sequence repeat (SSR) date palm primers were developed by Billotte et al. (2004). Zehdi et al. (2004) did genotyping of many Tunisian cultivars, and the comparison of theoretical geographic populations showed that each Tunisian oasis constitutes a single population. The identity probability of cultivars showed that the use of only three microsatellite markers can discriminate a cultivar from the rest of the population. Key cultivar identification was well established by Hamza et al. (2010) and Zehdi et al. (2004). In addition, the SSR markers have been used to study certain agronomic traits in Tunisia such as maturity period and fruit consistency (Hamza et al. 2011). The results detected genetic differentiation between subpopulations in terms of fruit consistency. The cultivar subpopulations with semisoft dates showed a significant genetic differentiation, which places them between soft and semidry date cultivars. This suggests that the continental date palm oases could be a set of populations with a different origin.

6.6 Cultivar Description

More than 200 named cultivars are grown in Tunisian oases (Ferchichi and Hamza 2008; Rhouma 1994, 2005). These cultivars exhibit differences in ripening time; three different classes are identified in that regard: early, mid-season, and late. Fruit consistency is the most important parameter that usually determines fruit quality and commercial interest in them. Traditionally, this trait was segregated into three different fruit classes: soft, semidry, and dry (Munier 1973). New research has established semidry fruit cultivar heterogeneity. As a consequence, semidry has been subdivided into two new clusters: semisoft and semidry with supportive morphological and genetic distinction (Hamza et al. 2009, 2011, 2012).

The application of the selected morphological characters on cultivars of continental oases (Hamza et al. 2009, 2011) showed a high correlation with maturity period and fruit consistency. In fact, the percentage of spiny midrib parts for early and soft cultivars was significantly smaller than for the others, and these cultivars also showed the highest percentages of solitary spines (Fig. 6.9). In addition, the percentage of bunch lengths without spikelets and the percentage of fructified spikelet lengths for the early cultivars were, respectively, the lowest and the highest in comparison with the mid-season and late-maturing cultivars.

6.7 Date Production and Marketing

The date harvest in Tunisia starts at the end of July for early-maturing cultivars and ends in the middle of December for the late cultivars. Earlier cultivars produce soft dates that must be transported in the early morning to local markets for consumption. Late-maturing cultivars like Deglet Noor have a long-season market. These cultivars are subjected to postharvest practices in order to prepare them for marketing.

Tunisian date production increased from 118,000 mt in 1999 to 190,000 mt in 2011. Concerning Deglet Noor cv., the production increased from 78,200 mt in 1999 to 137,000 mt in 2011. This increase was about 61 % for all date production and 76 % for Deglet Noor. The Nefzaoua region is the most important, accounting for about 59 % of the national date production (ODS 2011).

During the last quarter of the twentieth century, world date production has been allocated on average at a rate of 75 % for local consumption and 25 % for export. The latter began to increase significantly over the past decade, averaging a third of the total production (Fig. 6.10). At the same time, prices of Deglet Noor cv. dates have changed significantly. In fact, the farm gate price has increased from USD 604/mt in 1999 to USD 927/mt in 2008, while the price for local consumption increased from USD 1,327 to USD 2,039/mt during the same period. The export price was USD 1,434/mt in 1999 and climbed to USD 1,954/mt in 2008.

Date exports are an important component of the national economy. Average annual exports for the period 2002–2006 were approximately 41,000 mt, equivalent in value to USD 69.3 million per year. It should be noted that date exports for the 2008/2009 season exceeded 60,000 mt with a value of USD 140 million; those levels place Tunisia as the fourth largest world exporter of dates in quantity and the first in value.

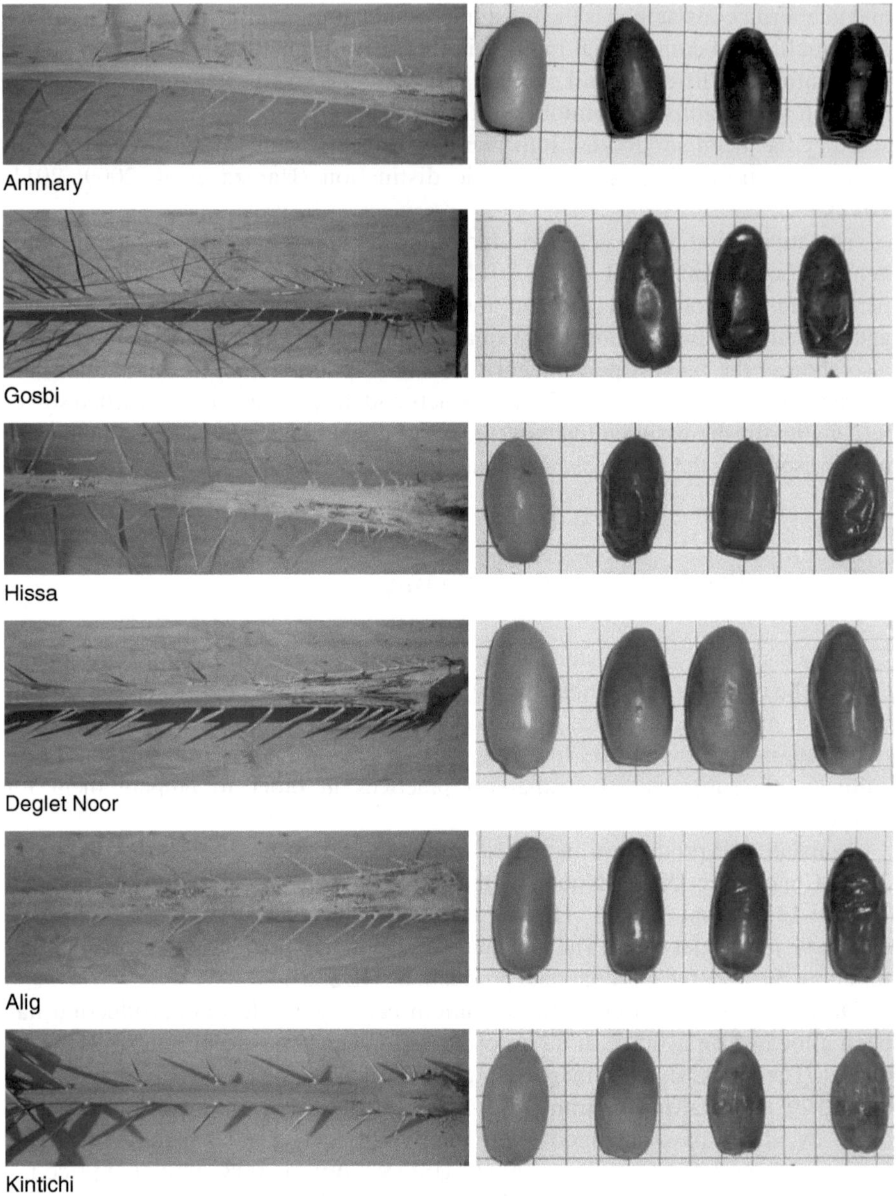

Fig. 6.9 Spiny midrib parts and fruits of Tunisian date palm cultivars. Early cultivars with soft dates (cvs. Ammary, Gosbi, and Hissa) and late cultivars with semisoft dates (cv. Deglet Noor), semidry dates (cv. Alig), and dry fruit (cv. Kentichi) (Photos by H. Hamza)

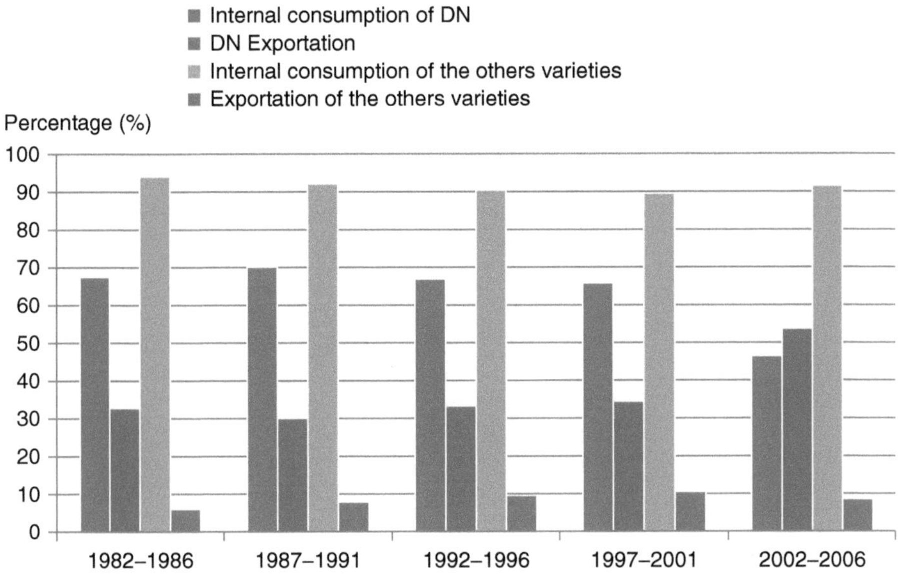

Fig. 6.10 Date palm production, internal consumption, and exportation (DN: Deglet Noor) as compared to the total of other cultivars (Source: Annuaires Statistiques Agricoles; Ministère de l'Agriculture 2010)

6.8 Processing and Novel Products

The main activity of date processing stations is date packaging and export. However, tons of cull dates are unfit for human consumption because of poor quality, dryness, or contamination. These culls are rejected by date processors but a small proportion is recycled as plant fertilizer or goes to animal feed. Cull dates are a rich source of carbohydrates essentially sugars and dietary fiber. Due to their richness, some studies have been carried out to valorize these dates and to develop new products as summarized in Fig. 6.11. In Tunisia, several research programs were established for this purpose, focused on:

(a) Date juice: dates are pitted and ground then the mix solubilized in hot water. After that, filtration is necessary to clarify the juice. The latter should be stored at a cold temperature. Chaira et al. (2009) perfected a cooking method for date palm juice. The application of a temperature of 80 °C for 90 min on pulpy cultivars gave the best results. It produces an energy juice drink rich in total solids, of low acidity, and with a high power to scavenge superoxide free radicals.

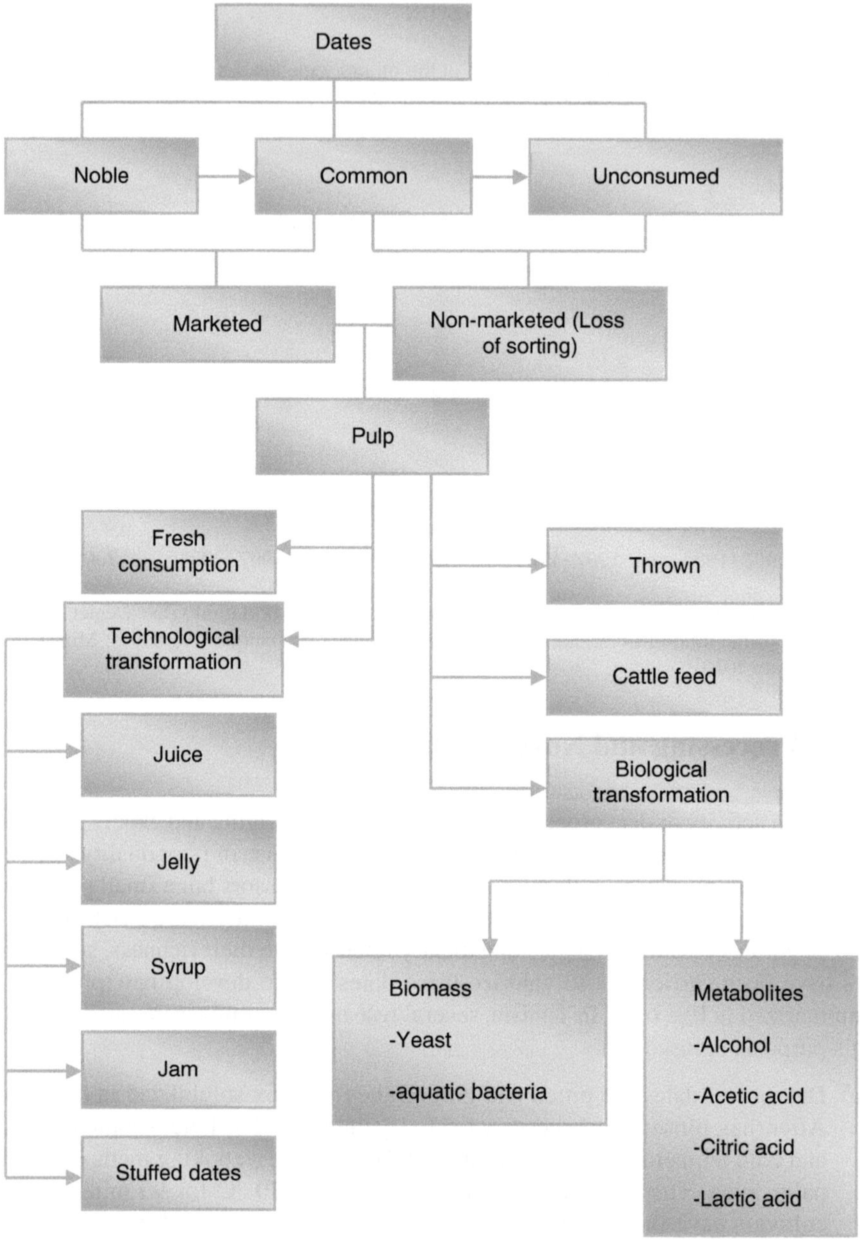

Fig. 6.11 Possible ways of date-pulp valorization (Besbes et al. 2006)

(b) Date syrup: date syrup, also called *rub al tamr*, is produced in southern Tunisia from certain cultivars. Extraction of date syrup is carried out as follows: date paste is mixed with water and boiled at 100 °C. The juice produced is passed through a cloth filter. To prepare date syrup, juice is concentrated at 100 °C. The date syrup is consumed directly or used as an ingredient in certain foods such as ice cream, beverages, confectionery, bakery goods, sesame paste/date syrup blends, jams, and butters. Date syrup is a high-energy food rich in carbohydrates and a good source of minerals; it also contains a very complex mixture of other saccharides, amino and organic acids, polyphenols, and carotenoids. These techniques have produced a high-commercial-value syrup by enzymatic treatment. This method has enhanced the total soluble solids and decreases the turbidity by using pectinase and cellulase. In addition enzymatic treatment improves the syrup clarity which becomes lighter. This new syrup was very appreciated by consumers at a large taste evaluation. In addition, results showed that the date syrup prepared after extraction with pectinase and cellulase mixture gave the lowest phenolic, flavonoid contents and antioxidant activity (Abbès et al. 2011, 2013).

(c) Date jam: date jam is unknown on the market of Tunisia despite the effort of some manufacturers trying to produce it. According to Besbes et al. (2009), jam is prepared by boiling ground fruit flesh with sucrose in water. The mix is cooked to about 65 °Brix. A significant effect of the date cultivar used was noted on the composition and physical characteristics of date jams (Besbes et al. 2009). Indeed, Alig cv. jam was richer in reducing sugars and was characterized by its greater firmness and water-retention capacity. Alig and Kentichi cv. jams presented a higher overall acceptability in comparison with quince and did not show any significant difference from Deglet Noor jam.

(d) Dietary fiber: fiber is extracted from date flesh by hot water and recuperated by centrifugation after dissolving the sugars (Elleuch et al. 2008). Extraction of dietary fiber (Fig. 6.12) was studied at different temperatures (40, 50, and 60 °C). A significant decrease in water-holding capacity, swelling capacity, and emulsifying capacity was noted at 60 °C (Borchani et al. 2012).

(e) Biomass production: according to the richness of date juice in simple sugar, it constitutes a favorable medium for biomass such as the production of biopesticide bacteria. To do this, an experimental design was devised. Indeed, given the low protein content of date juice, the addition of protein is necessary. Optimizing the composition date juice and fermentation parameters was performed. The main results obtained are the concentration of sugar in media and the concentration of protein added; the pH of media and temperature of fermentation have a significant effect on the yield and on the productivity of these biopesticide bacteria (Jemni et al. 2010).

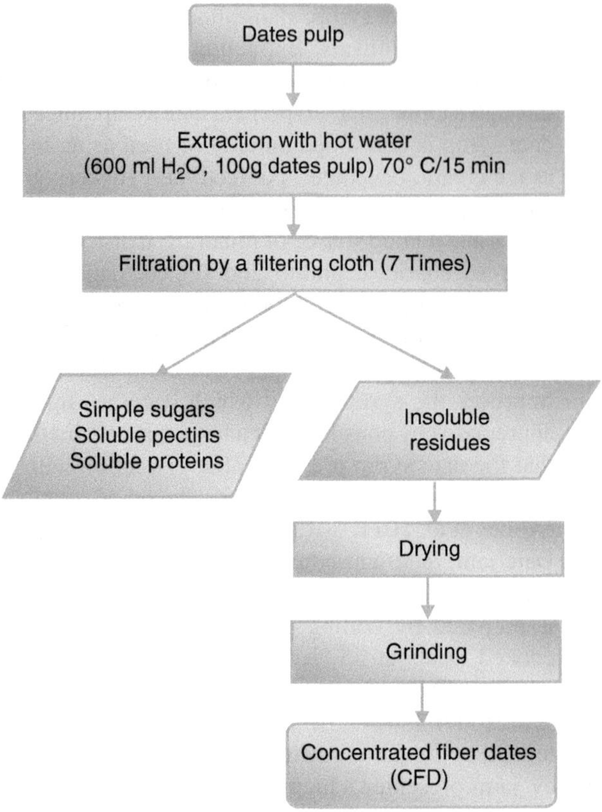

Fig. 6.12 Diagram of preparation of date-pulp fiber (Borchani et al. 2010)

(f) Metabolite production: the date juice has been used for the production of some metabolites such as vinegar which is obtained by double fermentation involving alcoholic and acetic acid at temperatures of 27–30 °C (Besbes et al. 2006) and curdlan production with a yield of 22.83 g/l (Ben Salah et al. 2011). The purified date by-product curdlan (DBP-curdlan) had a molecular weight of 180 kDa. Moreover, bioethanol, with a concentration of 25 %, was produced in Tunisia by fermentation of date juice by the yeast *Saccharomyces cerevisiae* using a sugar concentration of 200 g/l at 30 °C and natural pH (Louhichi et al. 2013).

There are many industries in Tunisia trying to produce some of these secondary products in addition to their main activity of processing dates. Syrup is the major product because of its ease of production. Projects to valorize dates are of particular importance for the national economy. The products have high added value and levels of integration and are important in terms of the extent of investment and job creation. However, marketing and the dietary habits of Tunisians are the main problems limiting the adoption of these products by consumers. Therefore, the major by-products currently are targeted for export and not for local markets. To promote the domestic market for date by-products, government grants to publicize and promote the products and to assist manufacturers would be a positive step.

6.9 Conclusions and Recommendations

At present, Tunisian oases enjoy an economic and social stability. However, they are threatened by unsustainability aggravated by climate change, overexploitation of water resources, and an increasing trend toward Deglet Noor cv. monoculture. Current research on oasis agriculture does not adequately address priority areas such as water needs and protection of date palms against diseases and pests which present the risk of biodiversity loss. Given these problems, comprehensive programs must be mobilized, to include: (a) halting of date-cultivation expansion in areas where water resources are overexploited; (b) enhanced proficient use of water resources; (c) improved understanding of the irrigation water needs of specific oasis crops; (d) collection, evaluation, and conservation of plant genetic resources; (e) improved opportunities for in vitro techniques in order to contribute to the propagation of cultivars under threat of extinction; and (f) strengthening of the role of general date palm research. Dates have always suffered from postharvest losses due to deformation, microorganism contamination, and carob moth and other insect infestations. Fumigation by methyl bromide is an effective method of protecting dates from the moth but its use will be prohibited in 2015. As a consequence, it is necessary to search for alternatives treatments. Valorization of lost biomass in the food industry and in bioenergy is also a significant issue which requires in-depth research.

References

Abbès F, Bouaziz MA, Blecker C et al (2011) Date syrup: effect of hydrolytic enzymes (pectinase/cellulase) on physicochemical characteristics, sensory and functional properties. LWT – Food Sci Tech 44:1827–1834

Abbès F, Kchaou W, Blecker C et al (2013) Effect of processing conditions on phenolic compounds and antioxidant properties of date syrup. Ind Crop Prod 44:634–642

Aldryhim Y (2008) Dubas bug. In: Capinera JL (ed) Encyclopedia of entomology, 4th edn. Springer, Dordrecht, pp 1254–1256

Baaziz M, Majourhat K, Bendiab K (2000) Date palm culture in the Maghreb countries: constraints and scientific research. Proceedings of the date palm international symposium, Windhoek, 22–25 Feb, pp 306–311

Baraud J (1985) Coléoptères Scarabaeidae. Faune du nord de l'Afrique, du Maroc au Sinaï. Paris France. Editions Paul Lechevalier. Encyclopédie Entomologie XLVI:652

Bchini H (2006) Quelques critères morphologiques de sélection indirecte des pollinisateurs a effet metaxénique chez la variété de palmier dattier 'Deglet Nour' de Tunisie (French). Plant Gen Res Newsl (IPGRI/FAO) 145:46–60

Ben Abdallah A, Stiti K, Lepoivre P, Du Jardin P (2000) Identification de cultivars de palmier dattier (*Phoenix dactylifera* L.) par l'amplification aléatoire d'ADN (RAPD). Cah Étud Recher Francoph/Agr 9:103–107

Ben Chaaban S, Chermiti B, Kreiter S (2011) *Oligonychus afrasiaticus* and phytoseiid predators' seasonal occurrence on date palm *Phoenix dactylifera* L. (Deglet Noor cultivar) in Tunisian oases. Bull Insect 64(1):15–21

Ben Salah M (1993) Description phénopomologique de 13 cultivars de palmier dattier des oasis tunisiennes. Rev Reg Arid 5:3–22

Ben Salah M (2012) Rapport d'expertise technique sur la biodiversite oasienne en Tunisie. RADDO & ASOC, Gabès

Ben Salah M, Hellali R (2004) Description phénopomologique et la distribution des variétés de palmier dattier (*Phoenix dactylifera* L.). Rev Reg Arid (spec num):64–70

Ben Salah R, Jaouadi B, Bouaziz A et al (2011) Fermentation of date palm juice by curdlan gum production from *Rhizobium radiobacter* ATCC 6466: purification, rheological and physico-chemical characterization. LWT – Food Sci Tech 44:1026–1034

Benkhalifa A (1999) Gestion de la diversité génétique du palmier dattier en Algérie. Paper presented at the workshop constitution et organisation d'équipes de recherche scientifique dans les domaines de foresterie et des arbres fruitiers, Marrakech, 13–15 Apr

Besbes S, Cheikh-Rouhou S, Blecker C et al (2006) Voies de valorisation des pulpes de dattes. Microbiol Hyg Aliment 18:3–7

Besbes S, Drira L, Blecker C et al (2009) Adding value to hard date (*Phoenix dactylifera* L.): compositional, functional and sensory characteristics of date jam. Food Chem 112:406–411

Billotte N, Marseillac N, Brottier P et al (2004) Nuclear microsatellite markers for the date palm (*Phoenix dactylifera* L.): characterization and utility across the genus *Phoenix* and in other palm genera. Mol Ecol Notes 4:256–258

Borchani C, Besbes S, Blecker C et al (2010) Chemical properties of 11 date cultivars and their corresponding fiber extracts. Afr J Biotech 9:4096–4105

Borchani C, Besbes S, Masmoudi M et al (2012) Influence of oven-drying temperature on physicochemical and functional properties of date fiber concentrates. Food Bioproc Tech 5:1541–1551

Bornet B, Branchard M (2001) Non anchored inter simple sequence repeat (ISSR) markers: reproducible and specific tools for genome fingerprinting. Plant Mol Biol Report 19:209–215

Chaira N, Mrabet A, Mtaoua H et al (2009) Characterization of date (*Phoenix dactylifera* L.) juices obtained from common varieties grown in Tunisia. Rev Rég Arid (spec num):69–76

Chebbi H (2011) First record of *Rhynchophorus ferrugineus* on *Phoenix canariensis* in Tunisia. Tunis J Plant Prot 6(2):149–153

Dhouibi MH (1991) Les principaux ravageurs des palmiers dattier et de la date en Tunisie.- Institut National Agronomique de Tunisie, Laboratoire d'Entomologie-Ecologie, Tunis

Djerbi M (1995) Précis de phoeniciculture. FAO, Rome

Dowling TE, Moritz C, Palmer JD, Rieseberg LH (1996) Nucleic acids IV: analysis of fragments and restriction sites. In: Hillis DM, Moritz C, Mable BK (eds) Molecular systematics. Sinauer Associates, Sunderland, pp 249–320

Drira N, Benbadis A (1985) Multiplication végétative du palmier dattier (*Phoenix dactylifera* L.) par réversion en culture in vitro d'ébauche florale de pieds femelle. J Plant Phys 119:227–235

Elleuch M, Besbes S, Roiseux O et al (2008) Date flesh: chemical composition and characteristics of the dietary fibre. Food Chem 111:676–682

FAO/IAEA (2001) Plant Breed Genet Newsletter

Ferchichi A, Hamza H (2008) Le patrimoine génétique phoenicicole des oasis continentales tunisiennes. Institut des Régions Arides, Medenine

Fki L, Masmoudi R, Drira N, Rival A (2003) An optimised protocol for plant regeneration from embryogenic suspension cultures of date palm, *Phoenix dactylifera* L., cv. Deglet Nour. Plant Cell Rep 21(6):517–524

Fki L, Bouaziz N, Kriaa W et al (2011a) Multiple bud cultures of 'Barhee' date palm (*Phoenix dactylifera*) and physiological status of regenerated plants. J Plant Phys 168(14):1694–1700

Fki L, Masmoudi R, Kriaa W et al (2011b) Date palm micropropagation via somatic embryogenesis. In: Jain SM, Al-Khayri JM, Johnson DV (eds) Date palm biotechnology. Springer, Dordrecht, pp 47–68

Fki L, Bouaziz N, Chkir O et al (2013) Cold hardening and sucrose treatment improve cryopreservation of date palm meristems. Biol Plant 57(2):375–379

Ghazouani W, Marlet S, Mekki I, Vidal A (2009) Farmers' perceptions and engineering approach in the modernization of a community-managed irrigation scheme. A case study from an oasis of the Nefzawa (South of Tunisia). Irr Drain 58:285–296

Groupement Interprofessionnel des Fruits (GIF) (2011) http://www.gifruits.com/

Hajji N (1997) The modernization of Tunisian oases: achievements and challenges. In: Jouve A (ed) The modernization of Mediterranean agricultural systems (in memory of Pierre Coulomb), vol 29, Options Méditerranéennes: Série A. Séminaires Méditerranéens. CIHEAM-IAMM, Montpellier, pp 257–264

Hamza H, Mokhtar E, Mahmoudi K et al (2006) Mesure de l'état de la biodiversité dans les oasis de Kébili. Actes du séminaire international: Gestion des ressources et applications biotechnologiques en aridoculture et cultures oasiennes: Perspectives pour la valorisation des potentialités du Sahara. Djerba (Tunisie) 25–28 Dec 2006. Revue des Régions Arides (spec num):194–200

Hamza H, Rejili M, Elbekkay M, Fercichi A (2009) New approach for the morphological identification of date palm (*Phoenix dactylifera* L.) cultivars from Tunisia. Pak J Bot 41:2671–2681

Hamza H, Vendramin GG, Ferchichi Ali (2010) Etude de la diversité de la population du palmier dattier (*Phoenix dactylifera* L.) dans les oasis continentales tunisiennes à l'aide des marqueurs microsatellites. Actes du 3ème Meeting International: "Aridoculture et Cultures Oasiennes: Gestion et Valorisation des Ressources et Applications Biotechnologiques dans les Agrosystèmes Arides et Sahariens". Djerba (Tunisie) 15-16-17/12/2009. Rev Rég Arid (spec num):7–11

Hamza H, Elbekkay M, Ben Abederrahim MA, Ferchichi A (2011) Molecular and morphological analyses of date palm (*Phoenix dactylifera* L.) subpopulations in southern Tunisia. Span J Agric Res 9(2):484–493

Hamza H, Benabderrahim MA, Elbekkay M et al (2012) Investigation of genetic variation in Tunisian date palm (*Phoenix dactylifera* L.) cultivars using ISSR marker systems and their relation with fruit characteristics. Turk J Biol 36:449–458

Hodgson RW (1932) La culture du dattier en Tunisie. Bulletin de la Direction Générale de l'Agriculture, du Commerce et de la Colonisation; 4ième trimestre 1932

Jemni M, Maaroufi A, Mejri S (2010) Optimisation d'un milieu de culture à base de sirop de dattes pour la production d'un bio pesticide bactérien. Rev Règ Arid 1(spec num):261–267

Jemni M, Otón M, Ramirez JG et al (2014) Conventional and emergent sanitizers decreased *Ectomyelois ceratoniae* infestation and maintained quality of date palm after shelf-life. Postharvest Biol Technol 87:33–41

Kearney TH (1906) Date varieties and date culture in Tunisia. USDA Bureau of Plant Industry Bulletin, No 92, Washington

Khoualdia O, Rhouma A, Hmidi MS (1993) Contribution to the bio-ecological study of the white scale *Parlatoria blanchardi*. Targ (Homoptera, Diaspididae) of date palm in Djerid (southern Tunisia). Ann Inst Nat Rech Agron Tun 66:89–108

Khoualdia O, Rhouma A, Marro JP, Brun J (1997) Premières observations sur *Oryctes agamemnon* (Col.:Scarabidae), nouveau ravageur du palmier dattier en Tunisie. Fruits 52:111–115

Louhichi B, Belgaib J, Benamor H, Hajji N (2013) Production of bio-ethanol from three varieties of dates. Renew Ener 51:170–174

Louvet J, Toutain G (1973) Recherches sur les fusarioses VIII. Nouvelles observations sur la fusariose du palmier dattier et précisions concernant la lutte. Ann Phytopath 4:35–52

Masmoudi-Allouche F, Chaâri-Rkhis A, Kriaâ W et al (2009) *In vitro* hermaphrodism induction in date palm female flower. Plant Cell Rep 28:1–10

Masmoudi-Allouche F, Meziou B, Kriaâ W et al (2010) *In Vitro* flowering induction in date palm (*Phoenix dactylifera* L). J Plant Growth Regul 29:35–43

Masmoudi-Allouche F, Gargouri-Bouzid R, Kriaâ W et al (2011) Investigation of genetic variability related to the *in vitro* floral hermaphrodism induction in date palm (*Phoenix dactylifera* L.). Afr J Biotech 10(52):10567–10574

MEDD (2010) Étude sur la gestion durable des écosystèmes oasiens. Ministère de l'Environnement et du Développement Durable. September 2010

Munier P (1973) Le palmier dattier, Techniques agricoles et productions tropicales. Maisonoeuvre et Larose edition, Paris

Murphy ST, Briscoe BR (1999) The red palm weevil as an alien invasive: biology and the prospects for biological control as a component of IPM. Biocon News Infor 20:35–46

Nagaoka T, Ogihara Y (1997) Applicability of inter-simple sequence repeat polymorphisms in wheat for use as DNA markers in comparison to RFLP and RAPD markers. Theor Appl Genet 94:597–602

Namsi A, Montarone M, Serra P et al (2007) Manganese and brittle leaf disease of date palm trees. J Plant Pathol 89(1):125–136

ODS (2011) Le Sud Tunisien en chiffres

Oihabi A (2000a) Date palm in North Africa and Near East: production and constraints. The date palm international symposium, Windhoek, Namibia, pp 312–313

Oihabi A (2000b) Technical report: date palm genetic resources in North Africa. The date palm international symposium, Windhoek, Namibia, pp 333–337

Omrani N, Burger D (2010) Improvement of irrigation efficiency in date palm oases: context and perspectives, case study Kebili, southern Tunisia. In: Meyer R, Burger D (eds) Low-input intensification of developing countries' agriculture – opportunities and barriers. Proceeding of the Kit-workshop, Karlsruhe, 8th Dec 2010, pp 103–114

OSS (2009) Observatory of the Sahel and the Sahara, the system aquifer of septentrional Sahara, the common management of a transborder basin. Serie of drafts, N°1

Othmani A, Bayoudh C, Drira N, Trifi M (2009) In vitro cloning of date palm *Phoenix dactylifera* L., cv. Deglet Bey by using embryogenic suspension and temporary immersion bioreactor (TIB). Biotechnol Biotec Eq 23: 1181–1188

Othmani A, Rhouma S, Bayoudh C et al (2010) Regeneration and analysis of genetic stability of plantlets as revealed by RAPD and AFLP markers in date palm (*Phoenix dactylifera* L.) cv. Deglet Nour. Int Res J Plant Sci 1(3):48–55

Othmani A, Mzid R, Bayoudh C et al (2011) Bioreactors and automation in date palm micropropagation. In: Jain SM, Al-Khayri JM, Johnson DV (eds) Date palm biotechnology. Springer, Dordrecht, pp 119–136

Palevsky E, Ucko O, Peles S et al (2004) Evaluation of control measures for *Oligonychus afrasiaticus* infesting date palm cultivars in the Southern Arava Valley of Israel. Crop Prot 23(5):387–392

Rahalkar GW, Harwalkar MR, Rananavare HD et al (1985) *Rhynchophorus ferrugineus*. In: Singh P, Moore RF (eds) Handbook of insect rearing, vol 1. Elsevier, New York, pp 279–286

Rhouma A (1994) Le palmier dattier en Tunisie I. Le patrimoine génétique, vol 1. IPGRI, Rome

Rhouma A (2005) Le palmier dattier en Tunisie I. Le patrimoine génétique, vol 2. IPGRI, Rome

Rhouma S (2008) Analyse de la diversité génétique chez le palmier dattier (*Phoenix dactylifera* L.) Étude transcriptomique de la maladie des feuilles cassantes. Université Tunis-El Manar, Tunis

Rouissi M (1969) Le fait migratoire au Djérid. Revue Tunisienne des Sciences Sociales. CERES. Actes du Colloque sur la démographie maghrébine

Sakka H, Zehdi S, Ould Mohamed Salem A et al (2004) Genetic polymorphism of plastid DNA in Tunisian date-palm germplasm (*Phoenix dactylifera* L.) detected with PCR-RFLP. Genet Resour Crop Evol 51:479–487

Sghaier S, Bahloul M, Gargouri Bouzid R, Drira N (2008) Development of zygotic and somatic embryos of *Phoenix dactylifera* L. cv. Deglet Nour: comparative study. Sci Hortic 116(2):169–175

Sghaier M (2010) Etude de la gouvernance des ressources naturelles dans les oasis, Cas des oasis en Tunisie. Union Internationale pour la Conservation de la Nature

Soltani R (2004) *Oryctes agamemnon* Burmeister (Coleoptera, Scarabaeidae): contribution à l'étude de sa biologie et de ses dégâts dans les oasis du Djérid au sud tunisien. Diplôme d'Etudes Approfondies. Ecole Supérieure d'Horticulture et d'Elevage de, Chott-Mariem

Soltani R (2010) The rhinoceros beetle *Oryctes agamemnon arabicus* in Tunisia: current challenge and future management perspectives. Tunis J Plant Prot 5(2):179–194

Trifi M, Rhouma A, Marrakchi M (2000) Phylogenetic relationships in Tunisian date palm (*Phoenix dactylifera* L.) germplasm collection using DNA amplification fingerprinting. Agron 20:665–671

Triki MA, Zouba A, Khoualdia O et al (2003) "Maladie des feuilles cassantes" or brittle leaf disease of date palms in Tunisia:biotic or abiotic disease? J Plant Pathol 85(2):71–79

Zaid A, de Wet PF, Djerbi M, Oihabi A (2002) Diseases and pests of date palm. In: Zaid A, Arias-Jimenez E (eds) Date palm cultivation, vol 156, FAO plant production and protection paper., pp 227–281

Zehdi S, Trifi M, Ould Mohamed Salem A et al (2002) Survey of inter simple sequence repeat polymorphisms in Tunisian date palms (*Phoenix dactylifera* L.). J Genet Breed 56:77–83

Zehdi S, Trifi M, Billotte N et al (2004) Genetic diversity of Tunisian date palms (*Phoenix dactylifera* L.) revealed by nuclear microsatellite polymorphism. Hered 141:278–287

Zehdi-Azouzi S, Rhouma S, Dkkhil-Dakhlaoui S et al (2011) Polymorphism and genetic relationship in date palm using molecular markers. In: Jain SM, Al-Khayri JM, Johnson DV (eds) Date palm biotechnology. Springer, Dordrecht, pp 407–425

Zouba A, Raeesi A (2010a) First report of *Ommatissus lybicus* Bergevin (Hemiptera: Tropiduchidae) in Tunisia. Afr J Plant Sci Biotech (spec iss):98–99

Zouba A, Raeesi A (2010b) First report of *Raoiella indica* Hirst (Acari:Tenuipalpidae) in Tunisia. Afr J Plant Sci Biotech (spec iss):100–101

Chapter 7
Date Palm Status and Perspective in Libya

Massimo Battaglia, Bashir Ghsera, Marta Mancini, Carlo Bergesio, Alessandro Camussi, and Milvia Luisa Racchi

Abstract Libyan date palm germplasm represents a heritage of great value both for agriculture and the environment. This chapter analyzes all the aspects of the date palm status in Libya starting from the historical aspect to the current status as well as the results of the recent International Union for the Protection of New Varieties of Plants cooperative program between Italy and Libya for improving and promoting date palm production in Libya, which at present is quite low. All the production is consumed locally, and the cultivation techniques and the processing industry are outdated and need to be modernized. The program strategy was driven by two main objectives: first, identification of high-quality dates through production protocols that ensure the consistency and quality of the final product; second, protection of the agrobiodiversity by promoting local date palm cultivars and strengthening traditional oasis management systems. For this purpose 18 Libyan cultivars, representing common genotypes in the central Libyan oasis of Al Jufrah, were studied in detail both from a morphological and genetic point of view. Cultivar descriptions were carried out on the basis of passport descriptors according to international standards to facilitate germplasm passport information exchange together with genetic fingerprinting. This was performed using 16 highly polymorphic simple sequence repeat (SSR) loci, which allowed setting up an efficient and unambiguous identification system. Clonal fingerprinting and cultivar identification are important elements for promoting typical local products and for linking a product to its place of origin. In this context, the Libyan germplasm represents an enormous richness that

M. Battaglia (✉) • C. Bergesio
Istituto Agronomico per l'Oltremare, IAO-MAE, Via Cocchi 4, 50131 Florence, Italy
e-mail: battaglia@iao.florence.it; carlo.berge@hotmail.it

B. Ghsera
Department of Plant Protection, Faculty of Agriculture, El Fatah University, Tripoli, Libya
e-mail: dr.bashlrgshera@yahoo.com

M. Mancini
Slow Food Foundation for Biodiversity, Bra, Cuneo, Italy
e-mail: martamancini@yahoo.it

A. Camussi • M.L. Racchi
Department of Agri-Food Production and Environmental Sciences, Genetics Unit, University of Florence, Florence, Italy
e-mail: alessandro.camussi@unifi.it; milvia.racchi@unifi.it

J.M. Al-Khayri et al. (eds.), *Date Palm Genetic Resources and Utilization: Volume 1: Africa and the Americas*, DOI 10.1007/978-94-017-9694-1_7

deserves to be exploited. Its characterization and valorization open new prospects for date palm breeding.

Keywords Agrobiodiversity • Al Jufrah Oasis • Cultivar • Date palm • Desertification • Libya • Molecular genetic diversity • Pomology

7.1 Introduction

The date palm, *Phoenix dactylifera* L., a dioecious perennial monocotyledonous species indigenous to the Middle East and perhaps to North Africa, is one of the world's first cultivated fruit trees (Govaerts and Dransfield 2005; Wrigley 1995). It was domesticated in Mesopotamia more than 5,000 years ago, and it has long been one of the most important plants of arid areas of northern Africa, the Middle East, and South Asia (Sauer 1993). The date palm has been called *the tree of life* because, in harsh environment where relatively few other plants can grow, it provides food as well as material for shelter, fiber, and fuel and it has been used for religious purposes (Carpenter 1975, 1981; Dowson 1982; Nixon and Carpenter 1978; Popenoe 1973). The Arabs have a saying, *the palm is our dear mother (al-umm al-hanuna).* Only its fruits, wood, and leaves make life in the desert possible. Without date palms, people could not survive in such a harsh environment. Whole civilizations have developed due to the date palm and Muslim populations living in the desert areas attach great importance to this tree and its fruit (Danthine 1937).

The 19th Surah of the Qu'ran tells of the birth of Isa (Jesus) and describes an image full of divine tenderness: the Lord, Allah, comes to the aid of Maryam (Mary) and eases her suffering during childbirth with the fresh, ripe fruits of the date palm:

> And the pains of childbirth drove her to the trunk of a palm-tree: She cried (in her anguish): "Ah! would that I had died before this! would that I had been a thing forgotten and out of sight!" But (a voice) cried to her from beneath the (palm-tree): "Grieve not! for thy Lord hath provided a rivulet beneath thee; "And shake towards thyself the trunk of the palm-tree: It will let fall fresh ripe dates upon thee. So eat and drink and cool (thine) eye. And if thou dost see any man, say, 'I have vowed a fast to ((Allah)) Most Gracious, and this day will I enter into not talk with any human being'." [Surah Maryam, 23–26, translation by Yusuf Ali]

The Muslim tradition has also passed down some Hadiths, sayings of the Prophet, which confirm the importance attributed to the date palm in the life of the universe and humans.

The Prophet Muhammad says that when the end of the world will come, whoever has a date palm offshoot in their hand must plant it, to ensure the continuity of life. When a woman gives birth, the first food that the newborn should eat is a date, because the first to do so was the Prophet Muhammad. In a house where there are no dates, it is said, the inhabitants will suffer from hunger.

Humans have spread the date palm far beyond its historical range, taking it to nearly all tropical and subtropical areas of the world (Zaid and de Wet 2002).

Recent research has revealed how the cultivated date palm is closely related to wild and feral date palm in the Near and Middle East and North Africa. These wild date palms are morphologically similar to domesticated cultivars, share the same climatic requirements, and can hybridize with the cultivars.

From a botanical point of view, wild and feral dates are considered the same species (Zohary and Hopf 1993). Wherever date palm is cultivated for fruit, a clear distinction is made between the traditional cultivars called *varieties* and cultivars that derive from a unique descended individual of seed, cloned thereafter by vegetative multiplication of offshoots.

Currently, cultivation techniques and the processing industry of dates in Libya are underdeveloped in the modern sense; consequently, date fruit production is quite low and entirely consumed locally. This chapter analyzes all aspects of the date palm status, starting from the past to the present day and provides the results of a cooperation program between Italy and Libya to improve and promote date palm cultivation and fruit production in Libya.

7.1.1 Historical Aspect of Date Palm Cultivation in Libya

Since ancient times, date palm cultivation has been widely practiced in Libya and has had a significant role in the livelihoods of the desert and semidesert areas. Early indications of date growing are known from the fifth century BC as reported by Herodotus who writes that the palms of Awjilah Gialo "proceeding westward, I meet the Nasamones who leave their cattle on the coast during the summer and go up the country to a place in the desert called Augila."

At that time the coastal areas of Libya enjoyed a much more favorable climate than at present. Pliny the Elder (77 AD) reports information gathered by the Roman military expedition in the Sahara led by Cornelius Balbus in the Fezzan and further south in Hoggar, Tassili, and confirms the information of Herodotus, indicating the existence of (date) palm trees in these areas to protect the desert.

The introduction and enhancement of the palms in Libya has benefited by cultivation and irrigation techniques of Egypt where date palm cultivation was developed earlier; many Libyan cultivars, in fact, can be considered of Egyptian origin. Favorable conditions existed for date growing in Libya; Scarin (1938), an Italian observer traveling in Western Libya, stated that *the basin has inexhaustible aquifers, at various depths*.

This kind of bold confidence would be misplaced today. Sokna, in southwestern Libya, still has the best water reserves, although the water table has dropped from 3–5 m to 150–200 m, while for more than 20 years, Waddan has been drawing on nonrenewable fossil water from depths of 1,500–2,200 m. This sulfurous water flows out of the ground at a temperature of over 70 °C and under very high pressure. It is allowed to cool in reservoirs before being mixed with fresher water and pumped to the fields for agricultural irrigation.

Historical photographs from the IAO (Istituto Agronomico per l'Oltremare) photograph archive representing date palm groves in the 1930s are shown in Figs. 7.1, 7.2, and 7.3.

Fig. 7.1 Libya, 1938. Old advertisement using an oasis background (IAO historical photograph archive)

Fig. 7.2 Libya – 1924. Farmers going to the market (IAO historical photograph archive)

Fig. 7.3 Libya – 1938. Date palm cultivation (IAO historical photograph archive)

7.1.2 Geographical Distribution of Date Palm in Libya

7.1.2.1 Coastal Area

Along the Libyan coast, north of parallel 32° N, the best date palm groves are found in the areas of Tripoli, Janzur of Zuara, Homs, Zliten, Misurata, and Tawurgha and Hisha (Fig. 7.4), where there is an abundance of water and reasonably good cultivation practices are followed. Date palm groves are also present to the north of Benghazi. Away from the coast, date palms are only sporadic and occur in the vicinity of wells or springs. Only a very limited percentage of the date palm trees can be considered productive.

For practical purposes, and to the contrary of the cultivation of the Saharan oases, it should be noted that because of the influence of marine moisture combined with lower temperatures, dates produced along the coast are of poor quality, have a high moisture content, and are not very sugary; they are consumed quickly and exclusively as fresh fruit. The main cultivars are Bukerary, Taboni, Lamsy, Blonde, Halaway, Bronzi, and Baudi.

7.1.2.2 Central Area

Located between 30° and 27° N lat., the Central Intermediate Zone includes oases of predesertic areas that run along the 29th parallel north, including Ghadames, Sokna, Hun, Waddan, Zellah, Al Fugha, Maradah, Jalo, Awjilah, and Giarabub (Fig. 7.4), where the best cultivars are grown. This area represents the most favorable climatic conditions for date palms in Libya.

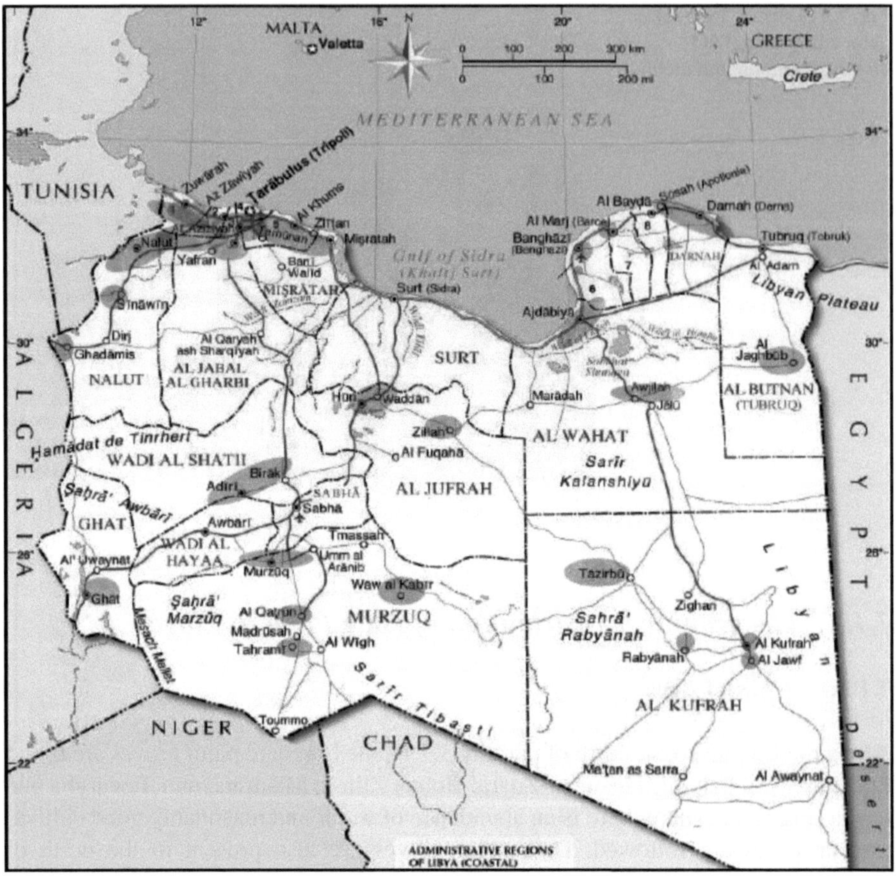

Fig. 7.4 The date palm productive areas in Libya (*shaded areas*)

The climatic conditions are represented by thermal units (°C/day) fluctuating between 1,944 ± 124 of Jalo and 1,569 ± 138 of Al Jufrah, by a relative humidity of 40–50 %, and a low rainfall of 10 mm during the period from August to October. The main cultivars in this area are Abel, Bestian, Deglet, Halima, Hamria, Kathari, Tagiat, and Saiedi.

7.1.2.3 Southern Area

South of the 27° N parallel, in the southern part of Libya, there are a series of Saharan oases which include Fezzan, Ghat, Sabha, Murzuk, Kufra, and Tazerbo (Fig. 7.4), where date palm cultivation has a different development in terms of production and quality.

The dates of these oases have a very high sugar content, which is usually above 70 %, and only rather limited moisture content; therefore, they are suitable for long-term storage, providing they are protected against insects and are well packaged. The dates of this third area are mainly of the dry type. The main cultivars are represented by Amjog, Emeli, Awarig, Tascube, Intalia, and Idaw.

7.1.3 Water Status in the Desert of Central Libya

The Al Jufrah region lies to the north of central Libya, around 200 km from Sirte on the coast. Al Jufrah oasis constitutes three adjacent oases (Sokna, Hun and Waddan) within a radius of aprox. 40 km and two smaller oases (Al Fugha and Zellah) approx. 200 km southeast and southwest, respectively, from the main nucleus. The presence of significant water resources and loose soil in the region has allowed the growth of different groups of date palm trees.

The large Al Jufrah basin (from the Arabic *jof*, meaning belly or hollow) stretches from west to east and is bordered to the south by the spurs of the volcanic Jebel Soda and the basaltic Black Mountains; to the northwest by the eroded slopes of the Jebel Machrigh; to the northeast by the Jebel Waddan, the Waddan Mountains; and to the east by the Harugi Mountains. The desert landscape is much more than just dunes and sand. Here it is a vast flat expanse of gravel (*serir*) or quite large pebbles (*hammada*), interrupted by hills shaped like truncated cones, smoothed by millions of years of erosion since their formation. The land is furrowed by *wadis*, riverbeds now permanently dry but once able to fill with water in just a few moments, becoming dangerous when sudden, heavy rains would cause flash flooding.

From above, the network of *wadis* looks like a maze of lines intersecting the whole landscape, their course easily traced by the vegetation concentrated along them, mostly made up of tamarisk and African acacia. Tree roots extend deep into the ground in search of water, and their leaves have become spines to limit moisture loss and for protection against animals desperate to feed on their greenery. The acacias are so hardy that on average they live for 200 years. The average elevation of Al Jufrah is around 220 m, but there is some difference (around 60 m) between the area of Sokna and the eastern part of the plateau, caused by ancient movements of the earth's plates. Springs flow around Sokna and feed the rest of the region, making it the most precious water reservoir in Al Jufrah.

The abundance of water just a few meters below the surface has permitted the cultivation of date palm trees, which in the past were rarely irrigated. There were 80,000–88,000 palms in Al Jufrah in the 1930s. Sadly, the Sokna plantations, probably the largest with around 35,000 trees, were partially destroyed during the constant clashes between Arabs and Berbers, who had one of their strongholds here. In fact the story of Al Jufrah has always been marked by the ongoing conflict between various ethnic groups with raids by nomads on more settled peoples, possessors of the region's only wealth: date palm groves and vegetable gardens.

Wild date palm groves, growing without human intervention, draw out water with their roots and still survive today, after hundreds of years. Whereas, the selected cultivars of date palms intentionally planted in groves are irrigated so that they fruit more abundantly and produce softer, more succulent fruits. While production from an irrigated tree can reach up to 80–100 kg, a nonirrigated palm produces on average 15–20 kg of fruit. The palms can be irrigated using the tradition system, through a network of channels made of soil surrounding every plant, or with modern drip irrigation, which uses much less water. The drip technique strategy can be used during the tree's initial growth.

The local people recall that up until the 1960s, the slopes of Jebel Soda were green with vegetation and, like the entire surrounding area, would become covered with grass as soon as it rained, allowing inhabitants of the nearby villages to raise animals other than camels. Today, however, the landscape has dried up, and 1952 remains imprinted in local memory. That year the peasants of Sokna rushed to the slopes of the Black Mountains after the extraordinarily abundant rains allowed them to plant even wheat and barley. In the new millennium, rains of a short period of time now come only every 3–7 years.

7.2 Cultivation Practices

Presented here are the best practices identified and shared with technical partners and date palm producers in the central area of Libya, during the realization of the program: Improvement and Valorization of Date Palm in Al Jufrah Oasis (Mancini 2010; Slow Food Foundation 2010). Some of the general information in this chapter is drawn from Mancini (2010).

Date palm production takes place in oasis zones whose arable soils are primarily sandy, and therefore highly permeable, and whose agricultural use is strongly influenced by the availability of irrigation. Water is drawn from strata of variable depths through wells and distributed using modern or traditional systems depending on the age of the plantation.

7.2.1 Plantation Establishment

Date palms are propagated through offshoots naturally produced by female and male adult plants. In order to be useful for a new planting, an offshoot must have a weight of 10–15 kg and a base diameter of at least 20–25 cm and no more than 35 cm when removed from the mother plant. Offshoots for propagation are taken between March and May. They are removed from the mother plant, generally after 4–5 years from planting depending on the cultivar, when the tips of the second leaf from the bottom of the offshoot start to dry up.

Offshoot planting in the ground must be performed during the same day in the afternoon or, at the latest, the day after they are removed. During cleaning of the offshoots, it is necessary to remove the external leaves and maintain at least five inner leaves by cutting them in half over the leaflets. Then the leaves are tied together with a leaf rachis.

To encourage root formation without weakening the plant, when the offshoot is planted, the foliage surface must be significantly cut back and covered to limit the action of the sun and water loss by transpiration. During this period it is necessary to keep the soil moist to guarantee absorption from the earliest development of the new root system. After planting, the date palm will be unproductive for around 5 years. The first significant production will only come after at least 8 years, and full productive maturity will be reached after around 20–25 years. Under specialized conditions, new plantations are laid out on a square grid with each plant at a minimum spacing of 6 by 6 m and a maximum of 8 by 8 m.

One month before digging holes for planting the offshoots, it is recommended to lay down the irrigation tubes underground between the rows and to irrigate the planting sites to facilitate the transplanting operations and marking the sites for holes (1 × 1 × 1 m) in areas with good sandy soil. Where a hard and calcareous underground layer is present, it must be broken up. It is not advisable to place the palms closer than 6 m from each other.

After planting the offshoot in the hole, it is suggested to raise the leaves off the ground to avoid any inward drainage and entrapment of water which could cause rot at the leaf base. The hole should be filled with the previously removed soil (except stones) and compacted around the palm and creating a basin around the trunk (diameter 1.5 m) to prevent water runoff. A male palm should be planted for every 20–25 female palms. Drip irrigation is recommended.

7.2.2 *Irrigation and Fertilization*

After planting, the soil must be kept constantly wet around the roots for the first 6 weeks, irrigating in the early morning (up to 10 am) and/or in the late afternoon (after 6 pm), providing 100 l/day water per offshoot and 200 l/day for adult palms, from February to October; but only once a week during winter to avoid early flowering.

The use of mechanical equipment to work the soil is very rare. The only soil work involves cutting weeds and repairing the furrows and basins constructed for irrigation (Arara 1975). Weeding is done manually, usually in January and February, before the date palms flower. The soil is fertilized with organic materials. Farms make use of their own manure, preferably well seasoned for at least 6 months, to fertilize the soil where a new palm grove is going to be planted. The use of compost heaps is common (Buys 1993). All plant waste produced by the farm is placed in a ditch, layered with soil and irrigated frequently to encourage the processes of decomposition and mineralization. The compost is then spread around the grove. The amount of fertilizer depends on soil composition and water quality and is best applied during the winter months from December to early March.

Chemical fertilizer such as DAP (diammonium phosphate) or NPK fertilizer (450–500 g/plant) is applied in November around the trunk, while the micronutrient fertilizer LINAFED (200–250 g/plant) is applied in two applications in spring or in early summer with the first application laid down north to south and the second east to west.

7.2.3 Disease and Pest Control

Protection of the date palms does not involve the use of any particular parasitical products, which is partly due to the climactic conditions, which discourage the development of multiple generations of entomophages and the proliferation of fungi (Bitaw and Ben Saad 1990; Edongali 1997; El-Alwani and El-Ammari 2007; Gariani et al. 1994; Martin 1958). During the winter, a natural copper-based antifungal treatment can be applied where necessary. It is possible to use natural parasitoids to organically combat entomophages, which attack the fruit, particularly during the ripening phase.

White scale (*Parlatoria blanchardi*) and mealybug (*Maconellicoccus hirsutus*) are the most dangerous pests of date palm and are effectively controlled with Dursban (150–200 ml/100 l) or Cyperkill 25 EC (150–200 ml/1,000 l) plus mineral oil during winter. The same products can be applied in June to control frond borers and *Ephestia* moths, when fruits start to mature up to 3 weeks before harvest. Damage by white scale is very serious on young palms 2–8 years of age, but even under severe attacks, the palm and its offshoots survive. Nymphs and adults suck the sap from the leaflets, midribs, and dates. Beneath each scale insect, a discolored area is created on the leaflet. Heavy infestation causes leaflets to turn yellow, and respiration and photosynthesis are nearly stopped resulting in early death of the infested leaves. Damage to fruits is easily noticed and makes the fruit unmarketable. The number of insect generations which can develop in 1 year varies from three to four depending on temperature conditions. All chemical treatments must be applied in early morning or late afternoon.

Red palm weevil (*Rhynchophorus ferrugineus*) and bayoud disease, caused by *Fusarium oxysporum* f. sp. *albedinis*, are prevented with quarantine controls on offshoot imports from Egypt, Algeria, or Morocco. From an economic point of view in Libya, damages caused by diseases and pests normally are not serious. This is due to the great biodiversity, the remoteness of the different crops from each other, and the minimal presence of intensive monoculture of cultivars.

7.2.4 Intercropping

The intercropping of alfalfa in date palm plantations has positive effects due to nitrogen fixation, microclimate improvement, and heat reflection/reduction from the soil but has high water requirements. Sowing is realized in early spring or autumn. Other crops like corn, garlic, onion, and other vegetables and cereals such as wheat, barley, and oats can easily be cultivated inside the basin surrounding the tree, which is irrigated regularly. An important benefit of intercropping is that it generates additional income to the farmers, especially in the period before the date palm begins to bear fruit.

7.2.5 *Pollination and Fruit Thinning*

From the end of February to mid-April, depending on temperature and humidity conditions and the cultivar, it is necessary to carry out artificial pollination, using, if possible, the same male cultivar of the female to be pollinated. The process consists of applying pollen to the female inflorescence by hand or aided by a pole once the flowers start opening. Hand pollination is normally repeated three times and depends on whenever the spathe is open and on the skill of the worker. In large modern farms, a compressor is helpful to better distribute the pollen (Wertheimer 1954).

Generally, fruit thinning is realized in May, leaving eight bunches (two for each cardinal point) per palm, choosing bunches in the middle part of the canopy.

7.2.6 *Harvest and Postharvest*

Date palm fruits are harvested from August to November (peak period in September/October) depending on cultivar and maturity stage, with manual selection of the best fruits on the stalk (for premium cultivars) or removal of the entire fruit bunch and selection on the ground. Manual fruit selection takes place until mid-September. Harvest timelines of the most important cultivars in the Libyan oases are indicated in Fig. 7.5.

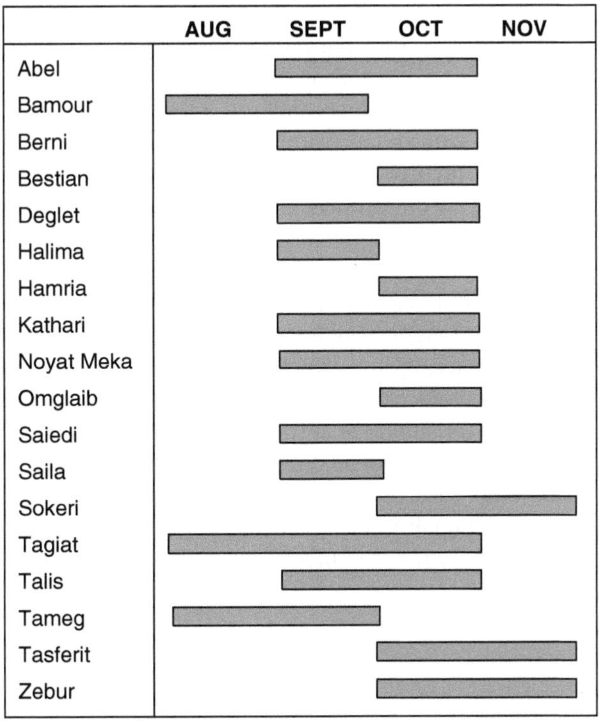

Fig. 7.5 Date palm cultivar harvest timeline in Libyan oases

To climb a date palm tree, which over the years can reach a considerable height, harvesters use the simplest of means: hands and feet. Tough calluses form on the skin after contact with the petiole bases where leaves have been cut from the trunk. Repeated experience shows that the most expert workers can climb even the tallest palms in just a few seconds. Many harvesters use a simple harness made from a strong rope woven from palm fibers, while others use a modern ladder. Some old and very tall palm trees have steps cut into the petiole bases for hand- and footholds, facilitating the climb to the top.

During the first stage of the harvest (September/early October), the harvesters climb each palm several times a week to remove the fruits from bunches as they gradually reach the appropriate ripeness, leaving unripe fruits to continue maturing in the sun. Fruit selection takes place directly on the tree during this first part of the season. In the later part of the harvest stage (end of October), when falling temperatures stop the fruit from ripening further, selection no longer takes place on the tree, but on the ground after the entire fruit bunch is detached and brought down. On the ground, the remaining fruit on the clusters are detached and classified as first, second, or third grades. Rejected fruits are fed to animals or used to produce alcohol in non-Muslim countries. Dates which have not fully ripened on the palm are left in the sun for around a week to mature before being packaged.

The traditional date cultivars of the Al Jufrah oases are primarily sold fresh, and commercialization must follow specific technical procedures to guarantee that the quality and safety of the product is maintained.

Immediately after harvest, it is necessary to wash them in potable water and dry the fruits collected at tamar stage (e.g., cvs. Saiedi, Tagiat, Abel, Bestian, Deglet). They then must undergo a rigorous sorting to remove impurities and/or imperfect fruits and any others unsuitable for sale. The sorting must take place in a hygienic facility to avoid any contamination from pathogens and dirt. Doing such work on the ground is not allowed. After processing, the fruits should be stored at −20 °C.

For direct consumption the dates are packaged in cardboard boxes in layers, ideally no more than three in depth. Packaging also must take place under hygienic conditions. Larger containers can also be prepared for distribution within the domestic catering industry. Packaged dates are kept refrigerated until sold, and a continuous cold chain must be maintained to guarantee refrigeration within the country or for shipment abroad.

The dates can also be dried, either whole or pitted. Drying is done in a hot oven. Also in this case, the fresh product must be kept refrigerated until being processed, to maintain quality and food safety. Pitted dates can also be pressed into a paste, used primarily in the confectionery industry.

7.3 Genetic Resources and Conservation

Some of the best dates in the world originate from Libya, and dozens of different cultivars have been preserved here. This extraordinary heritage is the legacy of centuries of agricultural history, carefully safeguarded by today's farmers to give hope for tomorrow's desert.

Compared to the dates that flood European supermarkets around Christmas, Libyan dates represent a superb celebration of biodiversity, with 95 different cultivars still being grown today. This incredible wealth has served as a highly effective natural defense for Libyan date plantations, which have remained free from pathogen attacks like bayoud disease. Pathogens have already devastated some of the widespread monocultures found in countries like Morocco (Djerbi 1983a, b, 1995).

Libya's date cultivars can be divided into three large groups: the soft fleshy-fruited coastal cultivars, which can be eaten fresh or refrigerated for months (cvs. Bronzi, Taluni, Baudi); the semidry cultivars from the central zone, mostly consumed fresh (cvs. Kathari, Abel, Tagiat); and the dry types from the oases of the south, less succulent and fleshy (cvs. Amjog, Emeli, Awarig, Tascube, Intalia, Tamjog). These latter cultivars are suited for drying and can be stored for up to 10 years, making them highly valued by the caravans that formerly crossed the desert.

Those familiar with the desert can distinguish the different date palm cultivars in various ways: the shape of the palm's foliage, the appearance of the leaves, and the length of the leaf spines and whether bunches of fruit are pendant or more upright. Even a European would immediately be able to see the striking diversity among the fruits when they are laid out next to each other, even before they enjoy the bewildering symphony of flavors on their palate. Kathari dates are greener than the yellowish Sokeri fruit; Tagiats have a tapered shape; the prized Halimas offer an incomparable concentration of pleasure that caresses the mouth without ever becoming cloying.

Libya is rich in sayings and legends associated with dates; one example is the old adage *if you plant Bernis, you'll eat Bernis*, referring to a particularly hardy cultivar able to guarantee food security. The date palm cultivars grown in Libya today are the same as those described by Italians between 1926 and 1930, showing how the impressive local genetic wealth representing in date palms has been wisely maintained and regenerated.

The Libyan government recently launched a major project to encourage, improve, and promote date production, creating new plantings in various desert and semiarid regions of the country, expanding research institutions in the sector, and supporting technical and scientific exchanges, with the aim of genetically, agriculturally, and biologically improving the crop in the arid regions. Single-cultivar plantations are not only more susceptible to possible parasitic epidemics but also at greater risk in

Fig. 7.6 Some of the main Libyan date cultivars of the central oases and their local use (IAO photograph archive)

the event of unusual weather patterns during certain key stages in the plant's life cycle, such as flowering and fruit setting, creating the possibility of serious production losses. Additionally single-cultivar crops are more vulnerable to market fluctuations dictated by the changing consumer preferences. Dates are also processed for the production of products such as fresh jam and syrup for local use (Fig. 7.6).

In other Maghreb countries, the gradual impoverishment of traditional cultivars, whose renewal and conservation is no longer ensured, has already led to the cultivation of a smaller number of selected cultivars. The complete abandonment of local cultivars of traditional crops such as date palm will inevitably lead to the reduction of genetic variability available to the species, a variability which comes from a long sequence of natural selection and constitutes the primary factor for environmental adaptation.

As in other North African countries, drought, salinity, desertification, and the old and less productive palm groves have created problems for Libyan date growers, but farmers recognize the importance of safeguarding the cultivars common at the local level and today have access to a heritage that is extremely valuable for the country's environmental and economic future. Typical date palm plantations in Libya of the present day are illustrated in Figs. 7.7, 7.8, and 7.9.

Fig. 7.7 Productive date palm plantation in a nonirrigated area in Libya (IAO photograph archive)

Fig. 7.8 New modern date palm plantation in an irrigated area in Libya (IAO photograph archive)

Fig. 7.9 A productive date palm plantation in an irrigated area in Libya (IAO photograph archive)

7.4 Plant Tissue Culture

At present in Libya, date palm propagation is by offshoots or seeds. Tissue culture, however, offers considerable advantages by comparison. It is important to initiate relevant research in Libya and adopt tissue culture for the following reasons:

(a) Propagation of healthy elite female cultivars (disease and pest-free).
(b) Large-scale multiplication of elite cultivars.
(c) No seasonal effect on plants because they can be multiplied under controlled conditions in the laboratory throughout the year.
(d) Production of genetically uniform plants.
(e) Clones can be propagated from elite cultivars already in existence or from the F1 hybrids of previous selections and seed-only originated palms.
(f) Insure an easy and fast exchange of plant material between different regions of the country or between countries without any risk of the spread of diseases and pests.
(g) Economically reliable when large production is required.

7.5 Cultivar Description

The morphological and pomological traits of the most important Libyan cultivars were recorded based on passport descriptors adopted by the International Plant Genetic Resources Institute, to establish international standards to facilitate germplasm information exchange (Alercia et al. 2001). Table 7.1 lists 18 important date palm cultivars grown in Libya and their main characteristics; Fig. 7.10 provides illustrations of them. Table 7.2 gives the nutritional composition of fruits of six cultivars which have been analyzed.

Table 7.1 Description of the date palm cultivars grown in Libya

Cultivar	Fruit characters	Pollination and harvest months	Distribution
Abel	Oval fruit, yellow with brown patches, smooth, tough, thick skin, hard flesh with a sweet but astringent flavor. High presence of fibers. Easy to harvest. Consumed at tamar stage	Pollinated in March and harvested in September/October by removal of bunches	Well adapted to Al Jufrah area. Widespread in Sokna, Hun, and Waddan and a limited number in Zellah
Bamour	Oval fruit, honey sometime red with pleated and blistered skin semidry flesh, medium presence of fibers. Consumed in rutab and tamar stages	Pollinated in March and harvested in August to September by selection of dates and by removal of bunches	Rare cultivar. It is only present in Waddan and few in Hun

Table 7.1 (continued)

Cultivar	Fruit characters	Pollination and harvest months	Distribution
Berni	Oval fruit, honey amber with smooth and blistered skin, soft flesh, medium presence of fiber	Pollinated in February/ March and harvested in October by removal of bunches	Rare cultivar of Sokna, Hun, and Waddan
Bestian	Oval fruit, honey with corrugated skin, altered skin color, semidry flesh, low sugar content, (most recommended for diabetes suffers), medium presence of fibers. Good for postharvest processing. Quick aging process. Consumed in tamar stage	Pollinated in February/ March and harvested in September/October by selection of dates and removal of bunches	Common cultivar in Waddan, Sokna, and Hun. It's one of the most widespread cultivars in Al Fugha and Zellah
Halima	Oval and subcylindrical fruit, amber with semidry flesh; the fruit is larger than average and pleasantly sweet without being cloying. Medium presence of fibers. Consumed in rutab and tamar stages. Considered a rare delicacy	Pollinated in March and harvested in September by selection of dates	Rare cultivar of Sokna, Hun, and Waddan
Hamria	Oval fruit, honey with tattooed corrugated skin, soft and semidry flesh. High presence of fibers. Consumed in rutab and tamar stages	Pollinated in February/ March and harvested in October by removal of bunches	Very abundant in Al Jufrah. Particularly in Zellah and Al Fugha
Kathari	Oval fruit, greenish yellow fruit, with tattooed corrugated skin, stubby with a thick hard skin, astringent and soft and semidry flesh, astringent. Though slightly, stays soft throughout the year. High presence of fibers. Consumed in rutab and tamar stages	Pollinated in March and harvested in September/ October by selection of dates and by removal of bunches	Well adapted to Al Jufrah area. It is one of the most widespread cultivars in Sokna, Hun, and Waddan. In Arabic Kathari means *green*
Libyan Deglet	Oval and subcylindrical fruit, amber with shiny skin, soft and brownish amber flesh. Mild sweet taste. Medium presence of fibers. Very suitable for long period conservation. Consumed in rutab and tamar stages	Pollinated in February/ March and harvested from September/ October by selection of dates and later removing the whole bunch	The most valuable date cultivar in Libya. It is one of the most widespread cultivars in Sokna, Hun, and Waddan
Noyat Meka	Ovate fruit, honey and dark brown with smooth skin, semidry flesh, medium presence of fibers. Consumed in rutab and tamar stages	Pollinated in March and harvested September/ October by selection of dates	Rare cultivar of Sokna and Hun

(continued)

Table 7.1 (continued)

Cultivar	Fruit characters	Pollination and harvest months	Distribution
Omglaib	Subcylindrical ovate fruit, honey sometimes red with blistered and pleated skin, semidry to soft flesh. High presence of fibers. Consumed in rutab and tamar stages	Pollinated in March and harvested in September/ October by removal of bunches	Rare cultivar of Sokna, Hun, and Waddan. In Arabic means "Mother of small Hearth"
Saiedi	Elongated oval fruit, translucent dark brown, with a thin tender skin and soft, syrupy flesh. Easy to harvest. Low presence of fibers. Consumed in rutab and tamar stages	Pollinated in March and harvested in September/ October by selection of dates	It is of ancient Egyptian origin but now one of Libya's most important cultivars. Widespread Al Jufrah Oases
Taila	Ovate fruit, honey with corrugated and tattooed skin, soft and semidry flesh, medium presence of fibers. Consumed in rutab stage	Pollinated in February/ March and harvested in August/September by removal of bunches	Present only in Sokna and Hun
Sokeri	Ovate subcylindrical fruit, honey with corrugated tattooed skin, semidry flesh, presence of fibers. Consumed in tamar stage	Pollinated in March and harvested from October/ November by removal of bunches	Common cultivar in Sokna, also exists in Hun and Waddan
Tagiat	Elongated oval fruit, dark brown with a smooth, thick, hard skin and soft flesh. Early and prolonged harvest period. All the maturity stages of the fruit can be used. The best cultivar to prepare paste, using tamar	Pollinated in February/ March and harvested from August/October by removal of bunches	One of the most widespread cultivars in entire locality of Al Jufrah Oases
Talis	Subcylindrical fruit, honey with pleated and corrugated skin, soft and semidry flesh, medium presence of fibers. Consumed in rutab and tamar stages	Pollinated in March and harvested in October by removal of bunches	It is present in Zellah
Tameg	Subcylindrical fruit, honey with pleated and corrugated skin, soft and semidry flesh, medium presence of fibers. Consumed in rutab stage	Pollinated in March and harvested from August/ September by selection of dates	Rare cultivar
Tasferit	Subcylindrical fruit, honey dark brown with blistered skin, semidry flesh, and medium presence of fiber. Consumed in tamar stage	Pollinated in March and harvested in October by removal of bunches	Presence in the dry farm plantations of Sokna, Hun, and Waddan
Zebur	Subcylindrical fruit, honey dark brown with pleated blistered skin, soft to semidry flesh, and medium presence of fiber. Consumed in rutab and tamar stages	Pollinated in March and harvested in October by removal of bunches	Rare cultivar

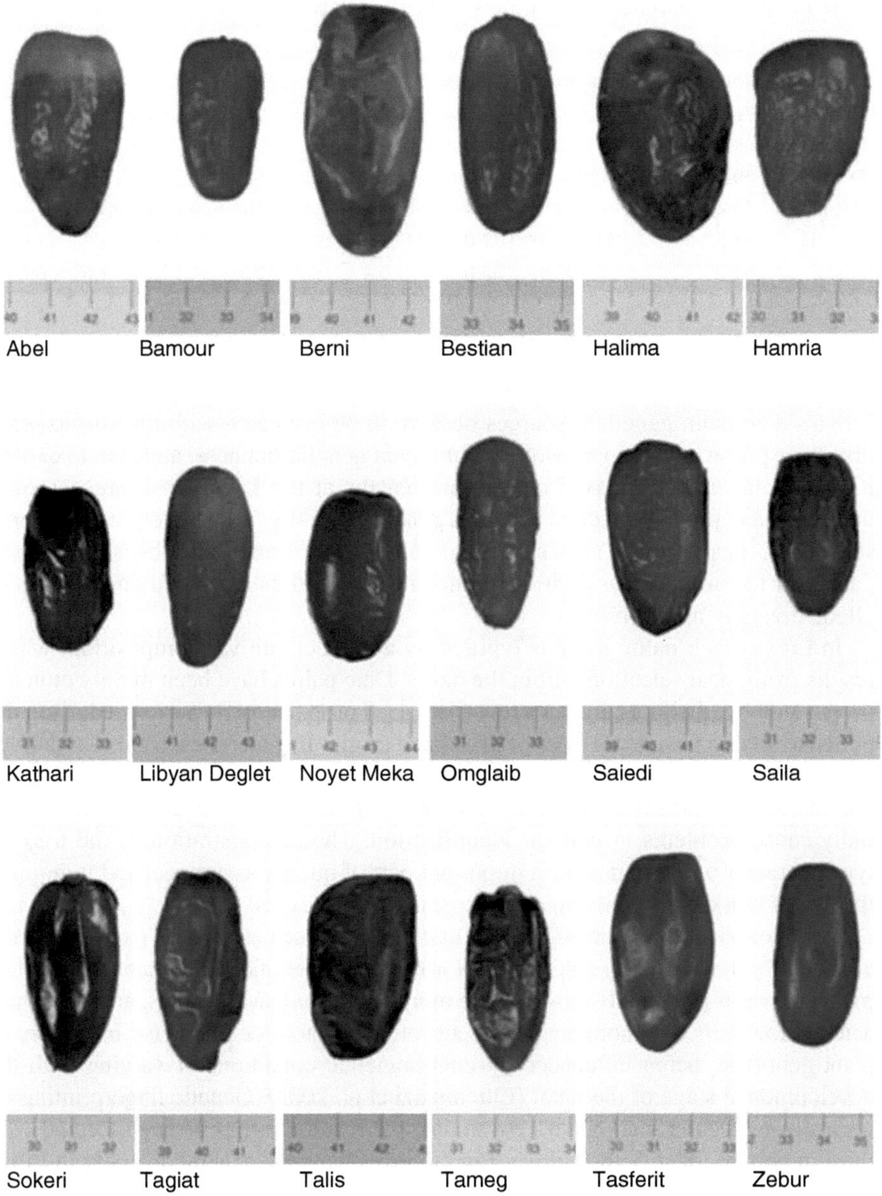

Fig. 7.10 Fruits of various date palm cultivars grown in Libya (Source: (IAO photograph archive))

Table 7.2 Nutritional composition of fruits of six date palm cultivars grown in Libya

	Sugars, %				Ions, mg/kg			
Cultivar	Total sugar	Fructose	Glucose	Sucrose	Mg	*K*	*Z*	Fe
Abel	71.3	35.1	36.0	0.2	594	7,013	3.2	7.6
Bestian	69.8	33.7	36.0	0.1	887	7,752	6.1	11.3
Deglet	69.2	20.3	22.0	26.9	561	7,314	1.7	5.7
Halima	73.4	34.9	38.4	0.1	610	6,010	3.9	7.4
Hamria	73.2	34.2	37.0	2.0	619	7,156	4.3	11.0
Saiedi	65.0	29.6	35.0	0.4	529	6,222	4.6	6.8

7.6 Cultivars Identification Using Molecular Markers

Libya's date palm genetic resources deserve to be evaluated with the aim to organize their preservation, to transmit a significant genetic richness, and also to exploit it. Molecular markers, based on polymorphisms at the DNA level, are currently used and have proved effective to assess genetic diversity. Microsatellites, or simple sequence repeats (SSR), represent a suitable tool for genotyping because of their particular features such as their codominant nature and their typically high levels of allelic diversity at different loci.

In Libya, each palm grove is typified by a distinct cultivar composition, which results from local selection within the oases. Date palms have been mainly clonally propagated by offshoots, in just a few cases seed propagation is performed using the pollen available from male trees of undefined origin. In general each cultivar derives from an individual seed, cloned thereafter by vegetative multiplication to ensure the identity and uniformity of the cultivar. However intra-cultivar variation could potentially cause problems in cultivar identification. The demonstration of the true-to-type character of the plants is an important part of quality assurance, and it requires the use of markers effective in distinguishing the cultivars.

Morphological traits and isoenzyme markers have been used in the past to describe and identify the date palm cultivars of North Africa. Identification of a particular date palm cultivar is principally based on the morphology of leaves, spines, and fruit characters. However, morphological traits are often variable or imprecise indicators of plant genotype, being influenced by environmental conditions or varying with the developmental stage of the plant (Elhoumaizi et al. 2002). Genetic fingerprinting by means of molecular markers of Libyan date palm cultivars has been performed with the aim both to identify the cultivars and to investigate the genetic diversity in Libya to improve production of this crop. For that purpose the parentage analysis of pollinator plants was also attempted to contribute to fruit quality breeding.

Eighteen cultivars, representing common genotypes in Al Jufrah oasis, were selected for their good fruit quality and were analyzed using 16 highly polymorphic microsatellite loci. Plant materials consisted of young leaves of adult trees randomly sampled in the localities of Sokna, Hun, Waddan, Zellah, and Al Fugha. The 18 cultivars are listed in Fig. 7.11, along with the number of female plants sampled in each locality. Dried leaf material was ground into a fine powder and then subjected

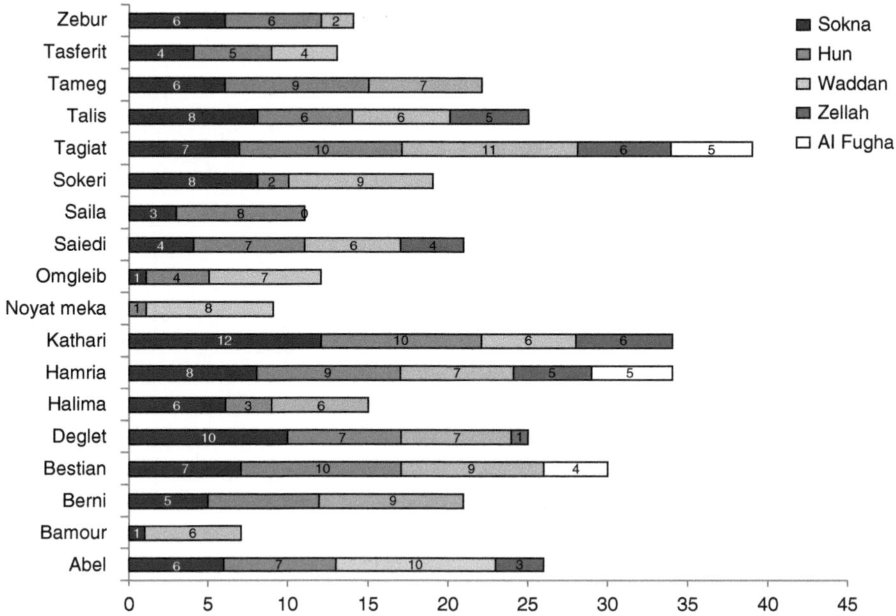

Fig. 7.11 Distribution of cultivar samples among the localities of Al Jufrah oasis, Libya (Modified from Racchi et al. 2013)

to DNA extraction. Sixteen date palm–specific primer pairs were selected for their polymorphic information content among the SSR loci, as developed by Billotte et al. (2004) and Akkak et al. (2009). Amplifications were performed in an Applied Biosystem Thermocycler (AB System, Germany), and PCR products were resolved on a MegaBACE 1000 (GE Healthcare, USA) sequencer (details in Racchi et al. 2013). A large number of SSR alleles were revealed with a mean of 6.88 per locus and allowed to detect a relatively high degree of genetic variability (Table 7.3). A high level of polymorphism was detected among cultivars as previously reported for cultivars in Algeria, Morocco, Tunisian, and Sudan using both isoenzyme and SSR markers (Bennaceur et al. 1991; Elhoumaizi et al. 2006; Elshibli and Korpelainen 2008, 2009; Zehdi et al. 2004a, b).

Each cultivar results from an empirical selection carried out by the farmers in the oases based on morphological characters and fruit quality; this fact justifies the presence at the same time of fixed alleles, 28 out of 110, due to random drift and the high level of heterozygosity due to a clonal breeding procedure for heterosis. Both number and frequencies of alleles vary among the localities due to a different presence of the cultivars in the localities. A good example is represented in Fig. 7.12 by locus mPdCIR10, which exhibits six alleles: the allele 154 is fixed in 9 out of 18 cultivars, while alleles at locus CAT11 are greatly polymorphic in Fig. 7.13. These loci well exemplify the different distribution among the oases; in fact while CAT11 alleles are present in all the oases, some CIR10 alleles are not equally distributed. The mean number of alleles varied from one cultivar to another.

Table 7.3 Microsatellite allelic data revealed by 16 SSR loci in female trees of 18 Libyan date palm cultivars

Locus code	Allelic range (bp)	Total alleles	Number of genotypes	H_{obs}	H_{exp}
PdCIR10	138–176	6	13	0.41	0.46
mPdCIR15	142–157	6	15	0.87	0.77
mPdCIR25	219–257	6	17	0.90	0.76
mPdCIR32	306–321	5	13	0.71	0.66
mPdCIR70	205–227	9	32	0.91	0.83
mPdCIR78	126–173	11	36	0.85	0.85
mPdCIR85	175–199	8	39	0.83	0.85
mPdCIR93	181–197	7	17	0.77	0.77
PDCAT1	103–123	4	10	0.23	0.63
PDCAT2	186–209	7	20	0.85	0.79
PDCAT6	142–172	7	17	0.82	0.71
PDCAT8	222–258	6	14	0.78	0.68
PDCAT11	154–177	6	20	0.75	0.79
PDCAT14	141–163	9	20	0.42	0.63
PDCAT17	131–157	6	14	0.45	0.63
PDCAT18	123–149	8	29	0.88	0.77

Modified from Racchi et al. (2013)
H_{obs} observed heterozygosity, H_{exp} expected heterozygosity at the HW equilibrium

The results, reported in Table 7.4, evidence the different genetic structure of the cultivars. All are characterized by negative values of the fixation index (F) due to an excess of heterozygotes respect to HW equilibrium, though at different level. In particular, cvs. Talis, Halima, Omglaib, Saiedi, Tagiat, Saila, and Zebur present $F = -1$, which indicates a strong heterotic selection at the base of the clonal breeding of these cultivars. On the other hand, an F value close to 0 is expected under random mating, as observed in Sokeri that is traditionally seed propagated.

An UPGMA dendrogram based on codominant genotypic distances of SSR loci is presented in Fig. 7.14. The observed cluster topology evidences the genetic diversity existing among cultivars that allow distinguishing them easily.

Codominant genotypic distances allow estimating the average similarity internal to each cultivar ranging from 0 to 20.98. Talis, Halima, Omglaib, Saiedi, Tagiat, Saila, and Zebur cvs. showed value 0, indicating no genetic difference within cultivar in agreement with the fixation index reported in Table 7.5. This result gives evidence that farmers have good skills, based on a long tradition, in clonal propagation. Nevertheless cases of misclassification can occur during propagation because of the difficulty to identify, in some cases, certain cultivars on the base of morphology. On the contrary, the high value (20.98) shown by cv. Sokeri indicating a high level of diversity among the palms relates to the practice of seed propagation of this cultivar.

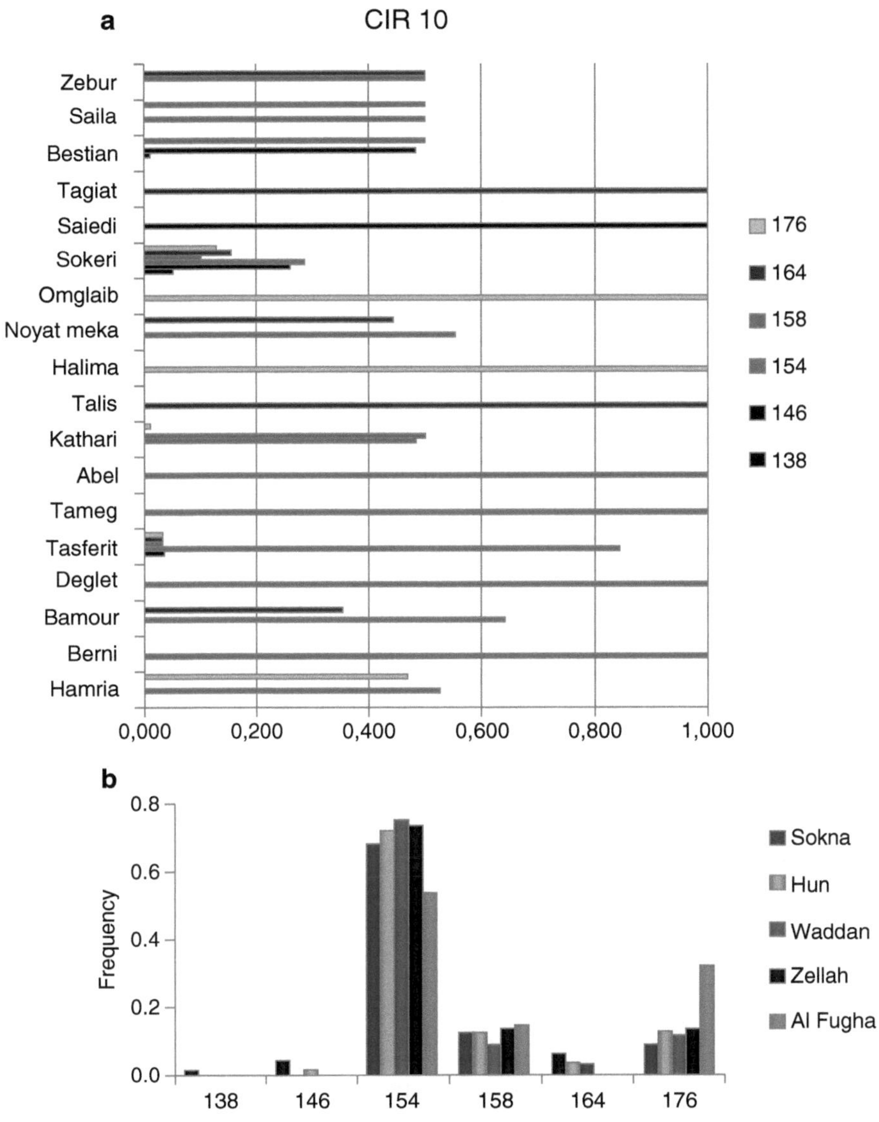

Fig. 7.12 Frequency distribution of alleles of CIR10 marker that varies both among (**a**) cultivars and (**b**) oases

An identification key (Fig. 7.15) was built using three microsatellite loci (mPdCIR78, mPdCIR93, mPdCIR25) and considering the 23 identified alleles: 10 alleles labeled (a1 to a11) for mPdCIR78 locus, 7 alleles (b1 to b7) for mPdCIR93 locus, and 6 alleles (c1 to c6) for mPdCIR25 locus.

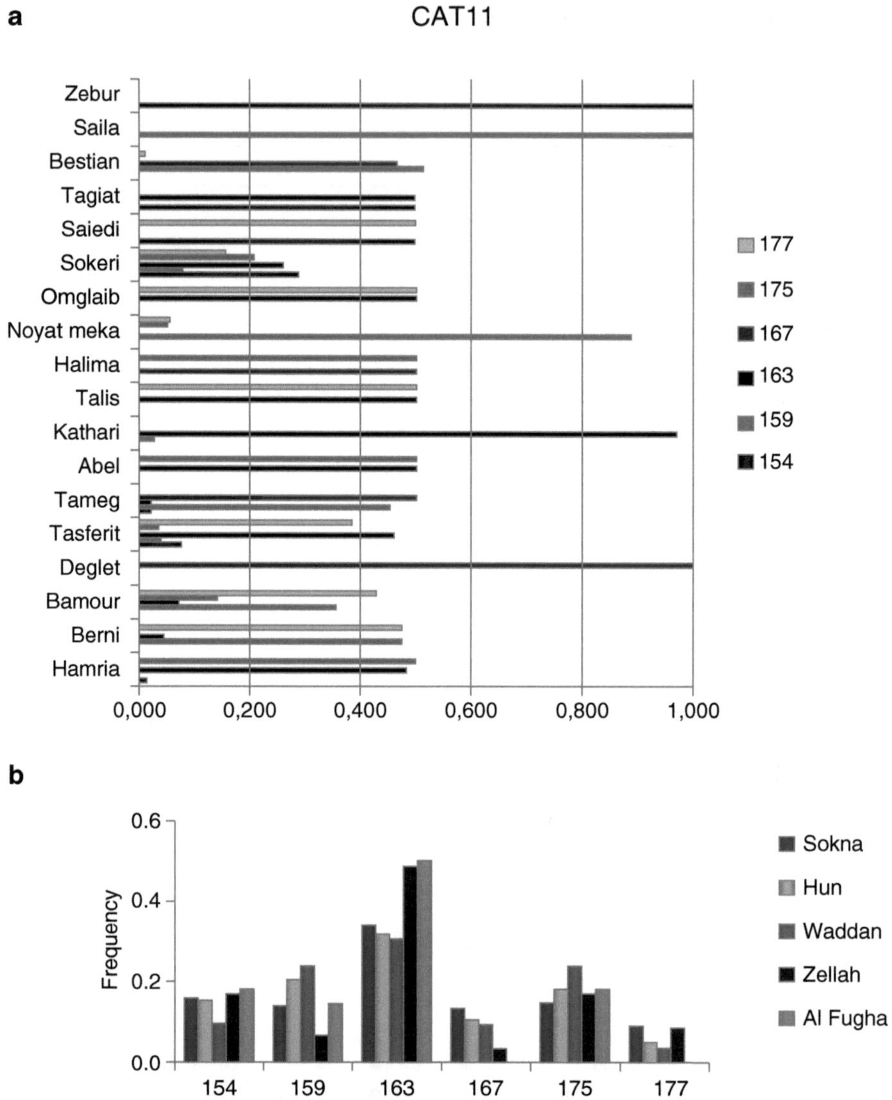

Fig. 7.13 Frequency distribution of alleles of CAT11 marker that varies both among (**a**) cultivars and (**b**) oases

The totality of local cultivars was univocally and easily identified on the base of their allelic profile. Similar result was previously obtained by Zehdi et al. (2006) in an analysis of 49 Tunisian accessions with three SSR loci.

Table 7.4 Genetic diversity indices for 18 Libyan date palm cultivars

Cultivar	*N*	Na	H_{obs}	H_{exp}	*F*
Abel	24	2.875	0.742	0.411	–0.553
Bamour	7	2.813	0.653	0.491	–0.268
Berni	21	2.625	0.622	0.348	–0.567
Bestian	33	2.750	0.629	0.339	–0.557
Deglet	25	1.688	0.625	0.314	–0.993
Halima	15	1.750	0.750	0.375	−1.000
Hamria	34	3.313	0.915	0.497	–0.799
Kathari	34	2.500	0.807	0.425	–0.747
Noyat Meka	9	2.563	0.576	0.365	–0.368
Omglaib	12	1.563	0.563	0.281	−1.000
Saiedi	21	1.813	0.813	0.406	−1.000
Saila	11	1.625	0.625	0.313	−1.000
Sokeri	19	5.125	0.708	0.656	–0.007
Tagiat	39	1.813	0.813	0.406	−1.000
Talis	25	1.625	0.625	0.313	−1.000
Tameg	22	2.750	0.685	0.380	–0.670
Tasferit	12	3.250	0.630	0.420	–0.240
Zebur	14	1.625	0.625	0.313	−1.000

Modified from Racchi et al. (2013)
N sample size, *Na* number of alleles, H_{obs} observed heterozygosity, H_{exp} expected heterozygosity at the HW equilibrium, *F* fixation index

The effectiveness of SSR in discriminating among all the accessions and cultivars examined confirms the usefulness of these markers for clonal fingerprinting and cultivar identification. Since each cultivar was identified by a unique profile, it is possible to generate an individual barcode using the multilocus genotype useful in the certification and the control of origin labels of date palm products.

Considering the SSR effectiveness in fingerprinting genotypes, we used them to assign male plants, sampled in each farm of the different localities within the Al Jufrah oasis. The method of maximum likelihood paternity assignment allowed assigning males to a single cultivar: 55 out of the 63 male plants were assigned to cultivars with strict confidence. The identification key applied to the 24 male plants presenting positive LOD score evidenced that each of them has at least one allele in common with the cultivar assigned by the parentage analysis (Table 7.5). The positive result obtained in identifying male trees further confirmed the suitability of SSR for genotyping and opens new prospects for date palm breeding. More detailed information about DNA extraction, amplification, and genotyping is reported in Racchi et al. (2013).

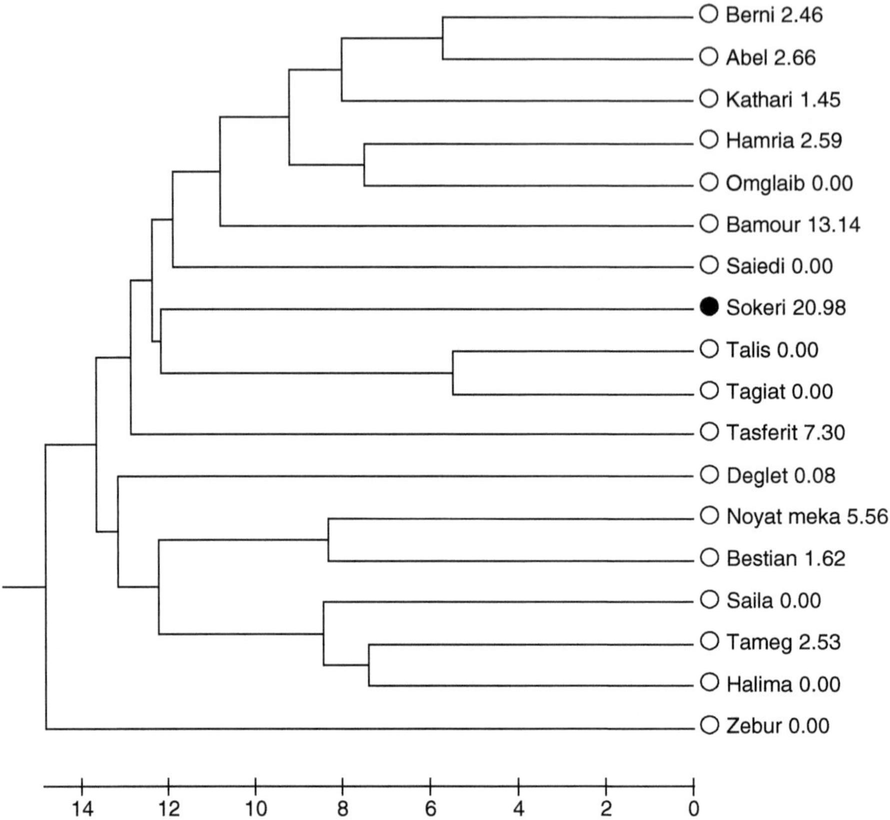

Fig. 7.14 UPGMA dendrogram based on codominant genotypic distances of SSR loci

The observed heterozygosity (H_{obs}) was calculated as the ratio of the number of heterozygotes and the number of samples, for each locus, and as an arithmetic average over loci. The expected heterozygosity (H_{exp}) assuming Hardy–Weinberg equilibrium was estimated as the difference from 1 of the sum of the squared frequency of the each allele. The statistic was computed for single locus and as average over loci. The fixation index or inbreeding coefficient (F) was computed as difference between H_{exp} and H_{obs} divided by H_{exp}.

The calculation of individual by individual genetic distance (GD) for SSR followed the method explained in Smouse and Peakall (1999). Genetic distance matrices for each locus were summed across loci under the assumption of independence. The matrix of distances among cultivars is obtained as an average of the individual distances between couples of cultivars, while the element of the main diagonal is the

Table 7.5 Assignments with positive scores of pair LOD value of parentage analysis performed on male plants sampled in Al Jufrah, Libya

Male ID	Candidate cultivar	Pair loci compared	Pair loci mismatching	Pair LOD score
SMM-Q-04	Kathari	15	0	9.47
SME-Q-03	Tasferit	16	0	0.13
SMT-Q-01	Tasferit	16	2	6.60
SMT-Q-02	Sokeri	15	1	1.26
SSA-Q-01	Deglet	15	1	9.87
SSA-Q-02	Deglet	16	2	1.77
SAK-Q-02	Bestian	16	1	6.97
SAK-Q-03	Sokeri	16	1	2.20
HSM-Q-02	Bestian	16	1	1.26
HRG-Q-01	Tagiat	16	1	3.08
H6I-Q-02	Deglet	16	2	1.05
H6I-Q-03	Tagiat	16	1	5.73
H6E-Q-03	Bamour	16	2	1.33
H3F-Q-03	Tameg	16	1	1.26
H5H-Q-02	Abel	16	1	3.45
H5H-Q-03	Hamria	16	2	2.16
H3H-Q-02	Bestian	16	1	1.12
WOE-Q-02	Abel	16	0	9.15
WBH-Q-01	Abel	16	1	3.33
WBH-Q-02	Sokeri	16	2	1.75
WFZ-Q-03	Tagiat	16	1	9.90
WHS-Q-01	Tagiat	16	1	3.80
W4B-Q-03	Hamria	16	1	1.82
WBB-B	Tasferit	16	0	1.60

Modified from Racchi et al. (2013)

average dissimilarities for all pair-wise comparisons internal to each cultivar. Male trees were assigned to cultivars by a maximum likelihood paternity assignment procedure through comparing genotypes of males and cultivars. To find the significant values of LOD scores, simulations were performed with 10,000 repeats, 0.01 as the proportion of loci mistyped and 61 individual profiles as probable cultivar candidate for each male tree. We used 95 % as strict and 80 % as relaxed confidence level. The LOD score is obtained taking the natural log (log to base e) of the overall likelihood ratio. Genetic variability measures and distance metrics were analyzed using GenAlEx 6.5 (Genetic Analysis in Excel; Peakall 2006; Peakall and Smouse 2012). Cluster analysis was performed using software MEGA version 5 (Koichiro et al. 2011); Cervus 3.0 (Kalinowski et al. 2007) was used for parentage analysis.

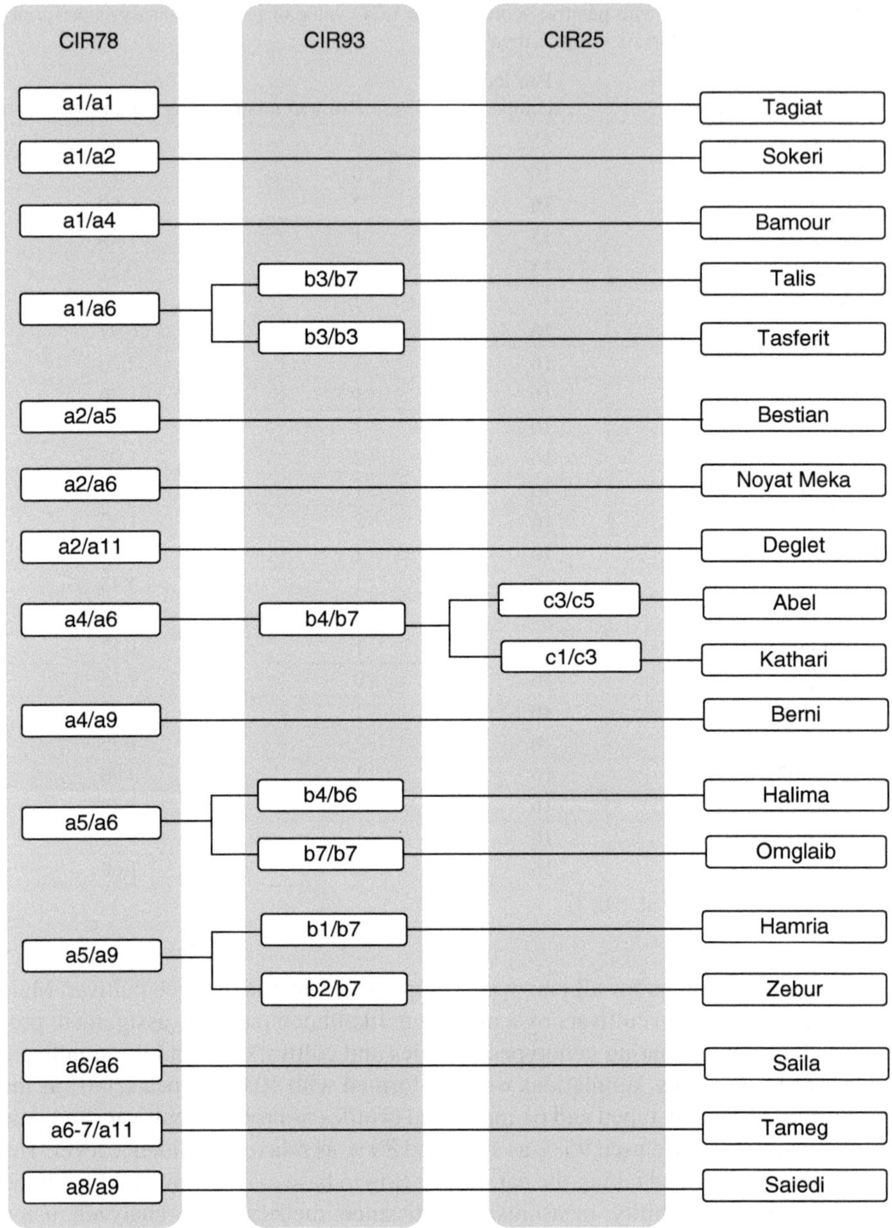

Fig. 7.15 Identification key of date palm cultivars based on 3 microsatellite loci (Modified from Racchi et al. 2013)

7.7 Date Production and Marketing

The date palm is a species with multiple uses (e.g., food, fodder, fuel, plaiting materials, construction) and adaptable to extreme environmental conditions. It has found wide use over the centuries in agroforestry-pastoral xerothermophilous regions, in oases along wadis, and in an environment where one can practice minimal irrigation, even with highly saline water (Dowson 1961; Dowson and Pansiot 1965).

In fact, the cultivation of date palms in Libya presents a very particular situation. The closure of the country because of well-known political events from the 1970s onward, impaired economic development and consequently the commercial development in the international markets of date palms, which represented in the other Maghreb countries the main factor responsible for the cultivar changes and for setting up new plantations (Glasner 1996). The internal market in Libya, in fact, less demanding and poorer than the European market, has allowed for maintenance of the cultivation of many local cultivars, whose fruits are also pleasingly consumed fresh.

The present level of date production in Libya is quite low. All production is consumed locally, and the cultivation techniques and the processing industry are not developed in the modern sense. Nevertheless, the domestic production of dates has increased significantly over the last 25 years concomitantly with the expansion of the cultivated area in dates. This has taken place mainly because of the Libyan government, which in recent years made significant investments in cultivation (FAOSTAT 2013). From 1994 to 2011, date production in Libya doubled, bringing the national production to more than 166,000 mt from about eight million trees. If on the one hand the general situation of date palm in Libya manifests some deficiencies when compared to the other North African countries producing dates in the Mediterranean areas, on the other hand, Libya, more than its neighbors, is a repository of high biodiversity, and it is still free from plant diseases and pests from outside. For this reason Libyan date palm germplasm represents a heritage of great interest both for agriculture and the environment. It is a source of genetic variability useful for genetic improvement, which is necessary to overcome current production limitations.

The cultivar component also significantly contributes to the quality characteristics of date production, based on local cultivars with a strong regional identity and high nutritional value. These characters of naturalness and authenticity make the Libyan dates potentially very attractive due to both the specific organoleptic characteristics and the potential for genetic improvement of cultivars which includes resistance to plant pathogens, adaptability to extreme environmental conditions, increase in production, and improvement of systems of conservation.

7.8 Processing and Products

Until a few decades ago, the date palm stem, cut into linear sections, was used to make support beams, doors, windows, and stairs in houses, while the woven branches, covered in lime, were used for roofing. The leaves and branches were also used to make fences to divide agricultural properties. A *zeriba* is a shelter built from palm fronds, where field tools can be stored. Also built out of palm leaves is the *cecabart*, a clever circular hut in which the hot air rises in the middle and leaves through a chimney-like opening, making it fresh and airy, rare, and precious in the desert. Still today, skilled craftsmen transform the leaves into mats, containers for storing food, everyday objects, incense, jewelry, hats, belts, and bags. The tough fiber can be woven into rope to make harnesses used by harvesters for support when climbing themselves up the palms. The wood not used for construction feeds the fires used to cook food and heating during the cold desert nights and winters.

The apical top of the palm stem can be tapped to extract a sap called *lagbi*, a thirst-quenching, sweet, and highly nutritious beverage. The cutting operation is very delicate, requiring great care so as not to cut into and injure the heart of the tree and kill it. In the past the *lagbi* was collected in a colocynth, a round gourd, typical of the desert which makes an excellent container when dried.

Date fruits are an essential staple food for both humans and animals. Following the way of thinking typical of rural zones, or anywhere with limited resources where nothing is thrown away, the date pits (and today the third-grade dates) are used to feed camels and goats, which give their milk an intense aroma. In Central and South Libya, date fruits are eaten fresh during the harvest season or pressed and mixed with other lesser ingredients to be conserved for leaner months. Dates have always been central to the diet of desert peoples. For the nomads and the animals which carried them across the desert, dried dates were the energy-giving food that could withstand the hottest temperatures (Ali et al. 1956). They were also a precious commodity to be bartered for grains grown along the coast.

7.9 Conclusions and Recommendations

The date palm is recognized for its unique capacity to grow, produce fruit, and accumulate a high quantity of important metabolites under farming conditions with highly restrictive temperatures and aridity. The species adapts well to semidesert conditions. It represents a fundamental economic and food resource in areas that are inhospitable to other plant species, and in fact it can even create a microclimate suitable for other plant species.

Generally, in the Maghreb countries, the gradual impoverishment of traditional cultivars, whose renewal and conservation is no longer ensured, has already led to the growth of selected cultivars. Abandonment of traditional crops will inevitably lead to the reduction of the genetic variability available to the species, a variability which derives from a long natural selection and constitutes the primary factor for environmental adaptation.

Drought, salinity, desertification, and the age of date palm groves have created problems for date palm cultivation, but farmers recognize the importance of safeguarding the cultivars common at the local level and today have access to a heritage that is extremely valuable for the country's environmental and economic future.

Single-cultivar plantings are not only more susceptible to possible parasitic epidemics but also are at greater risk in the event of unusual weather patterns during key stages in the plant's life cycle, such as flowering and fruit setting, creating the possibility of serious production losses. Additionally, single-cultivar production is more exposed to market fluctuations dictated by the changing preferences of consumers. In this context, the Libyan germplasm represents an enormous richness that deserves to be exploited. Its characterization and valorization open new prospects for date palm breeding, protecting the agrobiodiversity by promoting the local palm cultivars and strengthening traditional oasis management systems.

In the past, dates were the ideal staple food for desert crossings, providing energy for nomadic travelers and their animals, helping them withstand the extreme temperatures. Dates were also a precious commodity, bartered for the cereals grown along the coast. Today dates make a perfect breakfast food or a light snack; they are rich in sugars and fibers but have a very low fat content. More importantly, they are very rich in minerals, making them ideal as a source of rapidly available energy in cases of fatigue or physical debilitation.

Acknowledgments This publication represents the results of the international cooperation program for improving and promoting date palm production in Libya *Improvement and Valorization of Date Palm in Al Jufrah Oasis*, funded by the Directorate General for Development Cooperation of the Italian Ministry for Foreign Affairs and coordinated by the Istituto Agronomico per l'Oltremare (IAO) in Florence in collaboration with the Libyan Ministry of Agriculture, Slow Food Foundation for the Biodiversity, and the University of Agriculture, Florence.

References

Akkak A, Scariot V, Torello Marinoni D et al (2009) Development and evaluation of microsatellite markers in *Phoenix dactylifera* L. and their transferability to other *Phoenix* species. Biol Plant 53(1):164–166

Alercia A, Diulgheroff S, Metz T (2001) Multi-crop passport descriptors. FAO/IPGRI, Rome

Ali KT, Fine NC, Sarsam NH et al (1956) The use of dates and date pits for livestock feeding. Abu Ghraib Experiment Station, Dept. of Animal Husbandry, Ministry of Agriculture, Baghdad

Arara A (1975) Soils, irrigation and drainage of the date palm. 3rd FAO technical conference on improving date production processing and marketing paper N° a3, Baghdad, Iraq

Bennaceur M, Lanaud C, Chevallier MH et al (1991) Genetic diversity of the date palm (*Phoenix dactylifera* L.) from Algeria revealed by enzyme markers. Plant Breed 107:56–57

Billotte N, Marseillac N, Brottier P et al (2004) Nuclear microsatellite markers for the date palm (*Phoenix dactylifera* L.) characterization and utility across the genus Phoenix and in other palm genera. Mol Ecol Notes 4:256–258

Bitaw AA, Ben Saad AA (1990) Survey of date palm trees insect pests in Libya. Arab J Plant Prot 8(2):72–76

Buys AJ (1993) FSSA fertilizer handbook. Compiled for the Fertilizer Society of South Africa

Carpenter JB (1975) Notes on date culture in the Arab Republic of Egypt and the People's Republic of Yemen. Date Grow Inst Ann Rep 52:18–24

Carpenter JB (1981) Improvement of traditional date culture. Date Palm J 1(1):1–16
Danthine H (1937) Le palmier dattier et les arbres sacrés dans l'iconographie de l'Asie occidentale anciènne. Librairie Paul Geuthner, Paris
Djerbi M (1983a) Diseases of the date palm. FAO Regional Project for Palm and Dates Research Centre in the Near East and North Africa, Baghdad
Djerbi M (1983b) Report on consultancy mission on date palm pests and diseases. FAO, Rome, 28 pp
Djerbi M (1995) Précis de phoeniciculture. FAO, Rome
Dowson VHW (1961) Report to the Government of Libya on date production. FAO, Rome
Dowson VHW (1982) Date production and protection with special reference to North Africa and the Near East. FAO Tech Bull 35, pp 294
Dowson VHW, Pansiot EP (1965) Improvement of date palm growing. FAO Agr Dev Paper N°72
Edongali EA (1997) Diseases of date palms (*Phoenix dactylifera* L.) of Libya. Arab J Plant Prot 14(1):1–3
El-Alwani AM, El-Ammari SS (2007) Date palm fungal diseases in three Libyan oases. Dept of Botany, Fac Science, Univ Garyounis, Benghazi
Elhoumaizi MA, Saaidi M, Oihabi A et al (2002) Phenotypic diversity of date-palm cultivars (*Phoenix dactylifera* L.) from Morocco. Genet Resour Crop Evol 49:483–490
Elhoumaizi MA, Devanand PS, Fang J et al (2006) Confirmation of 'Medjool' date as a landrace variety through genetic analysis of 'Medjool' accessions in Morocco. J Am Soc Hortic Sci 131:403–407
Elshibli S, Korpelainen H (2008) Microsatellite markers reveals high genetic diversity in date palm (*Phoenix dactylifera* L.) germplasm from Sudan. Genetia 134:251–260
Elshibli S, Korpelainen H (2009) Biodiversity of date palms (*Phoenix dactylifera* L.) in Sudan: chemical, morphological and DNA polymorphism of selected cultivars. Plant Gen Res 7:194–203
FAOSTAT http://www.faostat.fao.org. Accessed Aug 2013
Gariani NK, Nuesery SM, Edongali EA (1994) Disease and pest outbreaks. Libya. Black scorch disease of date palms (*Phoenix dactylifera*) in Libya. Arab Near East Plant Prot Newsl 19:40
Glasner B (1996) Consultancy mission on date palm marketing and handling. First technical report, FAO, vol 41, pp 745–801
Govaerts R, Dransfield J (2005) World checklist of palms. Royal Botanic Gardens, Kew
Kalinowski ST, Taper ML, Marshall TC (2007) Revising how the computer program CERVUS accommodates genotyping error increases success in paternity assignment. Mol Ecol 16:1099–1106
Koichiro T, Peterson D, Peterson N et al (2011) MEGA5: molecular evolutionary genetics analysis using maximum likelihood, evolutionary distance, and maximum parsimony methods. Mol Biol Evol 28(10):2731–2739
Mancini M (2010) Desert fruit: the dates of Al Jufrah. Slow Food Foundation for Biodiversity, Florence
Martin H (1958) Pests and diseases of date palm in Libya. FAO Plant Prot Bull 6:120–123
Nixon RW, Carpenter JB (1978) Growing dates in the United States. US Dept of Agriculture, Agric. Information Bulletin N°207, Washington, DC
Peakall R (2006) GENALEX 6: genetic analysis in excel. Population genetic software for teaching and research. Mol Ecol Notes 6:288–295
Peakall R, Smouse PE (2012) GENALEX 6.5: genetic analysis in excel. Population genetic software for teaching and research – an update. Bioinformatics 28:2537–2539
Plini the Elder (77 AD) In: Naturalis historia, vol 3 Botanics, Book XIII, Einaudi, Torino 1984
Popenoe PB (1973) The date palm. In: Field H (ed) Field research projects. Coconut, Miami
Racchi ML, Bove A, Turchi A et al (2013) Genetic characterization of Libyan date palm resources by microsatellite markers. 3 Biotech. doi:10.1007/s13205-013-0116-6
Sauer J (1993) *Phoenix* – date palms. In: Historical geography of crop plants. CRC Press, Boca Raton, pp 182–186

Scarin E (1938) La Giofra e Zella. Le oasi del 29° parallelo della Libia occidentale. Sansoni Editore, Florence, p 34

Slow Food Foundation for Biodiversity (2010) http://www.libyandates.com/english/technical-staff-palm-libyan-dates.html

Smouse PE, Peakall R (1999) Spatial autocorrelation analysis of individual multiallele and multi-locus genetic structure. Heredity 82:561–573

Wertheimer M (1954) La pollination du palmier dattier (*Phoenix dactylifera* L.). Fruits 12:305–313

Wrigley G (1995) Date palm, *Phoenix dactylifera*. In: Smartt J, Simmonds NW (eds) Evolution of crop plants, 2nd edn. Longman, London, pp 399–403

Zaid A, de Wet PF (2002) Date palm cultivation, plant production and protection paper 156. FAO, Rome

Zehdi S, Sakka H, Rhouma A, Salem AOM, Marrakchi M, Trifi M (2004a) Analysis of Tunisian date palm germplasm using simple sequence repeat primers. Afri J Biotechnol 3:215–219

Zehdi S, Trifi M, Billotte N, Marrakchi M, Pintaud JC (2004b) Genetic diversity of Tunisian date palms (*Phoenix dactylifera* L.) revealed by nuclear microsatellite polymorphism. Hereditas 141:278–287

Zehdi S, Pintaud JC, Billotte N et al (2006) Etablissement d'une clé d'identification variétale chez le palmier dattier (*Phoenix dactylifera* L.) par les marqueurs microsatellites. Plant Gen Res Newsl 145:11–18

Zohary D, Hopf M (1993) Date palm, *Phoenix dactylifera*. In: Domestication of plants in the old world, 2nd edn. Clarendon, Oxford

Chapter 8
Date Palm Status and Perspective in Morocco

Moulay Hassan Sedra

Abstract In Morocco, the date palm is grown in several areas located on the southern flank of the Atlas Mountains along rivers and around water points. These areas are characterized by a pre-Saharan arid climate. The total number of 5.4 million date palms represents about 4.5 % of the palm patrimony of the world and is composed of 453 cvs. represented by nearly 2.8 million trees (51.8 %) and approximately 2.6 million *khalts* (48.2 %) originating from natural palm seedlings. Date production ranks first among fruit growing in the zones of oases and contributes 20–60 % of the income of oasis farmers. The diagnosis of date palm sector highlights the strengths, constrains, weaknesses, and opportunities of the sector. With the exception of some modern palm farms, the agricultural practices are generally traditional in oases. The Green Morocco Plan (PMV) designed and monitored by the Ministry of Agriculture and Fisheries (MAPM) designated this crop as a priority to give new impetus to the agricultural economy in oasis areas. Researchers investigated date palm protection especially to control bayoud disease, cultivar characterization, tissue culture micropropagation, and date valorization. Main development actions include distribution of over 1.5 million of vitro-plants since 1987 to farmers in order to restructure and reconstitute palm groves, encouraging investments and facilitating of new planting, encouraging research on water dynamics and participatory actions of water users, installation of refrigerated storage and some date processing, promoting domestic marketing and exports and ensuring of national coordination and networking of actors of date palm sector in the context of a Geographic Information System (GIS). The cvs. described were chosen for their level of representation. A promising future is expected for Moroccan date palm oases if all the actors reinforce the economic challenge and awareness of preservation of the environmental ecosystem.

Keywords Bayoud • Cultivars • Date palm • Genetic diversity • *Khalts* • Morocco • Biotechnology • Pests • *Phoenix dactylifera*

My.H. Sedra
Laboratory of Phytopathology, Genetics and Integrated Control,
Regional Centre of Agricultural Research of Marrakech,
National Institute of Agricultural Research (INRA), 533, Marrakech 40000, Morocco
e-mail: sedramh@gmail.com; mhsedra@yahoo.fr

J.M. Al-Khayri et al. (eds.), *Date Palm Genetic Resources and Utilization: Volume 1: Africa and the Americas*, DOI 10.1007/978-94-017-9694-1_8

8.1 Introduction

8.1.1 Moroccan Areas of Date Palm Cultivation

In the pre-Saharan and Saharan regions of Morocco, the date palm is grown in several areas located on the southern flank of the Atlas Mountains along the wadis (rivers) such as the Drâa and Ziz and around water points (e.g., springs, *khettaras* permanent wells) such as the plain of Tafilalet and palm groves of Figuig, Goulmima, and Alnif (Figs. 8.1 and 8.2). The provinces of Ouarzazate (Drâa Valley), Errachidia (Tafilalet and Ziz valleys), and Tata (Bani) alone account for nearly 90 % of the total palm trees. Other palm groves grow in areas of high elevation such as Tinghir and in North of Atlas' mountain areas such as the Marrakech palm groves. These growing areas are generally characterized by a pre-Saharan arid climate with very hot summers and cool and sometimes cold winters in northern areas, temperatures varying between 0 and 48 °C, very low and irregular annual rainfall oscillating between 50 and 150 mm. Winds are hot, dry, and often violent (40–50 km/h) and carrying sand in some places. The average evaporation is very high (about

Fig. 8.1 Map of main areas of the date palm in Morocco and some palm groves as seen from space (Source: Google Earth)

Fig. 8.2 Types of Moroccan palm groves: Tarnata located in the Drâa Valley, Ziz Valley, Figuig in Oriental, and Bani (Tissent and Tata localities) and example of recent grove extension located at Errachidia area. Development constraints: drought (**a**) and siltation and salinity (**b**, **c**)

3,358 mm/year) due to the combined effect of temperature, dry air, sunlight, and wind. In these areas, water resources, from the peaks of the High and Anti-Atlas, are both less frequent and irregular and are the most limiting factor. The soils are generally skeletal alluvial with aeolian sand deposits on the piedmonts, valleys, and plains.

8.1.2 Moroccan Date Palm Sector

8.1.2.1 Description

Morocco ranks seventh in the world in terms of area, eleventh in terms of date palm numbers, and twelfth in terms of date production (FAOSTAT 2011). At the national level of arboriculture, date cultivation is third, after the citrus and other fruit crops. In the Saharan regions, the date palm is the basic framework of the sustainability of the oasis ecosystem and plays both roles of great importance on ecological, social, and economic plans. Indeed, date palm cultivation, by its protective frame, creates a microclimate for underlying arboreal and herbaceous crops and ensures judicious protection against harsh external climate and the severity of its sudden changes. This protection helps to slow the phenomenon of desertification and mitigating the desert influences. At the socioeconomic level, the date palm sector contributes to the supply of dates to Moroccans. Annual national consumption of dates is about 3 kg/person and can reach 15 kg/person in production areas. With an average of around 100,000 mt annually, the date production ranks first among fruit oases and contributes 20–60 % of the income of oasis farmers. Table 8.1 summarizes data on the date palm sector in Morocco. The total number 5.4 million date palm represents about 4.5 % of the world's date population. It is composed of 453 cvs. (Sedra 2010a, 2011a, b, c, d, e, f) represented by nearly 2.8 million trees (51.8 %) and approximately 2.6 million *khalts* (48.2 %) from natural seedlings (Djerbi et al. 1986; Sedra

Table 8.1 Main data of the date palm sector of Morocco

Indicators	Value/quantity	Observations
Total number of palms	5 million	Majority of traditional palm
Total area	50,000 ha	Mean density of 108 date palms/ha
Number of cvs. enumerated	More than 453 cvs.	51.8 % of the total number
Number of *khalts* enumerated	More than 2.6 million	48.2 % of the total number
Mean annual production of dates	100,000 mt	Relatively low and diversified with approximately 30 % of dates are of merchantable quality
Average productivity of date palms	18–20 kg/tree	Relatively low
Cultivation practices properly performed	Relatively small number	Highly inadequate
Postharvest and marketing techniques	Relatively small number	Inadequate

In modern plantations, numerous practices are enough performed

1995, 2011b, c, d, f, 2013; Sedra et al. 1996). Only about 45 % of date palms produce fruit each year, and the average productivity of 20 kg/tree is low compared to the standard date-producing countries. This is due to numerous constraints cited below. With the exception of some modern palm farms, the postharvest practices are generally traditional in the oasis.

8.1.2.2 Strengths and Importance

(a) Moroccan date palm groves are a unique asset in the world by cultivar diversity and richness in seedling dates, each of which is genetically unique (Sedra 1995, 1997, 2011c, Sedra et al. 1996). Medjool cv. is the most famous and most sought after in the world. The commercial value of its fruits exceeds those of other cvs. exported in date-producing countries.
(b) Date production is very diverse in terms of fruit quality, maturity periods, and skills in conservation and transformation abilities. This permits meeting all levels of consumption and various uses.
(c) Opportunities for development and valorization of new genotypes of high performance and *terroir* products based on dates.
(d) Traditional character of oasis agriculture in favor of the establishment of biological date palm cultivation.

8.1.2.3 Abiotic and Biotic Constraints

(a) Arid climate, prolonged drought (Fig. 8.2), and decline of groundwater resources in quantity and quality
(b) Environmental degradation risk, mainly due to salinity and siltation (Fig. 8.2)
(c) Bayoud which impoverished national date palm oases of good quality cvs.
(d) Significant development of some palm enemies including white scale and inflorescence rot in several orchards around oases and date moths in places of traditional storage of dates

8.1.2.4 Weaknesses Hampering the Development

(e) Weakness and irregularity of productivity, mainly due to bayoud disease, traditional farming techniques, old and fragmented orchards and farms due to the effects of years of drought
(f) Predominance of low commercial value dates
(g) Significant shortfall in productivity and value of date production
(h) Inadequate facilities, packaging, processing, and storage of dates according to required standards as well as lack of date processing industries
(i) Weak farmers' organization and marketing and trading channels
(j) Exodus of farmers and/or their sons to urban centers in search of more remunerative work

8.1.2.5 Developments, Opportunities, and Government Actions

(a) Implementation, since 1986, of the National Plan for Restructuring and Development of palm groves
(b) Registration, in 2000, of the oases of southern Morocco in the Global Network of Biosphere Reserves of UNESCO
(c) Introduction in 2007 of the Act 01-06 concerning sustainable development of palm oases
(d) Development project in the sector launched by His Majesty the King Mohamed VI in 2009, hence the importance given to the sector in the Green Morocco Plan (PMV) designed and monitored by the Ministry of Agriculture and Fisheries (MAPM) that erected this crop as priority sector to give a new impetus to the agricultural economy in oasis areas
(e) Recent building up (2010) of the sector through the creation of the National Agency for Development of Zones of Oasis and Argan areas (ANDZOA)
(f) Establishment in 2010 of the National Federation of Dates Production (FNPD) and Interprofessional (FIMADAT) to contribute to the upgrading of this sector, in collaboration with their partners
(g) Integration of the palm groves under the Fruit Tree Project Millennium Challenge Account (MCA) Program 2008–2013
(h) Creation in 2013 of a national institution called the National Office of Agricultural Councils (ONCA) with regional and local representatives
(i) Existence of a framework to encourage investments in the sector
(j) Significant potential of communal lands may be object of investment
(k) Opening of the national economy to external markets

8.2 Cultivation Practices

8.2.1 Traditional Date Palm Groves

As mentioned above, farming techniques practiced in traditional oases are archaic and empirical. This mode of local knowledge inherited over several generations has evolved in recent decades with few peasant improvements that affect some practices including offshoots planting and date storage. At the level of the oasis village, irrigation water from the river is divided by small traditional lines respecting the water tower for each orchard. Date palm irrigation is realized by individual basins in which farmers grow crops under the trees. The use of chemical fertilizers is almost absent; organic manure is generally made for associated crops but with insufficient amounts, making relatively poor soil in terms of organic matter and nutrients. The traditional method of pollination is manual, attaching mature male spikelets between the female spikelets (3–5 spikelets male/female spathe). The date palms grow in clumps, planting distances are not respected, cleaning the palms after harvesting, pruning of date palm-mother, and offshoots are not removed at the appropriate times. This makes difficult passage in the field and agricultural manipulations under the palm trees. Generally, cleaning is done in case of need of palm leaves used for fencing and waste as firewood. These practices have decreased with the arrival of butane. Other techniques of fruit thinning and good harvest and postharvest practices are not followed. Fruit storage and processing techniques are traditional (e.g., soft dates pressed and flavored or not with aromatic and medicinal plants or semolina of dried dates). The control of palm pests and diseases is not practiced. Modern farming practices and postharvest technologies have not yet been adopted by small farmers for several reasons including uninformed farmers especially the elderly ones, insufficient financial resources, and lack of extension activities.

In recent years, the MAPM has made considerable efforts to support these farmers through awareness campaigns, close supervision, field schools, structuring social and professional organizations, and financial subsidies. A significant effect of these government actions was felt. To improve the effectiveness of extension services, the government created the ONCA, a national institution with regional and local representatives and programs.

8.2.2 *Modern Date Palm Farms*

In modern plantations located in pristine areas around traditional oasis and now covering more than 2,500 ha, good farming techniques are in part performed especially those related to the use of good cvs., planting, irrigation, fertilization, pollination, pruning of palms, and control of certain pests and diseases such as white scale and inflorescence rot. These plantations are recent and equipped with all modern facilities. The government is encouraging investors to reach a total of 19,000 ha of modern plantations by 2020 of which 17,000 ha for investors and 7000 ha (41.2 %) already under preparation until now. 5000 ha are planned to create small modern plantations on communal land to young farmers. The investment cost of creation and running modern orchards was evaluated in detail by Sedra (2003c). Sedra (2012, 2013) has developed a comprehensive and detailed bilingual (French/Arabic) guide of all the technical cultivation of date palms from the choice of terrain and cultivars to technical postharvest and marketing including control of pests and diseases.

8.2.2.1 Production of Offshoots

Good adult date palms produce many offshoots, most of which are aerial. The uprooting of these offshoots is essential for the growth of the parent plant, and separated offshoots can be used as an additional source of planting material. Since the Green Morocco Plan program (PMV) intends to plant 290,000 offshoots of good quality cvs., some nursery units have been established recently to create infrastructure to produce healthy offshoots for commercial purposes in accordance with regulations. These units do weaning and rooting of offshoots developed by agricultural research. To ensure rooting of small offshoots (<8 kg), they undergo the processing at the base site of weaning (indole butyric acid, AIB) hormone solutions at concentrations 10–15 mg/l (Zirari 2010) or gibberellic acid GA3 (350–400 ppm) (Sedra 2003c). This treatment stimulates the growth of new roots on offshoots and reinforces the nutrition of young date palms and ensures their proper growth.

8.2.2.2 Plantations

Investors first perform the analysis of soil and irrigation water to ensure the validity and appropriateness of the field site. The soil is plowed and leveled before planting to facilitate irrigation and maintenance of the plot. Cultivars are selected first Medjool cv. then Boufeggous cv. and selected varieties by INRA, e.g., var. Najda which is available in sufficient quantity. These cvs. are subsidized and recommended by the government.

The choice of planting density should take into account the potential of the soil, the size of the date palm leaves, plantation types, square or staggered planting with different distances, and the concern of mechanization and whether intercropping will be done. Densities are usually used 8 × 8 m (156 palms/ha) for cvs. Medjool and Boufeggous, 9 × 9 m (123 palms/ha) and 8 × 9 m (139 palms/ha) for var. Najda.

For newly weaned offshoots, the best planting season is from January to April. Plants rooted in bags can be planted from September to October. In areas with milder climate, planting can be done all year round. The volume of the planting hole is 1 m^3 for proper plant development. The soil removed is mixed with 10 kg of well-seasoned manure, 1 kg of ammonium sulfate (21 %), 1 kg of superphosphate (45 %), and 0.5 kg of potassium sulfate (46 %) to provide nutrients. The plant is positioned in the center of the hole so that the collar is exposed and then buried 30–40 cm depending on the size of the plant (offshoots or vitro-plants). The hole is filled, and a basin with a radius of 1–1.5 m is tailored to be enlarged every year. After the planting operation, shelters are made using dried palm leaves around young plants to protect them against sand-laden winds and intense sunlight and to reduce evaporation.

8.2.2.3 Irrigation

In general, gravity irrigation is used in small farms for most cvs. The dosage per date palm per month varies from 6 to 16 m^3 during cold periods and 17–25 m^3 in hot periods (Table 8.2). It also varies depending on the age of palms, their density, and irrigation method. In the case of gravity irrigation, the average annual irrigation application for 100 palms/ha varies from 11,000 m^3 for young to 16,750 m^3 for adult palms. These requirements increase when dates are grown in association with understory crops.

Drip irrigation system is increasingly being used in new plantations and large farms, which ensures the needs of plants and saves water. Indeed, the average doses of irrigation of plant vary during cold or hot climate from 85 to 110 l for young palms and 160–300 l for productive palms. The average annual irrigation dose for 100 palms/ha varies from 3,205 m^3 for young to 8,310 m^3 adult palms using drip irrigation.

Table 8.2 Dose of gravity irrigation and drip irrigation according to the age of palm trees and climatic conditions divided into cold period (November to mid-February) and hot period (mid-February to October)

Irrigation type and parameters	Young unproductive palm		Adult productive palm	
	Cold period	Hot period	Cold period	Hot period
Gravity irrigation				
Variation of the needs (m^3/palm/month)	6–10	11–17	9–16	17–25
Average per m^3/palm/month	8	14	12.5	21
Irrigation frequency	2–4 times per month	4–6 times per month	2–4 times per month	4–6 times per month
Annual average per m^3/ha (100 palms/ha)	11,000		16,750	
Drip irrigation				
Variation of the needs (l/palm/input)	50–120	150–170	150–170	200–400
Average per liter/palm/input	85	110	160	300
Average per period (m^3/ha) (100 palms/ha)	510	2,695	960	7,350
Irrigation frequency	Once per 2 days	Once daily	Once per 2 days	Once daily
Annual average per m^3/ha (100 palms/ha)	3,205		8,310	

Table 8.3 Quantities of organic and mineral fertilizers applied to date palm under gravity irrigation according to the age of date palms

Fertilizer	Dose kg/palm (average)[a]		Frequency and periods of fertilizer input	
	Young unproductive	Adult productive	Young unproductive	Adult productive
Manure or organic fertilizers	5–10 (7.5)	60–240 (150)	One land application every 6 months	One land application every 2–3 years, 1–2 weeks after harvest
Urea or ammonium sulfate (21 %) nitrogen (N)	0.1–0.2 (0.15)	2–3 (2.5)	Every month	After the winter season (Nov to Feb), during fruit setting (Apr to Jun), time of fruit development and coloring (June to Sept)
Super phosphate (45 %) Phosphorus (P_2O_3)	0.1–0.5 (0.3)	2–3 (2.5)		
Potassium sulfate (46 %) Potash (K_2O)	0.3–0.5 (0.4)	4–6 (5)		

[a]The recommended quantities of manure depend on the soil texture and soil organic matter content

8.2.2.4 Fertilization

Date palm nutrition palm is based on fertilization and irrigation. Balanced nutrition ensures a balance between growth and production. The fertilization with organic or mineral fertilizers plays an important role in increasing the productivity of trees and in improving the quality of the production. Fertilization at depth is realized at planting. Initial fertilizer quantities vary with the age of trees and the export level (Table 8.3). Organic fertilizer is applied at once after date harvest. Mineral fertilizer is spread around the trunk of the tree, within 1–2 m, and then buried in the depth of 10–30 cm.

For large date palm farms, fertigation aims to provide chemical fertilizers in dissolved form in localized irrigation "drip" at a concentration not exceeding 0.5 g of the fertilizer per water liter (equivalent average concentration: 0.05 % or 500 ppm). Watering with this enriched water is realized weekly (March to September); in contrast potassium fertilizers are applied 2–3 times, depending on the age of the palm and its development.

8.2.2.5 Pollination

The emergence and opening of female spathes takes place in late March to April depending on the region. Male pollinators selected are known locally for their high level of pollinator power or male pollinator varieties selected for their biological characteristics. Generally, 4 % of male date palms in the orchard are sufficient to produce the amount of spathes necessary to ensure good mechanical and manual pollination. In small farms, pollination is realized manually by deposition and fixation of mature male spikelets between the female spikelets (3–5 spikelets male/female spathe). In large and modern farms, the method of pollination is done by a semi-mechanical method on female inflorescences at favorable stage of floral receptivity by dusting from the ground using a pollinator by hand or a mechanical machine for pesticide treatment. This can only be achieved in plantations with sufficient planting distance. The last two methods are less costly and more effective than the traditional method. They require instead the preparation of mixture of pollen at 1 g+9 g talc. The pollen grains are stored in boxes, jars, or other suitable containers at a temperature of 4–7 °C in a refrigerator. The pollination period occurs in late March to April depending on the region and duration of female floral receptivity which varies from 1 to 13 days or more after the spathe opening, according to cultivar.

8.2.2.6 Thinning and Bending

As mentioned above, these techniques are generally not practiced in traditional oasis. On most modern farms, date palms are still young. These operations are recommended to increase the size of fruits, improve the quality, and restore a regular physiological balance to the tree by addressing the phenomenon of alternation. They must be done 3–4 weeks after fruit set: late April to May depending on the regions in accordance with the basis of 8–9 strands per fruit bunch on the tree. The number of fruit bunches recommended varies from 10 to 12 per tree depending on the age of the palm and maintenance level. Research in Morocco has showed that these practices have increased the size of the fruit which results in a rise of nearly 30 % of production with quality dates (Sedra and Zirari 1998). The techniques of thinning strands are recommended when the tree is bearing fruit bunches by equalizing the ends of strands, which corresponds to the removal of 15 % of strands or eliminating one-third of the strands located in the center of the fruit bunch.

The practice of fruit bunch arching aims to fix the bunches to ensure their distribution and uniform position around the palm tree and before fruit bunch lignification 3–4 weeks after fruit set. This agricultural technique is not yet practiced in Morocco.

8.2.2.7 Harvest and Postharvest

Harvest is an important step to ensure dates of good quality and suitable for marketing. On large farms when palms come into production, the protection of fruit bunches against possible attacks by birds, other pests, rain, and sand-laden winds is performed to improve fruit quality. Harvesting is done when 80 % of the fruit reach rutab and tamar stages and performed manually because the palm trees are not yet tall. In some semi-modern farms located along the borders of traditional oasis, date harvest is done using a ladder and rope. Mechanical harvesting using platform lifts is not yet practiced.

The following postharvest techniques are totally or partially practiced by some savvy producers and farmers' cooperatives: the harvest consists of cutting the fruit bunches and carefully placing them on clean sheets, collected dates sorted by size and by their state; dates are dried by exposure to the sun in boxes or in convection ovens then disinsected by heat treatment (70 °C for 2 h) followed by rinsing and drying before be packed in suitable containers (200, 250, or 500 g; 1, 2, or 5 kg). Necessary labeling information is placed on the packaging.

The practice of pruning or trimming date palms aims to eliminate all the organs of vegetative and reproductive apparatus in the process of drying or having only a very limited physiological activity. This cleaning operation is usually performed once a year after harvest in modern farms in expansion areas. Some cooperatives produce compost by grinding healthy dry palms.

8.2.3 Pest Control

Globally, the enemies of date palm cultivation are many and diverse. The damage they cause each year can reach 40 % of production varying according to date-producing country. More than 110 harmful enemies represented until now 135 species have been listed for date cultivation. There are 75 enemies (parasites and pests invertebrates) represented by 108 species, more than 11 animal and 8 diseases corresponding to physiological abnormalities or unknown origin (Sedra 2001a, 2003a, b, c, 2012). These enemies attack the date palm at all vegetative and reproductive organs. For competing weeds, more than 16 species can impact the development of palm (Sedra 2012). Some enemies are located in certain countries, and others are scattered in most date-producing countries. In Morocco, the date palm is threatened by pests and diseases that are not yet reported or reported in a particular locality such as the red palm weevil. These enemies cause enormous damage in other countries. The main dangerous pests and diseases require quarantine, routine monitoring, regular monitoring, and draconian phytosanitary measures in Morocco against the major date palm pests (Sedra 2003b, c, 2012).

In Morocco, the particular nature of the biotope where the date palm grows is exposed to a relatively small number of diseases and pests. Indeed, many enemies and competitive weeds have been identified in the Moroccan oases, some of which are less harmful and others generate a significant decrease in fruit production quantity and quality (Sedra 2012). Table 8.4 shows the main enemies and some significant competitive weeds of date palm in Morocco.

Table 8.4 Relative importance of major diseases and pests of date palm and in Moroccan oases

Names of enemies	Responsible agents	Importance of spread
Fungal diseases		
Bayoud (*Fusarium* wilt)	*Fusarium oxysporum* f. sp. *albedinis* Kill. & Maire[a]	++++++
Khamej or inflorescence rot	*Mauginiella scaettae* Calv.	+++
	Fusarium moniliforme J. Sheld	
Dry rot of inflorescences	*Thielaviopsis paradoxa* (Dade) C. Moreau	++
Black scorch	*Thielaviopsis paradoxa* (Dade) C. Moreau	++++
Heart and trunk rot	*Thielaviopsis paradoxa* (Dade) C. Moreau *Gliocladium vermoeseni*, Schr., *Botryodiplodia theobromae* Pat.	++
Diplodia disease	*Diplodia phoenicum* (Sacc.) H S Fawc. & Klotz	+++
Brown leaf spot	*Mycosphaerella tassiana* (De Not.) Johan.	++++
Bending head	*Thielaviopsis paradoxa* (Dade) C. Moreau and/or other	+++
Graphiola leaf spot	*Graphiola phoenicis* (Moug.) Poit.	+
Apical drying of leaves	*Alternaria* sp., *Chalara* sp.	++++
Insects and mites pests		
Red palm weevil	*Rhynchophorus ferrugineus* (Oliv.)	+
Parlatoria scale (white scale)	*Parlatoria blanchardi* Targ.	++++++
Old world date mite (local name *boufaroua*)	*Paratetranychus afrasiaticus* McGr.	+++
Date moths	*Ectomyelois ceratonia* Zell., others	++++++
White termites (local name *larda*)	*Microcerotermes diversus* Silv.	+++
Frond borer	*Apate monachus* Fabr., *Phonapate frontalis* Fahr., other species	+
Palm stem borer (long antennae)	*Phoracartha semipunctata* Fabr.	+
Desert locust (occasional pest)	*Schistocerca gregaria* Forsk.	+++ *some years*
Competitive weeds		
Quackgrass	*Elytrigia repens* (L.) Gould	+++
Bermuda grass	*Cynodon dactylon* (L.) Pers.	+++
Nutgrass	*Cyperus* sp.	++
Blady grass	*Imperata* sp.	++
Bindweed	*Convolvulus* sp.	++

[a]Often isolated in association with other fungi such as *Alternaria* and *Diplodia* or *Chalara*. +: number of + means the importance of spread

8.2.3.1 Major Harmful Diseases

Inflorescence Rot Locally called *khamedj*, inflorescence rot is caused by three pathogens according the different-colored symptoms. Inflorescence rot may be partial or total (Fig. 8.3). *Mauginiella scaetae* Mich. & Sabet, causes the white creamy

Fig. 8.3 Beginning of attack of inflorescence rot (**a**, **b**) at different stages: *1* free, *2* and *3* partially attacked by (*Mauginiella scaetae* Mich. & Sabet); it may be partial or total with symptoms and the appearance of blackening and charring (**c**) and dwarfism (**d**) of rachis and leaves caused by black scorch disease *Thielaviopsis paradoxa* (Dade) C. Moreau

symptoms, *Fusarium moniliforme* J. Sheld. provokes pinkish ones and in *Thielaviopsis paradoxa* (Dade) C. Moreau the symptoms are dry and brown inflorescence rot. The disease attacks both male and female palms. The infection begins during spathe formation from a primary bud and before its appearance on the tree. The development of the disease is favored by conditions including low temperature and high air humidity or rainy weather and injury of young spathes. Its dispersal is by rain, wind, contaminated pollen, insects, and cultivation activities involving the apical part of the tree. In case of severe attack, female and male rotand produce dry spathes and do not bear fruit or pollen. There are different levels of cultivar susceptibility to the disease; highly susceptible are cvs. Medjool and Boufeggous and several selected varieties. In order to control the disease, the following integrated management is recommended:

(a) Clean and incinerate infected inflorescences and tissue fragments.
(b) Avoid using pollen from contaminated spathes or collected from diseased male trees in order to prevent the disease spread.
(c) Preventive chemical treatments with fungicides after harvest (September to November) followed by another treatment before or at the beginning of the output spathes next year (December to March).
(d) Chemically treat palm tree very early at the onset of symptoms. If the symptoms are highly obvious, chemical intervention is not effective. Examples of fungicides used: *la bouillie bordelaise* (0.3–0.5 %), methyl thiophanate (0.2 %), thiram (0.2 %), and copper oxychloride (0.4 %).

Black Scorch This malady is caused by the fungus *Thielaviopsis paradoxa* (Dade) C. Moreau and leads to dwarfism and charring of rachis and leaves (Fig. 8.3). This pathogen attacks the area of the palm heart and secretes toxic substances leading to these primary symptoms and the appearance of blackening and charring. The importance of these symptoms depends on the level of date palm resistance and environmental conditions. This disease is controlled by pruning of infected palms, disinfecting of wounds resulting from pruning leaves, with a copper compound (e.g., copper oxychloride 0.4 %), incineration of infected parts of the palm, and spraying the tree with fungicide *la bouillie bordelaise* (0.3 %), methyl thiophanate (0.2 %), polyram thiram (0.2 %), and Mancozeb (0.2 %).

Bayoud, *Fusarium* Wilt *Geographical distribution* Date palm cultivation in Morocco has been suffering for more than a century from the effects of bayoud disease which is difficult to control. Fig. 8.4 illustrates the disease in an oasis where date palms were destroyed. In recent years, about 1,000 Moroccan oases have been affected. Results from satellite imagery, confirmed by field surveys, have permitted location of 60 foci of the disease alone in the area of Aoufous in Ziz Valley. Also causing significant damage in Algeria, the disease was found in Mauritania in 1999 (Sedra 1999, 2000b, 2002, 2003a, b, 2006b, 2007a, b, 2011f). The disease occurs in the majority of Moroccan date groves and has spread, in recent years, to areas beyond traditional oases (Sedra 2003b, 2006a, 2009). Sedra (2001b, 2003b, 2004, 2006a, 2011a, f) summarized the works on bayoud over several decades.

Damages In Morocco, the consequences of bayoud attack are harmful to oases and have a heavy negative impact on the country's palm patrimony. Thus, 10

Fig. 8.4 Bayoud disease focus indicating destroyed palm orchard (**a**), microconidia, macroconidia, and chlamydospores of causal agent *Fusarium oxysporum* f. sp. *albedinis* Kill. & Maire (**b**), typical symptoms (hemiplegic symptom) and different stages of their evolution on susceptible palms until death (**c**–**e**)

million of date palms were destroyed (i.e., two-thirds of the total), and several cvs. have disappeared, for example, Berni and Idrar. The best commercial cvs. (Medjool, Boufeggous, Bouskri, Jihel, Bourar Aziza Bouzid, and Bouittob) are significantly affected by this disease. Indeed, the groves have lost more than 50 % of the date palms including more productive commercial cvs., resulting in a loss of a vital source of income for oasis residents. Morocco became an importer of dates because the disease was difficult to control after its spread. In addition, bayoud not only reduced planting density but also decreased significantly the extent of annual associated plants that were protected by the palms and accelerated the process of desertification. Moreover, it was shown that the best commercial cvs. of Algeria (Deglet Noor), Tunisia (Deglet Noor, Boufeggous, Besser Lahlou, Gondi, Horra, Kenka, and Kentichi), and Iraq (Barhi, Halawy, Khastawy, Khadrawy, Sair, and Zahidi) also are susceptible (Sedra 1992b, 2003b, 2006a, 2011a, f). Recognized resistant cvs. are very few: in Morocco, Black Bousthammi, White Bousthammi, Iklane, Boufeggous Moussa, Sairlayalate, and Tadmainte have exhibited resistance since 1973 (Louvet and Toutain 1973; Saaidi 1992; Saaidi et al. 1981), and a seventh resistant cv. (Boukhanni) was selected 20 years later (Sedra 1992b, 1993a, 1995). In Algeria, the resistant cv. Takerbouchte is known (Bulit et al. 1967); Tirichine (1991) identified another resistant cv. named Akerbouch in the region of Mzab.

Causal agent, symptoms and internal dissemination Bayoud or *Fusarium* wilt is caused by the soil fungus *Fusarium oxysporum* f. sp. *albedinis* Kill. & Maire. Research on the parasite in Morocco has focused on the morphological, cultural, biological, biochemical, and molecular aspects of the organism (Amraoui et al. 2004, 2005a, b; Sedra 1992a, 1993a, b, 2003b, 2011f, 2012; Sedra et al. 1993) and development of rapid molecular methods for the pathogen identification (Fernandez et al. 1998; Sedra 2003b, 2006a, 2011f; Tantaoui et al. 1996). Under favorable conditions, the fungus spores (Fig. 8.4) germinate and attack the roots, develop in vessels, and colonize the trunk to infect palms in the apical part of the tree. External symptoms are characterized on the leaf by external hemiplegia character (one side dried) and dried palm leaves having the appearance of wet feathers. The disease can cause atypical symptoms corresponding to bilateral drying and browning extending the middle of the rachis. Internal symptoms are characterized by browning of vessels of the roots, trunk, and attacked leaves (Fig. 8.4). Infected leaves dry out one after the other until the death of the tree which may be between 6 months and 2 years depending on response level of the plant. Fig. 8.4 illustrates the different stages of the evolution of symptoms on susceptible palms until death. All organs of the tree can be infected except the spikelets and dates. The means and the modes of dissemination are numerous: e.g., offshoots, leaves, contaminated soil, decomposed parts of diseased palms, irrigation, root contact, labor tools, transplantation, contaminated manure, and sand-laden wind.

Host plants and pathogen variability The pathogen attacks other palm species such as the Canary Islands palm (*Phoenix canariensis* Chab.) and other known species as healthy carriers of the parasite: henna and alfalfa. Based on aggressivity levels and molecular markers, research has shown genetic variability in the pathogen population (Fig. 8.5) (Sedra 1993b, 2003b, 2006b, 2007b, 2008a, 2011f; Tantaoui et al. 1996; Sedra and Zhar 2010).

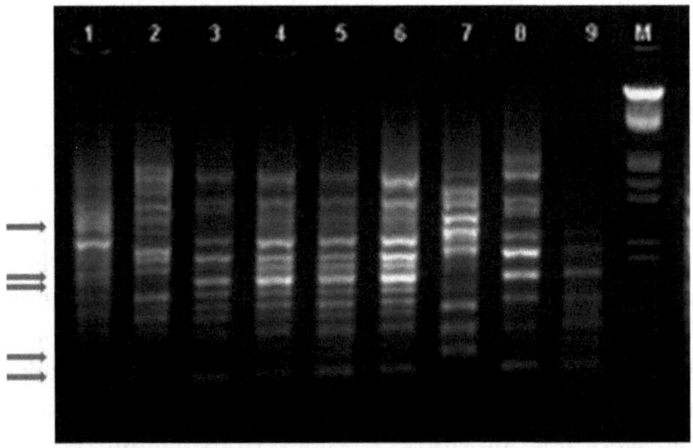

Fig. 8.5 Genetic variability within the pathogen population based on molecular markers using the technique of microsatellite ISSR-PCR of genomic DNA isolates of Moroccan and Algerian (from 1 to 9) using primer Mic43 $(AGG)_6$. Electrophoretic profile on a 1.8 % agarose gel of PCR products. M contains the DNA markers with sizes in Kb/Eco R1/Hind II/BAP. The arrows indicate the variability

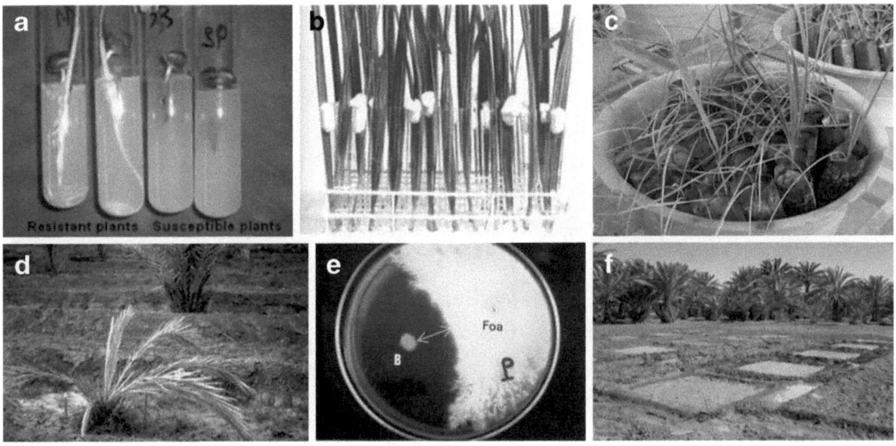

Fig. 8.6 Techniques of evaluation of plant resistance to bayoud in the laboratory in vitro selection toxins among seedlings (**a**) or detached leaflets (**b**) and selection in greenhouse (**c**) and in field (**d**) using the pathogen. (**e**) Selection of antagonist bacterium (*Pseudomonas* sp.) that inhibits the pathogen growth on medium. (**f**) Soil fumigation and solarization for pathogen eradication technique at Zagora Experiment Station

Strategy and control methods In 1967, investigations were initiated and planned; given the complexity of bayoud, control methods have focused on genetic control by selection of disease-resistant varieties, and since 1981, the research has been diversified. Numerous research studies have been developed in Morocco:

(a) Development of evaluation of plant resistance to bayoud in the laboratory with in vitro techniques using toxins and greenhouse and field using the pathogen (Fig. 8.6) (El-Fakhouri et al. 1996, 1997; Sedra 1994a, b; Sedra and Besri 1994; Sedra and Lazreak 2011; Sedra et al. 1993, 1998b, 2008a)
(b) knowledge of the mechanisms of resistance and infection (Amraoui et al. 2004, 2005a, b; Baâziz et al. 1994; Bendiab et al. 1993; El Hadrami et al. 1996; Ziouti et al. 1996)
(c) Soil suppressiveness or resistance to bayoud and microorganisms antagonistic to the parasite (Fig. 8.6) (Sedra 1993c, d, 2003b, 2006a, 2008b, 2010b; Sedra and Rouxel 1986, Sedra et al. 1994a, b)
(d) Soil fumigation and solarization for pathogen eradication technique (Fig. 8.6) (Essarioui and Sedra 2007, 2010)
(e) Characterization and varietal selection by conventional and molecular approaches (El-Houmaizi et al. 2002; El-Youssefi 1987; Sedra 1997, 2000a, 2001a, 2003b, 2012; Sedra et al. 1993, 1996)

In 1987, the first successful varieties such as Najda (INRA-3014) were selected, multiplied, and distributed to farmers to restore orchards devastated by bayoud (Sedra 1995, 2003b, c, 2005a, 2007d, 2010a, 2011a, f). New efficient and resistant varieties to bayoud (Al-Amal (INRA-1443) and Bourihane (INRA-1414) and

Fig. 8.7 Adult palm tree (from organogenic tissue culture) of selected var. Najda (INRA-3014) in date production in field (**a**, **b**). Examples of in vitro plantlets of var. Najda distributed by local services of Ministry of Agriculture, **c** planted by farmers to reconstruct a devastated orchard, **d** sparse as a result of bayoud

recently Sedrat (INRA-3003) and Darâaouia (INRA-1447)) were selected and some of which are being propagated (Sedra 2005a, 2007d, 2010a, b, 2011a, f). However, difficulties are expected, including the emergence of new races of the pathogen that could overcome varietal resistance used. Possible alternative methods of control based on pathogen eradication by soil solarization, fumigation by metam sodium, and the use of antagonistic microorganisms have been demonstrated, and research applications are in process (Essarioui and Sedra 2007, 2010; Sedra 1993a, b, 2003b, 2006a, 2008b, 2010b; Sedra 1994a, b). The strategy adopted in Morocco led to results mentioned above that are already in use in the field (Sedra 2003b, c, 2005a, 2007d, 2010a, 2011a, f, 2012, 2013). Indeed, since 1987, more than 800,000 plants of var. Najda (disease resistant) produced by tissue culture were distributed free to farmers to rehabilitate orchards devastated by bayoud (Fig. 8.7). The dates of this variety have been marketed for several years (Fig. 8.8).

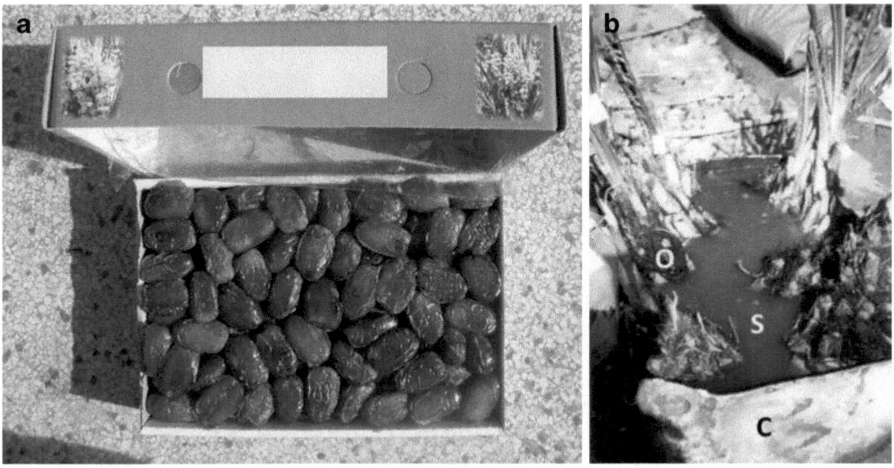

Fig. 8.8 (**a**) Dates of var. Najda packaged and distributed at different great souks of Morocco. (**b**) Chemical treatment of palm offshoots (*O*) by dipping in a systemic fungicide solution (*S*), e.g., hymexazole (0.2 %) or methyl thiophanate (0.2 %) contained in a vat (*C*)

8.2.3.2 Major Harmful Pests

Red Palm Weevil This insect (*Rhynchophorus ferrugineus* Oliv.) is spread over the five continents of the world. In North Africa, it is declared present in all countries except Algeria. As mentioned above, this pest was introduced in Morocco in 2008 in the northern city of Tangier. So far it has not been detected outside that city. The insect attacked hundreds of Canary Islands palm focusing on the apical portion of the trees (Fig. 8.9). The insect lays its eggs in wounds on the trunk and at the leaf base where the fiber is wet. The larvae burrow in and weaken the tree which dies quickly if the attack is severe. Figure 8.9 shows the damage caused by this pest, control methods, and the possibilities of its dispersal. Attack intensity varies from the presence of a few galleries to tree fall after the pest has drilled inside the trunk. In addition to bayoud, Moroccan palm groves cannot tolerate another destructive pest like red palm weevil. Indeed, recent contamination of palms by this pest in Tangier could represent, if the pest is not eradicated, a serious threat to ornamental palms (Fig. 8.9) in North Atlantic regions and ultimately date palm oases in North Africa (Sedra 2009, 2012, 2013). This represents a serious threat because there is currently no resistant palm varieties to this pest or 100 % effective direct control techniques. Alerted by the second scourge in a limited area in the north, Morocco should urgently develop its own strategy adapted to the reality on the ground and taking advantage of foreign experiences. Because of insect behavior and the nature of host range, this pest is classified as one of the most difficult to control pests in the world. Several attempts using IPM were tried (Sedra 2009, 2012, 2013); these efforts have reduced its incidence in some cases, but could not yet prevent its spread or eradicate the pest in a contaminated area.

Fig. 8.9 Symptoms of the attack of leaves (**a**) by red palm weevil of Canary Islands palm in Tangier City, attack focused on apical part of tree (**b**), that is covered after cleaning and pesticide treatment (**c, d**) possible spread of the pest from the North to South area because the presence of nurseries and city ornamental palm plantation, (**e**) example of pheromone trap used in Tangier to control pest spread

The control methods applied by ONSSA services in Morocco have given good results to limit spread of insect out of infested region at Tangier. In order to control this pest, the recommended Integrated Pest Management (IPM) consists of the following measures:

(a) Good tree maintenance and creation of a monitoring and control systems.
(b) Make certain to sound an emergency alert of the presence of new infestations to plant protection services or nearest agriculture services.
(c) Limit and treat with mastic insecticide injuries to the trees, especially those caused by farming activities.
(d) Insecticide treatment (malathion 0.2 % or imidacloprid or chlorpyrifos (0.1 %)) by spraying or injection of systemic insecticide and wrapping infested palms with plastic sheeting.
(e) Destruction of heavily infested and incurable palm trees by incineration and insecticide treatment of apparently healthy and surrounding date palms
(f) Accentuate monitoring in the surrounding area and quarantine the infested area in addition to strengthening of control points.

Date Moths Locally, this insect is called *soussa tmar*, and it is considered the most harmful to fruit quality. The moths are represented by three species with *Ectomyelois ceratoniae* Zell. the most important. The larvae are responsible for almost all damage of fruits and cause wormy dates (Fig. 8.10). The activity of the

Fig. 8.10 High infestation of leaf by white scale (**a**), wormy damaged dates by larvae of date moth (**b**), population of white termites (**c**) and spraying infested leaves with adequate pesticides, (**d**) to control pests

insect continues in warehouses and storerooms. Wet conditions in these locations promote significantly the invasion of moths. Damages in the field during fruit maturity range from 1 to 4 %, but the significant damages are in storage areas where it can reach 70 %. In order to control this insect, the following IPM is recommended:

(a) Practice good crop management in orchards to improve tree vigor.
(b) Protect fruit bunches by covering them just after fruit set or a week later using cloth or bags with very fine meshes to prevent insect entry (Fig. 8.10). A metal ring can be used to separate the dates and the net.
(c) Pack harvested dates immediately after harvest using clean cases and bags.
(d) Avoid mixing dates of the new harvest with those of previous harvests.
(e) Ensure clean warehouses and use lime anti-insect paint on the walls at the end and beginning of each season.
(f) If necessary, chemically treat date palms by two sprays at intervals of 2–3 weeks with malathion (0.15–0.3 %). The first application is done 8–10 days after fruit set.

(g) Disinsect dates for storage by fumigation under cover using nontoxic gases (carbon disulfide and carbon tetrachloride).
(h) Disinsect dates for storage by heat treatment of 55–60 °C for 1–2 h (e.g., Gonet oven type) for small quantities of dates.

White Scale The insect *Parlatoria blanchardi* Targ., called locally *al-qomila*, is the most important pest causing major damage in oases (Fig. 8.10) because of its high infestation levels and presence in all oases. In Morocco, there are 3–4 insect generations per year: in spring, summer, autumn, and winter. High density of date palms and absence of good pruning, the use of leaves for fencing orchards, and leaving large numbers of offshoots on the palms are the main factors that contribute to aggravation of insect attacks (Sedra 2003a, b, c, 2012, 2013). When the infestation is high, the tree is weakened, and the fruits become unfit for human consumption; attacks may lead to the death of young date palms. In order to control this insect, the following integrated management is recommended:

(a) Pruning of the tree and its offshoots and cutting and burning infested leaves.
(b) Spraying infested leaves with pesticide 3–4 times a year depending on the number of generations of the insect. Often, it is advised to carry out this operation after the fruiting season, at the beginning of winter using insecticides, dimethoate (0.15 %) and malathion (0.15 %), and spring using another insecticide like methidathion (0.15 %) (Sedra 1995, 2003b, 2012).
(c) The use of biological predators such as *Chilochorus bipustulatus* L. var. *iranensis* to supplement chemical control during the year. For this, it is necessary to establish small laboratories at the regional level in order to produce mass populations of predators and develop integrated control and guidance for the farmers (Sedra 2003a, c, 2012, 2013).

Date Mite Locally called *boufaroua*, this mite [*Paratetranychus* (*Oligonychus*) *afrasiaticus* Megr.] is considered harmful to fruit quality in some localities. Damage can be considerable on date bunches when the conditions are favorable. The pest spreads especially in areas affected by drought or lack of irrigation and oases with very high tree density. It attacks the fruit during its growth and feed on it leaving residue and dust. When the infestation is severe, fruits become useless (Fig. 8.10). Control is effected by spraying date bunches with sulfur powder (100–150 g/palm) or the use of the sulfur pesticide in water (concentration of 0.25 %) or other pesticides such as malathion (0.2 %). Spray 2–4 times depending on attack severity and duration of fruit development. Pesticides should not be used during the month before harvest to avoid pesticide poisoning through its aftermaths. For the success of this control process, farmers and agricultural extension agents play an important role.

White Termites Locally called *larda*, white termites (*Microtermes diversus* Silv., *Amitermes* sp., and *Reticulitermes* sp.) (Fig. 8.11) attack palm roots usually weakened by a lack of proper maintenance or a parasite infection. Termites build clay galleries from the surface up along the trunk exterior. They weaken date palms and often cause death, especially of young trees. In Morocco, the invasion and the development of these pests are increasing especially when drip irrigation is used and in

Fig. 8.11 Symptoms of attack of date bunches by mites (*Paratetranychus afrasiaticus* Megr.) (**a**). Damages of defoliant palm leaves caused by massive insect populations of Saharan locust in Tata area (**b**). Damages caused of dates in some palm groves around Ouarzazate and Tata by birds, sparrows (*Passer domesticus* L. and *P. hispaniolensis* Temm.) (**c**) and rats (*Rattus rattus* L.) (**d**)

recent plantations situated in farms neighboring traditional oases. The IPM recommended is as follows:

(a) Cull if necessary palms heavily attacked and burn them immediately.
(b) Clean the parts of palm trees of the infested clay galleries and treat the location with an insecticide (chlorpyrifos-ethyl 1.2 %) (Sedra 2012, 2013). This treatment can protect the palms for 2–3 years. Other insecticides registered in Mauritania may be used.
(c) Provide adequate and regular date palm maintenance (e.g., irrigation, hoeing to remove deep galleries and fertilizer) in order to reduce their susceptibility to termite attack. Good weed control is also recommended.

8.2.3.3 Control of Other Potential Enemies of Date Palm

Saharan Locust *Schistocerca gregaria* Forsk. is an occasional pest, and although attacks are very rare, they still threaten oases. In some years Morocco is alerted to a locust plague in the Saharan regions. Locust attacks are sometimes considerable on date palms. Massive insect populations defoliate the trees. Fig. 8.11 shows the damage caused by locusts in the Tata Oasis and the importance of invasion. Chemical control is by aircraft in partnership with affected countries. Usually malathion (0.2–0.3 %) is sprayed by aircraft or on the ground by all-terrain vehicles. Poison baits can be used in the areas of reproduction. In the case of widespread invasion in the oases, monitoring and control are performed by the Ministry of Agriculture and ONSSA.

Rodent Pests Several other animal enemies cause damage in some date producer countries (mollusks, birds, rodents, and herbivorous animals). In Morocco, animals such as cattle, sheep, goats, and camels sometimes wreak havoc on unfenced orchards. Moreover, rodents such as the black rat (*Rattus* L.) attack many agricultural products after harvest and most vegetables and fruits including dates. They cause considerable damages in some palm groves around Ouarzazate and Tata, for example (Fig. 8.11). The house mouse (*Mus musculus* L.) can also wreak havoc on stored dates. Preventive control against rodents is based on thinning clumps of date palms, flooding burrows and destruction of their small progenies, incineration of the remaining crops, and depriving rodents of suitable habitat and sources of nutrition. For rodents of stored products, this preventive strategy is the good construction of adequate storage facilities so as to prevent their introduction and proliferation inside. Poisoned bait prepared as pellets or a preparation must be placed in the evening near each burrow or side tunnel. These reduce their population to a lower threshold level. Placement of bait is done twice a year after the harvest of winter and summer in the oasis products. Rodenticides registered in Morocco are (a) for field mice, poisoned bait by strychnine sulfate (15 g/kg of food substrate, rat poison exclusively for services of ONSSA and the Ministry of Agriculture) or Vertox Pellets (Brodifacoum 0.01 % used at 5.1 mg/kg) and (b) for house mice, in local storage, 50 rat poison (Alphachioralose 95 % used at 100 g/kg).

Harmful Birds Some birds including the house sparrow (*Passer domesticus* L.), the Spanish sparrow (*P. hispaniolensis* Temm.), and the garden bulbul (*Pycnonotus barbatus* Desf.), found in some date palm groves, often attack fruit at rutab stage in early morning and before sunset causing significant damages (Fig. 8.11). Preventive control against birds requires proper bagging before the date maturity; if major attacks occur, noise emission repellents for more than 20 min to prevent birds from settling and feeding, and destruction of nests, eggs, and fledglings are effective. As in the case of locusts, if an invasion is widespread, monitoring and control are organized and supervised by the Ministry of Agriculture and ONSSA.

8.2.3.4 Competitive and Harmful Weeds of Date Palm

In Morocco, several weed species are harmful to date palm. Perennial plants are those that compete with date palms. Among these are: *Elytrigia repens* L., *Cynodon* sp., *Cyperus* sp. *Imperata* sp., *Convolvulus* sp., and others. Quackgrass (*Elytrigia repens* L.) is the most common and most difficult to control, spreading primarily by creeping rhizomes about 2–4 mm in diameter and able to invade quickly. Its spread is favored by tillage, regular irrigation, and lack of crop competition. Recommended methods of control involve hand weeding or with a small tiller, either requires repeated labor under dry conditions. Herbicides based on glyphosate are most effective to destroy quackgrass.

8.3 Genetic Resources and Conservation

Initial Moroccan scientific investigations on date palm were by the French in the 1950s in response to the damage caused by bayoud. A subsequent phase of research began in 1965 following the establishment of the Zagora Experiment Station. Afterward, the studies were expanded for oasis agricultural development. A research structure was created, Central Station of Saharan Agronomy (SCAS), in Marrakech with responsibility for research in date groves and experiments to integrate crops and livestock with date palm.

Research on date palm focused on varietal selection (cvs. resistant to bayoud), varietal profiles, and breeding and farming techniques. In 1980, palm tissue culture began in a Marrakech laboratory. During 1980–2000, scientists and technicians were recruited to strengthen the staff, and other experimental stations were added. In 1987, SCAS was reorganized into the National Program of Date Palm. Since then, research has focused on production, protection, breeding, biotechnology, and food technology. To strengthen the efforts in Marrakech, a regional INRA center was created in 2003 at Errachidia, near the oases, and a tissue culture laboratory (PALMINRA) established in 2011.

Data of date palm grove surveys, complemented by updated data during the past 30 years, observations, and investigations, have permitted determination of the distribution of the major cvs. and *khalts* in different regions of the country (Djerbi et al. 1986; Sedra 1997, 2003b, 2011d; Sedra et al. 1996). *Khalts* from spontaneous seedlings (hybrids with an unknown male parent) constitute about 48.2 % of total date palms. Certain *khalts* multiplied by farmers become cvs. and could be added to the total of 453 cvs. identified thus far. The quantitative and qualitative importance of the genetic material of date palms varies according to the growing areas.

8.3.1 Cultivars

Distribution of the major Moroccan date palm cvs. by region is shown in Fig. 8.12. The Drâa, Ziz-Tafilalet, and Bani regions are the richest in numbers of cvs. Some cvs. recur in different regions, while others are rare and limited to specific locations. The number of cvs. varies according to region. Among the total of 453 inventoried cvs., lesser represented cvs. in tree numbers, along with rare ones, are in the

Fig. 8.12 Importance and spread of main date palm cvs. according to regions and oases in Morocco

majority and exceed 400. Most cvs. are found in the Drâa Valley, Tafilalet, and Bani. The most dominant cvs. in each region are presented in Table 8.5. Boufeggous cv. alone is distributed in all regions even in small areas where irrigation is became practically impossible. In general, 12 cvs. are preferred by farmers: Medjool, Boufeggous, Jihel, Bouskri, Black Bousthammi, Bouslikhene, Outokdime, Bouittob, Ahardane, Aguelid, Taabdount, and Aziza Bouzid, along with the new selected var. Najda distributed in recent years. Some cvs. are famous in their regions of origin.

Table 8.5 Dominant and rare date palm cultivars present in the main palm groves of Morocco ranked in descending order of distribution

Palm grove or region	Dominant cultivars[a]	Rare cultivars
Drâa	*Jihel*, *Bousthammi Noire*, Iklane, *Boufeggous*, *Bouskri*, Aguelid, Bourar, Jaâfari, *Ahardane*	El Malha, Oum Tamarmaite, Tinbaba, El Kouhila, Acedame, Jaafari, Aloud-Ezzine, Boutoutobte Loubania
Tafilalet-Ziz	Bouslikhéne, *Boufeggous*, Racelahmer, Boucerdoune, *Medjool*	N'kila, Koul ou Skout, Aghtita, Timssassine, Chatouia
Bani	*Jihel*, *Boufeggous*, Iklane, *Bouskri*, *Bouittob*, Taranimt,	Aâdam Bouskri, Bouichane, Filal, Agueroui, Boudi ou Smen, Aâdam Ibrerin, Azbabay, Timachiouine
Anti-Atlas	*Boufeggous*, *Jihel*, Boulizeft	Boufezoua, Mouch
Saghro	*Boufeggous*, *Bousthammi noire*, Iklane, *Jihel*, *Bouskri*, Bouslikhéne	Mektoub, Tazzeggart, Maktoub
Oriental (East)	*Boufeggous*, *Assiane*, Afroukhten Tajent, Aghras, *Aziza Bouzid*, Hafs, Taabdount, Bouijjou	Amira, Lakmil, Taberchant
Between Saghro and High Atlas	*Boufeggous*, *Bouzeggar*, Azigzao, Outokdime, *Bouskri*, Hafs, Lahlaout	Agoujil, Ouarzaout, Takfout

[a]Cultivars in italics are those preferred by farmers in each region

This is the case for example, of cvs. Medjool and Bouslikhene in the Ziz Valley, Tafilalet, Jihel; Bouskri cv. in Drâa Valley, Anti-Atlas, and Saghro; Bouittob cv. in Bani, Bouzeggar, and Outokdime; higher-elevation cvs. in the region between Saghro and High Atlas; and finally, Aziza Bouzid and Taabdount cvs. in the East. Very rare cvs. are numerous, often localized in certain areas of one or a few known palm groves. Some examples of these cvs. are listed by region in Table 8.5. Given their very small numbers, these cvs. require preservation and improvement.

8.3.2 Seeding Dates (Khalts)

Surveys were carried out in date groves to identify quality seedling dates in two periods. In 1973–1974, Louvet and Toutain (1973) made a survey to locate individuals of quality and resistance to bayoud but limited their work to areas where the disease was present. More than 100 clones were selected and planted at the Zagora Experiment Station. In the second period (1979–1983), surveys covered all major date groves. This mass assessment covered more than 2.6 million palms and represented a collaborative effort among institutions at the national, provincial, and local levels (Djerbi et al 1986; Sedra et al 1996). More than 3,000 *khalts* were identified and evaluated on the criterion of fruit quality. Selected offshoots were transplanted to two experimental sites for further evaluation on agronomic criteria and response to bayoud. From the seventh year when normal *khalt* fruit production begins, genotypes were selected with performances that satisfy the needs of farmers (Sedra 2001a, 2005a, 2011a, f, 2012, 2013; Sedra et al. 1996).

The results obtained from the surveys showed that quantitative and qualitative importance of *khalts* varies according to region. The total number of selected *khalts* for their quality in all oases reached 2,337 genotypes. This represents 0.09 % of the total *khalt* population of 2.6 million (Djerbi et al. 1986; Sedra 1995, 1997). The importance of selected clones varies from 0.01 to 0.27 % depending on the oases. Fruit quality of these genotypes is much diversified and may exceed that of some commercial cvs. Offshoots were planted in experimental sites for evaluation. Given losses due to the non-recovery of offshoots, only 1,130 genotypes were tested at a rate from 1 to 7 palm trees per clone. Sedra (1995, 1997, 2003b) indicates that about 40 % of these clones were susceptible to bayoud. Although they exhibited extreme susceptibility to the disease, but produced good fruit, several clones, along with certain rare threatened cvs., have been preserved in the germplasm collections of Marrakech, far from bayoud contaminated areas (Sedra 2007d).

8.3.3 Selection and Breeding

The agro-morphological evaluation of Morocco's diverse date palm patrimony was assessed by Sedra et al. (1996), morphological characterization of cvs. initiated (El-Houmaizi et al. 2002; El-Youssefi 1987; Sedra 2001a, 2003b, 2012, 2013; Sedra et al. 1996), and agronomic, morphological, biochemical, and molecular descriptors developed (Sedra 2001a). The cultivar genotyping and diversity of a set of 45 accessions from Morocco and foreign countries were evaluated using molecular markers (Sedra et al. 1998a, b, 2011b, e). The main objective was selection of high-yielding types of good date quality with resistance to bayoud. Conventional selection was made from the following material.

8.3.3.1 Selection of Moroccan and Foreign Cultivars and *Khalts*

Selection for fruit quality was not a priority since there are excellent cvs. on national and international markets such as Medjool and Deglet Noor, but they are extremely susceptible to bayoud. The challenge was to select the cvs. that are both of good fruit quality and resistant to bayoud.

Varietal Behavior Towards Bayoud In general, the best cvs. are extremely susceptible to bayoud, whereas those producing fruit of lesser quality are the most resistant (Louvet and Toutain 1973; Saaidi 1992; Sedra 1990a, b, 1995, 2003b). Known cvs. with resistance are Black Bousthammi, White Bousthammi, Iklane, Tadmainte, Sairlayalate, Boufeggous ou Moussa, and Boukhanni. Cultivar Boukhanni, found only in a locality of the Drâa Valley, has shown resistance to experimental inoculation (Sedra 1994b, 1995, 2003b; Sedra et al. 1996). Furthermore, no positive correlation was found between the behavior of *khalts* towards bayoud and fruit quality (Sedra 1995). Among the 1,130 genotypes tested, 40 % were found to be susceptible. Many genotypes from this field work were selected and became varieties (Sedra 2003b, 2005a, 2011c). Some individuals of good *khalts* present in date groves are threatened with extinction by bayoud. Noted the loss of two famous

Moroccan cvs., Ildrar and Berni, in the early nineteenth century. It should be noted that all 14 commercial cvs. from Tunisia, Algeria, and Iraq proved sensitive to bayoud in field experimentation (Sedra 1992b, 1995).

New Selected Varieties New varieties were selected to enrich the national genetic patrimony and supply markets with quality dates (Sedra 2003b, 2005a, 2007a, 2010a, 2011a, b, e, f; Zaher and Sedra 1998). After obtaining var. Ayour (INRA-3415), Hiba (INRA-3419), Tanourte (INRA-3414), Al Baraka (INRA-3417), Tafoukte (INRA-3416), Mabrouk (INRA-1394), and Khair (INRA-3300) and exploiting the var. Najda (INRA-3014) (Sedra 1995, 2005a, b), seven new varieties were selected. These are Darâaouia (INRA-1445), Sedrat (INRA-3003), Al-Amal (INRA-1443), Al-Fayda (INRA-1447), Bourihane (INRA-1414), Mabrouk (INRA-1394), and INRA-3010 (not yet named) and two selected males Nebch-Bouskri (INRA-NP3) and Nebch-Boufeggous (INRA-NP4); these nine possess promise to control bayoud. Some of these varieties such as Sedrat (INRA-3003), Al-Amal (INRA-1443), and Darâaouia (INRA-1445) have high performance and are resistant to bayoud (Sedra 2003b, 2005a, 2007a, b, c, d, 2010a, 2011a, b, e, f, 2012, 2013). Other varieties recently characterized are less demanding of heat and have the advantage of being adapted to areas north of the Atlas Mountains such as the Marrakech climate due to their ability to produce mature dates (Sedra, unpublished results). Some varieties have been selected, multiplied by tissue culture, and distributed to farmers to restore orchards devastated by bayoud and restructure the palm groves to improve production (Sedra 2003b, 2005a, 2010a, 2011a, b, e, f, 2012, 2013). The level of conformity of produced plantlets in the field was assessed using phenological descriptors and molecular markers (Sedra 2005b). In Morocco, 14 selected varieties, including males even if multiplied by tissue culture technique or their multiplication is in progress, were protected and their use restricted by INRA. Laboratories, nurseries, and other persons are not permitted to valorize these varieties commercially without INRA approval.

8.3.3.2 Selection from Crosses

Date palm breeding by hybridization was initiated in 1972. Later selection schemes by Sedra (2003b) improved strategies and incorporated the addition of molecular and biochemical markers. During the period 1979–1983, more than 100 cultivar crosses were made using 20 males and 60 females of different origins and selected from different agricultural and biological traits. In 1998–2010, the number of crosses reached 1,000 producing over 1.2 million seeds. These crosses included those made between several types of known cvs. and males, between families and backcrossings, and between siblings. Important results (Sedra, unpublished) were obtained and related to: (a) procedures for the transmission of resistance, (b) parent effect and choice of good genitors, (c) cross effect and choice of progenies, and (d) segregation and genetic nature of resistance. Some descendants have served for selection and others for genetic studies. More than 100,000 descendants were tested at the seedling stage by artificial inoculation with the pathogen in the glasshouse (Saaidi et al. 1981; Sedra 1990a, 1995). The 8,000 young palms that have survived in the greenhouse to inoculation with the pathogen or are tolerant were planted in the field to assess their behavior towards bayoud under natural conditions and to evaluate their quality of fruit production in subsequent years (Sedra 1990a).

After several years of observations of more than 3,000 progenies, the percentage of plant mortality was 30–80 %, depending on the crosses. The first sort was performed on the basis of fruit quality of the surviving palms. Preliminary results permitted the selection of 40 descendants for their fruit quality ranging from good to very good. Selected descendants have been artificially reinoculated in the field by the pathogen according to a technique developed by Sedra (1994a) and compared with susceptible and resistant trees control towards bayoud. The evaluation of these 40 descendants from controlled crosses revealed 87 % of susceptible material (Fig. 8.13) (Sedra et al. 1996). Five showed resistance and good fruit quality (i.e., 0.005 % of the initial number). These selected varieties have been proposed for tissue culture (Sedra 1995, 2003b). The origins and some agronomic traits of selected varieties from crossing are presented in Fig. 8.13. Other palm hybrids from controlled crosses were subsequently selected (Zaher and Sedra 1998).

Fig. 8.13 (**a**) Descendants from crosses sorted at seedling stage by artificial inoculation with bayoud in a glasshouse. (**b**) Surviving date palms planted in the field to assess behavior towards bayoud and evaluate future fruit production. (**c**) Examples of selected date genotypes of *khalts*, No. 1221, 1355, 1414, and 1443, and from controlled crosses, B14/20 (Deglet Noor × American male DN3), C2/107 (Black Bousthammi (BSTN) x American male HLW1 (from cv. Halawy), A16/44 (Tadmainte local × male), C1/78 (Bouzeggar × local male), and B42/2 (Deglet Noor × local male). Note the diversity and differences in appearance compared to common resistant cvs. Black Bousthammi (BSTN) and Sairlayalate (SLY) and susceptible and good commercial cv Boufeggous (BFG)

Numerous research achievements of date palm breeding were obtained, some of which are summarized as follows:

(a) Some results showed that crosses between susceptible parents give 84–93 % susceptible descendants and on occasion 100 %, while descendants from crosses between resistant parents show only 5–10 % of attack. Crosses between resistant and susceptible parents often give descendants that present 35–70 % mortality. These results suggest that resistance is recessive in some cvs. and dominant in others.
(b) Djerbi and Sedra (1986) showed that the resistance of the date palm to bayoud appears to be additive and governed by minor genes, and it is possible to assess the behavior of parents by testing their descendants towards the disease.
(c) Some female cvs. such as Deglet Noor, Jihel, and Black Bousthammi transmit better to their descendants 2, 4, and 5 fruit qualitative characters (Zaher and Sedra 1995).
(d) Some male genitors transmit best resistance to descendants than others. The male Nebch-Boufeggous (INRA-NP4) selected gives an example.
(e) Some crosses (backcrosses) of cv. Medjool and Halawy (the best Iraqi cv.) gave descendants with the performance of date quality (e.g., fruit size, shape, and weight, pulp percentage in the fruit) higher than the parents. Some examples of these descendants are shown in Fig. 8.14.

Fig. 8.14 Examples of selected descendants from backcrosses with higher performance of dates quality than the parents: INRA-6001, INRA-6004, and INRA-6006 descendant from cross Medjool × American male MJH5 and INRA-6008 from cross Halawy × American male HLW3

Fig. 8.15 (**a**) Plantation and development of vitro-plants derived from selection using an irradiation and screening test with pathogen toxins. These selected mutants, and other plants as controls, are under field study at the Zagora Experiment Station, INRA, (**b**) collection of genetic material (e.g., cvs., descendants, or hybrids, *khalts*, vitro-plants) at the Marrakech (Saada) Experiment Station, INRA, for example

8.3.3.3 Selection by Mutagenesis

INRA Research in partnership with the International Atomic Energy Agency (IAEA) permitted in vitro propagation of some resistant mutants of the commercial bayoud susceptible cv. Boufeggous employing gamma irradiation and in vitro selection using pathogen toxins (Fig. 8.6) (Sedra 2011b, f; Sedra et al. 2008). Ten selected mutants are under field evaluation at the Zagora Experiment Station (Fig. 8.15) to confirm resistance to bayoud and evaluate their agro-morphological characters.

8.3.4 Genetic Diversity of Date Palm Patrimony

8.3.4.1 Agronomic Characteristics of Cultivars and Clones

Destination of Production Moroccan production is characterized by a predominance of cvs. of low to medium fruit quality. Indeed, cvs. of low fruit quality (e.g., Black Bousthammi, Iklane, Bouslikhene) represent 40 %; those of medium quality (e.g., Jihel, Bourar, Bouittob) represent 35 % and cvs. of good quality (e.g., Medjool, Boufeggous, Aziza Bouzid) 25 %. In recent years the destination of date production has been divided among direct human consumption (30 %), animal feed (20 %), and marketing (50 %) (Bouziane 2010).

Maturity Dates Different periods of fruit maturity are defined by the heat requirements of cvs. For major cvs., these data are reported as part of the cultivar description. A study of fruit maturity on a sample of 280 genotypes of clones (about 40 % of clones from the collection of *khalts* at Zagora Experiment Station) and 32 known cvs. revealed that mid-season clones are most frequent (60 %), followed by moderately precocious clones (31.8 %) (Sedra et al. 1996). Groups of precocious and late clones are each represented by 7 clones (2.5 %). In the group of cvs., Ahardane and Aguelid proved precocious (6.2 %), while 4 cvs. Bouzeggar, Iklane, Jihel, and Medjool were classified in the group of late cvs. (12.5 %). Most cvs. are represented by those that mid-season and moderately late fruit maturity is mid-season and moderately late (62.4 %). Other observations at palm groves have shown that certain cvs. (e.g., Bakria and Tagnanite) are very precocious, ripening fruit in July.

Appearance and Quality of Dates The main assessment criterion is general presentation of fruit (size, color, shape, texture) (Sedra et al. 1996). The appearance of soft dates degrades more rapidly than dry dates. Sedra et al. (1996) evaluated and found that cvs. Medjool, Boufeggous, Bourar, Ademou, Bouijjou, Mekt, and Jihel are characterized by a better appearance and fruit quality among a sample of 280 genotypes of clones (*khalts*) and 32 common cvs. By contrast, 40.6 % of cvs. studied are characterized by a low fruit quality, and one-fourth of the cvs. produce soft dates which are difficult to store under traditional conditions. Ademou, Bourar, Bouskri, Jihel, and Medjool cvs. showed the best performance for date conservation among a sample of 245 clones and 32 cvs. that were stored for 3 months in reed baskets at room temperature in Zagora Experiment Station. This study also showed that the production of 74 % of *khalts* has been devalued by the degradation of fruit, and 71.8 % cvs. are more or less suitable for conservation. This demonstrates the usefulness and the necessity of drying dates and storing them under cold conditions. Cultivars bearing black dates covered one-fourth of the lot. These cvs. are less valued at the market than brown or golden dates. The comparative study of some fruit characteristics of selected clones (*khalts*) and cvs. operated more traditionally indicates that the clones offer greater diversity and higher performance than those of most observed cvs. Over 80 % of clones have fruit characters whose value exceeds that of cvs. (Sedra et al. 1996).

Date fruit weight at maturity is an essential criterion on which to base the choice of producers and traders. Among 67 cvs. and selected varieties, the number of dates per kg varies from 40 to 294 (Sedra 2003b). Among 32 common cvs. studied, only Ademou, Boufeggous, Bouijjou, Bourar, Medjool, and Mekt bear fruits of sufficient individual weight such that 100 dates together exceed a weight of 1 kg. On the other hand, 8 cvs. have a pulp percentage greater than 92 %, while 6 cvs. produce dates with a length exceeding 3.8 cm and date width greater than 2.4 cm (Sedra et al. 1996). Among selected varieties with bayoud resistance and good fruit quality, Al-Amal (INRA-1443), Sedrat (INRA-3003), Darâaouia (INRA-1445), and others have high performances. These record a pulp percentage (90.9–95.2 %) and date weight (14.1–21.9 g) and are desirable light-brown fruits (Sedra 2003b, 2005a, 2011b, 2012, 2013); they are precocious to mid-season and have good storage qualities. Comparison and ranking of performance of those cvs. to 34 selected clones on date weight revealed that among the first 20 cvs. and selected clones, only cvs. Medjool and Boufeggous are, respectively, in the first and sixth place (Sedra 2003b). Indeed, Medjool cv. fruits are the heaviest since about 40 or fewer weigh 1 kg. Cultivars including Boufeggous ou Moussa, Ademou, and Jihel ranked, respectively, in the 22nd, 25th, and 26th place.

Table 8.6 Number of different sources of date palm genotypes preserved in INRA experiment stations

Genetic resources	Saada Menara	Tassaout	Zagora	Errachidia	Total (%)
Common cultivars and selected varieties	26	73	36	6	141 (3.1)
Progenies from crosses (male and female)	187		3,030		3,217 (70.9)
Female date palms from seed (*khalts*)	24	–	740	405	1,169 (25.8)
Male date palms from seed (*khalts*)	3	–	8	–	11 (0.2)
Total	240	73	3,814	411	4,538

8.3.5 Germplasm Banks of Genetic Resources

As part of the INRA strategy, the preservation of genetic material is an important action for the future. In this framework, collections have been installed at five sites in the experimental fields of INRA (Errachidia and Zagora in the oasis region, Saada and Menara in Marrakesh and Tassaout located north of the Atlas Mountains. These collections include 4,538 genotypes represented by over 8,000 palms (Sedra 2007d). This material is kept for the preservation of national patrimony against bayoud attack and prolonged drought effects and for the use of genotypes in the date palm breeding program. Table 8.6 shows different sources of genetic material. The descendants of more than 100 crosses represent the largest part with 3,217 genotypes (70.9 %) and common cvs. (3.1 %). Figure 8.10 cited above shows an example of genetic resources collection.

The characterization and profiling of Moroccan varieties encounter several difficulties; first, in naming varieties or cvs. and second as famers' selection continues over time of new varieties from local *khalts*. However, it is clear that Moroccan date palm oases have very important genetic potential that is not being exploited to the desired level. Productivity remains below potential due to low-tech empirical methods in practice on the majority of date farms.

For several decades, INRA, in collaboration with its partners, has developed and adapted technologies to the oasis system and improved date palm cultivation. Indeed, several new varieties were selected, some of which were valuable and others are in process. This action aims to support the Moroccan Green Plan whose main objective is strengthening the proper development of the date sector.

8.4 Plant Tissue Culture

8.4.1 Role and Importance

Despite its relatively high cost, in vitro culture is the most powerful tool for rapid production of hundreds of thousands of true-to-type plantlets. In Morocco, somatic embryogenesis is used for research purposes and date palm breeding, while organogenesis was developed by researchers in 1984 for large-scale commercial

Fig. 8.16 Different stages of in vitro propagation of date palm using the technique of organogenesis. (**a**) Offshoots weaned, cleaned, and ready for use, (**b**) removal of explants (apex, leaves, leaf bases, axillary buds) to culturing (initiation phase of organogenic tissue), (**c**) production budding strains, (**d**) elongation phase and rooting plants, (**e**) transplanting into individual tubes of complete seedlings, (**f**) transfer of plants in pots in the greenhouse for acclimatization, (**g**) development of plantlets in bags, (**h**) developed plants at different stages in bags of different sizes, (**i**) hardening phase of plants under shelter shade before planting, (**j**) planting of vitro-plant, (**k**) adult palm in offshoots and date production

micropropagation. The main steps of production plants using organogenesis are illustrated in Fig. 8.16. This new technology is essential for: (a) mass multiplication of plants of desired cvs. and varieties; (b) increased multiplication within a reasonable time period of selected clones having high agronomic performances that are represented by only a few specimens in order to repopulate devastated orchards by bayoud, to restructure the traditional date groves, and finally to create new farms and oases; and (c) saving rare cvs. threatened by extinction due to factors of genetic erosion of abiotic or biotic origins. It is the same for selected good quality individuals but extremely susceptible to bayoud and (d) producing plants free of diseases.

8.4.2 Research and Development

Organogenesis is preferred since it ensures maximum conformity and consistency of vitro-plants and minimal risk of genetic variation. Indeed, the comparison of some descriptors and agro-morphological and molecular characters of plantlets produced by organogenesis and their mother trees for three Moroccan cvs. detected no polymorphism or variability (Sedra 2005b). In the case of somatic embryogenesis, somaclonal variations in tissue obtained from in vitro Egyptian cvs. were detected using molecular markers and isozymatiques compared to the mother trees (Saker et al. 2000, 2006). Genetic variations reach approximately 4 % of analyzed plants representing 70 regenerated plants. Although relatively more profitable than organogenesis, embryogenesis applied in some countries has led to in the production of some stunted deformed and nonproductive adult plants; Namibia, Yemen, and Saudi Arabia are good examples.

As mentioned above, INRA Morocco has focused its research on organogenesis (Abahmane 2011; Anjarne et al. 1995, 2010; Beauchesne et al. 1986; Benjama et al. 1996; Rhiss et al. 1979). This technique of multiplication from offshoot heart material has been adapted to over 40 Moroccan cvs. and selected varieties. This technique has also been developed and applied on explants derived from inflorescences of several varieties (Abahmane 2011). So far, INRA has signed partnership agreements with three private national laboratories: Domaines Agricoles of Meknes, Issemeghy Biotechnology, Palmagro Biotechnology in Agadir, Invitro Palm Biotechnology in Casablanca, Maghreb Palm in Inzegane and Oasis Biotechnologie in Erfoud. The principle of the agreements is based on the provision of private laboratories budding strains produced by organogenesis at INRA and the laboratories being responsible for continuing the process according to the technique developed by INRA for producing conformed plantlets suitable for planting. Tens of thousands of budding stems of cvs. and selected varieties were provided by INRA, and over 800,000 plantlets were produced and planted in date groves between 1987 and 2008 with no reliable claims of problems recorded and 400,000 plantlets during in 2009–2013 under the Green Morocco Plan (PMV). The PMV program provides for the production of 2.9 million plantlets by 2020 to reconstruct, restructure Moroccan palm groves, and create new date farms on 19,000 ha. More than 4,000 ha were planted to 2013. All these recent operations will be implemented within the framework of a partnership among INRA, ANDZOA, MAPM, and Regional Directions of Agriculture (DRAs, ORMVAs, DPAs) and private commercial laboratories. Some new other private commercial laboratories will open their doors in the next few years.

Plants produced and delivered to farmers must have a number of criteria including (a) certified and authentic plants, (b) plantlets free of all pests and diseases, (c) plants having at least one pinnate leaf blade pinnate and plants with only juvenile leaves that are very susceptible to environmental stress, and (d) plants 6–12 months of age after acclimatization and with regular potting and not root-bound.

Research in tissue culture continues to improve the efficiency of organogenesis and its adaptation to new selected varieties and difficult cvs., improve organogenesis of inflorescences tissues especially for multiply selected hybrid dates without offshoots, improve the use of mycorrhizal fungi to facilitate acclimation of young plantlets immediately after transfer tubes to pots, and develop molecular techniques to assess conformity levels.

8.5 Cultivars Identification

8.5.1 Morphological Descriptors

The presence of the same cultivar in several regions bearing different names and/or the same name for different cultivars resembling each other in terms of morphological characters makes exact enumeration difficult. Cultivars often bear regional names, often derived from significant fruit characters such as color, shape, size, maturity of fruit, etc. The name may also come from the fruit or the tree resembling something from nature such as *chicken egg*. The owner of the date palm, the garden, or the village or region of origin may also provide a cultivar name.

Given the many and varied origins of cultivar names, identification based upon them is unreliable. Several studies have been done in Morocco to contribute to the identification of cultivars based on morphological characters. The problems of synonymy and homonymy led to the establishment of research strategies to evaluate genetic diversity created by farmers over time. The evaluation of phenotypic diversity constitutes a basic first step in a program to improve germplasm management of crops. Methods based on morphological traits, of either vegetative or fruit parameters, are of some benefit in evaluation of date palm genetic resources (El-Houmaizi et al. 2002; El-Youssefi 1987; Sedra 2001a, 2007d; Sedra et al. 1996). Sedra (2001a) enumerated 342 date palm descriptors of which 105 are descriptive characters (tree, inflorescence, fruit), 132 agronomic characters (e.g., maturity, pollination, resistance), 62 chemical characters (e.g., chemical structure), 6 biochemical scorers (e.g., enzymes), and 37 RAPD molecular markers. Statistical analyses using the Static and Systat programs on a sample of more than 90 date palm cvs. grown in North Africa permitted an appreciation of the interrelationships observed between all quantitative and qualitative characters to determine those that proved to be (a) discriminative, (b) highly discriminative, (c) very highly discriminative, and (d) correlated to highly discriminative quantitative characters (Sedra 2001a, 2003a; Sedra et al. 1996). The main discriminative qualitative characters were found to be consistency of pinnae, density of leaf crown, rigidity of spines, tournure of the palm, curvature of the palm, color of the leaf base, and shape of the fruit. The main discriminative quantitative characters are percentages of spiny and leaf base parts, length of central spine, pinnae and spikelet, angle formed by center spine with leaf rachis, number of spathes per tree and their length, width of fruiting bunch (central free zone), angle formed by dorsal pinnae, total number of flowers of central spikelet of the spathe, weight of fruit and seed, and length of part of lower spikelet without flowers.

8.5.2 Molecular Descriptors

Initial molecular marker studies of Moroccan date palms involved a restricted set of cultivars (Aît Chitt et al. 1995; Corniquel and Mercier 1994). The first molecular studies on large populations were conducted on 43 Moroccan date palm accessions and some cvs. from Tunisia and Iraq. The studies used PARD markers which revealed relatively low polymorphism and lack of evident organization observed among the date palm cvs. grown in Morocco (Sedra et al. 1998a, b). This outcome could be related to the mode of introduction and maintenance of germplasm (Sedra et al. 1998a, b). The foundation of the germplasm is somewhat limited. The fact that the cvs. from Tunisia and Iraq did not markedly diverge from the genetic diversity present in Morocco suggests a narrow genetic diversity of populations from which the present cvs. have been derived and maintained over several centuries (Sedra et al. 1998a, b). Exchange of cvs. between plantations and periodic development of new recombinant cvs. through sexual reproduction and seedling selection also may have played an important role. In addition, selection by farmers concerns mainly end-use quality-related genes which may represent only a small fraction of the date palm genome. Sedra (2007d, 2011b, e) selected several hundred RAPD and ISSR primers that allowed genetic diversity analysis, genotyping, and identifying of several markers candidates to distinguish partially or totally between resistant and susceptible cvs. of date palm to bayoud. A total of 550 RAPD primers 10-Decamer were tested on date palm DNA, and 170 of them were selected permitting identification of more than 300 polymorphic markers which are able to detect polymorphism and analyze genetic diversity and genotype date palm cvs. The percentage of polymorphism may reach 70 %, and 1–5 polymorphic bands per primer were generally generated. Some examples of selected RAPD primers are 5′-CACCGTATCC-3′, 5′-CCGAACACGG-3′, 5′-CATCCGTGCT-3′, 5′-GAGGGAAGAG-3′, 5′-CCGCTACCGA-3′, and 5′-CTGGGGATTT-3′ (Sedra 2007d, 2011b, e). For ISSR primers, the PCR analysis applied to 45 tested primers (short sequences) allowed the selection of 21 primers that showed a high rate of polymorphism among 45 date palm cvs. studied. These primers have revealed more than 80 polymorphic markers. The polymorphism level varies from 43 to 100 % (average 80 %) according to primers, and the average of 6.1 polymorphic bands per primer was generated. Some examples of selected ISSR primers are $(AC)_{10}$, $(AG)_{12}$, $(TA)_{14}$ (Fig. 8.17), $(AGT)_5$, $(GATA)_4$, $(AAG)_8$, $(GGGA)_4$, $(CTCACA)_4$, $(TCC)_5G$, and $(ATG)_5AG$ (Sedra, 2007d, 2011b, e). For both RAPD and ISSR markers, the results were reproducible.

The genetic relationship and genotyping of 45 Moroccan cvs. and selected clones and some Tunisian cvs. (cited in Table 8.7) were determined using 79 ISSR markers

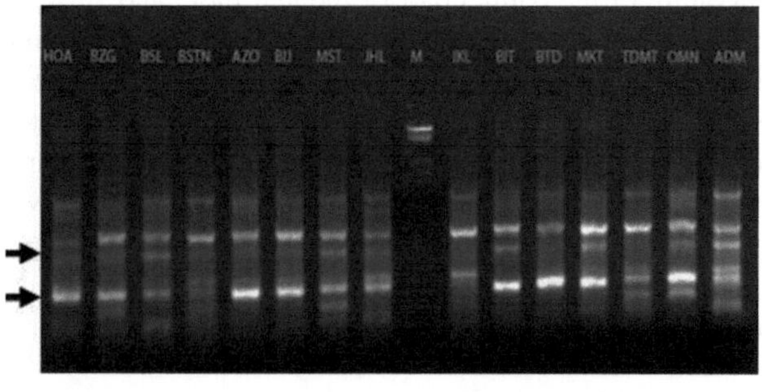

Fig. 8.17 Examples of DNA polymorphism detected between accessions of date palm. Ethidium bromide-stained agarose gel (1.4 %) of amplification fragments produced with microsatellite ISSR primer MIC 19 $(TA)_{14}$. M (lane 9): fragments of standard molecular weight marker (λ/Eco R1/Hind III. BAP). The arrow indicates the polymorphic marker. 16 Moroccan cvs: Houa (HOA), Bouzeggar (BZG), Bouslikhene (BSL), Black Bousthammi (BSTN), Azigzao (AZO), Bouijjou (BIJ), Mestali (MST), Jihel (JHL), Iklane (IKL), Bouittob (BIT), Boutemda (BTD), Mekt (MKT), Tadmainte (TDMT), Oum N'hal (OMN), and Ademou (ADM) (see Table 8.7). The *arrows* indicate the discriminative markers

(Sedra 2011e). The cluster analysis based on genetic distance between genotypes could not distinguish the group of Tunisian cvs. among all Moroccan ones, but it allowed totally regrouping the selected clones that have some similarity with Tunisian cvs. (Fig. 8.18). In fact, according to the dendrogram, the genotypes of the palm date palm studied can be clustered in 2 large groups (A and B). Group A is subdivided in two subgroups A1 and A2; subgroup A1 contains all Moroccan selected clones and the Tunisian cv. Ftimi (FTM). Cultivar Okhtftimi (OFTM) is clustered in the subgroup A2. The group B regrouping of many different cvs. is subdivided in two subgroups B1 and B2. Because of its excellent fruit quality and high susceptibility to the bayoud, cv. Medjool (MJH) alone represents the subgroup B2. The subgroup B1 contains several sub-subgroups of which certain ones contain some susceptible and resistant cvs. to bayoud with fair or moderate fruit quality as IKL, BSTN, AZO, BZG, and BSL. For others sub-subgroups, similar results were obtained. The male genotypes: NP3, NP4, A18, and B18 were clustered in different subgroups with female cvs. The results showed that the ISSR technology is very effective in identifying different varieties of date palm, but the markers used in this case cannot cluster the sex genotypes or resistant cvs. (Sedra 2011e).

Table 8.7 Name and main characteristics of date palm genotypes studied

Code	Name	Resistance to bayoud*	Code	Name	Resistance to bayoud[a]
BSTN	Black Bousthammi	R	JHL	Jihel	HS
BSTB	White Bousthammi	R	BSK	Bouskri	HS
TDMT	Tademainte	R	BRR	Bourar	S
IKL	Iklane	R	MST	Mastali	S
SLY	Saîrlayalate	R	DN	Deglet Noor	HS
BFGM	Boufeggous/Moussa	R	FTM	Ftimi	S
HFS	Hafs	S	OFTM	Okhtftimi	S
BSL	Bouslikhéne	MR	NJD	Najda (selected clone)	HS
BZG	Bouzeggar	MR	Sed 1	Selected clone	R
AIB (ISY)	Aissa-Iyoub	S	Sed 2	Selected clone	R
BHZ	Belhazit	S	Sed 3	Selected clone	R
AZO	Azigzao	MR	Sed 4	Selected clone	R
BIT	Bouittob	S	Sed 5	Selected clone	R
BTD	Boutemda	S	Sed 6	Selected clone	R
ADM	Ademou	S	Sed 7	Selected clone	R
BIJ	Bouijjou	S	Sed 8	Selected clone	S
OTK	Outokdime	MR	Sed 9	Selected clone	S
AGL	Aguélid	MR	Sed 10	Selected clone	R
HOA	Houa	S	NP3	Selected male palm	R
AHD	Ahardane	HS	NP4	Selected male palm	R
BKN	Boukhanni	R	A18	P10C-A18 (male palm)	S
MJH	Medjool	HS	B18	P10C-B18 (male palm)	S
BFG	Boufeggous	HS			

[a]Resistance phenotypes: *HS* highly susceptible, *S* susceptible, *MR* medium resistant, *R* resistant. Source: Sedra et al. (1996, 1998a, b); Sedra (2001a, 2003b, 2005a, 2011a)

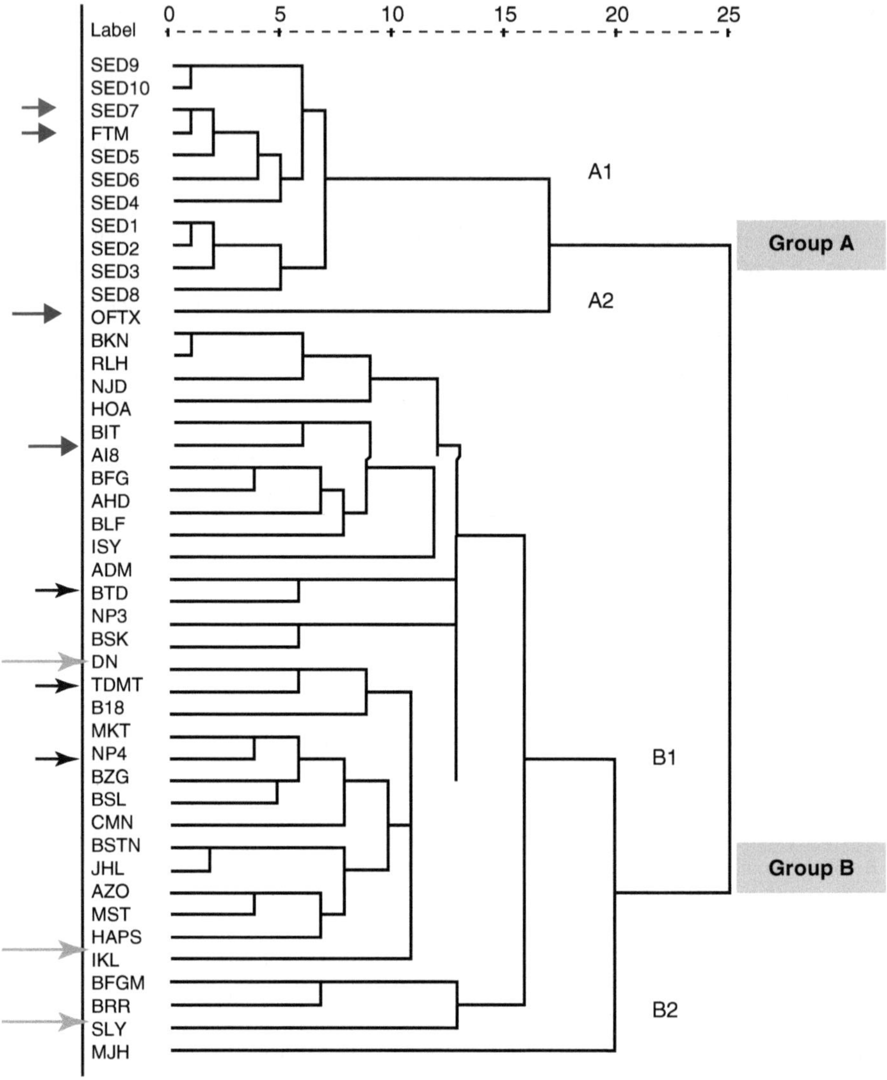

Fig. 8.18 Dendrogram of the 45 date palm genotypes (31 cvs., 10 selected clones, and 4 males) listed in Table 8.7 generated by group average clustering analysis (UPGMA) using RAPD and ISSR-based genetic distance. 79 ISSR polymorphic markers revealed by 13 primers. The polymorphism level is 79.8 %

Preliminary studies revealed the relationship between 31 quantitative and qualitative descriptors of date palm tree and 34 selected RAPD molecular markers (from m1 to m34) (Sedra 2007c). Molecular markers can be linked to genes of interest such as disease resistance and sex, allowing indirect selection of the desired genotypes. All of the molecular techniques (RAPD, AFLP, SSR, and ISSR) have been applied to evaluate the genetic diversity and identification of date palm cvs. (Adawy et al. 2005; Sedra 2007d; Sedra et al. 1998a, b) and for genetic comparison and the identification of vitro-plants obtained by tissue culture techniques from adult mother palms (Saker et al. 2006; Sedra 2005b).

8.5.3 *Marker-Assisted Selection*

Advances in selection for sex and agronomically important traits, such as fruit quality or disease resistance, are difficult due to the date palm's long generation time. The Moroccan experience in date palm genetic improvement by crosses and mass selection confirms this reality (Djerbi et al. 1986; Louvet and Toutain 1973; Saaidi et al. 1981; Sedra 1995, 1997, 2003b, 2005a, 2011a, f). The resistance character to bayoud disease seems the least complicated of the characters in date palm. Relatively fast methods were developed for the assessment of resistance to bayoud in different stages of plant growth (i.e., seedlings, young plants, and adult trees) in the laboratory and the field (Sedra 1994a, b; Sedra and Besri 1994; Sedra et al. 1993). These methods of conventional breeding and hybridization programs are limited practically and financially when it is necessary to evaluate a very large number of trees. Recent molecular research results allowing gender determination in young date palm will enhance breeding approaches and facilitate marker-assisted selection (MAS) and genetic association studies (Cherif et al. 2013).

8.5.3.1 Bayoud Resistance Markers

In the approach based on biochemical and plasmid mitochondrial DNA markers, many markers that are correlated to resistance in date palms have been reported, such as isozymes (Bendiab et al. 1993; Bennaceur et al. 1991), polyphenolics (El Hadrami et al. 1996), and mitochondrial plasmid-like DNAs (Benslimane et al. 1994, Trifi 2001). However, the correlation between phenotype and the described marker has not been clearly established. The selected primes revealed by RAPD and ISSR techniques (Sedra 2011b, e) were tested on 16 cvs.: 7 resistant (IKL, BSTN, BSTB, TDNT, BFGM, SLY, and BKN) and 9 susceptible. Seven molecular markers were candidates to be associated with resistance to bayoud disease (Sedra 2011b, e). Their estimated molecular weight (kb) varied from 0.1 to 1.5 kb, and they have been detected in different resistant cvs. The RAPD marker UBC-145-1.22 is present in 5 resistant cvs. among 7 studied (Sedra 2011b, e). The RAPD-UBC-578-1.50 is present in 5/6 resistant cvs. The ISSR marker Mic19-1.37 kb revealed by the primer $(TA)_{14}$ is detected in the 6 resistant cvs. studied (Sedra 2011b, e).

Other markers are only detected in one or a few resistant cvs. Moreover, the reliability of molecular techniques was very comparable to the biological technique using artificial inoculation of plants that needed 4 months at least (Sedra 2011e). These markers have been revealed in the majority of resistant hybrids (young plantlets) derived from crossing a resistant parent with the other susceptible parent. The biological trial conducted in the greenhouse (Fig. 8.19) presents these markers, therefore can be transmitted to the progeny. These results suggest that the resistance could be encoded by different genes. In fact, the first technique can select 55.0 % of resistant plants which is equivalent to 91.7 % of resistant plantlets identified by the pathogen (Sedra 2011e). Moreover, these markers allowed rapid screening to reduce large tested population and can be used at very early stage of plantlet growth. This interesting approach permits reduced high trial cost of biological test screening of large numbers of plantlets. Prospectively, it will be interesting to select a precise minimal number of markers associated with bayoud disease resistance in date palm and to sequence them.

Fig. 8.19 Selection of resistance of date palm plantlets (issued from the cross Black Bousthammi (resistant female) × P10C-A18 (susceptible male)) to bayoud disease by biological test using inoculation technique with the pathogen, *Fusarium oxysporum* f. sp. *albedinis* in the greenhouse. The *arrow* indicates the samples of leaflets that were taken for DNA extraction

8.5.3.2 Sex Identification Markers

Previous studies using tens of RPAD and ISSR primers permitted selection of several primers which revealed polymorphic markers that completely distinguish females and males. Among them, 19 RAPD and 4 ISSR primers were tested on numerous other genotypes: 28 males and females (list 3) and allowed generating total of 178 polymorphic markers among 258 markers revealed. This number of selected polymorphic markers for sex distinguishing is also very high (48 RAPD and 27 ISSR markers) (results not yet published). In order to reduce the interesting markers, the third trial using these 14 males and 14 females has permitted to select only 10 RAPD and 2 ISSR markers that distinguish males and females. The primers RAPD-536 and RAPD-580 had, respectively, revealed 3 and 2 markers, whereas other primers had only generated one marker each. These markers allowed globally about 61.8 % of polymorphism. The estimated molecular weight (pb pair bases) varied from 564 to 3,342 pb. The best examples of these primers are RAPD-553 (5′-TTCGAGATCG-3′), Mic7-ISSR ($(GACA)_4$), and Mic8-ISSR ($(GT)_8$). The markers RAPD-553-564pb, Mic7-ISSR-432pb, and Mic8-ISSR-2027pb were the most interesting markers which can distinguish 100 % of males used. However, total selected markers were necessary for population analysis. The dendrogram presents two large groups 1 and 2 totally formed each by males and females (Fig. 8.20).

However, it was noted that two cvs. (IKL) and (TDMT)) were clustered nearly with the some male genotypes. This can be explained by the presence of a common genetic origin or a near relationship link between these cvs. with the male genotypes. The detected molecular markers are capable of distinguishing 28 genotypes (males and females) with an error of 2/28 (0.07 %). Prospectively, it will be interesting to select a precise minimal number of markers associated with sex identification in date palm and to sequence them.

These research results open new doors to explore in the use of molecular technologies in the development of a breeding program to select rapidly new varieties desired by farmers and more demanded by different markets. They also may provide an area in research and allow a construction program of a date palm genetic map.

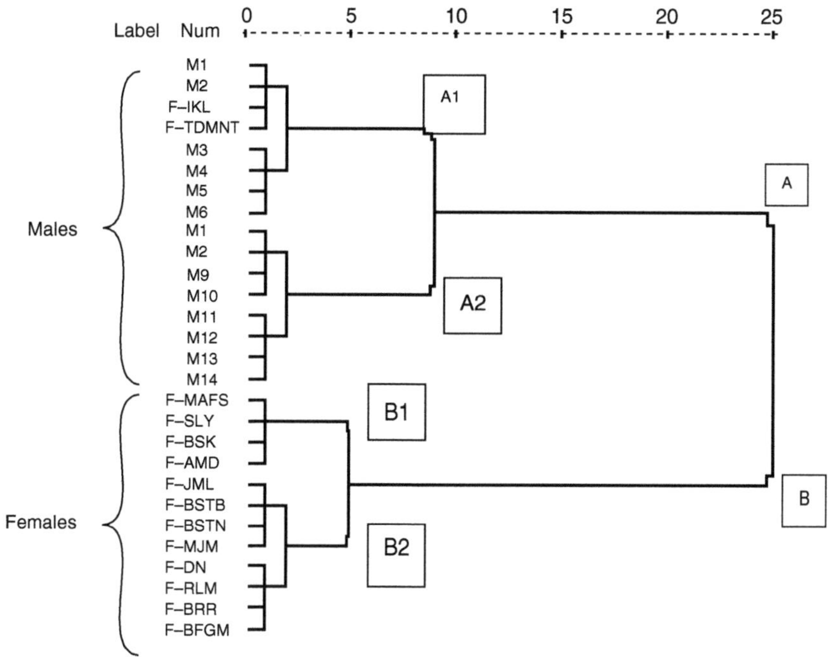

Fig. 8.20 Dendrogram of the 28 date palm genotypes (14 males that begin by M) and 14 female cvs. by F) listed in Table 8.7 generated by group average clustering analysis (UPGMA) using RAPD and ISSR-based genetic distance. Analysis realized by 10 RAPD and 2 ISSR polymorphic markers. The polymorphism level is 61.8 %

8.6 Cultivars Description

A total of 453 date palm cvs. have been enumerated in Moroccan oases using local names, the majority of them being considered less representative and rare. A representative sample of 51 cvs. were selected for description here, chosen for their level of representation and recent finds: 19 most representative and popular, 15 poorly representative, 5 rare representing different date-growing areas, 10 new selected varieties, and 2 selected males. Descriptions, accompanied by photographs (tree, leaf, fruit bunch, fruit), indicate the main characters, mostly agro-morphological, which are important for farmers, producers, and investors in the date palm sector. The characters of value to producers and farmers in making a choice of cultivars are (a) length of leaf, (b) length of spadice bearing fruits, (c) number of dates per kg, (d) percentage of pulp in fruit, (e) fruit color at tamar stage, (f) heat requirements for date maturity, and (g) behavior towards bayoud disease. The description of some cultivars is presented in Table 8.8 grouped based upon the degree of their representation in the country. Tree and fruit traits of most representative cultivars are shown in Fig. 8.21, and those of some poorly representative cultivars are shown in Fig. 8.22. Also presented are the characteristic features of fruits of other cvs. grown in Morocco grouped as rare cvs. (Fig. 8.23), selected date palm varieties (Fig. 8.24).

Table 8.8 Description of some date palm cultivars grown in Morocco classified based on their representation

Cultivar	Length of leaf	Behavior to bayoud	Pollination period	Maturity and consistency	Fruit shape and color	Commercial assessment	Possible uses	Distribution
1. *Most representative cultivars*								
Aguelid	Moderate	Moderately resistant	Beginning of March	Early, semisoft	Cylindrical, light brown	Good, valued for earliness	Fresh, processed	Drâa, Bani, Saghro, Anti-Atlas
Ahardane	Short	Very susceptible	Beginning of March	Early, semisoft	Cylindrical, light brown	Moderate, valued for earliness	Fresh, processed	Drâa, Anti-Atlas, Saghro, Bani, Oriental
Azigzao	Short	Susceptible	End of March	Moderately early, semidry	Ovoid opposite, light brown	Weak, locally appreciated	Fresh, processed	Tafilalet, between Saghro and High-Atlas areas, Ferkla, Gheris, Saghro
Aziza Bouzid	Long	Susceptible	Beginning of March	Mid-season, semisoft	Cylindrical, light brown	Relatively good, very appreciated in area of origin	Fresh, processed	Oriental (East)
Black Bousthammi	Very long	Resistant	End of March	Moderately late, soft	Ovoid, black	Weak	Fresh, pressed, processed	Bani, Drâa, Tafilalet, Saghro, Anti-Atlas
Boucerdoune	Very short	Susceptible	End of March	Mid-season, dry	Rotund, dark brown	Moderate	Fresh, dry processed	Tafilalet, between Saghro and High-Atlas areas, Gheris, Saghro, Oriental
Boufeggous	Very short	Susceptible	End of March	Mid-season, soft	Ovoid, dark brown	Good	Fresh, processed	Both areas
Bouijjou	Short	Susceptible	End of March	Mid-season, dry	Cylindrical, light brown	Moderate	Fresh, dry processed	Tafilalet, Oriental, Guir

Bouittob	Short	Susceptible	End of March	Moderately late, dry	ovoid opposite, light brown	Moderate	Fresh, dry processed	Anti-Atlas, Bani
Bourar	Moderate	Very susceptible	End of March	Moderately late, semidry	Elliptic, dark brown	Good	Fresh, processed	Drâa, Bani, Saghro, Tafilalet
Bouskri	Short	Very susceptible	End of March	Moderately late, dry	Ovoid opposite, dark brown	Moderate to good	Fresh, processed	Bani, Drâa, Saghro, Todra, Oriental, Tafilalet, between Saghro and High-Atlas areas, Anti-Atlas
Bouslikhene	Moderate	Moderately resistant	End of March	Moderately early, semidry	Ovoid opposite, light-dark brown	Weak to moderate	Fresh, processed	Tafilalet, Saghro
Bouzeggar	Very long	Moderately resistant	End of March	Late, semisoft	Cylindrical, black	Weak	Fresh, processed	Drâa, between Saghro and High-Atlas areas, Ferkala, Gheris
Hafs	Short	Susceptible	End of March	Mid-season, semisoft	Cylindrical, dark brown	Weak	Fresh, processed	Drâa, Oriental, Tafilalet, between Saghro and High-Atlas areas
Iklane	Moderate	Resistant	End of March	Late, semisoft	Cylindrical, black	Weak	Fresh, processed	Anti-Atlas, Bani, Drâa, Saghro
Jihel	Moderate	Very susceptible	End of March	Late, dry	Ovoid opposite, light brown	Good	Fresh, processed	Drâa, Bani, Anti-Atlas, Tafilalet, Saghro, between Saghro and High-Atlas areas

(continued)

Table 8.8 (continued)

Cultivar	Length of leaf	Behavior to bayoud	Pollination period	Maturity and consistency	Fruit shape and color	Commercial assessment	Possible uses	Distribution
Medjool	Short	Very susceptible	End of March	Late in Drâa Valley and mid-season in Tafilalet, semisoft	Ovoid elongated, brown-dark brown	Excellent	Fresh	Ziz, Tafilalet, extended to Drâa Oriental and Bani areas
Outokdime	Very short	Moderately resistant	Beginning of April	Moderately late, dry	Ovoid elongated, light brown	Moderate	Fresh, processed	Between Saghro and High-Atlas areas, Todra
Ras Lahmer	Moderate	Susceptible	End of March	Mid-season, dry	Ovoid, light brown	Moderate	Fresh, processed	Bani, Drâa, Saghro, Tafilalet, between Saghro and High-Atlas areas
2. *Poorly representative cultivars*								
Ademou	Long	Susceptible	Beginning of March	Mid-season, dry	Ovoid opposite, light brown	Good	Fresh, processed	Drâa and Tafilalet
Aissa-Iyoub	Moderate	Susceptible	Beginning of April	Moderately late, semidry	Cylindrical, light brown	Good	Fresh, processed	Oriental and Drâa
Belhazit	Very short	Susceptible	Beginning of April	Mid-season, semisoft	Ovoid, brown	Moderate	Fresh, processed	Tafilalet
Boufeggous ou Moussa	Very short	Resistant	End of March	Moderately early, soft	Ovoid elongated, black	Weak	Fresh, processed	Bani
Boukhanni	Very long	Resistant	Beginning of March	Moderately early, semisoft	Cylindrical, dark brown	Moderate	Fresh, processed	Drâa
Boutemda	Short	Susceptible	End of March	Moderately early, semidry	Ovoid, dark brown	Moderate	Fresh, dry processed	Bani
Houa	Short	Susceptible	End of March	Mid-season, semidry	Ovoid elongated, brown	Moderate	Fresh, dry processed	Drâa

Mah-Labaid	Short	Susceptible	End of March	Moderately late, soft	Ovoid elongated, dark brown	Fairly good	Fresh, pressed, processed	Anti-Atlas and Bani
Mekt	Short	Susceptible	End of March	Mid-season, soft	Ovoid elongated, black	Weak	Fresh, processed	Drâa, Bani, Saghro, Anti-Atlas
Mestali	Moderate	Susceptible	End of March	Moderately early, semisoft	Ovoid, dark brown	Weak	Fresh, processed	Drâa
Oum N'hal	Moderate	Susceptible	End of March	Moderately late semisoft	Ovoid elongated, light brown	Weak to moderate	Fresh, processed	Anti-Atlas and Drâa
Sairlayalate	Moderate	Resistant	End of March	Moderately late, semidry	Cylindrical, light brown	Moderate to fairly good	Fresh, processed	Bani
Sbaa Sultane	Long	Susceptible	End of March	Moderately early, soft	Cylindrical elongated, golden brown	Moderate to fairly good	Fresh, processed	Drâa and Tafilalet
Tadmainte	Long	Resistant	End of March	Mid-season, semidry	Cylindrical, dark brown	Moderate	Fresh, processed	Anti-Atlas, Bani, Drâa, Oriental, Saghro
White Bousthammi	Moderate	Resistant	End of March	Mid-season, soft	Ovoid, dark black	Weak	Fresh, pressed, processed	Anti-Atlas, Bani
3. *Rare cultivars*								
Acedame	Moderate	Susceptible	End of March	Moderately early, soft	Cylindrical, dark brown	Moderate	Fresh, processed	Drâa
Al Aoud-Ezzine	Long	Unknown	Mid-March	Moderately early, semisoft	Ovoid elongated, brown	Moderate, little sweet	Fresh, processed	Drâa
Al Jaafaria or Jaafari	Long	Unknown	End of March	Moderately early, semidry	Ovoid elongated, brown	Moderate	Fresh, processed	Drâa

(continued)

Table 8.8 (continued)

Cultivar	Length of leaf	Behavior to bayoud	Pollination period	Maturity and consistency	Fruit shape and color	Commercial assessment	Possible uses	Distribution
Boutoutobte Loubania	Long	Unknown	End of March to beginning of April	Mid-season, semisoft	Ovoid elongated with bumps, light brown	Moderate	Fresh, processed	Drâa
Chatouia or Chatoui	Short	Unknown	End of March	Very late, maturity of date at the end of autumn semisoft	Ovoid elongated, light brown	Good	Fresh, processed	Drâa, Tafilalet
4. *Selected female varieties*								
Al-Amal (INRA-1443)	Long	Resistant	Mid to end of March	Moderately early, semisoft	Cylindrical, light brown	Good fruit, moderate storage quality, advised for rebuilding bayoud-devastated orchards	Fresh, thicket with fines, processed	Originally from Drâa, massive multiplication and diffusion in progress
Al-Fayda (INRA-1447)	Long	Resistant	Mid-April	Moderately early, dry	Ovoid elongated, light brown	Good fruit and storage quality, advised for rebuilding devastated orchards	Fresh, processed	Originally from Drâa, massive multiplication and diffusion in progress
Ayour (INRA-3415)	Very long	Susceptible	Beginning of March	Mid-season, soft	Ovoid, blackish brown	Good fruit and storage quality ability, advised for bayoud-free zones, good commercial value	Fresh, processed	Good, can be valued in zones free of bayoud, good commercial value

Bourihane (INRA-1414)	Moderate	Resistant	End of March	Moderately early, semisoft	Ovoid opposite, light brown	Good fruit and storage quality, advised for rebuilding devastated orchards by bayoud	Fresh, processed	Originally from Drâa, massive multiplication and diffusion in progress
Darâaouia (INRA-1445)	Very long	Resistant	End of March	Moderately early, semisoft	Ovoid opposite, light brown	Good fruit and storage quality, advised for rebuilding devastated orchards by bayoud	Fresh, thicket with fines, processed	Originally from Drâa, massive multiplication and diffusion in progress
Hiba (INRA-3419)	Very long	Susceptible	End of March to beginning of April	Moderately late, semidry	Cylindrical, light brown	Good fruit and storage quality, advised for bayoud-free zones, good commercial value	Fresh, processed	Originally from Drâa, massive multiplication and diffusion in progress
Khair (INRA-3300)	Moderate	Susceptible	End of March to beginning of April	Late, semisoft	Ovoid opposite, light brown	Good, advised for bayoud-free zones, good commercial value	Fresh, processed	Originally from Drâa, massive multiplication and diffusion in progress
Mabrouk (INRA-1394)	Long	Resistant	End of March to beginning of April	Mid-season, semidry	Cylindrical, light brown	Good fruit and storage quality, advised for rebuilding bayoud-devastated orchards	Fresh, processed	Originally from Drâa, massive multiplication and diffusion in progress

(continued)

Table 8.8 (continued)

Cultivar	Length of leaf	Behavior to bayoud	Pollination period	Maturity and consistency	Fruit shape and color	Commercial assessment	Possible uses	Distribution
Najda (INRA-3014)	Long	Resistant	End of March	Mid-season, semisoft	Cylindrical, light brown	Good fruit and storage quality, advised for devastated orchards	Fresh, processed	Originally from Drâa, massively multiplied and diffused in all oases, avoid planting in trays in altitude or marginal areas
Sedrat (INRA-3003)	Very long	Resistant	End of March	Moderately early, semidry	Ovoid elongated, light brown	Good fruit and storage quality, advised for devastated orchards	Fresh, processed	Originally from Drâa, massive multiplication and diffusion in progress
5. *Selected male varieties*								
Nebch-Boufeggous (INRA-NP4)	Long	Resistant	March	Very high, weak part of deformed or aborted grains	Very high	Valued commercially, advised for rebuilding devastated orchards	Fresh, processed	Originally from Drâa, massive multiplication and diffusion in progress
Nebch-Bouskri (INRA-NP3)	Moderate	Resistant	March	Very high, weak part of deformed or aborted grains	Very high	Valued commercially, advised for rebuilding devastated orchards	Fresh, processed	Originally from Drâa, massive multiplication and diffusion in progress

Fig. 8.21 The characteristic features of tree, leaf, and fruits of the most representative date palm cvs. grown in Morocco

Fig. 8.21 (continued)

Fig. 8.21 (continued)

Fig. 8.21 (continued)

8.7 Dates Production and Marketing

Date production techniques in the oases are traditional and result in low yields, while in modern farms the use of appropriate technology has improved the yield quantity and quality. Traditional production, which represents the majority, is now characterized by a predominance of cvs. of low to medium quality. Among 5 million trees, only average of 45 % are productive because of insufficient care of trees due generally to weak application of adequate farming practices and alternation phenomena of production. The national average of date production during the past 5 years is 90,000 mt. This insufficient production requires Morocco to import dates, most recently 40,000 mt in 2013. The average yield of 20 kg/tree or about 2 mt/ha

Fig. 8.22 The characteristic features of fruits of some poorly representative date palm cultivars grown in Morocco

Fig. 8.23 Characteristic features of fruits of some rare date palm cvs. grown in Morocco

Fig. 8.24 Characteristic features of fruits of some selected date palm varieties grown in Morocco

is still low compared to Tunisia (2.6 mt/ha) and Algeria (3.4 mt/ha) and the standard date-producing countries.

The destination of the production is shared between consumption 51 % (33 % human and 18 % animal) and marketing 48 %. The remaining 1 % represents the total loss of production. Annual consumption is around 15 kg/inhabitant in production areas against 3 kg/inhabitant at the national level. In oases, postharvest techniques focus on the collection, sorting, and packing, the remaining products intended for sale at low prices or for human and animal consumption. Storage of bulk dates in farm houses is still the most common practice before the sale to intermediaries. In some cases, traders store the dates in mountain areas where the climate is cool (e.g., villages of Aguelmous and Timahdit).

Marketing problems are numerous, and dates are situated in different links in the chain of production and marketing. Detailed diagnosis of date marketing have been established (Chetto et al. 2005; Harrak and Chetto 2005). There are 30 medium and large souks (markets) according to volume sold most of which are to be developed to make them well organized and respect the standards and conditions of hygiene and food safety. In addition to marketing in the local souks, dates are shipped to larger cities and other urban centers of Morocco to satisfy demand. Internal trade of dates has significant potential because of low current consumption pattern, the growing demand trends, and significant rate of population growth. Given the lack of domestic production, imports of dates in recent years averaged 30,000 mt/year. The contract programs between the government and the private sector will contribute to the promotion of exports of quality dates through the implementation of communication campaigns and participation in trade fairs and exhibitions in target markets. For the valorization and commercialization of products, appropriate future action concerns the development of marketing platforms to improve the quality, hygiene, and safety of marketed dates. With financial support from international donors, the recent installation of 14 new units for date refrigerated storage and packaging near the southern oases has encouraged producers to organize economic interest groups (EIG) under the supervision of the government. This has provided an important take-off for products valorization. In fact, institutional structures and organizations that conduct the date valuation and its by-products are still few in number, and volumes are lower than the opportunities.

Under the PMV program of the MAPM, thus far 23 organizations (EIG) have been strengthened and added at sites of production and urban centers including women's organizations in charge of processing dates and cooperative storage, processing, and marketing. In order to improve product quality and remediate marketing channels, the PMV has provided, since its inception in 2009, to protect the different products (e.g., fresh dates, raw date, jam, syrup, powder) by labels. Date labeling was carried out by development of distinctive signs of origin and quality of Moroccan dates (Harrak 2011). SDOQ is a collective system accessible to producers and manufacturers of agricultural products with a specific quality allowing the recognition and protection of local products. The MAPM established SDOQ in 2009. Some cvs. were protected by a label Protected Geographical Indication (PGI), examples: *Medjool of Tafilaet* in 2010 (Fig. 8.25) and Geographical Indication (GI) *Dates Bouittob of Tata* in 2013 (Harrak 2013). This labeling, initiated in 2010, involves 30 rural and urban associations of date producers in the region of Tafilalet. The recovery process is now engaged and should continue to guarantee conditions for success through monitoring, coaching, training, and the establishment of appropriate regulations, well-trained staff, and laboratories for quality control of the products.

Fig. 8.25 Labeling of Medjool dates in Tafilalet area by Protected Geographical Indication

8.8 Processing and Novel Products

Obtaining good quality dates is driven by quality improvement at all stages through which the product passes: production techniques including harvesting and postharvest technology in addition to processing technologies for products other than fresh dates. Marketing of dates is still mainly through intermediaries who buy production in the field or through the different major souks of production areas (Drâa, Tafilalet, Bani, and Figuig). In rural bazaars, sales of dates are loose, whole plans and dates traditionally packaged (e.g., bags, cartons). These conditions negatively affect their quality. Currently some traders use appropriate packaging (Fig. 8.26). There is no genuine integrated date industry but rather methods of date packaging and processing realized by cooperatives that produce relatively insufficient volumes of production to satisfy demand. The trade souks were not designed to accommodate these products and to keep consumers insufficiently informed. As mentioned above, there are 30 souks most of which require upgrading. To promote date products, it is necessary to put in place units of valorization in different production areas, where the volume of production is justified. Thus as mentioned above, recent installation of 14 new refrigerated storage and date processing units (capacity of 100 to 400 mt annually) in different areas of production out of 23 units planned by 2020. Forty other units of processing and refrigerated storage are expected to be installed in the next few years in collaboration with private investors and the MAPM. It will also be useful to upgrade local souks and conditions of sale.

Moroccan date oases have an ancestral know-how of preservation and processing of dates into various products (e.g., juice, syrup, vinegar, flour, paste), developed over time by local dwellers. These products are characterized, in addition to their nutritional and organoleptic qualities, by therapeutic properties due to the incorporation in their preparation of aromatic and medicinal plants. Nevertheless, this traditional knowledge is under valorized and risks being forgotten. Similarly, traditional

Fig. 8.26 Some examples of appropriate packaging of date in Morocco for the best commercial cvs. (Medjool, Boufeggous, Najda, Jihel)

preparations of dates, using the experience of oases women, require archiving, valorization, and promotion of their consumption by their integration into Moroccan cuisine and modern culinary art (Harrak and Zirari 2006).

The main research achievements to improve and further improve the Moroccan date postharvest sector are as follows:

(a) Determination of typicity and quality criteria of Moroccan dates. This concerns development of nutritional, technological, sensory, and commercial qualities such as physical, physicochemical, biochemical, and sensory date characterizations in addition to date technological suitability. About 60 quality criteria have been studied including: sensory criteria, weight criteria, dimension criteria, importance of the pulp, quality index, Brix, pH, acidity, sugar content, protein content, ash, minerals, cellulose content, and aroma compounds (Harrak and Boujnah 2012; Harrak and Hamouda 2005; Harrak et al. 2005; Lebrun et al. 2007). This led to the establishment of the first date sensory analysis panel of INRA in Morocco.
(b) Archiving, analysis, and improvement of date palm traditional knowledge of Moroccan oases. In order to safeguard this traditional knowledge, detailed information was collected on the various traditional preparations of different Moroccan oases including illustrations of practical demonstrations of some of these preparations done by oases women. More than 30 of traditional preparations were identified. These are often associated with medicinal and aromatic

Fig. 8.27 (**a**) Improved traditional date juice Tassabount, (**b**) date paste, (**c**) date jam with no added sugar, (**d**) date butter (Photos: Hasnaâ Harrak, INRA Marrakech, Morocco)

plants which provide the properties of flavoring, preservation, and medication (Zirari et al. 2006). These traditional date products can be justifiably regarded as local or *terroir* products and deserve to be promoted (Harrak 2013; Harrak and Zirari 2006; Sedra 2003c). Examples of improved traditional products are shown in Fig. 8.27.

(c) Development of novel products and adaptation of modern industrial date fruit processing of Moroccan dates (e.g., jam, paste, juice, syrup, flour) (Harrak and Boujnah 2012; Harrak and Jaouan 2010).
(d) Adaptation of preservation processes to Moroccan dates (Harrak 2011).
(e) Literature and exploratory studies for improving date valorization and commercialization (Harrak and Chetto 2005).

In these topics, the research and development achievements permit technology transfer and technical assistance to date valorization units and training in the fields of date postharvest valorization for the benefit of technicians, farmers, associations, and cooperatives in Moroccan oases: Drâa, Tafilalet, Guelmim, Tata, Figuig, and Marrakech (Harrak 2010a, b)

8.9 Conclusions and Recommendations

In terms of varietal richness, Moroccan oases constitute an ecological and environmental asset; the current handicap is represented by underdeveloped packaging, packing, and marketing of date products. Except for certain cvs. such as Boufeggous or Medjool, there is no significant marketable volume for the same cultivar. This diversity must be better managed and used in the manufacture of various products like fresh fruit, date syrup, paste, and powder for different markets, as well as for long-term development of quality fresh dates to benefit Morocco. Recently, a number of small units of processing, packaging, and storage have emerged, and more refrigeration units continue to be established. But the volumes of processed products are still insignificant.

The groundwater is not of equal importance throughout the oases; the adequacy of the volume of available water resources and the maintenance of traditional oases and expansion of plantations must be well managed and thoughtful, hence the need for a good combined use of surface water resources (management of dams and flood waters) and groundwater and of appropriate associated crops with date palm. Given the adverse effects of climate change, the spread of drip irrigation by installing a network of interconnected basins should be considered, supplied from dams and collective pumping stations. Thorough research on water dynamics and participatory actions of water users through partnerships are necessary to achieve the objectives. To increase profitability, it is useful to develop other activities including handicrafts and ecotourism to generate additional income for the oasis populations and therefore to contribute to control poverty and ensure a level adequate living for oasis populations and mitigate the rural exodus. In this context, the oasis environment must be preserved to assure sustainable management of natural resources and adequate development of products and by-products.

To achieve these objectives, it is recommended, in addition to the aspects mentioned above, to implement the following:

(a) Implement the Green Morocco Plan into all development activities of date palm sector.
(b) Educate the populations to be aware and responsible.
(c) Strengthen research activities and research and development of appropriate technologies and capitalize on the knowledge and research findings.
(d) Pursue development of procedures for labeling products, under Moroccan law 25-06 on Distinguishing Marks of Origin and Quality.
(e) Encourage investment to ensure adequate and locally efficient production which is favorable for exports.
(f) Promote domestic marketing and exports by launching campaigns and awareness.
(g) Strengthen operations against desertification, silting, and soil and water salinization.
(h) Ensure the protection of date palms against harmful enemies (pests and diseases) of quarantine by applying phytosanitary measures under the regulations in force.

(i) Refresh the data by conducting baseline studies (e.g., population census, inventory, and characterization of date groves, foci map of bayoud, hydrological maps).
(j) Facilitate the formulation of legislation for the establishment of subsidies for the acquisition of facilities and equipment for packaging and promotion of products and by-products of date palm (e.g., dates, paper, wood, compost).
(k) Ensure national coordination and networking of actors of date palm sector in the context of a Geographic Information System (GIS). This tool permitted the mastery of inventories of localities and oases and their monitoring. Thus, GIS of three important oasis Tarnata, Aoufous and Erfoud was made.

Acknowledgments I gratefully acknowledge all persons for their help and providing me with information.

References

Abahmane L (2011) Date palm micropropagation via organogenesis. In: Jain SM, Al-Khayri JM, Johnson DV (eds) Date palm biotechnology. Springer, Dordrecht, pp 69–90

Adawy SS, Hussein E, Ebtissam HA et al (2005) Genomic diversity in date palm (*Phoenix dactylifera* L.) as revealed by AFLP in comparison to RAPDs and ISSRs. Arab J Biotech 8: 99–114

Aît Chitt M, Mantell SH, Thangavelu M, Ainsworth CC (1995) Cloning date palm (*Phoenix dactylifera* L.) DNA and characterization of low, medium and high copy DNA sequences. Elaeis 7:57–63

Amraoui H, Sedra MH, Hamdaoui A (2004) Etude des sécrétions protéiques et enzymatiques du Fusarium oxysporum f. sp. albedinis, agent causal de la fusariose vasculaire du palmier dattier. Al Awamia 109/110:47–62

Amraoui H, Lazrek HB, Sedra MH et al (2005a) Chromatographic characterization and phytotoxic activity of *Fusarium oxysporum* f. sp. *albedinis* and saprophytic strain toxins. J Phytopathol 153:203–208

Amraoui H, Sedra MH, Hamdaoui A (2005b) Mise en évidence d'enzymes à activité antifongique chez le palmier dattier: dosages des activités chitinases et β-glucanases, comme réaction au *Fusarium oxysporum* f. sp. *albedinis*, agent causal du bayoud. Al Awamia 116:18–34

Anjarne M, Bouguerfaoui M, Aitchitt M, Cheikh R (1995) Production de vitroplants de palmier dattier par la technique d'organogenèse in vitro: L'expérience marocaine. Proc Journées Internationales sur le palmier dattier dans l'agriculture oasienne des pays méditerranéens. Elche, Espagne, pp 25–27

Anjarne M, Bougerfaoui M, Abahmane L (2010) *In vitro* propagation as tools for the development of Moroccan date palm groves. Proc. 4th Int date palm conference, Abu Dhabi, 15–17 Mar, p 50

Baâziz M, Brakez Z, Sedra MyH (1994) Correlation between phenol-oxidizing enzymes and bayoud disease resistance in date palm cultivars. 5th Arab Congress of Plant Protection, Fès, 28 Nov–3 Dec

Beauchesne G, Zaid A, Rhiss A (1986) Meristematic potentialities of bottom of young leaves to rapidly propagate date palm. Proceeding of the second symposium on the date palm. March 3 6/1986, Date Palm Research Center, King Faisal University, El-Hassa, pp 87–94

Bendiab K, Baâziz M, Brakez Z, Sedra MH (1993) Correlation of isoenzyme polymorphism and bayoud-disease resistance in date palm cultivars and progeny. Euphytica 65:23–32

Benjama A, Charkaoui B, Samson R (1996) Effets de certains antibiotiques et antiseptiques sur les bactéries contaminant les cultures in vitro de tissus de palmier dattier. Al Awamia 93:53–61

Bennaceur M, Lanaud C, Chevalier MH, Bounaga N (1991) Genetic diversity of date-palm (*Phoenix dactylifera* L.) from Algeria detected by enzyme markers. Plant Breed 107:56–69

Benslimane AA, Rode A, Quetier F, Hartmann C (1994) Characterization of two minicircular plasmide-like DNAs isolated from date-palm mitochondria. Curr Genet 26:535–541

Bouziane M (2010) Le marché des dattes au Maroc: prévisions et stratégies de développement. Packinfo 89:50–52. Juillet/Aôut 2010, Maroc

Bulit J, Louvet J, Bouhot D, Toutain G (1967) Recherches sur les fusarioses. I. Travaux sur le bayoud, fusariose vasculaire du palmier dattier en Afrique du Nord. Ann Epiphytie 18:231–239

Cherif E, Zehdi S, Castillo K et al (2013) Male-specific DNA markers provide genetic evidence of an XY chromosome system, a recombination arrest and allow the tracing of parental lineages in date palm. New Phytol 197(2):409–415

Chetto A, Harrak H, El-Hachami N (2005) Le marketing des dattes au Maroc: Défaillances Préférences et Attentes. INRA, Rabat

Corniquel B, Mercier L (1994) Date palm (*Phoenix dactylifera* L.) cultivar identification by RFLP and RAPD. Plant Sci 101:163–172

Djerbi M, Sedra MyH (1986) Résistance to Fusarium oxysporum f. sp. albedinis in Phoenix dactylifera: evaluation of new screening method and performance of American high quality backcrossed males. Proceeding of the second symposium on the date palm. March 3–6/1986, Date Palm Research Center, King Faisal University, Al-Hassa, pp 400–412

Djerbi M, Aouad L, El Filali H et al (1986) Preliminary results ou selection of high quality bayoud resistant lines among natural date palm population in Morocco. Proceeding of the second symposium on the date palm. March 3–6/1986, Date Palm Research Center, King Faisal University, El-Hassa, pp 386–399

El Hadrami I, Ramos T, Macheix JJ (1996) Caractérisation de nouveaux dérivés hydroxycinnamiques amines chez *Phoenix dactylifera* L.: relation avec le brunissement des tissus et la résistance des cultivars au bayoud. Polyphénols Commun 2:341–342

El-Fakhouri R, Lazrek HB, Bahraoui E et al (1996) Preliminary investigation on a peptidic toxins produced by *Fusarium oxysporum* f. sp. *albedinis*. Phytopath Mediterr 35:11–15

El-Fakhouri R, Lotfi F, Sedra MH, Lazrek HB (1997) Production et caractérisation chimique des toxines sécrétées par *Fusarium oxysporum* f. sp. *albedinis*, agent causal du bayoud. Al Awamia 93:83–93

El-Houmaizi MA, Saadi M, Oihabi A, Cilas C (2002) Phenotypic diversity of date-palm cultivars (*Phoenix dactylifera* L.) from Morocco. Genet Res Crop Evol 49:483–490

El-Youssefi B (1987) Mise au point sur la culture du Palmier dattier au Maroc et contribution à l'étude de la caractérisation variétale. Rapport de stage à l'INRA- Marrakech

Essarioui A, Sedra MH (2007) Effet de la solarisation et du Metam sodium sur les champignons telluriques et possibilité de lutte contre *Fusarium oxysporum* f. sp. *albedinis*, agent causal du bayoud du palmier dattier. Al Awamia 121(122):123–139

Essarioui A, Sedra MH (2010) Biocides, soil solarization and fumigation to control *Fusarium oxysporum* f. sp. *albedinis* inciting bayoud disease on date palm. Acta Hort 882:520–533

FAOSTAT (2011). http://faostat.fao.org

Fernandez D, Ouinten M, Tantaoui A et al (1998) Fot1 insertions in the *Fusarium oxysporum* f. sp. *albedinis* genome provide diagnostic PCR targets for detection of the date palm pathogen. Appl Environ Microbiol 64:633–636

Harrak H (2010a) Développement d'une indication géographique des dattes d'Akka au Maroc. Deuxième séminaire international d'Antalya Indications géographiques, dynamiques socio-économiques et patrimoine bio-culturel en Turquie et dans les pays méditerranéens. 15–19 décembre 2010, Antalya

Harrak H (2010b) Assistance technique de l'INRA aux coopératives des dattes dans les oasis marocaines. Article de vulgarisation. Packinfo 89:46–47

Harrak H (2011) Distinctive sign of origin and quality in Morocco: legal system and typical products. 5th edition of training program InterGI – a way forward for local development 6–17 June 2011, Geneva

Harrak H (2013) Evaluation de la qualité des pâtes de dattes traditionnelles Tassabount et Toummit, produits de terroir des oasis de Tata. Programme de Recherche. Bilan et perspectives.

Programme MCA-Maroc, Projet Arboriculture Fruitière, Consortium National de Recherche en Agriculture (IAV Hassan II, INRA, ENA, ENFI), pp 84–85
Harrak H, Boujnah M (2012) Valorisation technologique des dattes au Maroc. INRA Eds, Rabat
Harrak H, Chetto A (2005) Etude de la commercialisation des dattes, de l'industrie dattière et de l'exploitation des sous-produits du palmier dattier et des dattes au Royaume du Maroc. INRA Eds, Rabat
Harrak H, Hamouda A (2005) Etude de quelques critères de qualité des principales variétés de dattes marocaines. Proceeding of the international symposium: sustainable agricultural development of oasian systems. 07–10 Mar 2005, Erfoud, pp 554–557
Harrak H, Jaouan F (2010) Processing dates of low market value into flour: evaluation of quality and storage stability. Acta Hort 882:593–602
Harrak H, Zirari A (2006) Préservation et valorisation du savoir-faire traditionnel de transformation des dattes: produits de terroir des oasis marocaines. Colloque international: Avenir des oasis face à la désertification, 14–17 Sep 2006, Errachidia, p 8
Harrak H, Reynes M, Lebrun M et al (2005) Identification et comparaison des composés volatils des fruits de huit variétés de dattes marocaines. Fruits 60(4):267–278
Lebrun M, Billot C, Harrak H, Self G (2007) The electronic nose: a fast and efficient tool for characterizing dates. Fruits 62(6):1–6
Louvet J, Toutain G (1973) Recherches sur les fusarioses VII. Nouvelles observations sur la fusariose du palmier dattier et précisions concernant la lutte. Ann Phytopathol 5:35–52
Rhiss A, Poulain C, Beauchesne G (1979) La culture in vitro appliquée à la multiplication végétative du palmier dattier (*Phoenix dactylifera* L.). Fruits 34:551–554
Saaidi M (1992) Comportement au champ de 32 cultivars de palmier dattier vis à vis du bayoud: 25 années d'observations. Agron 12:359–370
Saaidi M, Bannerot TG, Louvet J (1981) La sélection du palmier-dattier (*Phoenix dactylifera* L.) pour la résistance au bayoud. Fruits 36:241–249
Saker MM, Bekheet SA, Taha HS et al (2000) Detection of somaclonal variation in tissue culture-derived date palm plants using isozyme analysis and RAPD fingerprints. Biol Plant 43:347–351
Saker MM, Adawy SS, Mohamed AA, El-Itriby HA (2006) Monitoring of cultivar identity in tissue culture-derived date palms using RAPD and AFLP analysis. Biol Plant 50:198–204
Sedra MyH (1990a) Preliminary results on the evaluation of the resistance to the bayoud of the clones (khalts), cultivars and some hybrids of the date palm trees selected on the fruit quality criterium. 8th Congr. of Mediter. Phytopathol. Union, Agadir, p 529
Sedra MyH (1990b) Observations on some phenotypical criteria that are eventually in relation with the date palm tree resistance to Bayoud. 8th congr. of the Mediter. Phytopathol. Union, 28 Oct–3 Nov 1990. Agadir, p 527
Sedra MyH (1992a) Variabilité dans le pouvoir pathogène des isolats et souches du *Fusarium oxysporum* f. sp. *albedinis*, agent de la fusariose vasculaire (bayoud) du palmier dattier. Séminaire sur les Interactions Plantes-Microorganismes. IFS-ORSTOM, Dakar, pp 204–213
Sedra MyH (1992b) Evaluation and selection of the resistant good cultivars and clones of date palm to the bayoud disease. Arab Soc Plant Prot 10(2):155–160
Sedra MyH (1993a) Caractéristiques morphologiques et culturales du *Fusarium oxysporum* f. sp. *albedinis*, agent causal de la fusariose vasculaire (bayoud) du palmier dattier. Al Awamia 83:209–222
Sedra MyH (1993b) Remarques sur la stabilité et la relation entre les caractères morphologiques et le pouvoir pathogène du *Fusarium oxysporum* f. sp. *albedinis*, agent du bayoud du palmier dattier. Al Awamia 82:39–52
Sedra MyH (1993c) Search results on the receptivity of some Moroccan date palm grove soils to *Fusarium* wilts. January 17–20/1993, Date Palm Research Center, King Faisal University, Al-Hassa, pp 1–14
Sedra MyH (1993d) Role of some physical-chemical and microbiological characteristics in the observed resistance of Marrakech date palm grove soils to *Fusarium* wilts. January 17 20/1993, Date Palm Research Center, King Faisal University, Al-Hassa, pp 42–54

Sedra MH (1994a) Mise au point d'une méthode pour l'évaluation rapide de la resistance au bayoud de plantules du palmier dattier issues de semis. Al Awamia 86:21–41

Sedra MH (1994b) Evaluation de la résistance à la maladie du bayoud causé par *Fusarium oxysporum* f. sp. *albedinis* chez le palmier dattier: recherche d'une méthode fiable d'inoculation expérimentale en pépinière et en plantation. Agronomie 14:445–452

Sedra MH (1995) Triage d'une collection de génotypes de palmier dattier pour la résistance au bayoud causé par *Fusarium oxysporum* f. sp. *albedinis*. Al Awamia 90:9–18

Sedra MyH (1997) Diversité et amélioration génétique du patrimoine phénicicole marocain. Proceeding du Séminaire National sur Ressources Phytogénétiques et Développement Durable Actes Editions IVA Hassan II, Rabat, pp 283–308

Sedra MyH (1999) Report of international expert consultation in Mauritania. Bayoud disease status, cultivation, development and project proposal for the protection and development of research in Mauritania. FADES/FAO Project: UFT/MAU/025/MAU, Oct,-Nov. 1999, Technical assistance for date palm protection in Adrar State

Sedra MyH (2000a) Biological and genetic characteristics of bayoud resistant Moroccan date palm cultivars and strains. Agric Water 20:55–66

Sedra MyH (2000b) Report of international expert consultation on sanitary state of date palm groves in Mauritania and discovery of bayoud disease. Suggestions and recommendations for a prevention program and methods of control. Training a team of prospectors in palm groves and field diagnostic methods. AOAD, Al-Khartoum, 6–21/7/2000

Sedra MyH (2001a) Descripteurs du palmier dattier (*Phoenix dactylifera* L.). Imprimerie Al Watania, Marrakech

Sedra MyH (2001b) Bayoud disease of date palm (*Phoenix dactylifera* L.), extension, damage, diagnosis and prevention of practical recommendations for the spread and development of the disease. Bulletin guidelines (4). Research network and the date palm development. ACSAD, Damascus

Sedra MyH (2002) Report of international expert consultation in Mauritania. FADES/FAO Project: UFT/MAU/025/MAU, 6 Nov–4 Jan 2002, Technical assistance for date palm protection in Adrar State

Sedra MyH (2003a) Date palm cultivation, characterization and classification of main Mauritanian varieties. Edit AOAD, Al-Khartoum

Sedra MyH (2003b) Le bayoud du palmier dattier en Afrique du Nord. FAO, RNE/SNEA-Tunis. Imprimerie Signes, Tunis

Sedra MyH (2003c) Le palmier dattier base de la mise en valeur des oasis au Maroc. Techniques phoénicicoles et création d'oasis. Edit. INRA Maroc, Imprimerie Al Watania-Marrakech

Sedra MyH (2004) Le bayoud (fusariose vasculaire) du palmier dattier en Afrique du Nord. Situation actuelle et stratégies de lutte. Journées nationales sur la protection des Plantes. Association Marocaine de la Protection des Plantes (AMPP), Rabat

Sedra MyH (2005a) Caractérisation des clones sélectionnés du palmier dattier et prometteurs pour combattre la maladie du bayoud. Proceeding of the international symposium: sustainable agricultural development of oasian systems. 07–10 Mar 2005, Erfoud, p 72–79

Sedra MyH (2005b) Phenological descriptors and molecular markers for the determination of true-to-type of tissue culture-derived plants using organogenesis of some Moroccan date palm (*Phoenix dactilyfera* L.) varieties. Al Awamia 113:87–101

Sedra MyH (2006a) Bayoud disease on palm date (*Phoenix dactylifera* L.): its appearance, losses, spread, development conditions, diagnosis and control methods, achievements, prospects and practical recommendations. Regional Project research for early detection of bayoud disease on date palm and development of its control technologies. AOAD/Imprimerie Al-Watania, Al-Khartoum/Marrakech

Sedra MyH (2006b) Final report of international expert consultation in Mauritania. Project: FAO/ UFT/MAU/025/MAU technical assistance for date palm protection in Adrar State, 27 Aug–17 Sep 2006

Sedra MyH (2007a) Discovery of bayoud disease on date palm tree in Mauritania: its importance, kind and control strategies. Fourth symposium on date palm King Faisal University, Hofuf, 5–8 May 2007

Sedra MyH (2007b) Bayoud disease of date palm in North Africa: recent distribution and remarks about its characterization, diagnosis and origin. Fourth symposium on date palm King Faisal University, Hofuf, 5–8 May 2007

Sedra MyH (2007c) Report of international expert consultation in Mauritania. Training of Mauritanian staff for Phytopathology laboratory operation at Atar city affiliated to the Ministry of Agriculture and Rural Development and research and development program of date palm for short and medium term. Project OADA 17–23 June 2007

Sedra, MyH (2007d) Selection of morphological characteristics and molecular markers and their use for identification and distinguishing between date palm varieties and the plants issued from tissue culture. Fourth symposium on date palm, King Faisal University, Hofuf, 5–8 May 2007

Sedra MyH (2008a) Pathogenic and molecular characterization of *Fusarium oxysporum* f. sp. *albedenis* strains, causal agent of the bayoud disease of date palm. Proc. 10th International Fusarium Workshop and Fusarium Genomics Workshop. Alghero, Sardinia, 30 Aug–2 Sept, p 15

Sedra MyH (2008b) Antagonistic activity of some microorganisms isolated from suppressive soil and possibility of their use to control bayoud disease of date palm and other wilt diseases. Proc. 10th International *Fusarium* Workshop and *Fusarium* Genomics Workshop 2008, Alghero, Sardinia, 30 Aug–2 Sept, p 20

Sedra MyH (2009) Bio-écologie du charançon rouge (*Rhynchophorus ferrugineus* (Olivier)) du palmier et questions sur la stratégie de lutte. Proceedings of the international workshop on the strengthening national strategy for research, eradication and control of red palm weevil (*Rhynchophorus ferrugineus* Oliver), 9 Oct 2009, Rabat, pp 37–50

Sedra MyH (2010a) Les nouvelles variétés du palmier dattier sélectionnées pour combattre le bayoud. Packinfo 89:42–45. Juillet/Aôut 2010, Maroc

Sedra MyH (2010b) Evaluation of soil receptivity of date palm groves in Arab countries to *Fusarium oxysporum* f. sp. *albedinis*, causal agent of bayoud disease of date palm. Acta Hort 882:515–526

Sedra MyH (2011a) Development of new Moroccan selected date palm varieties resistant to bayoud and of good fruit quality. In: Jain SM, Al-Khayri JM, Johnson DV (eds) Date palm biotechnology, Springer, Dordrecht, pp 513–533

Sedra MyH (2011b) Molecular markers for genetic diversity and bayoud disease resistance in date palm. In: Jain SM, Al-Khayri JM, Johnson DV (eds) Date palm biotechnology. Springer, Dordrecht, pp 533–550

Sedra MyH (2011c). La palmeraie Marocaine: un site unique au monde, des potentialités à mieux valoriser pour un défi du futur. Pack Agro 101:36–40

Sedra, MyH (2011d) Présentation des principales variétés du palmier dattier au Maroc. INRA, Atlas du palmier dattier au Maroc. Rabat, pp 34–100

Sedra MyH (2011e) Genetic diversity analysis, Moroccan cultivar genotyping and rapid screening for bayoud disease resistance in date palm using molecular techniques. Proc. first symposium international sur le palmier dattier en Algérie 13–14 Nov 2011, pp 271–288

Sedra MyH (2011f) Le Bayoud (fusariose vasculaire) du palmier dattier en Afrique du Nord: situation, acquis de la recherche et applications. Proc. first symposium international sur le palmier dattier en Algérie 13–14 Nov 2011, pp 59–75

Sedra MyH (2012) Le guide du phéniciculteur: mise en place et conduite des vergers phoénicicoles. Edition INRA Maroc, Imprimerie Nadacomdh, Rabat

Sedra MyH (2013) Le guide du phéniciculteur: mise en place et conduite des vergers phoénicicoles. Edition INRA Maroc, Imprimerie Nadacomdh, Rabat

Sedra MH, Besri M (1994) Evaluation de la résistance au Bayoud du palmier dattier causé par *Fusarium oxysporum* f. sp. *albedinis*: recherche d'une méthode de discrimination des vitroplants acclimatés en serre. Agron 14:467–472

Sedra MH, Lazreak HB (2011) *Fusarium oxysporum* f. sp. *albedinis* toxin characterization and use for selection of resistant date palm to bayoud disease. In: Jain SM, Al-Khayri JM, Johnson DV (eds) Date palm biotechnology. Springer, Dordrecht, pp 253–271

Sedra MH, Rouxel F (1986) Résistance des sols aux maladies. Mise en évidence de la résistance d'un sol de la palmeraie de Marrakech aux fusarioses vasculaires. Al Awamia 66:35–54

Sedra MH, Zhar N (2010) Genetic variability analysis of populations of *Fusarium oxysporum* f. sp. *albedinis*, causal agent of bayoud disease of date palm and other *Fusarium oxysporum* using molecular techniques. Acta Hort 882:491–504

Sedra MyH, Zirari A (1998) Rapport final des travaux de la convention établie entre l'Office Régional de Mise en Valeur Agricole de Ouarzazate (ORMVAO) et l'Institut National de Recheche Agronomique (INRA) durant les années 1997–98 sur la conduite de la culture du palmier

Sedra MH, El Fakhouri R, Lazrek HB (1993) Recherche d'une méthode fiable pour l'évaluation de l'effet des toxines secrétées par *Fusarium oxysporum* f. sp. *albedinis* sur le palmier dattier. INRA-Rabat Maroc. Al Awamia 82:89–104

Sedra MH, El Filali H, Benzine M et al (1996) La palmeraie dattière marocaine: evaluation du patrimoine phoénicicole. Fruits 51:247–259

Sedra MyH, Lashermes P, Trouslot P et al (1998a) Identification and genetic diversity analysis of date palm (*Phoenix dactylifera* L.) varieties from Morocco using RAPD markers. Euphyt 103:75–82

Sedra MyH, Lazrek HB, Lotfi F, Rochat H (1998b) *Fusarium oxysporum* f. sp. *albedinis* toxin isolation and use for screening of date palm plants for resistance to the bayoud disease. Proc. XXV Int Hort Congr (IHC), Brussels, 2–7 Aug

Sedra MyH, Lazrek HB, Amraoui H, Nour S (2008). Pathogen toxins of bayoud disease on date palm: in vitro selection and biological and specificity activities. Proc. 10th International *Fusarium* Workshop and *Fusarium* Genomics Workshop. Alghero, Sardinia, 30 Aug–2 Sept

Tantaoui A, Ouinten M, Geige JP, Fernandez D (1996) Characterisation of a single clonal lineage of *Fusarium oxysporum* f. sp. *albedinis* causing bayoud disease of date palm (*Phoenix dactylifera* L.) in Morocco. Phytopathology 86:787–792

Tirichine M (1991) Caractéristiques des palmeraies du M'zab et de Metlili. Ressources génétiques du palmier dattier – comportement vis-à-vis du bayoud. Bulletin du Réseau Maghrébin de Recherche sur la Phoéniciculture et la Protection du palmier dattier, PNUD/FAO 1(3):7–10

Trifi M (2001) Polymorphisme et typage moléculaire de variétés tunisiennes de palmier (*Phoenix dactylifera* L.): relation avec la résistance au bayoud. Thèse de Doctorat d'Etat, Université de Tunis El Manar, Faculté des Sciences de Tunis, Tunis

Zaher H, Sedra MH (1995) Etude de quelques caractèes morphologiques et pomologiques chez trois croisements du palmier dattier. Al Awamia 91:25–39

Zaher H, Sedra MyH (1998) Evaluation of some date palm hybrids for fruit quality and bayoud disease resistance. Proceedings of date palm research symposium, Réseau de Recherches et du développement du palmier dattier. Marrakech, 16–18 février 1998, pp 181–192

Ziouti A, El Modafar C, El Mandili A et al (1996) Identification des acides caféoylshikimiques des racines du palmier dattier, principaux composés fongitoxiques vis-à-vis de *Fusarium oxysporum* f. sp. *albedinis*. J. Phytopathology 144:197–202

Zirari A (2010) Effects of exogenous indole butyric acide (IBA) on rooting and leaf grown of small date palm offshoots (*Phoenix dactylifera*) derived from adult vitroplants of Najda cultivar. 4th international date palm conference, Abu Dubai, 15–17 Mar 2010, p 52

Zirari A, Harrak H, Ouberhou A et al (2006) Cuisine marocaine -délices de dattes. Nouiga Eds, Tanger

Chapter 9
Date Palm Status and Perspective in Mauritania

Moulay Hassan Sedra

Abstract The date palm population in Mauritania is composed of 2.4 million trees distributed in 217 oases in three representative states: Adrar, Tagant, and Laassaba and small number in two *hodhs* (basins). The number of cultivars varies from 75 to 250. About 65 % of date palms annually produce an average of 20,000 mt. of fruit. The Mauritanian people practice, in most cases, two types of agricultural activities, camel husbandry during 7–8 months of transhumance and traditional agriculture which includes the date palm during the rest of the year. Natural constraints including the silting of date palm groves, salinization of soil and water, and the traditional agricultural techniques used by farmers create a development impediment to the Mauritanian date palm sector. This sector is still traditional and lacks modern formal plantations to bring about true date commercialization. In order to develop the date sector, the Mauritanian government, with international cooperation, established several programs based on social organizations, equipment development in oases, building of capacity of human resources, and strengthening of agricultural research and extension services. It established the Project of Sustainable Oasis Development (PDDO) in 2002 to preserve the fragile but valuable oasis ecosystems and stem a rural to urban population exodus. Few studies were done to control major date palm pests or to analyze the genetic diversity of date patrimony using morphological characters or molecular markers. A morphological study covered 22 main cultivars in the three important regions. The correlation between numerous quantitative and qualitative characters was evaluated, and several discriminating criteria were selected. The grouping association of cultivars was identified by cluster analysis using 75 characters. All cultivars studied were identified, and the variability was evaluated. It was recommended to preserve this material and to improve it by mass selection and breeding programs in order to select new date palm cultivars of better production quality and wider adaptability.

Keywords Bayoud • Cultivars • Genetic diversity • Mauritania • Pests

My.H. Sedra
Laboratory of Phytopathology, Genetics and Integrated Control, Regional Centre of Agricultural Research of Marrakech, National Institute of Agricultural Research (INRA), 533, Marrakech 40000, Morocco
e-mail: sedramh@gmail.com, mhsedra@yahoo.fr

J.M. Al-Khayri et al. (eds.), *Date Palm Genetic Resources and Utilization: Volume 1: Africa and the Americas*, DOI 10.1007/978-94-017-9694-1_9

9.1 Introduction

9.1.1 Importance of Date Palm

The date palm (*Phoenix dactylifera* L.) is one of the oldest cultivated plants in the world. It is a tree of great interest because of its high productivity, fruit nutritional quality, and adaptability to Saharan regions. Indeed, the date palm is the essential element in the oasis agrosystems in arid or semiarid areas. The world population of date palm trees is estimated at 100 million distributed over an area of about 600,000 ha, of which 422,000 ha (70 %) are in the Arab world. This area represents 5 % of planted land in Arab countries with more than 62 million date palms that produce 4.5 million mt or 67 % of annual global production. The best-quality dates are sought on the market and especially on the international level. The average productivity of date palms varies from 18 to 250 kg/tree, according to cultivar and tree age, environmental conditions, cropping patterns, and care in each of the date-producing countries.

In North Africa, the total number of date palms in the Maghreb countries is estimated at 17.5 million; Mauritania accounts for only 10.7 %. Date palm cultivation in Mauritania was revived at the time of the first migration of Arabs from the Arab Peninsula to North Africa. This crop has been a relatively successful development during the seventh and tenth centuries AD in two areas in particular, Azougui and Wadane located in the Wilaya (State) of Adrar in the northeast. From there, it was dispersed into the central and southeast regions. In Mauritania, all parts of the date palm are used, in particular fruits for human consumption; waste, leaflets, and seeds as food for domestic animals; trunk wood for construction; and the rachis and leaves used for building materials and furniture.

Mauritanian date palms number about 2.4 million, variably distributed in 218 inventoried oases. These oases cover a total area of 12,000 ha (Sedra 1999a, 2003a). This area is divided into five main regions administratively delimited by Wilayas (i.e., states) (Fig. 9.1a): Adrar, Tagant, and Laassaba states and two *hodhs* (basins) with densities of 171, 354, 502, and 150 date palms per ha, respectively. In 2011, the total date palm area was 8,638 ha, making Mauritania rank the 6th in that category, but with only 0.3 % of world production (FAOSTAT 2011).

Date palm oases exist near the Saharan rivers and in narrow valleys between sand dunes and rocky mountains, with some aligned plantations in flat areas (Fig. 9.1b–e). They are generally located in a hostile desert environment characterized by drought, low or infrequent rainfall, high temperatures, and strong winds carrying sand. The oasis ecosystem plays an extremely important role in the management of large-scale dry lands. Indeed, dry land populations in Mauritania manage natural resources that have allowed them to survive and adapt to difficult conditions. These populations practice, in most cases, two types of agricultural activities, focused on camel husbandry, during 7–8 months of transhumance, and traditional agriculture based on date palm for the rest of the year. Oasis agriculture is poorly practiced and production is relatively low in quality and quantity.

Fig. 9.1 (**a**) Map of states of Mauritania and main regions of date palm cultivation. Different types of date palm grove structures: oases beside Saharan rivers, (**b–d**) in narrow valleys between the sand dunes and rocky mountains. (**e**) Some aligned plantations in flat valley areas

Indeed, the average productivity of date palm does not exceed 18 kg per tree and harvested dates are not properly handled or valued and predominantly of poor- to medium-quality cultivars. In this situation, Mauritania continues to import some quantities of dates.

About 65 % of the date palms produce 18,000–23,000 mt of dates with an annual average of 16,000 mt in the period 1992–2003 and 21,440 mt in 2011; in the future, it may reach 60,000 mt. The portion of the Adrar region in production is generally estimated at 75 % (MDRE/NSO 2003). The average productivity of 18 kg of dates per palm is low compared to the Maghreb countries. Productivity increased in 2011 to 24.8 mt/ha (FAOSTAT 2011). Mauritania ranks 5th in the Maghreb and 19th worldwide (FAOSTAT 2011). Human population of the oases exceeds 250,000, approximately 14 % of the total Mauritanian population. The region of Laassaba is the most populous with 47 % of total oasis population, followed by Tagant (25 %), Adrar (24 %), and the *hodhs* (4 %). In general, the oases are located in areas characterized by hostile Saharan climate.

According to old reports and studies, there were reported to be 75–250 date cultivars in Mauritania. This imprecision is due to the absence of an exhaustive survey in all oases. In 1999, a detailed study for the identification and classification of cultivars showed that the Mauritanian patrimony includes more than 185 known cvs. of which 22 were morphologically characterized based on 47 quantitative and 28 qualitative vegetative and reproductive aspects (Sedra 1999a, 2003a). The best adapted cvs. exhibiting good production are Lamdina, Ahmar, Tinterguel, Tijeb, Oum Arich, Tiguedert, Soukani, Mahboula, Tinwazid, Mriziga, and Al-Falha, where the first 7 cvs. are most common (Sedra 2003a, b).

9.1.2 Constraints of Development of Date Palm Sector

9.1.2.1 Natural Constraints

Most Mauritanian oases are located in narrow and difficult valleys with a delicate environment, huge dunes, and/or rugged rocks; in addition to a hot, dry hostile climate and frequent prolonged drought (Fig. 9.2a). These natural constraints create important variations within the oases especially in the surrounding areas where transport and communication between oases is difficult. Other constraints include

Fig. 9.2 Constraints of date palm cultivation in Mauritania. (**a**) Effect of drought, (**b**) silting of palm groves, (**c**) losses by floods, and (**d**) low maintenance and care of date palms

the silting of date groves (Fig. 9.2b), floods causing loss of land and date palms (Fig. 9.2c), and salinization of soil and water aggregated by low maintenance and care of palms (Fig. 9.2d).

9.1.2.2 Technical Constraints

The agricultural practices of date palm farmers are traditional and undeveloped, handed down from one generation to another. This is due to inadequate field supervision and extension services to improve cultivation practices as well as the lack of operational field stations for applied research and demonstrating proper cultivation of date palm in association with other crops and livestock.

Despite Mauritanian Government projects supporting oasis development, the organization of inhabitants, and more efficient water use, many oases still have technical constraints that can be summarized as follows:

(a) Dispersion and isolation of oases and production units as well as transhumance populations.
(b) Significant deterioration of the ecological environment of the oases.
(c) Old plantations and orchard land ownership fragmentation due to inheritance.
(d) Cultivars are very unproductive due to inefficient farming techniques.
(e) Deficiency of managers and specialized engineers and technicians in date palm cultivation topics.
(f) Inadequate or lack of a practical awareness program for date farmers and their initiation into use of modern palm farming techniques.
(g) Weak organization and profession representation and difficult access to sources of specialized financing.
(h) Inadequate implementation of oriented and/or integrated control of programs against diseases and pests of date palms designed and adapted in Mauritania (Sedra 1999b, 2002, 2003a, b, 2006a, b, 2007a, b, c). This has led to increased risk to known and unknown disease spread, particularly bayoud which is the most serious threat of destruction to the Mauritanian date palm patrimony (Sedra 1999b, 2002, 2003a, b, 2006a, b, 2007a, b, c).
(i) Inadequate balanced and economic management of available water.
(j) Inadequate or total absence of utilization of genetic resources of cultivated date palms in the oases, particularly the low valuation of date fruits and other derived products in order to improve and diversify oasis products using conventional, modern, and agroindustrial methods.

9.2 Cultivation Practices

The Mauritanian date palm sector is still traditional, lacking formal plantations for commercialization. The government established the Project of Sustainable Oasis Development (PDDO), in 2002 to preserve the fragile but valuable oasis ecosystems and slow the rural to urban exodus that had begun to increase. The International Fund for Agricultural Development (IFAD) contributed some tens of millions of dollars in cooperative work with the Ministry of Rural Development. The main actions were focused on organizing oasis farmers and supporting emergence of a civil society able to sustain oasis participatory management associations (AGPOs) and make collective investments. Among the notable actions, the creation of small field schools and the plantations of fruit trees and vegetables interspersed with the date palms to demonstrate sustainable land management techniques. Other actions have concerned protection against soil erosion, efficient irrigation, and a diversification of farm income. In the Adrar region, the program focused on developing date palm groves with recommendations for improved practices of planting density, pollination, drip irrigation, organic fertilizer, harvest, date transporting, and commercialization in Nouakchott City. Farmers were advised to use solar-powered pumps to draw irrigation water. Because of their adaptation to Saharan and Sahel climate, the date palm and the camel are the twin pillars of the economy of these regions.

9.2.1 Cultivation

Good cultivation practices and proper maintenance of the date palms plays an important role in the expression of production performance of cvs. and their true morphological and phenological characteristics. Indeed, the measurements of many vegetative, productive, quantitative and qualitative palm characters are detrimentally influenced by poor and inadequate agricultural practices (Sedra 2001a, 2003a). As previously mentioned, date cultivation practices are traditional in most Mauritanian oases. Plantations are perpetuated by offshoots, generally without frequent watering or protection against strong sunlight and hot strong winds carrying sand, leading to low productivity. Farmers continue to allow seedling dates to grow in the fields (Fig. 9.3a).

Irrigation systems differ from region to region. In some oases, date palms are rarely irrigated, the trees drawing on ground water close to the surface. For example, part of Aouja Oasis, Laassaba State, shows open areas in the center of the oasis because of the difficulty of the date palms to survive solely from ground water (1–2 m deep). Orchards are flood irrigated through traditional channels. Date groves with pumps deliver water by large pipes to fill the basins around the trees without calculating the amount used (Fig. 9.3b). The use of drip irrigation is rare. Generally, the date palms only benefit from fertilizers applied to associated crops. This is practiced by spreading organic manure; mineral fertilizers are rarely applied. Pollination is done by placing a few male spikelets into the female inflorescences.

Male trees are chosen and identified by farmers; pollen from these palms can serve several farmers in the same village. Thinning to reduce strand numbers is generally not practiced by date farmers; protection against the bird attack is usually done by cloth bags (Fig. 9.3c). Figure 9.3d shows the traditional drying process of

Fig. 9.3 (**a**) Growth of young natural seedling dates mentioned by the *red arrows*, (**b**) irrigation of palm trees using large pipes to fill the basins around the trees, (**c**) protection by covering date bunches using bags (Toues Al-Alia Oasis, Tijigja valley), Tagant State, and (**d**) example of drying process of dates by traditional method in Taqadat Iriji Oasis, Laassaba State

dates. For date palm protection, insect pest control is not practiced, except for the use of sulfur against mites in some oases. This is due to low agricultural support of extension services and lack of financial resources and of managers and technicians specialized in palm cultivation. In Mauritania, date palm research undertaken by IFAC (Institut Français de Recherches Fruitières Outre-Mer) is the best known and used as a reference for date palm growing, despite some gaps. A few research projects have been undertaken by the University of Nouakchott in collaboration with institutions in Senegal, France, and other North African countries.

9.2.2 *Pests and Diseases*

9.2.2.1 Quarantine

Like many countries that produce dates, Mauritania is exposed to potential pest attacks. The main threats are listed in Table 9.1. The most serious is red palm weevil (*Rhynchophorus ferrugineus*) that has caused high losses in many countries.

Table 9.1 Principal pests and diseases threatening date palm oases in Mauritania

Insects and diseases	Responsible agent	Nearest infested countries to Mauritania
Insects		
Red (or Indian) palm weevil	*Rhynchophorus ferrugineus*	Spain (Canary Islands), Morocco, Libya
Green scale	*Asterolecanium phoenicis*	Sudan
Dubas	*Ommatissus binotatus-lybicus*	Libya and Gulf countries
Butterfly	*Paysandisia archon*	France and Southern Europe
Diseases		
Bayoud disease	*Fusarium oxysporum* f. sp. *albedinis*[a]	Morocco, Algeria
Bacterial blight	*Erwinia chrysanthemi*	USA and recently reported in Saudi Arabia
Brittle leaf disease	Undetermined	Algeria
Lethal yellowing	Mycoplasma	Caribbean, Central America

Source: Sedra (2003b, 2012, 2013)

[a]Often isolated in association with other fungi such as *Alternaria, Diplodia*, or *Chalara*

To prevent the introduction and development of pests and diseases into Mauritania, the government has passed legislation governing the importation of the date palms, phytosanitary measures, and establishment of control stations at major border points and a national diagnosis laboratory in Nouakchott. The following measures are taken into consideration:

(a) Respect the regulations on the import of plant material potentially harboring these maladies, plants, seeds, and contaminated or polluted products.
(b) Notify immediately the proper plant protection services and/or the Ministry of Agriculture of plant anomalies and cases of disease.
(c) Ensure infected orchards make the appropriate treatments of curative and preventive control to prevent the spread of insects or diseases, according to the advice of specialized services.
(d) Ensure routine inspections and regular monitoring of date palms in oases and in urban areas (ornamental palms).

9.2.2.2 Control Strategies

Several date palm pests and diseases have been identified and studied during the IFAC cooperation. Biological control against white scale (*Parlatoria blanchardi* Targ.) by predatory beetles has been successful. Indeed, releasing vast numbers of lady beetles predatory of white scale in oases reduced the pest significantly, but these operations were suspended for lack of human and financial resources. Later, consulting missions to over 70 oases in different regions drew attention to the development of certain diseases and pests (Sedra 1995, 2003a, b, 2007a, b, c). The

Table 9.2 Relative importance of major diseases and pests of date palm in the main regions of Mauritania

Names of enemies	Responsible organism	Region (or state)		
		Adrar	Tagant	Laassaba
Fungal diseases				
Bayoud (*Fusarium* wilt)	*Fusarium oxysporum* f. sp. *albedinis*[a]	+	±	Unknown
Khamej or inflorescences rot	*Maugeniella scaettae* *Fusarium moniliforme*	+	+	+
Dry rot of inflorescences	*Thielaviopsis paradoxa*	±	±	±
Black scorch	*Thielaviopsis paradoxa*	++	++	++
Heart and trunk rot	*Thielaviopsis paradoxa, Gliocladium vermoeseni, Botryodiplodia theobromae*	++	++	++
Diplodia disease	*Diplodia phoenicum*	++	++	++
Brown leaf spot	*Mycosphaerella tassiana*	+	+	+
Bending head	*Thielaviopsis paradoxa*	++	++	++
Graphiola leaf spot	*Graphiola phoenicis*	–	–	++
Apical drying of leaves	*Alternaria* sp., *Chalara* sp.	++	+	+
Faroune (black)	*Thielaviopsis paradoxa*	++	++	++
Faroune (white)	Undetermined	++	++	++
Omphalia root rot (takakte)	*Omphalia tralucida* *Omphalia pigmenta*	++	++	++
Pests (insects and mites)				
Parlatoria scale (white scale)	*Parlatoria blanchardi*	+++	+++	+++
Old world date mite (boufaroua)	*Paratetranychus afrasiaticus*	++	++	+
Date moths	*Ectomyelois ceratonia* and other	+	+	+
White termites (larda)	*Microcerotermes diversus*	+	+	+
Frond borer	*Apate monachus, Phonapate frontalis,* other species	+	+	+
Palm stem borer (long antennae)	*Phoracantha semipunctata*	+	+	+
Desert locust (occasional pest)	*Schistocerca gregaria*	+	+	+

Source: Sedra (1995, 2003a, b)
[a]Often isolated in association with other fungi such as *Alternaria* and *Diplodia* or *Chalara*

harmful level of these maladies varies according to the general condition of the groves and the degree of date palm care and oasis maintenance. Table 9.2 shows the relative importance of major palm diseases and pests in different states (Sedra 1995, 2003a, b). Some diseases like white or black *faroune*, black scorch, bending heart, and heart and trunk rot are the major fatal diseases which are devastating Mauritanian date groves, and considerable sanitary problems are caused by white scale and apical drying of the leaf in some orchards, and the attack of a harmful mite on date production. Bayoud disease represents a serious threat of expansion into and destruction of Mauritanian date groves (Sedra 1995, 1999a, b, 2003a, b, 2011c).

9.2.2.3 Main Diseases Development and Control

Faroune Disease This disease is one of the most serious in Mauritanian oases; it is widespread and impacts fruit production. Symptoms are characterized by dwarfism of trunk and leaves with spines and leaflets growing irregularly. Then yellowing appears on inner dwarf leaves. The disease leads to abnormal growth of buds stopping tree growth over a long period and even death. The cause of this disease is unknown and no cure yet exists. There are two types of this disease: white described above and black *faroune* which causes the same symptoms as well as the emergence of some blackening or charring on dwarf leaves (Sedra 1995, 2003a, b, 2007b, c, 2008b, 2012, 2013). There are three hypotheses as to the cause of the disease:

(a) Aggravated occurrence of white *faroune* is related to the care level of the date palms (Fig. 9.4a). The disease often appears in oases where there is insufficient irrigation water and drought. Under such conditions, the palm reacts by reducing the length of the leaves and trunk and bud and spathe production. This leads to a lack of mineral nutrition that causes some of the problems that lead to the physiological disorders in leaflet and spine growth and to yellowing of sensitive internal leaves (Fig 9.4b–d). In this case, the diseased palm tree can be treated by proper agricultural services including soil tilling to aerate roots and a fertilizer program.
(b) The frailty of date palms due to nutritional problems can lead to infections with fungi, nematodes, or other parasites in the soil which provoke the symptoms. White *faroune* resembles Al-Wijame disease of Saudi Arabia and Libya. The causes are complicated because the effect of soil conditions on disease aggravation. Experimental treatment is three injections of tetracycline into the trunk spaced over 6 months (20–30 g/tree).
(c) Black *faroune* (Fig. 9.4e) is caused by the fungus *Thielaviopsis paradoxa* which leads to dwarfism and charring of rachis and leaves. Possibly, white *faroune* is a stage of black *faroune* where the fungus is still in the heart of palm and secretes toxic substances leading to these primary symptoms and the appearance of blackening and charring. The importance of these symptoms depends on the level of date palm resistance and environmental conditions. Control of black *faroune* disease is known; in Morocco by pruning of infected palms, disinfecting of wounds issued from pruning leaves with a copper compound, incinerating infected parts of the palm, and spraying the tree by fungicide *bouillie bordelaise* (0.3 %) and polyram Thiram (0.2 %) that gave good results for healing diseased trees. It is also necessary to fungicide-treat any wounds of upper parts of the palm.

Takakt Disease or Trunk/Heart Rot During field work, a few date palms infected by this disease were observed. Symptoms are characterized by trunk base or heart rot (Fig. 9.5a–c). This leads to leaf dryness and death. Symptoms of brown spots of rot may be accompanied by liquid that has a characteristic odor and the existence of some species of ants abnormally on the tree during a trunk injury, and it is probably due to the presence of the liquid. Laboratory analysis of a sample of the rotten tissue permitted isolation of several different fungi, but it was difficult to establish a relationship between their presence and causes of the disease due to lack of a quality sample (Sedra 1995). However, the analysis of the other samples in 1999 showed the presence of some unknown bacteria and fungi known to infect date palms and cause these symptoms (Sedra 1999a, b, 2003a, b), for example, *Thielaviopsis paradoxa* and *Botryodiplodia theobromae*.

Fig. 9.4 Focus of disease of white *faroune* disease. (**a**) Characterized by slowdown of growth leading to trees and offshoot decline, (**b–d**) development and yellowing of leaflets and decline of young tree, and (**e**) black *faroune* disease (black scorch) of leaves and their dwarfing

Fig. 9.5 (**a**) Disease of bending heart, (**b**, **c**) heart and trunk rot (takakt), (**d**) apical drying of leaves, (**e**) *Parlatoria scale* (white scale), and (**f**) date mite (*Paratetranychus* (*Oligonychus*) *afrasiaticus*)

Apical Dry Leaf Disease The disease is known in its initial phase by dryness of apical date palm leaves (Fig. 9.5d); symptoms then evolve to include all the leaf and can affect its base. When conditions are favorable to the disease (drought and lack of irrigation water), it spreads to all leaves and can infect the heart of the tree area and thus constitutes a threat to its life. This disease often spreads in the oases because of poor care of date palms. Fungi isolated from infected leaves are *Alternaria* and *Chalara* and others, unidentified (Sedra 1995, 2003b). For control, it is recommended to remove and burn infested leaves and prune the trees. In the case of aggravation of the disease in the head of the tree, the use of fungicides like Methylthiophanate (0.2) and Thiram (0.2) is necessary.

9.2.2.4 Main Pests and Control

During field work, symptoms of date palm insect attacks were observed, such as on the stem, leaf, and rachis by borers, but apparently the damages they cause locally are relatively minor. White scale and date mite are the most widespread and destructive pests of date palms in Mauritanian oases (Sedra 1995, 2003a, b). Occasionally, other pests like white termites and the Saharan cricket cause damages locally.

White Scale The insect *Parlatoria blanchardi* (Targ.) is the most important pest causing major damage in oases of Mauritania (Fig. 9.5e) because of its high infestation levels and presence in all oases. High density of date palms and absence of good pruning, use of leaves for fencing orchards, and leaving large numbers of offshoots growing on the palms are the main factors that contribute to aggravation of insect attacks (Sedra 1995, 2003a, b). When the infestation is high, the tree is weakened and the fruits become deformed; attacks may lead to the death of young date palms. Previous research studies show that control of white scale using biological predators, such as *Chilochorus bipustulatus* var. *iranensis*, had good efficacy to reduce damage caused by the pest. However, this control program ceased in 1975 because of technical and financial circumstances; this again permitted spread and severe outbreaks of the pest. In order to control this insect, the following integrated management is recommended:

(a) Cutting and burning infested leaves.
(b) Good and beneficial pruning of the tree and its offshoots.
(c) Spraying infested leaves with pesticide three or four times a year depending on the number of generations of the insect. Often, it is advised to carry out this operation after the fruiting season, at the beginning of winter and spring. Some pesticides have given good results in Morocco: Dimethoate (0.15 %) and Malathion (0.15 %) (Sedra 1995, 2003b).
(d) The use of predators to supplement chemical control during the year. For this, it is necessary to establish a small laboratory at the regional level with participation of federal farmer offices or nongovernmental organizations under the supervision of researchers and technical specialists in order to produce mass populations of predators and adjust the period of control of chemical and biological and develop integrated control and guidance for the farmers (Sedra 2003a).

Date Mite Locally, this mite is called *taka*, and it is considered the most harmful to fruit quality. Damage can reach up to 80 % of production. *Paratetranychus* (*Oligonychus*) *afrasiaticus* (Megr.) attacks the fruit during its growth and feeds on it leaving residue and dust. When the infestation is severe, fruits become useless (Fig. 9.5f). The pest spreads especially in areas affected by drought or lack of irrigation and oases with very high tree density. High relative humidity (65–75 %) and moderate temperature (20–25 °C) are the favorable conditions for the spread of the pest. Control is realized by spraying date bunches with sulfur powder (100–150 g per tree) or the use of the sulfur pesticide in water (concentration of 0.25 %) or other pesticides such as Malathion (0.2 %). This is one of the easiest processes: spraying two to four times depending on attack severity and duration of fruit development. Pesticides should not be used during the month before harvest to avoid pesticide poisoning through its aftermaths. For the success of this control process farmers and agricultural extension agents play an important role.

White Termites White termites (*Microtermes diversus*) Amitermes and Reticulotermes attack palm roots usually weakened by a lack of proper maintenance or a parasite infection. Termites build clay galleries from the surface up along the trunk exterior. They weaken date palms and often cause death. The invasion and the development of these pests are increasing especially when drip irrigation is used. The Integrated Pest Management (IPM) recommended the following:

(a) Cull if necessary palms heavily attacked and burn them immediately.
(b) Clean the parts of palm trees of the infested clay galleries and treat the location with an insecticide (chlorpyrifos-ethyl 2 %) (Sedra 2012, 2013). This treatment can protect the palms for 2–3 years. Other insecticides registered in Mauritania may be used.
(c) Provide adequate and regular date palm maintenance (e.g., irrigation, hoeing to remove deep galleries, and fertilizer) in order to reduce their susceptibility to termite attack. Good weed control is also recommended.

Saharan Locust The Saharan locust (*Schistocerca gregaria* Forsk.) is an occasional pest that infrequently attacks but still threatens oasis. Some years, Mauritania is alerted to a locust plague spreading in the oases. Massive locust attacks cause considerable damage, often characterized by total defoliation of date palms. The locusts invade large areas sometimes extending over several countries. Chemical control is performed under partnership with the countries infested and the assistance and coordination of FAO. Usually, the insecticide Malathion (0.2–0.3 %, 200–300 ml/hl) is sprayed from aircraft or from all-terrain vehicles (Sedra 2012, 2013). Poisoned bait can be used in the areas of insect reproduction. In widespread invasion in the oases, monitoring and control are performed by the protection service of the Ministry of Agriculture.

9.2.2.5 Discovery of Bayoud and Study of Pathogen Strains in Mauritania

Spread and Characterization of Bayoud Disease The common symptoms of stiffness and dryness along palm rachis generally characterize bayoud disease (Fig. 9.6). But these symptoms are not enough to diagnose bayoud disease because other diseases can cause the same symptoms such as rotting of the leaf base due to *Botryodiplodia theobromae*. This disease is considered the most serious disease of date palm, especially in Morocco and Algeria, where it has destroyed more than 13 million date palms of the best quality and the best known commercial cultivars (Sedra 2001b, 2003a, 2006a, 2011c). In Mauritania, during initial field surveys (Sedra 1995, 1999b, 2000, 2007a), some exterior symptoms that looked like bayoud symptoms in Morocco and Algeria were found in the Ammaria and Chanker oases not far from Atar City.

Bayoud disease has caused damages and losses in some date palm orchards in Mauritania. Samples were taken from infected palms (Fig. 9.7) and analyzed in a Moroccan laboratory, and the result showed the isolation of two fungi together: *Fusarium oxysporum* and *Gliocladium vermoeseni*. The same fungi was isolated from the Canary Islands palm (*Phoenix canariensis*) in Morocco, and laboratory

Fig. 9.6 Typical external symptoms of bayoud disease on the leaves of palm tree in Morocco, right (**a**), and in Mauritania, left. (**a**, **b**) Atypical symptoms are characterized by dry brown central blade rachis with green sides (*left*) or browning of rachis sides (*right*), (**c**) symptoms of rotting of leaf bases similar to the symptoms of bayoud disease in (**a**, **d**) internal symptoms of bayoud disease, some infected vascular tissue indicates the passage of the fungus in the trunk, and (**e**) samples of infected leaves showing internal symptoms. The *yellow arrows* mention the evolution of symptoms (**a**) and the types and localities of symptoms in (**d**–**f**). The *red arrows* mention the types and localities of symptoms (**b**, **c**)

experimentation showed that the fungus *F. oxysporum* can cause wilting of seedlings of date palm Canary Islands palm, and *G. vermoeseni* is able to provoke seedling rot. According to references, this latter fungus frequently infects ornamental palms in several countries due to its rapid spread and high sporulation on infected leaves when the conditions are appropriately spread by wind and on infected pruning and harvesting tools.

For the fungus *Fusarium oxysporum* isolated from infected palm leaves, the pathogenic test applied to date palm seedlings confirmed the presence of vascular wilt disease (bayoud) (Sedra 1995, 2007a, b). During surveys, other disease foci were found in oases of Wadane Valley and Atar and confirmed by laboratory diagnostic. In 1999, during technical consultations by FAO in the areas of Adrar State, Sedra (1999b) collected several dozen infected palm leaf samples (Fig. 9.7) and isolated many strains of *F. oxysporum* that have proven to cause wilt disease of date palm seedlings. The isolation process of the presence of fungi associated with

Fig. 9.7 Foci of bayoud disease of date palm in Mauritanian oasis (**a**, **b**) (near Atar City, Adrar state in the north). Existence in the infected orchard of henna crop *Lawsonia inermis* (*left*) and the lucerne (*Medicago sativa*) (*right*), mentioned by the *red arrows*; both crops are silent carriers aiding spread of the disease, (**b**) infected date tree with bayoud disease cut and destroyed by the farmer, mentioned by the *red arrows,* (**c**) sampling of infected leaves from diseased tree, the *red arrows* mention the symptoms of brown vascular tissue of rachis (**d**) isolation of *Fusarium oxysporum* from infected leaves using artificial and selective medium (Komada) (5) and the difficulty of isolating of pure *Fusarium* colonies on nonselective medium (PDA) (1), (**e**) isolation of other mixed fungi like *Botryodiplodia* or *Thielaviopsis*, and (**f**) isolation of pure culture of the pathogen from infected vascular tissue of the trunk

F. oxysporum such as *Thielaviopsis paradoxa* and *Botryodiplodia theobromae* and other unknown fungi was carried out through the use of selective medium. The *Fusarium* was often isolated in pure culture from infected vascular trunk tissue (Fig. 9.7). Advanced studies on bayoud and associated rot diseases in Mauritania are needed to establish a clear relationship between them and suggest how, together, to prevent their spread and to develop control methods. From 2001, several teams of technicians and engineers carried out extensive date palm disease diagnostic field surveys of the most important oases in the Adrar region and other areas as a part of the FAO project (1999–2000), AOAD project (05/14–18/2005), FAO project (09/1–6/2006), and AOAD project (June 2007).

Within the framework of the regional bayoud disease project, managed by the Arab Organization of Agricultural Development (AOAD), technology using aerial surveys for early detection system of bayoud disease was realized to orient field surveys in the oases of Atar. According to result analysis and satellite image processing, green color intensity increases wherever date palms exhibit more growth and are healthy. Results from satellite imagery and confirmed by field surveys have permitted location of some foci of the disease in the area of Atar City and in other oases. Table 9.3 indicates the localities that were confirmed with Moroccan laboratory analyses (Sedra 2001b, 2003a, b, 2004, 2007a). During field visits to the oases of Adrar in northern Mauritania (Atar and Wadane) (Sedra 1995, 1999a, b, 2000, 2002), it was observed that most of the date palm found infected with the fungus in July 1999 had already been destroyed by the disease in November 1999. Date palm

Table 9.3 Contaminated date palm oases with bayoud disease in Mauritania

Area	Contaminated oases	Detection year	References
Adrar State	Atar, Wadane	1995	Sedra (1995, 1999a, b, 2001b, 2003a, b, 2004)
	Ain Taïae, Anterqante, Taouz, Tayart, Tounqat oujeft	1999	Sedra (1999a, b, 2001b, 2003a, b, 2004)
Tagant State	Tichitt	2002	Sedra (2002, 2003a, b, 2004)

cvs. infected are Soukani, Tijeb, Lamdina, and probably Ahmar, which is the most common in Mauritania, but the source or origin of these cvs. is unknown. It should be noted that cv. Touatia cultivation which flourished in the oasis of Wadane may have its origin and its source from the Algerian oasis called Touatee. Another cv. named Bousakar is among the Mauritanian cvs. (Sedra 2003a) that produce dry dates and suitable for storage and that have been carried by travelers in caravans of Saharan traders from Morocco or Algeria. So, it is probable that the origin of Mauritanian Bousakar cv. is the Algerian Bousoukar cv. or the Moroccan Bouskree cv.

It is also probable that bayoud disease was introduced into Mauritania by the transport of offshoots of the cvs. mentioned by caravans over time. In addition, it was observed that henna plants, grown under the date palms, serve as a host for but are immune to bayoud; imported henna seedlings may have spread the disease from Morocco and Algeria. Bayoud may have been in Mauritania for a long time but that farmers did not know its symptoms. The presence of bayoud may also not be obvious if date palms die of multiple causes. For example, diseases like bending head or heart rot of other causes could lead to tree death. Despite these theories, nothing confirms that the bayoud disease in Mauritania originally came from Morocco (Sedra 2007a, b, 2011c). It will be necessary to conduct research using molecular markers to identify the origin of the fungus strains isolated in Mauritania.

Comparative studies of strains from Mauritania and Morocco showed a partial compatibility in some morphological characters to strains prevalent in Morocco and also in Italy but with a lower pathogenic compatibility with Mauritanian strains (Sedra 2004, 2007b). However, these latter strains are relatively similar to the fungus strains that infect ornamental palms (Canary Islands palm) and henna (Sedra 2003a, b, 2004, 2007b, 2011c). It seems that Mauritanian strains are old and not as genetically advanced compared with Moroccan strains that have become very lethal to date palms. This poses different hypotheses on the origin of bayoud disease (Sedra 2003a, b, 2007b, 2011c). Some studies analyzing DNA of the *Fusarium* strains have helped to clarify these hypotheses. In fact, the study of genetic diversity of the strains based on molecular random amplified polymorphic DNA (RAPD) and intersimple sequence repeat (ISSR) microsatellites markers revealed a non-negligible polymorphism on the one hand within the (f. sp.) albedinis and on the other hand a genetic proximity between some *F.o.* f. sp. albedinis and *F. oxysporum* strains. This permits formulation of a new hypothesis on the origin of *Fusarium oxysporum* f. sp. *albedinis* (Foa), suggesting that the parasite could derive from saprophyte strains in individual countries without needing to disseminate from an original strain. A dendrogram of 45 *Fusarium* strains from different Arab countries generated by group average clustering analysis using combined RAPD and ISSR microsatellite-based genetic distance regroups all *Fusarium* strains in only one group (group II) except for an original Foa195-MA which may be a particular strain (Sedra and Zhar 2010). The Mauritanian pathogen strain of *Fusarium oxysporum* is clustered in a subgroup containing the pathogen strains from Morocco and Algeria and some strains isolated from soils in Morocco and Egypt.

The Bayoud Threat Publicity about the dangers posed by bayoud and other date palm diseases was well received on a large scale in Mauritania. Indeed, several actions have been taken: publication and distribution of five leaflets and posters to raise awareness, production of ten awareness panels, and improvements in major roads. Indeed, several actions have been taken: (i) publication and distribution of five leaflets and posters to raise awareness, production of ten awareness panels, and improvements in major roads, (ii) an awareness seminar on the danger and spread of bayoud in 2002 at Atar for 100 participants (e.g., farmers, NGOs) and an awareness day in 2006 (AOAD project) for 50 farmers and technicians about bayoud and other sanitary and technical problems, (iii) developing a brochure on the spread of bayoud (AOAD project), seven television and radio announcements on bayoud, disease, and pollination techniques (Sedra 2008a).

The threat of bayoud has prompted policymakers and date palm farmers to seek responsible and appropriate actions to ensure effective protection of date palms in the oases. Protection against bringing in infested plants makes it necessary to set up border posts and interregional phytosanitary controls to protect the crops of the oases. A ministerial order by the Ministry of Rural Development (MDR), No. R0000928/MRDE, dated May 8, 2002, was issued to create phytosanitary control posts; the first station at Nouakchott has been established.

The pressures exerted by and persistence of these disease constraints have contributed to a steady deterioration of oases and loss of some cultivars of interest, as well as especially good spontaneous seedling date palms, contributing to the depletion of the date palm gene pool in different regions. To improve date palm cultivation practices and ensure good fruit production and healthy palms, Sedra (1995, 2003a, b, 2008b) recommended agricultural techniques and cultural care adapted to Mauritanian oasis conditions. An integrated production and protection program was initiated gradually over recent decades through national programs including PDDO, Project of Protection of Date Palm in Adrar State (PPDA) with the support of the Islamic Bank (IB), International Fund of Arab Development (IFAD), and Arab Fund for Economic and Social Development (FADES) and assistance from international agricultural organizations such as FAO, AOAD, and Arab Center for the Studies in Arid Zones and Dry Lands (ACSAD).

Summary Strategy to Control Bayoud Disease In order to limit the spread of the disease in Mauritania, it is necessary to conduct the following operations:

(a) Direct control by eradication of the disease foci and application of quarantine and legislation to compensate farmers for necessary date palm destruction.
(b) Indirect control aiming to fully inform farmers and set forth relevant plant protection interests in agricultural extension. In Mauritania, within the framework of several projects supported by international organizations (e.g., FAO, AOAD, ACSAD), training sessions for technicians and engineers were held and several field days and awareness notices through audiovisual media and distribution of posters and leaflets extension (Sedra 2006a, b, 2007d, 2008a).
(c) Enacting strict laws and creation of a national headquarters for monitoring the movement of plants within the country and in border crossing points, airports, and sea ports. Some headquarters monitoring ports have been created.
(d) Create a modern plant pathology laboratory and experiment station to deal with all aspects of date palm, especially integrated control measures against the main diseases and pests. The laboratory project was achieved in Atar City but needs to become fully operational; creation of a station has begun (Sedra 2006a, b, 2007c, 2008a, b).
(e) Formation of a team of technicians and engineers, with a framework to strengthen their abilities and competencies. Many teams have trained in the field and at the laboratory in Atar City.
(f) The need for coordination among all stakeholders in the date palm oases: institutions, concerned ministry, farmers, producers, private sector, and NGOs.
(g) Creation of a working system and partnership with appropriate software to control bayoud disease and diminish its spread.

These proposals and achievements are detailed in the reports of the technical consultation final reports of national projects (PDP, PPDA, etc.) and regional project for the early detection of bayoud disease and development of technologies, financed by FADES with technical collaboration of FAO and AOAD (Sedra 2003a, 2006a, b, 2008a, b).

Use of Pathogen Toxins for Evaluation of Cultivars Resistance to Bayoud A training course was held on palm protection and particularly, bayoud disease led by Dr. Sedra and funded by AOAD in the Phytopathology Laboratory at Atar in northern Mauritania. Trials were established to assess the susceptibility and resistance of five Mauritanian date palm cvs. (Lamdina, Ahmar, Tijeb, Wandghad, Tiguedert) using toxins extracted in the Phytopathology Laboratory in Morocco from the liquid culture of the fungus causing bayoud disease (*Fusarium oxysporum* f. sp. *albedinis*). Figure 9.8 illustrates the experiment that showed some preliminary results within 6 days, exhibiting differences of cultivar behavior toward the toxins (Sedra 2008a).

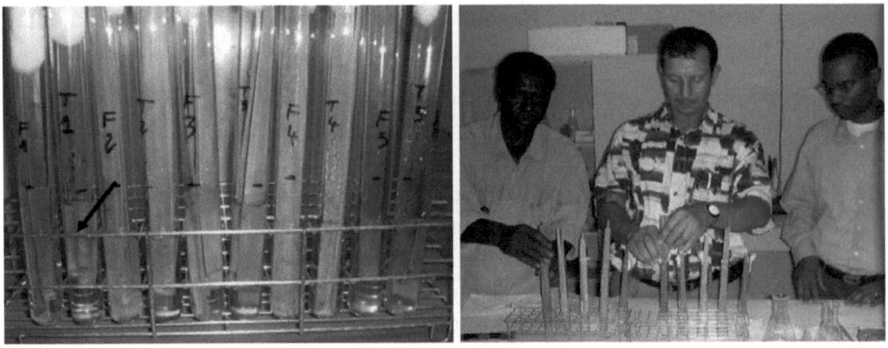

Fig. 9.8 Laboratory experiment to assess the susceptibility and resistance of five Mauritanian date palm cultivars (Lamdina, Ahmar, Tijeb, Wandghad, Tiqedert) using toxins derived from the fungus culture, causing bayoud disease (*Fusarium oxysporum* f. sp. *albedinis*). The *arrow* shows the symptoms caused by the toxin. Expert consultation, AOAD (6/15–23/2007), Dr. Sedra (*center*) and trainees in the Phytopathology Laboratory at Atar City

9.3 Genetic Resources and Conservation

There is no doubt that the Mauritanian date palm cvs. are numerous. Munier (1955) confirmed more than 200 cvs. in studies carried out by the IFAC; however, those studies lacked precision in light of modern science. The traditional approach to classification of date palm cvs. is based on the shape and color of fruits, among date growers in Mauritania, as is the case in many countries producing dates. Local names are the fundamental basis for designating cultivars by farmers. Based on this rule, Sedra (1999a, 2003a) compared some cvs. having the same name in different areas by comparing their morphological characteristics. The study also set out to identify the most important cultivars and phenotypes, widespread and rare, and to ascertain any genetic relationships between them. This study relied on conventional methods developed for Moroccan cvs. (Sedra et al. 1996, 2001a) and did not use modern molecular markers (Sedra et al. 1998; Sedra 2011a, b) due to their high cost. Later, Ould Mohamed Salem et al. (2008) used microsatellite markers to analyze the genetic diversity of several Mauritanian date palm cvs.

9.3.1 Mauritanian Cultivars and Oasis Areas

Sedra (1999a, 2003a) was concerned with cultivar characterization and classification and included three areas representing three states: Adrar (Atar City) in the northeast of the country, Tagant (Tidjikja City) in the middle, and Laassaba (Kifa City) in the south-central. In consultation with local officials, certain oases were selected for inclusion in a study according to their importance within the region, as well as the existence of farmer organizations in the form of communities or associations (Table 9.4).

Table 9.4 Oases and date palm cultivars chosen for phenotyping and clustering studies

Region (state)	Chosen oases	Number of cultivars		Local chosen cultivars
		Known	Chosen	
Adrar	Oued Wadane	81	2+ (1)	Soukani (SKAN), Tinwazid (TNZD), Tiguedert 2 (TGD2)
	Taïerte	17	3	Tijeb (TBJ1), Amcherssi (AMRS), Lamdina (MDNA)
	Oued Seguelil	14	3	Boujira (BGRA), Salmadina (SMDN), Al-Falha (AFHA)
Tagant	Toues Al-Alia	66	3+ (2)	Basbrek (BSBK), Alfat Al-Bahoua (ABAH), Alfat Foum Agadir (AFAG), Tijeb 2 (TBJ2), Mahboula 2 (MHB2)
	Al-Huitate	39	2+ (1)	Tifred (TFRD), Louted (LTED), Tinterguel 2 (TGR2)
Laassaba	Al-Aouja	12	3	Tiguedert (TRG1), Mahboula (MBH1), Ahmar Dli (AHML)
	Al-Andaouda	13	3	Oum Arich (OMAA), Lakhdira (LKD), Tinterguel (TGD1)
	Taqadat Iriji	26	3	Ahmar Danca (AHMJ), Mriziga (MRZG), Al-Hanaouia (HANN)
Total		268	22+ (4)	

9.3.2 Genetic Resources of Mauritanian Date Cultivars

Based on field visits and research with groups of farmers in the areas covered by the study, it appeared that genetic resources are apparently quite diversified in Mauritania. More than 185 cv. names (famous, not famous, or rare) were enumerated only in 8 oases, representing 3 states (Sedra 1999a, 2003a). Some oases include more cvs. than others. The total number could be higher if the surveys were extended to include the remaining oases. There is no doubt that the cultivar numbers continue to increase because of the presence of spontaneous seedling dates, which can become new cvs. Date palm farmers are interested by maintaining, propagating, and naming these cvs according to their origin, owner, tree, or fruit descriptors. It was observed that most of the first cultivar names in Fig. 9.9 of cvs. (about 19 %) include *alfat*, which means seed, for example, Alfat Laararem (seed of the cv. Laararem), Alfat Al-Aqah, and Alfat Al-Chorafae (seed of cv. of an honorable person). Some seedling dates became cultivars after a long period of their propagation by offshoots like Alfat Foum Agadir and Alfat Al-Bahoua (Sedra 1999a, 2003a). When the mother tree is unknown, the seed offspring are generally called *daughters of earth*. Generally in Mauritania, date palms are not intentionally propagated by seed.

Mauritanian date cultivars are often given names that may refer to (a) fruit color, such as cv. Ahmar (red); (b) size and shape of fruit, such as cv. Bazoul al-Baqara (cow's udder); (c) character of spikelets, such as cv. Oum Az-Zaouane, for its large number of long spikelets; (d) the discoverer or owner, such as cv. Alfat Ouled Laghlal; (e) area of origin, such as Touatia from Touata region (Algeria); (f) an

Fig. 9.9 Importance and spread of date palm cultivars according to regions (states) and oases in Mauritania

Table 9.5 Periods of fruit maturity of main and chosen cultivars studied in Mauritania

Cultivar categories	Period of fruit maturity	No. of cultivars (%)	Representative cultivars
Early maturity	First weeks of June	3 (14)	Alfat Al-Bahoua, Lakhdira, Mahboula
Middle maturity	Mid-June to mid-July	7 (32)	Tijeb, Ahmar Danca, Alfat foum Agadir, Al-Hanaouia, Foum Agadir, Basbrik, Tinouazid
Late maturity	Mid-July to beginning of August	12 (54)	Tinterguel, Soukani, Oum Arich, Tiguedert, Louted, Amchekhsi, Mriziga, Tifred, Lamdina, Salmadina, Ahmar Dli, Boujira
Total		22	

animal name, such as Oumat Al-Jamel (mother of camel); and (g) fruit quality, such as cv. Bousakar, which means dates rich in sugar. (h) Some cultivars have diminutive names due to the similarity of their dates with other cultivars but of a smaller size, such as cv. Touba and cv. Twiba; (i) other names, such as cv. Lamdina Al-Ghilalia or Al-Qilalia. Cultivars also vary in the onset of maturity (Table 9.5).

9.3.3 *Spread and Importance of Cultivars*

Mauritanian date cultivars vary in certain characteristics, abundance of production, and fruit quality. There are up to 185 named cultivars in the important oases (Sedra 1999a, 2003a) and their dispersal varies according to regions and oases (Fig. 9.9). Table 9.6 indicates cultivar importance by states and date-growing areas in Mauritania (Sedra 1999a, 2003a). Several cvs. just retain their existence in some oases such as Al-Aassia, Al-Kafha, and Tajnanate in Laassaba State and Admenmej, Al-Khaouara, and Al-Maskha in Adrar State.

According to farmers, the commercial value of date fruit varies by cultivar. In descending order of importance, cvs. Tiguedert, Mahboula, Lamdina, Alfat Al-Khadra, Tinterguel, Oum Arich, Basbrik, Soukani, Tijeb, and Ahmar Dli are the most mass marketed and can be stored in traditional ways by farmers. This group of cvs. is generally the most widespread and best known in the oases studied, with some more prevalent in certain major important oases but less widespread in others The most valuable are cvs. Ahmar Dli, Mahboula, Lamdina, Oum Arich, Tinterguel, and Tiguedert, especially if the fruits are of good quality from palms receiving adequate care (Sedra 1999a, 2003a). Most of the dates produced in Laassaba State (southern Mauritania) are soft or semidry, for example, cvs. Mahboula, Oum Arich, and Lakhdira; this is due to relatively high humidity in this region. The opposite is found in Adrar where many cvs. produce dry or semidry dates (e.g., cvs. Tifred and Amcherssi). Tagant includes several oases producing dates of varying consistency from soft to dry.

Table 9.6 Importance and the best-quality cultivars in main oases and their classification in States of Laassaba, Tagant, and Adrar, cvs. arranged in descending order of importance

State and oasis	Tree distribution	Best date quality cultivars
Laassaba State		
Al-Aaouja	Tinterguel, Tiguedert, Ahmar, Mahboula	Timzad, Ahmar, Mahboula
Andaouda	Tinterguel, Ahmar, Oum Arich, Lakhdira	Tinterguel, Tiguedert, Oum Arich, Ahmar
Taqadate Iriji	Tiguedert, Ahmar	Tinterguel, Ahmar, Oum Arich
Tagant State		
Toues Al-Alia	Tinterguel, Oum Arich, Tiguedert, Ahmar Danca, Al-Alfat Al-Khadra	Tinterguel, Oum Arich, Tiguedert, Al-Alfat Al-Khadra, Al-Alfat foum Agadir, Mahboula, Lamdina
Al-Huitate	Tinterguel, Ahmar Danca, Oum Arich, Mriziga, Louted, Tifred	Ahmar Dli, Tinterguel, Tifred, Tiguedert, Oum Arich, Mahboula, Ahmar Danca, Touatia, Tijeb
Adrar		
Oued Wadane	Soukani, Idmanmej, Tinterguel Al-Kahla, Ahmar, Al-Khouara, Ajib An-Nada, Timzad As-Safra, Tinqdir	Soukani, Tiguedert, Tinwazid, Tinterguel Al-Kahla, Ahmar
Taïerte-Azougui	Ahmar, Tijeb, Lamdina, Amchekhssi, Tiguedert, Adeqate, Mahboula, Tifred, Tamchkarte	Ahmari, Tifred, Tiguedert, Amchekhsi
Oued Seguelil	Ahmar, Tijeb	Ahmar, Tiguedert, Al-Falha, Lamdina

Prior to 1999, except for a cultivar inventory that is not exhaustive and the characterization of some cultivars, no research has been cited to determine the true aspects of genetic improvement and conservation of the date palm resource patrimony at the national level except for a small group of cultivars realized during the IFAC collaboration. The first research work on the Mauritanian date palm patrimony and cultivar characterization was initiated by Sedra (1999a, 2003a). In his reports, Sedra (2002, 2003a) proposed to conduct surveys extended to all oases in order to refresh data of Mauritanian date palm patrimony, accurately establish a cultivar inventory, and assess the levels of genetic diversity, with the objective of establishing a national collection of cultivars at the Atar Experiment Station.

Carrying forward the research by Sedra referenced above, Ould Salem et al. (2008) expanded morphological descriptions of most Mauritanian cvs. and developed 34 quantitative and 27 qualitative characters. These were used to evaluate the correlation between the characters and to select several discriminating criteria, using 22 main cvs. from the important regions. The grouping association of cultivars was identified by cluster analysis. Some cvs. like Ahmar Janca and Tijib constitute small separated group according to the analyzed criteria. In preliminary work, Bodian et al. (2012) characterized 5 early bearing Mauritanian cvs. (Ahmar, Medina, Tijeb, Edaghed, Tiguidert) and males (Emacine) using microsatellite markers. These cvs. were clustered in 92 samples from 5 oases (Aoujeft, Chinguitti, Ksar

Torchane, Terjit, Teyeret) and analyzed with 16 SSR markers. The results reflected a great genetic diversity between cvs., and the dendrogram shows an individualization of cvs. in spite of a certain degree of similarity (Bodian et al. 2012). Genetic variability is very low for cv. Ahmar and moderate for cv. Medina but very high for cv. Tijeb. Three cluster grouping samples of different cultivars were obtained, but no genetic specificity was observed for male cv. Emacine.

Ould Mohamed Salem et al. (2007) also analyzed 10 Mauritanian date palm cultivars and progenies of two controlled crosses according to the identity of mitochondrial plasmid-like DNAs. The PCR technique used on total genomic DNA and appropriate primers was designed to amplify either a 373 bp or 265 bp fragments corresponding to the S- and the R-plasmid, respectively. The results showed that five cultivars out of ten studied have exhibited the R-plasmid suggesting their resistance to bayoud disease. Similar results were found in Morocco (Benslimane et al. 1996) and Tunisia (Trifi 2001) but apparently, the correlation between the date-palm phenotype and the described marker has not been clearly established (Sedra 2011a, b) because both of the tested cultivars revealed resistance by markers that have been shown to be susceptible to bayoud, according to Saaidi (1992) and Sedra (1992). This observation is reinforced by the existence of intra-cultivar variability revealed in the cv. Ahmar mentioned by Ould Mohamed Salem et al. (2007). For progenies of two controlled crosses, the analysis suggested the strict maternal transmission of the date palms' mitochondrial genome.

Sané et al. (2005) studied the growth and development of date palm seedlings under drought and salinity stresses. They founded different tolerance in cvs. Nakhla Al-Hamra and Tijeb to osmotic stress induced by polyethylene glycol (PEG) or NaCl. But the seedlings are hybrids obtained from crosses and do not genetically represent the cultivar parents. The breeding and genetic improvement program of date palm through controlled crosses aims to identify good male genitors (pollinizers) that have a positive metaxenic effect on fruit quality.

9.4 Plant Tissue Culture

Date palm propagation is usually done by offshoots that are often exchanged between farmers, families, and friends, often in a traditional trade. There are no approved nurseries producing date palm offshoots or commercial laboratories producing plantlets in Mauritania. Research on in vitro culture of date palm has been done at the university but is academically oriented and not reaching the level of applied technology development. Under the PPDA project, the laboratory created at Atar was supported by FAO and AOAD to implement several basic in vitro culture facilities to encourage research on tissue culture of date palm in Mauritania (Sedra 2006b, 2008a). It is urgent to make this laboratory operational by providing equipment, qualified technicians, and adequate funding to mass produce plantlets of the best cvs. in Mauritania within the framework of a vision to rehabilitate and restructure the existing date palm groves.

9.5 Cultivars Identification

9.5.1 Cultivars Diversity Based on Morphological Descriptors

Sedra (1999a, 2003a) studied 22 cvs. and 4 duplicate cvs. in different regions and observed that the proportion of the total of 75 distinctive characteristics varied from 10.7 % for cv. Tijeb, 34.7 % for cv. Mahboula, and 36.0 % for cv. Tinterguel to 44.0 % for cv. Tiguedert. This difference can be explained by the fact that the cvs. observed in different oasis have different origin, but they have the same name due to their similarity, especially cv. Tijeb, or there is a family relationship between them. It could also be that the majority of different characteristics have been affected by different environmental conditions. Symbols of quantitative and qualitative characteristics (Sedra 1999a, 2003a) are explained in Table 9.7.

Statistical analysis, considering 22 and 4 duplicate cvs., generally showed only 34 quantitative characteristics that have importance in making distinctions in cultivar classification. Table 9.8 lists the most important distinctive characteristics intra-cultivar in two different oases. The number of quantitative and qualitative characteristics is, respectively, 22 and 5 (Tinterguel), 8 and 0 (Tijeb), 22 and 4 (Mahboula), and 26 and 7 (Tiguedert) (Sedra 1999a, 2003a). In these studies, 47 quantitative characteristics were ranked according to their importance for Mauritanian cvs. Through the results, the percentage of the pinnae area (ppe) is the best distinctive quantitative characteristic. Table 9.9 presents top ten of the important characteristics.

On the other hand, data analysis showed the positive relations between characteristics (by pair) and their relative importance. Over this relationship between two traits reach up to about 100 % plus any of these traits may represent and compensate other traits. Perhaps the relationship is intuitive, such as the number of spines on the right (ned) and the total length of the spines in the right side (sed). Characteristics that the proportion of their relationship (by pair) varies from 70 to 99.4 %: 88.3 % between the number of pinnae (ntp) and length of its area on leaf rachis (lpf), 98.4 % between the total length of spines in the right side (sed) and left side (seg), and 86.9 % between the length of top spine (leh) and length of central spine (lem).

9.5.2 Cultivar Classification

9.5.2.1 Analysis of Quantitative Characteristics

Statistical analyses of the differentiating characteristics demonstrated a very high relationship between them where the relationship r factor rises to 0.99. This relationship may be negative, such as the percentage of pulp fruit with the ratio of weight seed in fruit. Analysis using the principal components analysis (PCA) method on quantitative traits permitted the focusing and shorthand which can be interpreted by three axes responsible for explaining about 50 % of the total variance (Sedra 1999a, 2003a).

Table 9.7 Symbols of quantitative and qualitative characteristics and traits

Symbols of quantitative characteristics and traits
Main phenological quantitative characteristics of vegetative organ
Vigor of palm trunk (cm) *val*, production of offshoots at trunk base (average number) *prp*, production of offshoots above base of trunk (average number) *pra*, number of leaves *npn*, length of leaf (center of crown) (cm) *ltp*, length leaf rachis *lrp*, length of pinnae area (cm) *lpf*, percentage of pinnae part (%) *ppf*, density of pinnae area on rachis (%) *dip*, length of spiny part (cm) *lpe*, percentage of spiny part (%) *ppe*, length of the base leaf part (cm) *lpb1*, percentage of the leaf base part (%) *ppb*, total number of pinnae *ntp*, length of terminal pinnae (cm) *lph*, thickness of terminal pinnae (cm) *eph*, length of central pinnae (cm) *lpm*, thickness of central pinne (cm) *epm*, length of lower pinnae (cm) *lpb*, thickness lower pinnae (cm) *epb*, angle formed by terminal pinnae (°) *apt*, angle formed by anterior pinnae (°) *apv*, angle formed by dorsal pinnae (°) *apd*, angle formed by dorsal pinnae with rachis (°) *apr*, total number of spines *nte*, density of spines on leaf rachis (%) *die*, number of spines on the right side *ned*, number of spines left side *neg*, length of top spine (cm) *leh*, thickness of top spine (cm) *eeh*, length of central spine (cm) *lem*, thickness of central spine (cm) *eem*, length of lower spine (cm) *leb*, thickness of lower spine (cm) *eeb*, total length of spines on right side (cm) *sed*, total length of spines on left side (cm) *seg*, angle formed by center spine with leaf rachis (°) *aer*
Main phenological quantitative characteristics of reproductive organ
Number of fruit bunches *nsp*, length of fruiting bunch (central free zone) (cm) *lsp*, width of fruiting bunch (central free zone) (cm) *lms*, weight of 100 fruit weight (g) *pcd*, proportion of fruit pulp (%) *ppd*, fruit length (cm) *lf1*, width or diameter fruit (cm) *ef1*, weight of 100 seeds (g) *pcg*, proportion of seed in fruit (%) *pgd*, length of seed (cm) *lg1*, width or diameter seed (cm) *eg1*
Symbols of qualitative characteristics and traits
Main phenological qualitative characteristics of vegetative organ
Inclination degree of palm leaf *cpa*; inclination degree of palm leaf summit *cep*; color of the base of leaf *cbp*; softness of spines *re1*; color of spines *ce1*; color of top spine *cee*; presence of dropper on spine *gp1*; existence of individual spines in right side *pesd*; presence of individual spines in left side *pesg*; presence of individual spines in right and left sides *pes*; presence of spine grouping *par*; groups of spines and frequency *rge*; place, grouping, and order of spines in right side *poed*; place, grouping, and order of spines in left side *poeg*
Main phenological qualitative characteristics of reproductive organ
Color of fruit bunch *csp*, shape of fruit *ff1*, color of the fruit in stage *balah csb*, color of date (final phase) *csm*, shape of fruit rind (external wall) *aep*, thickness of fruit rind *eep*, shape and place of fruit crown *fca*, pulp fruit *cf1*, fruit structure *tf1*, shape of seed *fg1*, color of seed *cg1*, surface of seed *sg1*, shape of seed valley or slot *fsg*, place of germination point of seed *sfg*, shape of germination point of seed *apg*, length of thread link between seed nucleus and fruit crown *lpg*, adhesion of the inner wall of date *atg*, bosses on seed *prg*

Table 9.8 Most important distinctive characteristics intra-cultivar in two different oases

Cultivar	Quantitative characteristics	Qualitative characteristics
Tinterguel	pr1, lpb, ppb, ntp, eph, nte, ned, neg, leh, eeh, eem, sed, seg, aer, apr, nsp, lms, pcd, ppd, pcg, lg1, pgd	dcf, cbp, cp1, re1, eep
Tijeb	lpb, ppb, dip, apr, pcd, pcg, pgd	–
Mahboula	val, lpf, ppf, lpe, ppe, lpb, ppb, lph, eph, lpm, epb, nte, ned, neg, leh, lem, eem, pcd, ppd, pcg, lg1, pgd	ff1, eep, fg1, sfg
Tiguedert	npn, lpf, ppf, ntp, eph, nte, lpm, epb, dip, nte, die, ned, neg, eeh, lem, eem, leb, eeb, sed, seg, aer, lms, pcd, ppd, ef1, pcg, pgd	re1, ff1, fca, cg1, sfg, lpg

Table 9.9 Evaluation of variation range of quantitative characters and their importance in Mauritanian date palm phenotyping based on top 10 among 47 characters

Rank	Variation factor	Symbol	Quantitative character
1	124,0	ppf	Percentage of the pinnae area
2	58,0	lpb1	Length of leaf base area
3	56,2	seg	Total length of spines in left side
4	54,3	die	Density of spines on leaf rachis
5	53,9	ppb	Percentage of leaf base area
6	53,6	sed	Total length of spines in right side
7	45,5	pcg	Weigh of 100 seeds
8	42,7	eeb	Thickness of down spine
9	41,5	leb	Length of down spine
10	40,9	eem	Thickness of central spine

9.5.2.2 Analysis of Qualitative Characteristics

The analysis of 28 qualitative characteristics (symbols are defined in Table 9.7) showed differences in their contributions to characterize cvs. (Table 9.10). Some permit characterization of numerous cvs., others only a half or few of cvs. studied, and other ones did not characterize any cultivar.

9.5.2.3 Classification of Cultivars according to Qualitative Characteristics

A total of 13 qualitative characters only were used for the cvs. ordering process with reference to its clustering and the impact of some characters and their contribution to constitution of the tree nodes and cultivar clustering. It has shown that some qualitative characteristics contribute more than others in the formation of a large number of tree nodes to classify cvs. such as the fruit color in stage *balah* (csb: 36 %), thickness of fruit rind (eep: 36 %), spine color (cee and ce1: 32 %) and presence of individual spines in the spines area (pes: 32 %), color of palm leaf base (cbp: 28 %), and other characters that do not contribute at all, such as the length of accumulated spines (leg) that are almost 100 % equal in the case of all cvs. (Sedra 1999a, 2003a). Based on these attributes, the cv. classification permits a gathering composed of five groups (Table 9.11).

9.5.2.4 Clustering Based on Quantitative Characteristics

Data analysis using Euclidean distance and 47 quantitative traits permitted constitution of four groups of cvs. and other cvs. separately in three individual group sets each of which is formed by a cv. (Sedra 1999a, 2003a). Cv. Ahmar Danca (AHMJ)

Table 9.10 Contribution of qualitative characteristics to characterize date palm cultivars

Distinguished cultivars	Qualitative characteristics
Large number of cultivars	dcf, cee , ce1, aer, par, csb, aep, sfg, fsg, fg1, tf1, ff1, csp
Half of studied cultivars	cp1, re1, csm, eep, cf1, prg, cg1
Few of studied cultivars	Pes, fra, lpg, atg,
Any cultivar	sg1, fca

Table 9.11 Division of date cultivars according to their classification based on quantitative traits

Group	No. of cultivars	Cultivars
1	1	Ahmar Dli (AHML)
2	5	Alfat Foum Agadir (AFAG), Louted (LTED), Oum Arich (OMAA), Tijeb 1 (TBJ1), Tijeb 2 (TBJ2)
3	16	Alfat Al-Bahoua (ABAH), Al-Falha (AFHA), Amcherssi (AMRS), Boujira (BGRA), Al-Hanaouia (HANN), Mahboula 1 (MHB1), Mahboula 2 (MHB2), Mriziqa (MRZG), Soukani (SKAN), Salmadina (SMDN), Tifred (TFRD), Tiguedert 1 (TGD1), Tinterguel 2 (TGR2), Tiguedert 2 (TGD2), Tinwazid (TNZD), Tinterguel (TGR1)
4	2	Ahmar Danca (AHMJ), Basbrik (BSBK)
5	2	Lakhdira (LKD), Lamdina (MDNA)

is unique in group 1 due to longer distance with other cvs. and to several quantitative characteristics: longer leaf palm (ltp), longer leaf rachis (lrp), the widest angle formed by central spine with leaf rachis (aer), the longest (lf1) and widest (ef1) fruit, smaller proportion of seed in fruit (pgd), as well as several characters with either greater or lesser value.

9.5.2.5 Clustering Based on Qualitative Characteristics

Statistical analysis using 28 qualitative characteristics permitted clustering of the cvs. into 11 groups (Sedra 1999a, 2003a). Four groups are composed of one cv. each: Ahmar Dli, Tiguedert 1, Boujira, and Alfat Foum Agadir. The grouping of spines did not permit arranging and classifying Mauritanian cvs., but for Moroccan cvs., it was found that combination of spine shape and their grouping and arrangement are among the elements and characteristics that distinguish cvs. (Sedra 2001a). This spine arrangement suggests an arrangement of a series of spines starting from the base of the leaf palm on the left (G) or on the right (D) (Sedra 2001a). As mentioned above, in view of the difficulty of traditional methods and the effort it takes, the traditional definition of farmers' cultivars depends mainly on fruit description and continuous preference over the years for local cvs. This inherited experience in the oases may fail in the identification of some palm individuals derived by seed from these similar cvs. in their fruit. The study using 47 quantitative and 28 qualitative characteristics has permitted identification of several characteristics that can distinguish between the palms of Mauritanian date palm cvs. or group of cvs.

In general, cultivar classification and arrangement appear gradual and relative and their clustering is not in very strong groups and of sufficient Euclidean distances except that some cvs. stand out by quantitative and qualitative traits such as Ahmar Danca and Lamdina based on quantitative traits and cv. Tijeb based on qualitative traits (Sedra 2003a). This arrangement and classification suggests that Mauritanian cvs. converge generally with each other and have a uniform origin or some close and inherited assets. Likely, the two cvs. Ahmar Dli and Lamdina have different origins to each of the other cvs. outside groups. Therefore, the rumor among farmers reports that cv. Lamdina was originated from seed collected in palm farms of Medina in Saudi Arabia. This study (Sedra 1999a, 2003a) only a launching of rapprochement analysis to Mauritanian cvs, and it is necessary to follow up this study to include other Mauritanian and non-Mauritanian cvs. in order to make comparisons. It is necessary to use modern methods based on enzymatic fingerprints and DNA analysis and also to search for genetic fingerprinting traits associated with quantitative and qualitative characters, and all this in order to assess the genetic diversity and genetic relationship between cvs. integrated with traditional methods.

Ould Mohamed Salem et al. (2008) carried forward vegetative description of most Mauritanian cvs. described by Sedra (1999a, 2003a). In fact, other related morphological variability of Mauritanian date palm cultivar work as revealed by vegetative traits was subsequently found by Ould Salem et al. (2008). These authors assume that the studied cultivars are characterized by a high level of genetic diversity while Sedra (2003a) using 22 cvs. and 47 quantitative and 28 qualitative standards descriptors, found that the level of this diversity was not very high, although Ould Mohamed Salem et al. (2008) used 12 date palm ecotypes and only 18 of the same descriptors. Bodian et al. (2012) revealed, using molecular markers, the existence of genetic variability between individuals of the same cv. according to oasis sources. This genetic differentiation remains low in Ahmar cv. contrary to Tijib cv. which showed a high genetic differentiation. In comparison of 4 cvs. and their duplicated samples, Sedra (2003a) has generally found no fundamental differences between the trees, but these differ by several less-important characters. Sedra (2007d) selected numerous morphological characteristics and molecular markers that can be used for identification and distinguishing between date palm cultivars and the plants issued from tissue culture.

The morphological characteristics of the parts of the date palm subject to controls and mathematical principles governing their quantity, quality, and gathering and its engineered may be useful to the researcher in order to study response to environmental factors such as light reflection and temperature change inside the oasis (Sedra 2003a). This response may suggest practical recommendations for each cv., for example, on the determination of adequate distances between palm trees when creating a field or a new oasis.

9.6 Cultivar Description

The cultivars studied, listed below, are of the best quality and most popular in Mauritania. They are from palms which have received at least minimal care: Tijeb, Lamdina, Al-Falha, Mahboula 2, Tinterguel 2, Tiguedert, Mriziga, Ahmar Danca, Oum Arich, Al-Hanaouia, and Ahmar Dli. Sedra (2003a) created identification cards for 22 cultivars common to the date-growing areas of Mauritania. These cards contain details of the dimensions and characteristics for each cv. (75 quantitative and qualitative characters), accompanied by photographs (tree, leaf and spines, fruit bunch, fruit). These attributes for each cultivar are shown in Fig. 9.10 and summarized as follows:

Ahmar Danca This cv. is present in Adrar and Laassaba states and is very widespread in Tagant. Main characters are: mid-season bearing, good fruit, and moderate to good storage quality. Ahmar Danca resembles cv. Ahmar Dli in several ways, with larger characters than those found in most other cvs. It has the longest leaf and leaf rachis, widest angle formed by the central spine with leaf rachis, and largest fruit size (comparable to fruits of cvs. Tinwazid, Lamdina, and Al-Falha) and a high percentage of fruit pulp (83.8 %). Ahmar Danca differs, compared to cv. Ahmar Dli, by a shortness in the total length of spines on the left and right sides of the leaf and a low pinnae density on the rachis. Weight of the fruit is 10.22 g.

Ahmar Dli This cv. is present in Adrar and Laassaba states and is widespread in Tagant. Main characters are: late season bearing, good fruit, and moderate to good storage quality. It is characterized by high offshoot production; relatively large fruit size, especially by the highest fruit pulp content (86.05 %) compared to other cvs. studied; and the highest values for the following property: angle formed by dorsal pinnae with leaf rachis. However, the trunk is thin and the proportion of spine area on the leaf rachis is very moderate. Weight of the fruit is 7.8 g.

Al-Falha The cv. is present in Adrar State (Oued Seguelil), widespread there, mid-season bearing, good fruit, and moderate to good storage quality. This cv. is characterized by the longest leaves compared to other cvs., except for some, such as cvs. Ahmar Danca, Amcherssi, and Boujira whose leaf length is similar to Al-Falha's, but differs by the longest area of spines on the leaf rachis and the longest fruiting bunch. The cv. features the lowest angle formed by dorsal pinnae, fruit and seed length and fruit weight (12.8 g), relatively high percentage of seed of the fruit (43 %) and small bosses (protuberances) rarely appear on the seed surface; percentage of fruit pulp (57 %).

Alfat Al-Bahoua This cv. is present in Adrar and Laassaba states and is widespread in Tagant; it is very precocious in fruit production and has average quality of fruit and moderate storage quality. It is characterized generally by moderate features, except for the following distinguishing characteristics: low number of pinnae, short central pinnae, and relatively long lower spines. Further notes are that the palms of

Fig. 9.10 The attributes of date palm cultivars grown in Mauritania

Fig. 9.10 (continued)

Fig. 9.10 (continued)

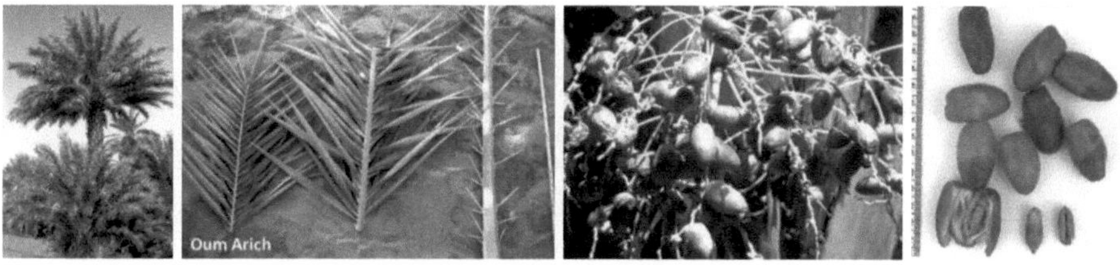

Oum Arich

Salmadina

Soukani

Tifred

Tiguedert 1

Fig. 9.10 (continued)

Tiguedert 2

Tijeb 1

Tijeb 2

Tinterguel 1

Fig. 9.10 (continued)

Tinterguel 2

Tinwazid

Fig. 9.10 (continued)

this cv. bears small numbers of offshoots and fruit bunches compared to other cvs. As with cv. Lakhdira, cv. Alfat Al-Bahoua is early season bearing, weight of fruit is 8.47 g, and percentage of fruit pulp is 67.7 %.

Alfat Foum Agadir The cv. is present in Laassaba and Tagant states, moderately common, mid-season bearing, has good fruit but the seed is large, and has moderate storage quality. This cv. has features like some other cvs. such as Lamdina and Tijib, by several short or small dimensions in comparison with all cvs. For cv. Alfat Foum Agadir, these comprise trunk circumference, leaf number, length and amount of pinnae area, pinnae density on leaf rachis, length and thickness of central spine, total length of spines on the left and right, and finally the proportion of fruit pulp. This cv. is also characterized by fruit shape which is often pear shaped, weight of fruit is 5.95 g, and percentage of fruit pulp is 42.4 %.

Al-Hanaouia This cv. is present in Laassaba State, widespread in oasis of Taqadat Iriji, mid-season fruiting and has good fruit and storage quality. The cv. is distinguished by a dense crown of leaves, despite a moderate number of them, and thinness of pinnae in the central region and short and thin fruit bunches and areas of short spines. Dimensions of other properties are generally moderate. The cv. produces a large number of offshoots, weight of fruit is 8.22 g, and percentage of fruit pulp is 79.3 %.

Amchekhssi This cv. is present in Adrar State, very widespread in Taïert and Oued Seguelil and other oases, late season bearing and has average fruit quality and moderate to good storage quality. The main characters are lengthy leaf and top pinnae, but comparatively small features with respect to pinnae density on leaf rachis, angles formed by dorsal pinnae between them and apical, shortness of spines lower on rachis, shortness of total length of spines on left and right sides and low angle formed by the middle spine with leaf rachis; the weight of the fruit is 7.64 g and percentage of fruit pulp is 60.6 %.

Basbrik This cv. is present in Tagant and Laassaba states and is widespread and mid-season bearing and has relatively good fruit and moderate storage quality. The cv. is characterized by short leaves and pinnae, smallest area of leaf base, higher leaf density on rachis compared to other cvs., in addition to shortness and small total length of spines on left and right sides and relatively shorter fruiting bunch, slim trunk and relatively high offshoot production in addition to oval shape of the fruit at the base; the weight of the fruit is 8.7 g and the percentage of fruit pulp is 74.8 %.

Boujira This cv. is present in Adrar State and widespread in Oued Wadane; it originated from Chenguiti region in Adrar, and it is late season bearing and has relatively good fruit and moderate storage qualities. The cv. is characterized by a huge trunk and long leaves and pinnae and large number of pinnae, presence of many brown-black spots on leaf base, the lowest angle formed by dorsal pinnae and leaf rachis, long fruiting bunch (part free of fruits), relatively high weight of seed relatively (4.08 g), weight of fruit (11.17 g), and percentage of fruit pulp (63.65 %).

Lakhdira This cv. is present in the states of Adrar, Laassaba, and Tagant, very widespread, and very precocious bearing and has good fruit and good storage quality. It is characterized by some relatively large dimensions like the percentage of spiny area, length of lower pinnae, angle formed by terminal pinnae, and length of upper and lower pinnae. The cv. is distinguished by the shortest fruiting bunch, color of fruit brown-greenish that gives its name, and great length of fruit and seed; weight of fruit is 7.34 g and percentage of fruit pulp is 73.8 %.

Lamdina This cv. is present in Adrar, Laassaba, and Tagant states; it is very widespread in Adrar and is late season bearing and has good fruit but low to moderate storage quality. It is characterized by relatively thick trunk and smaller measurements for many characteristics such as leaf and pinnae length, angle formed by pinnae, especially the apical ones and the number and length of spines. The cv. features include relatively unusual leaves and long fruit bunch and heavy fruit weight (14.47 g) and very long thin seed, in addition to the *wade* form of the seed slot that is relatively wide, percentage of pulp fruit (65.93 %).

Louted This cv. is present in Tagant State, widely grown, late season in fruit, relatively good fruit with moderate storage quality. It is characterized by the thinnest tree trunk compared with cvs. studied and less base area of palm leaf without spines, comparable to cvs. Basbrik and Tijeb. Some characters are high, such as: percentage of the pinnae area, length of top and middle pinnae, fruit bunch numbers and length. Immature fruit is yellow violet (*balh*) in color and often horned, fruit weight is 2.64 g, and percentage of pulp fruit is 64.7 %.

Mahboula This cv. is present in Adrar, Laassaba, and Tagant states, very widespread in the latter and very precocious bearing and has good fruit but poor to moderate storage quality. In Aouja Oasis (Laassaba), cv. Mahboula 1 is characterized by numerous criteria that have large dimensions as compared to other cvs.: total length of leaves, percentage of spiny area, especially the length of leaf base area (36.7 %) and its proportion to the leaf (9.73 %), total number of leaves (194.7), angle formed by apical pinnae (104.4°), angle formed by dorsal pinnae with leaf rachis (53.3°), high percentage of fruit pulp (81.5 %), low weight of seed (9.75 g), and its proportion in fruit (14.9 %). However, the same tree, named Mahboula 2 by farmers in the oasis Toues Al-Alia (Tagant) did not possess all properties and characteristics mentioned. Mahboula 2 is characterized by the shortest spine area (44.5 cm) and the low proportion on leaf of this spine area (330.0 cm) and its percentage (88.6 %), longer top and central leaves, lowest number of spines (18.7), and this number is quite different to the spine number of Mahboula 1, which is 29.2. The tree (Mahboula 2) also differs by long spines and fruit weight (11.08 g). Statistical analyses show differences between Mahboula 1 and 2. Explanation is that one cv. is original and the second is from the first through the seed that led to denominate it like Alfat Mahboula as the farmers did and because of its resemblance with known cv. Cv. Mahboula remains one of the best Mauritanian date quality cvs., and as to the beauty of its trees.

Mriziga This cv. is present in Tagant and Laassaba states, very widespread, and late season bearing and good fruit and moderate to good storage quality. The cv. is characterized a very large number of leaves and this makes the crown dense, short and thin terminal pinnae, low angle formed by terminal pinnae, few spines, and relatively short, in addition to the low angle formed by the spine center with leaf rachis. Despite moderate fruit weight of date (7.92 g), the proportion of fruit pulp is very high (80.5 %).

Oum Arich This cv. is present in Tagant and Laassaba states, very widespread in the latter, and late season bearing and has good fruit and poor to good storage quality. The cv. is characterized by large dimensions of certain features: the largest number of pinnae, the highest density of pinnae on rachis due to the small percentage of wide pinnae, especially in the middle and lower leaf, high angles between the front or dorsal or upper pinnae, large number of spines, longer spines at leaf top and the greatest total length of spines on the left and right sides of leaf, very large number of fruit bunches, great density of fruiting bunch in the middle of the zone of fruits, fruits crowded and low in weight (5.78 g) and shortest in length, and proportion of fruit pulp (62. %). Occasionally, oval bumps appear on seeds.

Salmadina This cv. is present in Adrar State and widespread in Oued Seguelil and other oases, originates from Chenguiti region (Adrar), is late season fruiting, and has relatively good fruit and low to moderate storage quality. The cv. has moderate dimensions of morphological characteristics of the leaves and fruits.The cv. shows the smallest measurement of length and thickness of pinnae and especially the upper leaves, fruiting bunch relatively long and thin in the middle; generally leaf crown is dense and not airy. The weight of fruit is 12.22 g and the percentage of pulp fruit is 59.2 %.

Soukani This cv. is present in Adrar, Laassaba, and Tagant states and very widespread in Adrar, originates from Oued Wadane, is late season fruit bearing, and has relatively good fruit and moderate storage quality. Generally the cv. is characterized by the largest dimensions, especially with regard to the characteristics of the pinnate leaves and spines. The cv. presents high percentage of leaves that are wide, especially in the middle and lower portion, giving a highly dense crown, high angle of the upper leaves, large number of spines, longer spine at leaf apex and the largest total length of the spines on both the left and right sides of leaf. Seed represents moderate percentage (40.97 %) of the fruit size, so about 59.03 % of pulp fruit that is the moderate proportion in comparison with other cvs., fruit has cylindrical shape sometimes with a kind of ring in the fruit center, fruit weight is 8 g.

Tifred This cv. is present in Adrar and Tagant states, widespread in Oued Wadane (Adrar), and late season bearing and has relatively good fruit and good storage quality. The cv. has moderate features, except for the large number of pinnae, relatively high angle formed by the center spine with leaf rachis, smaller density of spine groups on rachis, smaller number spines, low fruit weight (5.83 g), and shorter fruit compared with other cvs. The cv. is characterized by oval-opposite rectangular fruit shape. Percentage of fruit pulp is 57.6 %.

Tiguedert This cv. is present in Adrar State, widespread in Oued Wadane, and late bearing and has good fruit and good storage quality. Like cv. Mahboula, clear differences were observed between the trees. Tiguedert 1 in Aouja Oasis (Laassaba) and trees of Tiguedert 2 in Oued Wadane Oasis (Adrar) a distant oasis with different relative humidity conditions. According to their identification cards, numerous characteristics distinguish Tiguedert 1 and Tiguedert 2. The most important ones are number, proportion, and density of pinnae, angle formed by dorsal pinnae with the rachis, total number and density of spines, total length of spines on both the right and left sides, proportion of fruit pulp, seed weight and its proportion in the fruit, as well as some of the characteristics like fruit shape, fruit bunch color, and spine softness. Based on only the fruit shape, for cv. Tiguedert 1, the fruit is smaller at the top like skinny pear. The difference between Tiguedert 1 and Tiguedert 2 cannot be due to different oases because the properties have not been affected by environmental factors; for example, the spine numbers which vary from 40 (Tiguedert 1) to 42.7 (Tiguedert 2) and angle formed by the central spine with leaf rachis with which varies from 24.9° (Tiguedert 1) to 49.3° (Tiguedert 2) and proportions of pulp fruit and seed which differ between Tiguedert 1 and Tiguedert 2 vary, respectively, from 72.3 to 48.7 % and from 27.7 to 51.3 %. It appears that one cv. of these two cvs. is original and the other is from seed of original, fruit weight of both (7.67 g). In comparison with other cvs., Tiguedert is characterized by several small dimensions, for example, thinness of lower pinnae (0.43 cm) and shortness of lower spine (2.6 cm), low angle formed by center spine with leaf rachis, and low fruit weight (4.14 g) and seed weight (1.02 g), in addition to the fruit shape mentioned above. Cv. Tiguedert is well-known commercial date producer and its fruit is consumed by farmers and urban populations. It is known for long storage quality among growers and traders because fruit dries and its pulp hardens and can be milled. These dates are consumed by travelers during their travels in the desert. This property is well-known for the cv. Bouskri in Morocco.

Tijeb This cv. is present in Adrar, Laassaba, and Tagant states, very widespread in Adrar and Tagant, and mid-season fruit bearing and has relatively good fruit and moderate storage quality. Cv. is characterized by fruit color and shape, short leaf and central and low pinnae, low angle formed by terminal pinnae and by dorsal pinnae with rachis, thinnest fruiting bunch in the free-fruit zone. The comparison of trees of cv. Tijeb 1 (Taiert Oasis, Adrar) and trees of cv. Tijeb 2 (Toues Al-Alia Oasis, Tijigja Valley, Tagant), did not reveal any fundamental difference between the two in important characteristics not affected by environmental conditions. This is confirmed by the dendrogram showing that trees of Tijeb 1 and 2 are similar and clustered in one group; the fruit weight varies is 6.67 g (Tijeb 1) and 10.58 g (Tijeb 2), and percentage of fruit pulp are respectively (46.13 % and 75.5 %).

Tinterguel Cv. Tinterguel 1 (Andaouda Oasis, Laassaba) is notable among the rest of the cvs. by high density of pinnae with leaf rachis, thickness of central pinnae, and wide angles: the first angle formed by top pinnae and the second angle by central spine with leaf rachis (61.3°), its characteristics are: short fruiting bunch (56.7 cm), lightest fruit (5.67 g), and long seed (5.44 cm). Compared to the cv. Tinterguel 1, the trees of cv. Tinterguel 2 (Lahouitate Oasis, Tagant) differ only by some characteristics such as length of top spine (19.7 cm) for Tinterguel 1 and (9.1 cm) for Tinterguel 2, but there is no substantial differences between the followed characteristics: total length of the spines on the right and left sides, length and width of fruit bunch and angle formed by central spine with leaf rachis, but there is no substantial differences between these last characteristics. The late maturing cv. is one of the most popular in Mauritania due to its adaptation in most oases and good fruit quality and storability. Weight of fruit is 5.39 g (Tinterguel 1) vs. 6.17 g (Tinterguel 2); percentage of fruit pulp is both 48.7.

Tinwazid This cv. is present in Adrar, Laassaba, and Tagant states, very widespread in Adrar, and mid-season bearing and has good fruit, and moderate to good storage quality. The cv. is characterized by thick trunk, large number of leaves and relatively few offshoots, and greater density of pinnae on rachis despite fewer numbers, but this is due to its width and thickness record. The fruit is thicker compared with other cvs., in addition to its great length and weight of fruit (12.4 g). The seed weighs almost more than a third of the entire weight of the fruit, in addition to great length and thickness of the seed. Percentage of fruit pulp is 60.8 %.

9.7 Dates Production and Marketing

In Mauritania, the date palm sector is still traditional and will remain so until large plantation are established leading to date marketing and commercialization. Annual date production is estimated at 20,000 mt, to which is added a small quantity of imports (1,000 mt) from Algeria and 500 mt from Tunisia. Around 60 % of dates are consummated between June and August, during the *guetna* (Mauritanian Arabic name for the date harvest season) that coincides with the season of fruit ripening. This is well known for some of cvs. consumed at the immature stage of *balah*, for example, cvs. Alfat Al-Bahoua, Timzad, and Bazoul Chlouha in Tagant State. Taste is one of the essential elements of date quality. All date fruits, at whatever stage of maturity, that are not suitable for human consumption are used by farmers as feed to livestock. Uneaten good dates are dried for consumption throughout the year. According to Mauritanian tradition, dates are consumed within families before meals, especially in rural areas. The consumption is also important during the period of Ramadan.

9.8 Processing and Novel Products

Modern post-harvest operations of dates are almost absent. Dates are usually processed the traditional way. These treatments were limited in sorting dates and setting traditional packaging in cartons or bags or sold in bulk. Soft dates are sometimes pressed and held for auto consumption or intended for marketing. Some dried dates are traditionally crushed and mixed with other human foods. Dates of poor quality are intended for animal feed.

9.9 Conclusions and Recommendations

In Mauritania, the rural sector remains the most important sector of the national economy. This country has been strongly affected by the effects of drought for several decades. Because of drought, the Mauritanian oases suffer from silt, water scarcity, and a lack of soil fertility, following the absence of alluvial flows. Promoting irrigation with water control has reduced the sensitivity of agricultural production to drought. In order to develop the date palm sector, the Mauritanian government, with international cooperation, established several programs based on social organizations, equipment development in oases, building of capacity of human resources and strengthening of agricultural research and extension services. It established PDDO in 2002 to preserve the fragile but valuable oasis ecosystems and stem a rural exodus. Several programs are underway to upgrade the date palm sector through investment in the oasis, strengthening and developing research actions, supervision of farmers, and support for social organizations. To ensure sustainable development of oasis, it is recommended to:

(a) Restructure palm groves adjusting densities and introducing varieties with higher commercial value
(b) Develop the micropropagation of good quality cvs. by reinforcing the laboratory at Atar City and encourage creation of private laboratories
(c) Install collective wells and basins for supplying irrigation water
(d) Generalize gradually drip irrigation to save water and ensure adequate fertilization of trees
(e) Develop integrated production and protection techniques of date palm
(f) Install conservation industrial or semi-industrial units, processing and packaging of dates promoting the private sector and social organizations
(g) Provide ongoing staff training and mentoring up of palm farmers
(h) Establish a national research program and research and development in the medium and long term
(i) Benefit from the experiences of the leading countries in topics of research and development and of international cooperation

Acknowledgement I gratefully acknowledge all persons for their help and providing me with informations.

References

Benslimane AA, Hartmann C, Ouenzar B, Rode A (1996) Intramolecular recombination of a mitochondrial minicircular plasmide-like DNA of a date-palm mediated by a set of short direct repeat sequences. Curr Genet 29:591–593

Bodian A, Ndoye Ndir K, Diedhiou PM et al (2012) Analysis of genetic diversity of date palm (*Phoenix dactylifera* L.) cultivars from Mauritania using microsatellite markers. Int J Sci Adv Technol 2:36–43

FAOSTAT (2011) http://faostat.fao.org

MDRE/NSO (2003) Agrégats de la Comptabilité nationale et indicateurs socio-économiques 2002

Munier P (1955) Le palmier dattier en Mauritanie. Annales Inst. Fruits et Agrumes Colon, No. 12

Ould Mohamed Salem AS, Rhouma S, Zehdi M et al (2007) Molecular characterization of Mauritanian date-palm cultivars using plasmid-like DNAs markers. Biol Plant 51:169–172

Ould Mohamed Salem AS, Rhouma S, Zehdi S et al (2008) Morphological variability of Mauritanian date-palm (*Phoenix dactylifera* L.) cultivars as revealed by vegetative traits. Acta Bot Croat 67(1):81–90

Saaidi M (1992) Comportement au champ de 32 cultivars de palmier dattier vis à vis du Bayoud: 25 années d'observations. Agronomy 12:359–370

Sané D, Ould Kneyta M, Diouf D et al (2005) Growth and development of date palm (*Phoenix dactylifera* L.) seedlings under drought and salinity stresses. J Biotechnol 4(9):968–972

Sedra MyH (1992) Evaluation and selection of the resistant good cultivars and clones of date palm to the Bayoud disease. Arab Soc Plant Prot 10(2):155–160

Sedra MyH (1995) Report of international expert consultation in Mauritania. Phytosanitary problems of date palm in Mauritania and proposals of control methods, 8–16 June 1995. Research network and the date palm development. ACSAD, Damascus

Sedra MyH (1999a) Report of international expert consultation in Mauritania. Identification, characterization and classification of date palm cultivars in Mauritania, 7–23 July 1999. AOAD, Al-Khartoum

Sedra MyH (1999b) Report of international expert consultation in Mauritania. Bayoud disease status, cultivation, development and project proposal for the protection and development of research in Mauritania. FADES/FAO Project: UFT/MAU/025/MAU, Oct–Nov 1999, Technical assistance for date palm protection in Adrar State

Sedra MyH (2000) Report of international expert consultation on sanitary state of date palm groves in Mauritania and discovery of Bayoud disease. Suggestions and recommendations for a prevention program and methods of control. Training a team of prospectors in palm groves and field diagnostic methods, 6–21 July 2000. AOAD, Al-Khartoum

Sedra MyH (2001a) Descripteurs du palmier dattier (*Phoenix dactylifera* L.). Edit. INRA Maroc, Imprimerie Al Watania, Marrakech Maroc

Sedra MyH (2001b) Bayoud disease of date palm (*Phoenix dactylifera* L.), extension, damage, diagnosis and prevention of practical recommendations for the spread and development of the disease. Bulletin guidelines (4). Research network and the date palm development. ACSAD, Damascus

Sedra MyH (2002) Report of international expert consultation in Mauritania. FADES/FAO Project: UFT/MAU/025/MAU, 6 Nov–4 Jan 2002, Technical assistance for date palm protection in Adrar State

Sedra MyH (2003a) Date palm cultivation, characterization and classification of main Mauritanian varieties. Edit AOAD, Al-Khartoum

Sedra MyH (2003b) Le Bayoud du palmier dattier en Afrique du Nord. FAO, RNE/SNEA-Tunis. Editions FAO, Imprimerie Signes, Tunis

Sedra MyH (2004) Le Bayoud (fusariose vasculaire) du palmier dattier en Afrique du Nord. Situation actuelle et stratégies de lutte. Journées nationales sur la protection des Plantes. Association Marocaine de la Protection des Plantes (AMPP). Rabat

Sedra MyH (2006a) Bayoud disease on palm date (*Phoenix dactylifera* L.): its appearance, losses, spread, development conditions, diagnosis and control methods, achievements, prospects and practical recommendations. Regional Project research for early detection of bayoud disease on date palm and development of its control technologies. AOAD/Imprimerie Al-Watania, Al-Khartoum/Marrakech

Sedra MyH (2006b) Final report of international expert consultation in Mauritania. Project: FAO/UFT/MAU/025/MAU Technical assistance for date palm protection in Adrar State, 27 Aug–17 Sept 2006

Sedra MyH (2007a) Discovery of bayoud disease on date palm tree in Mauritania: its importance, kind and control strategies. In: Fourth symposium on Date Palm King Faisal University, Hofuf, 5–8 May 2007

Sedra MyH (2007b) Bayoud disease of date palm in North Africa: recent distribution and remarks about its characterization, diagnosis and origin. In: Fourth symposium on Date Palm, King Faisal University, Hofuf, 5–8 May 2007

Sedra MyH (2007c) Report of international expert consultation in Mauritania. Training of Mauritanian staff for Phytopathology laboratory operation at Atar city affiliated to the Ministry of Agriculture and Rural Development and research and development program of date palm for short and medium term. Project OADA 17–23 June 2007

Sedra MyH (2007d) Selection of morphological characteristics and molecular markers and their use for identification and distinguishing between date palm varieties and the plants issued from tissue culture. In: Fourth symposium on date palm, King Faisal University, Hofuf, 5–8 May 2007

Sedra MyH (2008a) Final report of regional project research for early detection of bayoud disease on date palm and development of its control technologies. AOAD, Al-Khartoum

Sedra MyH (2008b) Report of international expert consultation in Mauritania. FAO project: TCP/MAU/3104 (D), sector strengthening of plant protection in Mauritania, 12–22 Sept 2008

Sedra MyH (2011a) Molecular markers for genetic diversity and bayoud disease resistance in date palm. In: Jain SM, Al-Khayri JM, Johnson DV (eds) Date palm biotechnology. Springer, Dordrecht, pp 533–550

Sedra MyH (2011b) Genetic diversity analysis, Moroccan cultivar genotyping and rapid screening for bayoud disease resistance in date palm using molecular techniques. In: Proceeding of the first symposium international sur le palmier dattier en Algérie, 13–14 Nov 2011

Sedra MyH (2011c) Le Bayoud (fusariose vasculaire) du palmier dattier en Afrique du Nord: situation, acquis de la recherche et applications. In: Proceeding of the first symposium international sur le palmier dattier en Algérie, 13–14 Nov 2011

Sedra MyH (2012) Le guide du phéniciculteur: mise en place et conduite des vergers phoénicicoles. Edition INRA Maroc, Imprimerie Nadacomdh, Rabat (in French)

Sedra MyH (2013) Le guide du phéniciculteur: mise en place et conduite des vergers phoénicicoles. Edition INRA Maroc, Imprimerie Nadacomdh, Rabat (in Arabic)

Sedra MyH, El Filali H, Benzine M et al (1996) La palmeraie dattière marocaine: Evaluation du patrimoine phoénicicole. Fruits 51:247–259

Sedra MyH, Lashermes P, Trouslot P et al (1998) Identification and genetic diversity analysis of date palm (*Phoenix dactylifera* L.) varieties from Morocco using RAPD markers. Euphyt 103:75–82

Sedra MyH, Zhar N (2010) Genetic variability analysis of populations of *Fusarium oxysporum* f. sp. *albedinis*, causal agent of bayoud disease of date palm and other *Fusarium oxysporum* using molecular techniques. Acta Hort 882:491–504

Trifi M (2001) Polymorphisme et typage moléculaire de variétés tunisiennes de palmier (*Phoenix dactylifera* L.): relation avec la résistance au bayoud. Thèse de Doctorat d'Etat, Université de Tunis El Manar, Faculté des Sciences de Tunis

Chapter 10
Date Palm Status and Perspective in Sub-Sahelian African Countries: Burkina Faso, Chad, Ethiopia, Mali, Senegal, and Somalia

Mohamed Ben Salah

Abstract Date palm (*Phoenix dactylifera* L.) is present in sub-Sahelian African countries. Although date production is usually conducted using poor cultivation operations, date palm genetic resources are rich and highly diversified. This is due to years of date palm propagation by seed; hence, the populations of males are as numerous as females in cultivated oases. Generally, Sahelo-Sudanian and Sahelian climate and the predominant agroecological conditions are favorable for the cultivation of date palm. However, climate changes can render more areas in the sub-Sahelian borders better suited for date palm cultivation in the future. In some Sahelian countries such as Mali, Niger, Chad, and Somalia, date palm cultivation and knowledge of techniques are more extensive than in other countries of the region. In Senegal, Burkina Faso, Cameroon, and Ethiopia, date palm is an alternative crop and development efforts are in progress. Internationally known cultivars are recommended for cultivations in this region based on the reported market success in southern African countries especially South Africa and Namibia. Some local genetic resources can also be used for date palm expansion. Nonetheless, training is needed in date palm cultivation and protection. The main objectives of current date palm agricultural development projects are the diversification of production and the development of irrigated crops. This chapter focuses on the distribution in six sub-Sahelian countries of date palm and highlights the major obstacles facing date palm cultivation and presents recommendations to ameliorate the situation. Some other sub-Sahelian countries are discussed in separate chapters in this book.

Keywords Cultivation • Date palm • Development • Oases • Perspectives • Production • Sub-Sahel

M. Ben Salah
Aridology and Oases Crops Research Laboratory (LACO),
Research Regional Center on Oases Agriculture,
Degache 6021, Tunisia
e-mail: bensalah_mohamed@yahoo.com

J.M. Al-Khayri et al. (eds.), *Date Palm Genetic Resources and Utilization: Volume 1: Africa and the Americas*, DOI 10.1007/978-94-017-9694-1_10

10.1 Introduction

Date palm (*Phoenix dactylifera* L.) has been grown in sub-Sahelian African countries since the Neolithic civilization but with some gaps. This is illustrated by evidence of Sudanese date plantations around Neolithic sites. Dollé et al. (1990) estimated the total number of date palms in sub-Sahelian countries to be about 2 million trees which represent about 2 % of the total date palm number in the world (Table 10.1).

The latitudinal limits of date palm distribution in the Northern Hemisphere as reported by Zaid and de Wet (2002) are between 10° N (Somalia) and 39° N (Elche/Spain or Turkmenistan) (Table 10.2) with the most favorable areas between 24° and 34° N (Fig. 10.1).

The area of the sub-Sahel, the poorest of the African continent, is a vast region located between the isohyets 100 and 600 mm with a geographic extension of 400–600 km north to south and nearly 6,000 km from east to west and encompassing seven countries (Mali, Burkina Faso, Niger, Chad, Senegal, Cameroon, and Mauritania) with a combined population of approximately 50 million. Agriculture occupies less than 50 % of the gross domestic production, of which crop production is 25 %, livestock 17 %, and forests and other agricultural practices 8 %.

All of the countries of the sub-Sahelian center (Niger, Nigeria, Chad, and Cameroon), with 142.5 million inhabitants and a high proportion of Muslims (58 %), represent an important market for date production (Granry 1999).

Table 10.1 Date palm population's estimation in sub-Sahelian countries

Country	Number of date palms
Chad	1,360,000
Niger	500,000
Mali	20,000
Somalia	114,000
Djibouti	10,000
Senegal	6,000
Total	2,010,000

Source: Dollé et al. (1990)

Table 10.2 Southern latitude limits of date palm cultivation in sub-Sahelian Africa

Country	Region/district	Parallel
Tanzania	Tabora	5° S
South Africa	Henkries Fontein	29° S
	Kakamas	27° S
	Klein Pella	27° S
Namibia	Naute/Keetmanshoop	26° 57′ S
	Hardap/Mariental	24° 33′ S
	Aussenkehr/Karasburg	28° 24′ S
	Eersbegin/Kunene	20° 09′ S

Source: Zaid and de Wet (2002)

Fig. 10.1 A map of Africa showing the date palm-growing countries of the sub-Sahelian region

This chapter begins with a presentation of the sub-Sahelian agricultural production context and reports the distribution of date palm in different countries including Chad, Mali, Senegal, Burkina Faso, Somalia, and Ethiopia; Niger is discussed in a separate chapter in this book. The potentialities of date palm production as traditional practices and available genetic resources are presented. Next, the major obstacles facing date palm cultivation are highlighted in these countries classified by technical and socioeconomic constraints. This review presents some considerations to enhance date palm development in the region at three plantation scales: extended oases, valley plantations, and small-scale orchards. Finally, some recommendations are presented to ameliorate the date palm situation especially in the choice and propagation of cultivars and for pest and disease control.

10.1.1 Distribution and Climatic Conditions

From the climatic point of view, the sub-Sahelian region is subdivided into four ecological zones that largely influence agricultural production. They are hyperarid, arid, semiarid, and dry subhumid regions. In the majority of the Sahelian countries, the agriculture system is very fragile, characterized by an active nomadism and unprofitable agricultural production. The sub-Sahelian date palm population is estimated at 6,530,000. Sahelian countries annually produce approximately 105,000 mt of dates, representing about 5 % of world date production (DMP-IPALAC 1997).

In the southern African countries, date palm is present in South Africa, Namibia, Mozambique, and Tanzania. Conditions are better there than in the sub-Sahel for date palm cultivation in relation to water availability.

In the sub-Sahel, the year is generally divided into a hot season which extends from April to October (maximum temperature 42 °C) and cold season from November to March (minimum temperature 13 °C). The climate is hyperarid, with little or no rainfall. Indeed, rainfall recorded in 24 relatively wet years in the Sahel (1937–1961) was on average 26 mm/year (Rahme 1991). The sand-laden winds are an adverse factor in general and occur some 70 days per year. In the Saharan portions of the Sahelian countries, the winds are nearly always oriented northeast to southwest. High temperatures in combination with intense winds cause high evapotranspiration rates. The winds activate sand dunes without vegetation and form crescent-shaped dunes called *barchans*.

These barchans threaten not only the farms and natural resources such as Ayn Galakka in Chad but also urban infrastructure. Thus, the dunes invaded toward the Chadian city of Faya and the new airport was quickly covered with silt (Rahme 1991). Sand-buried plant matter forms humus and soil rehydrating because the sand dune limits evaporation. When the dune has passed, the peasants reinstitute farming with more fertile soils.

10.1.2 Economic Aspects

Date palm is a promising crop for the Sahel. The dry and hot regions in the north are suitable for the production of dry and semidry cultivars; whereas, the more humid south is suitable for soft cultivars.

The introduction of date palm cultivation to the sub-Sahelian region can result in very significant economic and environmental gains. The FAO (1997) proposed commercial date palm introduction into Mauritania, Senegal, Mali, Burkina Faso, Niger, Cameroon, and Chad as a regional project effort aimed at benefiting the middle- and low-income producers. Discussing the opportunities of this project, experts presented the following advantages:

(a) The date fruit is well known and highly cherished by the mostly local Muslim population.
(b) There is a strong local market for dates which is presently limited by the high prices of mostly imported dates.
(c) Dates are highly nutritious. One kilogram of date has more than 3,000 cal as compared with cooked rice with 1,800 cal/kg and bananas with 970 cal/kg. Dates are also very rich in minerals and fiber (Zaid and de Wet 2002).
(d) Dry and semi-dry dates can be kept well and can be stored for a long period of time, over a year for some cultivars.

(e) There are regional and international markets for dates.
(f) In intercropping farming systems, date palm trees protect the more sensitive irrigated crops from wind, sandstorms, and scorching sun. The very high rate of transpiration from date leaves together with its shading effects results in a significant cooling of the microenvironment within the date grove and an increase of relative humidity. This effect allows the production of heat-sensitive vegetable and fruit species during the hot season.

The introduction of date cultivation to the sub-Sahelian countries will have a tremendous economic benefit to the region by: (a) ensuring food security, (b) generating income to the farm community, (c) increasing exports, and (d) protecting the environment (FAO 1997).

10.2 Cultivation Areas in the Sub-Sahel

10.2.1 Chad

Date palm cultivation in Chad is concentrated in the northern part of the country near the Libyan border. BET (Borkou, Ennedi, and Tibesti) is the northernmost prefecture of Chad. It is a desert region which covers an area of 600,350 km^2 or almost half of the country. It is divided into three subprefectures: Borkou (57,000 inhabitants); Belluno Faya-Largeau/Ennedi (53,000), with its capital Fada; and Tibesti (10,000), capital in Bardai. The oases of Borkou are located in a geologic depression (Fig. 10.2a). Different hydrogeological studies carried out recently (Aureli 2011) confirm the existence of deep fossil groundwater reserves. This deep water feeds the water table by capillary action and reaches the surface along geologic fractures in the form of springs.

In Chad, large areas of date palm cultivation are found in Borkou and Kanem (Table 10.3). The total number of palms is about 1,110,000, including 85,000 irrigated plants. In addition to the Borkou Region, there are large plantations to the level of the Kanem Region but in smaller areas. In the region of Kanem and Lake Chad, there are nearly 20,000 date palms planted in 17 valleys and wadis without irrigation (Baroin and Pret 1993).

10.2.2 Mali

The area of date palm cultivation in Mali is located in the north, the seventh region. The village of Tessalit already existed upon arrival of the Arabs and the Islamization of the area (Ag Sindibla 1988). The oldest report of the number of date palms is found in Munier (1973) which cites about 50,000. He noted the existence of date palms in the Nioro du Sahel and at other sites such as Bourem on the Niger River.

Fig. 10.2 Date palm cultivation in sub-Sahelian African countries. (**a**) Date palm production in Faya, Chad (Source: http://www.ndjamena-matin.com/), (**b**) Date palm plantation in Burkina Faso (Source: http://www.les-palmeraies-du-burkina-faso.com/)

Table 10.3 Date palm number estimates in Chad

Region	Number of trees
BET (Borkou, Ennedi, and Tibesti)	1,300,000
Kanem	20,000
Other regions	630,000
Total for Chad	1,950,000

Source: Baroin and Pret (1993), Pret (1990)

Lenormand (1988) in his mission reported the number of date palms in the region of Kidal to be 2,320. Togo (1988) reported a number of 4,302 palms spreading over an area of 100 ha. The number of date palms in Mali (Table 10.4) as evaluated by PSARK (Food Security Program and Income of the Kidal region) financed by the IDB (Islamic Development Bank) and OPEC (Organization of the Petroleum Exporting Countries) is 4,435 palms (IFAD 1989, Ben Salah 2001). The number of date palms has steadily declined. Moreover, growing date palm is often subject to the effects of climate including drought as well as occasional exceptionally strong water flow in the wadis which can destroy plantations. The inhabitants of Tessalit reported that the old Tessalit Palm Grove was entirely swept away by the water flow in the Tessalit wadi during one of its floods.

10.2.3 *Senegal*

Date palm cultivation for fruit is not well known in Senegal although the palm is commonly planted in settlements, in mosque esplanades, along avenues, and in public squares. Agricultural statistics show that date imports have increased more than 75 % over the past 5 years, stabilizing at about 500 mt per year since 1999 (Ben Salah 2005). This represents an outflow of foreign currency of more than XOF

Table 10.4 Date palm population estimates in Mali

Region	Estimated numbers (Lenormand 1988)	Estimated numbers (Togo 1988)	Number of date palms
Tessalit + Hambouber	2,050	1,680	1,660
Intedeini	50	–	5
Abanko	–	98	13
Intabdok	20	–	73
Aguelhok-Sawani	–	90	20
Telabit-Tahist	150	568	562
Tassijdimt	–	501	709
Tagrabat	–	449	448
Tanezrouft	50	517	487
Tefeinak	–	19	117
Tin-Zaouatene	–	203	225
Kidal ville + Intekoi	–	177	278
Matelme	–	–	215
Sorene	–	–	118
Abeibara	–	–	5
Total	2,320	4,302	4,435

Source: Ben Salah (2001)
(–): not available

(West African CFA Franc) one billion in 5 years. Date palms are reportedly grown in the area of northern Senegal along the banks of the Dagana River in Bakel. Munier (1973) estimated their numbers at 5,000–6,000. In the town of Bakel itself, their presence is also reported but in smaller numbers and with less interesting fruit quality.

Analyses of vegetation show that date palms are present in Saint Louis. At the town of Richard Toll, an old experiment of the ISRA (Senegalese Institute of Agronomic Research) station contains a collection of introduced date palm cultivars. Near Saint Louis, at Savoye, one producer released offshoots of introduced cultivars Tamar Akhal, Tijip, and Mdina, from Atar, Mauritania. Those cultivars are commonly grown in Assaba and Kankossa in southern Mauritania, under similar conditions to those of northern Senegal (Ben Salah 1994).

In the Lake of Guiers and Louga areas, different gardens surrounding the lake have some date palm trees planted by locals which produce edible dates.

In Keur Momar Sarr, an experimental plot was established and equipped by the Israeli organization IPALAC (International Program for Arid Land Crops) and the USAID (United States Agency for International Development) in 1994. Two cultivars, Barhi and Medjool, were introduced. In Diourbel, date palm is present in the city. An old introduction of 20 years ago was made by the caliph of Touba.

Palm trees in very good vegetative state still exist in the residence of the sheikh of Touba, in the Great Mosque of Touba esplanade, as well as in a small plot in front of Ainou Rahma. The general vegetative behavior of date palm seedlings in Touba is acceptable but fruit quality is poor. Children gather the fruits before they mature.

In the Niayes zone, there are only a few date palms in isolated plots and in home gardens. The climate of the area was classified by Toutain (1967) as Senegalese or sub-Canarian Sahelo-side climate which marks the northern regions on the south bank of the Senegal River and is not very favorable to the maturation of the dates.

10.2.4 Burkina Faso

Date palm cultivation is not well known in Burkina Faso where more than 83.6 % of the population lives in rural areas. Agriculture is highly impacted by climatic fluctuations, especially drought cycles. The cultivated area of the country is estimated at 35,000 km^2 or 13 % of the total area, including 2.7 % under irrigation. The Sahelian portion (200–600 mm rainfall) of the country covers an area of about 61,000 km^2 (±22 % of the total area). The main date palm cultivation sites in Burkina Faso are in Djibo, Dori, and Oudalan.

The Burkina Faso government encourages farmers to plant date palms to prevent the advance of the desert. One national program has been established, supported by the FAO, and includes 200 farmers with the aim to plant and maintain date palm plantations (Fig. 10.2b). The program, based on national and regional studies, covers the northern provinces of Seno, Soum, Oudalan, and Yahga. Local inhabitants of that region will benefit from the program through the manufacture of baskets with date palm leaves and selling fruits. The 2-year program lays the foundation for a larger-scale effort to plant more date palms.

10.2.5 Somalia

According to Chazée (1990), oases production activities are not so old in time. Arab inhabitants from Djibouti and Yemen introduced date palm cultivation about 200 years ago.

Dowson (1963) indicated that the date palm grows all around the Red Sea, Gulf of Tajura, Southern Arabia, Eastern end of the coast, and also East coast southward from Cape Guardafui to south of Mogadiscio. He estimated the total number of date palms to be about 90,000. Date palm numbers in Somalia were estimated by Chazée (1990) to about 100,000 covering an area of 350 ha with a fruit production of less than 900 mt. Additional dates consumed in the country are imported from Iraq. Date palms grow in the far north of the country, around the mountains, in an arid region where the annual rainfall is less than 200 mm.

The Somalian date palm production areas are Ambouli, Douda, Gobad, and Tajoura. Six local cultivars are planted: Nemahan (40 %), Sahcari (20 %), Suqadari (14 %), Masili (13 %), Faqur (10 %), and Farad (10 %). Production based solely on date palm production represents about 24 % of the area. The average of production estimated by Dowson (1963) is nearly 4 kg per palm tree. Production systems including intercropping with date palm represent just 6 % of the total. Poor maintenance of the palms is common. For example, a mere 5 % of farmers apply manure in the groves. Practically 100 % of production is consumed locally within the family or shared with the extended family, and about 6 % is commercialized, also locally.

10.2.6 Ethiopia

According to Hussein (2010), the Afar Region in Ethiopia is suitable for date plantations. Located within the Danakil Depression, the region experiences the harshest climate in Ethiopia with temperatures reaching 50 °C. The area is home to the Afar people who are primarily seminomadic. Historically the Afars established date palm farms in areas around the Awash Delta and the Afambo Region. These farms produce date fruits mainly for local consumption. The Awash River is the main source of irrigation in the Afar Region; it runs from the Ethiopian highlands, fed by a number of tributaries, through the Afar desert land. There is currently a shortage of date fruit in Ethiopia where the demand reaches its peak during the month of Ramadan. About a third of the Ethiopian population is Muslim, representing a large potential market for date fruits. Labor cost in Ethiopia is significantly low compared to other date-producing countries. Dates are therefore attractive to produce locally at low cost to meet the high Ethiopian demand and possibly for export.

Dates have the potential to improve human nutrition and provide food security for the nomadic Afar community. Jobs filled by local farmers will create a source of income and help develop and improve the community as a whole (Gérard et al. 1996). Large-scale date plantation development in this desert land aims to make a significant contribution toward preparation for impending climate change (Shabani et al. 2012).

10.3 Agriculture Systems

10.3.1 Traditional Agriculture System

A 1990 survey of farming systems carried out in the Borkou oases in northern Chad by Barrow (1998) identified the following three systems: The first one is extensive system largely dependent on two dominant activities, extensive camel rearing in the plains to the south of the BET and extensive date farming in the oases, thus, a system based on mobility. The second is composed by two semi-intensive agropastoral systems distinguish the majority of the region's farmers, which combine within the palm groves, crop agriculture, date farming, and livestock, as part of a risk-avoidance diversification strategy. The first of these two agropastoral systems is based on subsistence consumption by the farmers, and the second on commercialization. The third is a system located on the periphery, where extensive date farming is practiced but based mainly on fruit growing rather than either agricultural or pastoral products.

10.3.2 Proposed Agriculture System

New development models proposed for date growing in the sub-Sahelian countries are adapted to local production conditions and are under experimentation in Niger, Mali, and Senegal. Four models are proposed for new production units of date palm.

10.3.2.1 Intensive Model

This proposed intensive model combines date palm culture with certain other fruit trees such as pomegranate (*Punica granatum*), guava (*Psidium guajava*), jujube (*Ziziphus mauritiana*), papaya (*Carica papaya*), and other understory vegetables like okra, pepper, tomato, peanut, and hibiscus (bissap) for a beverage. Irrigation is by a modern drip system. The area of this model is 1 ha. It has the advantage of providing income to farmers before date fruit production begins, which takes at least 3 or 4 years, after which loaned equipment is returned.

10.3.2.2 Semi-intensive Model

This model is similar to the previous one from the point of view of crop composition as well as objectives. The only difference is that it uses conventional irrigation. The area proposed for this model is also 1 ha.

10.3.2.3 African Market Garden (AMG) Model

The AMG model was inspired by the small gardens operated on a limited scale in some areas of Africa (Leroux 2009). This may include small farmers with access to irrigation water (wells, sump) or associations, schools, mosques, etc. This system combines date palm with food and vegetable crops excluding other fruit trees. The proposed system includes drip irrigation at low pressure. The area proposed for this system is limited to 1,500 m^2.

10.3.2.4 Pastoral Model

The last proposed model is agropastoral but similar to the semi-intensive model; it is established in areas where the groundwater is shallow (2–5 m). This model is proposed for nomads who move from a region to another but return to oasis locations for some months each year. Agricultural fields are located near the water source and must be protected from livestock (especially goats) when they come for watering.

10.4 Cultivation Practices

10.4.1 Propagation

In general, date palms growing in sub-Sahelian countries, with perhaps the exception of certain oases in Chad and Niger, do not originate from offshoots. The main method of propagation is by seed. As a consequence, there are a large number of male palms, which is not the case in oases cultivated with offshoots. No selection of superior genetic material is made. This situation results in a wide diversity of fruit morphotypes and contributes to a richness of genetic resources but makes identification and selection of new planting material complicated.

10.4.2 Pollination

Assisted pollination practices are generally negligible and not well known. Natural pollination is by wind, enhanced by the large number of male date palms in the grove. Some ineffective pollination practices in the sub-Sahel include the belief that it is sufficient to place some flowering male strands at the base of the trunk to complete pollination.

10.4.3 Date Handling and Conservation

No specific methods of date fruit harvest and storage are made for dates except for dry dates in Saharan parts of sub-Sahelian regions where nomadic traditions are the same as in Saharan portions of Algeria, Mauritania, and Libya. In the sub-Sahel, date imports are generally of dry dates from Algeria, and the storage practices are adapted for them. Dates are simply packaged in jute bags which allow for ventilation.

10.4.4 Protection Against Pests

Some insect pests are present in the region especially white scale (*Parlatoria blanchardi* Targ.) which is a serious threat to young date palms and date mites (*Oligonychus afrasiaticus* McGr.) which feed on the fruits in the sub-Sahelian Saharan regions. In more humid areas, other pests such as red scale (*Phoenicococcus marlatti* Cock.) are present; these scale insects attack the leaves and their threat is exacerbated by neglecting to prune infested leaves.

Termites and ants attack the date palm, colonizing the trunk and the base of leaves where they nest and multiply; these pests are frequent in the oases of Sahel. They can cause the death of date palms already debilitated by the abandonment of cultivation practices. Birds also cause damage to the production by feeding on soft dates and causing damage, preventing them from reaching maturity.

10.5 Genetic Resources

10.5.1 Local Genetic Resources

The quality of local date fruit production is not very good, but the palms are adapted to local environments giving an advantage to local genetic resources. In Somalia the date palm local cultivars are reported by Chazée (1990)to be the following: Nemahan cv. represents one-half the total numbers, with the other major cultivars being Faqur, Farad, Sahcari, Suqadari, and Masili.

In northern Mali, there is a morphotype identified by Ben Salah (2005) and Togo (1988) as suitable for offshoot propagation, called Tiber. Local farmers have used Tiber cv. to compare the performance of imported Algerian cvs. Tegaza (red fruit) and Telemsou (yellow fruit).

In northern Niger, some seedling dates have local names; the most widespread in the Kawar, Air, and Ingall areas are designations as *Agues*, *Klaw*, *Yilbodam Guewas*, *Alarka*, and *Almadeina* (Lenormand 1988). The differentiation lies in the appearance of harvested fruit: soft, semidry, or dry. In southern Niger, the date fruit distinction is solely on the basis of color. There the dates are called Dan *Fari* and *Dan Dja* whether for human or animal consumption. More details on date palm cultivation in Niger are described in another chapter.

In Chad dry cultivars are the most widespread in the Borkou area where Bornow is most predominant. This cultivar represents about 70 % of the date palms of the Borkou Depression in northern Chad. Combined with four other cvs., Kougoudou, Waserdow, Koîdow, and Aribo, they represent 85 % of the total date palm trees in Chad (Arditi 1995; Pret 1990).

Local date palm genetic resources should be collected and evaluated. There is also the need to establish research stations in the region for collecting, evaluating, and preserving local plant genetic resources

10.5.2 Introduced Genetic Resources

A few dates have been imported by development projects such as the three well-known commercial cultivars Medjool, Barhi, and Deglet Noor. But others need to be introduced for more experimentation to choose better-adapted cultivars. Certain cultivars in Senegal were proposed for introduction into Mali such as Zemli, Saggai, and Shishi (Ben Salah 2001, 2005). As well, Ashal Al Hassa from Saudi Arabia, Zaghloul from Egypt, and Rochdi from Tunisia. Those cultivars appear suited to the relatively wetter regions of the Sahel, more so than Medjool, Barhi, and Deglet Noor cvs.

10.6 Agricultural Constraints

10.6.1 Technical Constraints

There are at least three main technical constraints facing the promotion of commercial date palm production:

(a) The inexistence of high-performing cultivars. Most of the presently planted date palms were propagated from seeds. About 50 % of the plants are male and the quality of the female trees is very poor. Most of the European trade in date fruit is based on three high-quality cultivars: Medjool, Barhi, and Deglet Noor. To guarantee future exportation of dates, they should be introduced to the region. Presently there are no cultivars suitable for the higher rainfall regions (800–1,000 mm) of the Sahel and new ones should be introduced.
(b) The lack of technological know-how about date palm cultivation. This is true for all aspects of date palm cultivation and management such as planting methods, removal of offshoots and production of rooted offshoots, pollenization, fruit thinning and postharvest handling, and, most importantly, irrigation management.
(c) Inadequate irrigation systems. The prevalent irrigation system (hand irrigation) is not suitable for dates. Methods of surface irrigation are not widely distributed and have many inherent constraints.

10.6.2 Socioeconomic Constraints

The main socioeconomic constraint for large-scale cultivation of date palms for small sub-Sahelian producers is lack of a proper credit system. Prospective date palm farmers need a source of credit to allow them to purchase irrigation equipment, date plants, and agricultural inputs.

Training for date palm cultivation is also necessary. Exchange visits to oases in North Africa (Morocco, Algeria, and Tunisia), in Arab Gulf countries (Saudi Arabia, UAE, Yemen, Oman), or southern African countries (South Africa and Namibia) would provide practical examples rather than theoretical training for farmers and technicians.

10.7 Possibilities for Date Palm Development in the Region

Other than encouraging the planting of date palms in towns, esplanades, avenues, and squares, sub-Sahelian date palm-based agriculture can be composed of both small-scale production systems and large plantations.

10.7.1 Creation of Extended Oases

Opportunities exist to extend existing oases in Mali (Tilemsi and Tamessana), Senegal (Bakel), Chad and Niger (northern), and Afar (Ethiopia), where adequate water resources are present. Soil quality and adequate water resources are the necessary conditions for extended development of oasis sites. Water wells on sites of dozens of hectares are necessary along with prospect drilling that will be sufficient for the development of date palm plantations. Lazarev (1988) suggested key conditions for introduction of oases in the Sahelian environment. A critical factor is related to the combination of climatic conditions and soil quality in response to the agroecological requirements of date palms and associated crops. These larger systems must utilize pressurized drip irrigation and modern production technologies.

10.7.2 Integrated Wadis

Small-scale date growing can be carried out by integrated management of local wadis such as El Wedj and Aguelhok in Mali and in similar areas in northern Niger and Chad. Most valley plantations should ensure the availability of shallow groundwater capable of maintaining plantations through irrigation during the dry season.

Retention can be improved by the works of water conservation and soil and even water storage as practiced in the arid regions of North Africa (Jahiel and Morou 1990). It is possible to integrate semi-extensive livestock with the agricultural production of date palms and vegetables, as well as forage production, with areas seeded in times of rainfall with the use of water from shallow wells during the dry season.

10.7.3 Small-Scale Date Palm Plantation

Small-scale date palm plantations, inspired by small gardens operated with limited resources in some areas of Africa, can be a good model for expanding date palm cultivation in areas where soil and water conditions are favorable.

The DMP-IPALAC (1997) proposed preparing date palm farms of some 500 m^2 AMG units to be set up in four Sahelian countries as pilot farms. They proposed that nongovernmental organization technical staff and regional extension workers construct the units. After they become operational, the program plan is to provide the pilot farmers with quality date palm cultivars. The pilot AMG units are also proposed for use to train local farmers.

10.8 Approaches to Alleviate Agricultural Constraints

10.8.1 Identification of Suitable Date Palm Cultivars

In the sub-Sahel, it is necessary to identify adapted genetic material and cultivation practices for three distinct sets of climatic conditions:

1. A rainy season that extends from mid-June until the middle of October: This season is characterized by high humidity, moderately high temperatures and small temperature fluctuations between day and night, and heavy monsoonal rains.
2. The cool and dry season: This season extends from the beginning of September until mid-March. This is the main vegetable production season.
3. The hot and dry season: It starts in mid-March and ends in mid-June with the onset of the rainy season. Very high day temperatures typify the season, many times exceeding 40 °C, combined with dry winds and very low humidity. There is a need to identify vegetable cultivars that are suitable for production in each of the above seasonal scenarios.

10.8.2 Testing Date Palm-Based Production Units

Sub-Sahelian date palm-based agriculture should be composed of both small- and larger-scale production systems. The larger systems will use pressurized drip irrigation and modern production technologies.

Some of these systems will be composed of pure date stands, whereas others will be mixed stands of date palms with other fruit trees, vegetables, or forage crops. Research and demonstration plots have to be established as technical support for potential producers. In these units, the combinations of date palms with other fruit trees, vegetables, and forages must be tested and evaluated.

10.8.3 Establishment of Date Palm Acclimatization Nurseries

The best solution to the lack of improved date palm planting material is to develop regional laboratories for in vitro date palm propagation. According to the DMP-IPALAC (1997), there are two tissue culture laboratories in the region. One is situated in Bamako, Mali, and the other in Dakar, Senegal. One of them can be designated to become the regional laboratory for tissue culture-propagated date palm.

The staff of the selected laboratory has to be trained in the methods of tissue culture propagation of date palms and the laboratory must be equipped for commercial production of plantlets.

Whether date palm plantlets are produced in a regional laboratory or purchased from other certified tissue culture laboratories in another country, acclimatization can be conducted in the country where planting will occur. Date palm plantlets are generally 1 year of age when transported. The acclimatization operation can normally take up to 16 months to produce a date palm plant ready for field planting.

10.8.4 Development of Techniques for Pest and Disease Control

In general, crops grown under irrigation are highly susceptible to diseases and pests. This is particularly true for small-scale irrigated production systems of the Sahel where there are little space for crop rotation and lack of funds and knowledge for pest eradication.

Training on date palm pest management in particular and on the intensified oasis production is necessary for technicians and farmers. Numerous pests and diseases can attack date palm and fruit trees in the sub-Sahel where conditions of humidity are favorable for their development and reproduction. Some of the pests and diseases are white scale (*Parlatoria blanchardi* Targ.), Graphiola leaf spot (*Graphiola phoenicis* (Mong.) Poit.), *belaât* (*Phytophthora* sp.), Diplodia disease (*Diplodia phoenicum* (Sacc.) H.S. Fawc. & Klotz), date mite (*Oligonychus afrasiaticus McGr.*), almond moth (*Ephestia calidella*), black scorch disease (*Thielaviopsis paradoxa* (De Seynes) Höhn.), and Khamedj disease or inflorescence and leaf rot (*Mauginiella scaettae* Cav.), fruit rot, mealy bug, and termites (Djerbi 1997).

Efforts under way to establish effective integrated pest control management for the protection of the date fruits from infestation on the tree, during transportation and in storage, should be intensified, because losses from insect pest infestation are still substantial and in certain instances may destroy the entire crop. Some success has been realized combating *Parlatoria blanchardi* L. in northern Mali by biological control using *Chilocorus bipustulatus* var. *iranensis*, a practice followed in Mauritanian oases. This success has to be replicated throughout the sub-Sahelian countries.

Red Palm Weevil (*Rynchophorus ferrugineus* Oliv.) is considered the most dangerous pest, and Bayoud caused by *Fusarium oxysporum* f. sp. *albedinis* the most dangerous date palm diseases. Bayoud is easily spread by the transport of infected plants, offshoots or plant parts harboring the fungus. Attention should be considered to those two pest and disease.

10.9 Conclusions and Recommendations

Date palm growing is a component of the sub-Sahelian agriculture landscape and is present in different sub-Saharan parts of the region and is proposed to be extended to other wetter regions in the Sahel. The date palm is adapted to the Sahel and its planting can contribute to combating desertification and, above all, contribute to the creation of microclimates favorable to the development of under-canopy crops such as other fruit trees, fodder, and market garden vegetables. Marketing provides income and fruit and tree by-products for use in livestock feed and handicrafts. Currently, date palm cultivation in the sub-Saharan countries suffers from a lack of effective propagation, pollination, protection procedures, and very low fruit quality. However, the region does contain extensive date palm genetic resources.

The first condition for the maintenance of the current oases production and for development of oases in the sub-Sahel is mobilization of water resources and their rational management. It is also necessary to select superior local genetic material, if it exists, and to enlarge the choices of introducing cultivars. Mobilizing human resources to build capacities in date palm production and protection through a training framework is necessary to assist farmers and agronomists to achieve success in such projects.

References

Ag Sindibla I (1988) La phoeniciculture, contraintes et perspectives d'avenir en 7ème région. Mémoire de fin d'Etudes d'Ingénieurs en Sciences appliquées. Institut Polytechnique Rural de Katibougou, Mali

Arditi C (1995) Le commerce des dattes du Borkou (Tchad). Cah Sci Hum 31(4):849–882

Aureli A (2011) Les ressources en eau une responsabilité partagée. Geosciences 13. Juillet 2011:107–108

Baroin C, Pret PF (1993) Le palmier du Borkou, végétal social total. Journal des africanistes tome 63. Fascicule 1:5–20. doi:10.3406/jafr.1993.2369

Barrow SA (1998) Monograph of Phoenix L. (*Palmae*: *Coryphoideae*). Kew Bull 53:513–575. doi:10.2307/4110478

Ben Salah M (1994) La culture du palmier dattier en République Islamique de Mauritanie. Rapport de consultation. Organisation Arabe de Développement Agricole: OADA

Ben Salah M (2001) Réhabilitation de la culture du palmier dattier dans la région de Kidal-Mali. programme de sécurité alimentaire et de revenus de la zone de Kidal (PSARK). Report of con-

sultation. Islamic Development Bank (IDB)- OPEC (Organization of the Petroleum Exportation Countries)
Ben Salah M (2005) Promotion of the production of date palm in Senegal. Project TCP/SEN/3003. Report of consultation, Food and Agriculture Organization: FAO
Chazée L (1990) L'agriculture d'oasis et les palmeraies en Somalie. In: Dollé V, Toutain G (eds) Les systèmes agricoles oasiens, vol 11, Options Méditerranéennes: Série A. Séminaires Méditerranéens. Centre International de Hautes Etudes Agronomiques Méditerranéennes: CIHEAM. Paris-France, Montpellier, pp 67–75
Djerbi M (1997) Précis de phoeniciculture. FAO, Rome
Dollé V, Ferry M, Toutain G (1990) Situation des systèmes oasiens en régions chaudes. In: Dollé V, Toutain G (eds) Les systèmes agricoles oasiens. Options Méditerranéennes: Série A. Séminaires Méditerranéens, vol 11. Centre International de Hautes Etudes Agronomiques Méditerranéennes: CIHEAM. Paris-France, pp 7–18
Dowson VHW (1963) Report to the Government of Somalia on date production. Report N° 1731. Project SOM/TE/PL. FAO, Rome
Food and Agriculture Organization: FAO (1997) Date for the Sahel. A program for the introduction and cultivation of date palms. Experts report. Rome, Italia
Gérard D, Ferry M, Toutain G (1996) Introduction d'une agriculture oasienne dans le système pastoral traditionnel Afar de la vallée de l'Awash (Ethiopie). Opt Med A/28:193–194
Granry E (1999) Les oasis du centre sahélien. In: Ferry M, Bedrani S, Greiner D (eds) Agroéconomie des oasis. Groupe de recherche et d'information pour le développement de l'agriculture d'oasis (GRIDAO). Centre de Coopération Internationale en Recherche Agronomique pour le Développement (CIRAD). Montpellier, France
Hussein AH (2010) Potential of date palm plantation in Afar Region, Ethiopia and its market. Acta Hortic 882:43–48
International Fund for Agricultural Development: IFAD (1989) Programme de sécurité alimentaire et des revenus dans la zone de Kidal-Mali. Rapport d'évaluation. Annexe 5 Agriculture d'Oasis
Jahiel M, Morou A (1990) Evolution géographique de la phoeniciculture en relation avec la désertification. In: Dollé V, Toutain G (eds) Les systèmes agricoles oasiens. Options Méditerranéennes: Série A. Séminaires Méditerranéens, vol 11. Centre International de Hautes Etudes Agronomiques Méditerranéennes: CIHEAM. Paris-France, pp 59–66
Lazarev G (1988) L'oasis une réponse à la crise de pastoralisme dans le Sahel. Actes du séminaire Les systèmes Agricoles oasiens. Tozeur (Tunisie) 19–21 Nov 1988. pp 77–89
Lenormand C (1988) Compte rendu de la mission effectuée dans le Nord du Mali. Mars 1988
Leroux X (2009) Jardin potager africain. La recherche publique au secours du développement. Partie II. http://www.lepanoptique.com/sections/politique-economie/la-recherche-publique-au-secours-du-developpement-agricole-sahelien-partie-ii/. Accessed 15 Sep 2013
Munier P (1973) Le palmier dattier. Ed. Maisonneuve et Larose, Paris
Pret PF (1990) Les cultivars de palmier dattier (*Phoenix dactylifera* L.) de la dépression du Borkou (Tchad). Groupe de Recherche et d'Echange Technologique: GRET, Paris
Rahme J (1991) The date palm and the wellspring: Mandaeism and Jewish mysticism. Aram Newsl, pp 209–223
Shabani F, Kumar L, Taylor S (2012) Climate change impacts on the future distribution of date palms: a modeling exercise using CLIMEX. PLoS One 7(10):e48021. doi:10.1371/journal.pone.0048021
The Desert Margin Program: DMP and (Hosted by International Crop Research Institute for the Semi-Arid Topics: ICRISAT Niamey, Niger) International Program for Arid Land Crops: IPALAC (Ben Gurion University, Israel) (1997) Dates for the Sahel. Introduction of date palm-based agriculture to the Sahel
Togo I (1988) Inventaire des palmeraies et évaluation de la variabilité génétique des cultivars des palmiers dattiers dans la région de Kidal au Nord du Mali. Mémoire de fin d'études pour l'obtention du diplôme d'ingénieur d'état en agronomie saharienne. Institut de l'Hydraulique te de l'agronomie saharienne, Ourgla, p 56
Toutain G (1967) Le palmier dattier, culture et production. Al Awamia 25(4):23–151
Zaid A, de Wet PF (2002) Date palm cultivation, vol 156, Plant production and protection paper. Food and Agriculture Organization: FAO, Rome

Chapter 11
Date Palm Status and Perspective in Niger

Sahidou Abdoussalam and Dov Pasternak

Abstract Niger's economy is dominated by the agricultural sector. About 80 % of the population lives in rural areas producing mainly rainfed grains and pulses for subsistence living. Land degradation and high population growth have undercut economic development. Food security remains a serious problem in Niger. Date palms are cultivated in three main zones: the Aïr Mountains and the Ingal valley in the north, the Bilma plateau (Djado-Kawar-Agram) in the northeast, and the Damagaram and Manga depressions of the southeast stretching from Zinder to the Diffa regions. The number of date palms in Niger is about 720,000 producing 8,000 mt of dates annually. Average fruit yield, 10 kg per tree, is considered very low. Fruit quality is good in the northern traditional zones, where locally selected cultivars are grown, and poor in the southeastern marginal zones. Niger imports about 5,000 mt of dates per annum. In the traditional regions of date production, there are many locally selected cultivars, some of good quality. The International Crops Research Institute for the Semi-Arid Tropics (ICRISAT) introduced to the Sahelian region of Niger and neighboring countries some 10,200 tissue culture-propagated date palms of the Barhi cultivar. At the ICRISAT research station near Niamey, there are currently ten international cultivars under investigation. In spite of the great potential of date palm cultivation in Niger, very little research of consequence has been carried out, and currently there are no concentrated efforts to encourage large-scale date cultivation.

Keywords Cultivars • Cultivation practices • Date palm • Double flowering • Drought • Land degradation • Niger • Tissue culture

S. Abdoussalam (✉)
Systems and Crops Diversification Program,
International Crops Research Institute for Semi-Arid Tropics (ICRISAT-Niger),
Niamey BP12404, Niger
e-mail: s.abdoussalam@icrisatne.ne

D. Pasternak
Consultant, International Crops Research Institute for Semi-Arid Tropics (ICRISAT-Niger),
Niamey BP12404, Niger
e-mail: dov.pasternak@gmail.com

J.M. Al-Khayri et al. (eds.), *Date Palm Genetic Resources and Utilization: Volume 1: Africa and the Americas*, DOI 10.1007/978-94-017-9694-1_11

11.1 Introduction

The Republic of Niger is a landlocked, sub-Saharan country lying between 0° 16′ and 16° E long. and 11° 1′ to 23° 17′ N lat. Niger's 5,000 km border is shared with Libya and Algiers to the north, Mali and Burkina Faso in the west and southwest, Nigeria and Benin in the south, and Chad in the East. The country occupies an area of 1,267,000 km^2 and has a population of about 17 million people mostly concentrated in the Sahelian and the Sudano-Sahelian regions of the country. About 80 % of the population live in rural areas and derive their income from rainfed agriculture and animal husbandry.

Niger is a developing country and is consistently one of the lowest ranked in the United Nations' Human Development Index (HDI); in 2012 it was ranked 186th out of 186 countries. The economy that is dominated by the agriculture sector centers on subsistence crops and on livestock. Agriculture contributes about 40 % of GDP and provides livelihood for about 90 % of the population. Droughts, land degradation, and a high population growth rate have undercut the economy. Food security remains a problem in Niger.

The climate is typified by a rainy season stretching from June to September and a dry season starting in October and ending in May. Rainfall varies from less than 100 in the north to 700–800 mm per year in the south. The climatic zones of Niger can be divided into Saharan with precipitation less than 200 mm/year, Saharo-Sahelian with an average rainfall of 200–400 mm/year, Sahelian with an average annual precipitation of 400–600 mm, and Sudano-Sahelian with rainfall ranging from 600 to 800 mm/year. Most of the country is situated in the Sahara and Saharo-Sahelian regions. The agroecological zones in Niger, based on average annual rainfall, are shown in Fig. 11.1.

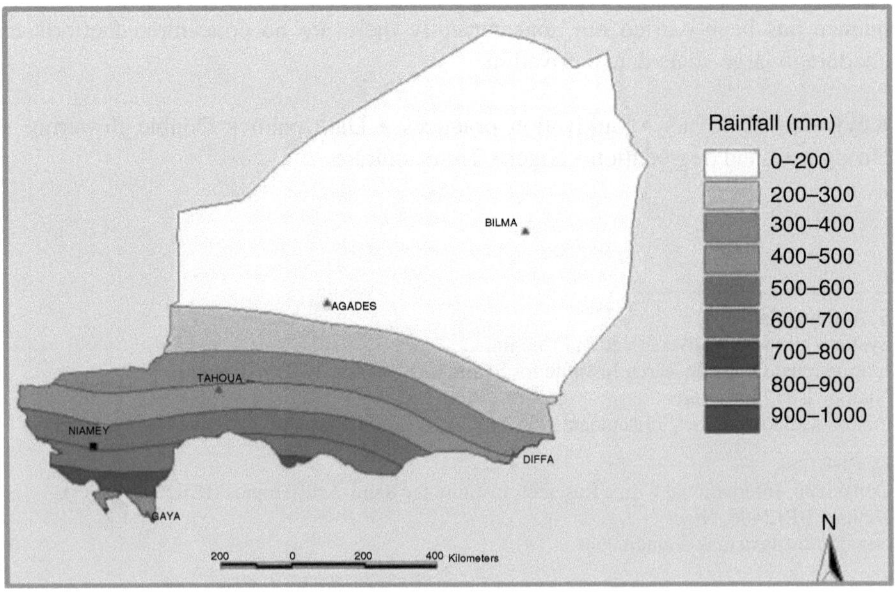

Fig. 11.1 Agroecological zones of Niger based on rainfall (FAO data)

The cultivable area (±6 million ha) represents about 3 % of the total area of Niger. Recurrent droughts in recent decades have accelerated expansion of the desert which increased from 66 to 77 % of the area of the country. Temperatures are high and suitable for date palm cultivation (Cook 1956; Laborie 1999).

Date palm cultivation has huge advantages for Niger and the rest of the Sahel. Date palm is unique in its ability to be cultivated with confidence (in appropriate sites) given the climatic conditions of the Sahel. The date palm can grow and produce on shallow ground water which is abundant, yet hardly utilized, in many areas of the Sahel. Furthermore, the date palm's water requirements can be met by irrigation from shallow aquifers and perennial sources of surface water; therefore, its yield should not be affected by temporary droughts. The date fruit is well known and highly appreciated in all Muslim, sub-Saharan African countries. And last, date palm cultivation in the Sahel will contribute to food security, desertification control, income generation, saving of imports, export earnings, and mitigation of climate change by growing this high-temperature-tolerant species and in general support the development of irrigated agriculture.

This chapter describes the date palm production zones in Niger, the local date palm germplasm, the results of research, and the new interventions by the International Crops Research Institute for the Semi-Arid Tropics (ICRISAT).

11.2 Cultivation Practices

Date palm propagation is carried out by offshoots in the traditional zones of date palm cultivation (Djado-Kawar-Agram and Aïr), while in the marginal zones (Manga and Damagaram), seed propagation is used. Contrary to the circumstances in the marginal zones, good management of date palm is well known by farmers in the traditional zones, but due to lack of labor, it is not regularly carried out. Pruning and manual pollination are practiced only in the traditional zones. In the traditional zones, the harvest is well organized but fruits are dropped to the ground and picked up later, compromising the quality of the dates. In the Aïr region, fruit maturation requires 180 days; flowering begins in January with full maturity in July (Jahiel 1993).

Average annual rainfall in the Aïr Mountains varies from 50 to 150 mm per year. Runoff flows in wide temporary riverbeds creating shallow water tables alongside the rivers. Depth of the water table ranges from 1 to 20 m. These water sources are used to irrigate a number of crops such as corn, forage, millet, potatoes, garlic, tomatoes, and onions. In addition, since the 1980s, improved cultivars of various citrus species have been planted. Delou is the common traditional method of irrigation used in the Aïr Mountains (Fig. 11.2). Animals pulling a rope and pulley raise leather buckets of water from wells. In recent years the Aïr Mountains farmers started producing large quantities of off-season onions which are exported to West African countries in the months of September–November, a period when onion prices are very high there. In many instances, annual crops are intercropped with date palms planted at a spacing of 10×10 to 20×20 m.

Fig. 11.2 Delou, a traditional method of irrigation in the Aïr Mountains

Farmers in the Bilma region have a long tradition of date palm cultivation. They apply all advanced date palm cultivation techniques: manual pollination, bunch and fruit thinning, palm pruning, etc. The palm groves at Bilma are older than those of the Aïr Mountains and are in a state of advanced degradation, probably as a result of the slowdown of economic activity resulting from the diminishing caravan trade. Because of climatic perturbations, many oases are not cultivated and trees are not producing. In addition to dates, the region also produces alfalfa for forage and vegetables for local consumption.

In some locations where the soil is fertile and the water is less saline (such as Fosso), one can find 30–40-year-old date palm trees in reasonably good condition and yielding 20 kg/palm, higher than in other locations. Spacing is 9 m between plants. Date palms, many times mixed with vegetables, are produced around the center of topographic depressions. Millet is the main crop of the region which is mostly a grazing area. In most cases, the date palms reach the water table and do not need to be irrigated. The average annual rainfall in the depression zones is around 400 mm. The dates planted in the depressions originate from seeds. Plantation maintenance is not carried out and therefore about 50 % of the date plants are male. In many instances offshoots are not removed, each palm thus forming a very dense canopy. Modern cultivation techniques are not known and thus not applied. Agricultural extension to introduce standard practices of date palm cultivation would have a very large positive impact on date yields in the depressions.

In the Diffa region, one can observe double flowering of dates during the year. Some 10–25 % of palms will flower in October and again in February. But almost all date palms flower in February–March. Flowering apparently is related to a period of cooler temperatures. Temperatures decrease during the rainy season (July–September), thus inducing flowering in October, and again during

December–January (winter) bringing about a February flowering (Alassane 1988; Jahiel 1989).

Dates in all regions of the country are attacked by white aphids (*Parlatoria blanchardi* Targ.) and spiders (*Oligonychus afrasiaticus* Mac Gr.). Desert locusts (*Schistocerca gregaria*) are a major threat because they attack the female inflorescence. Usually farmers do not take control measures against insects and the damage can be significant. Other date palm pests include birds, monkeys, rodents (rats and mice), termites *(Microcerotermes diversus)*, and ants.

Weeds growing in date palm plantations act as competitors for nutrients and water and serve as hosts for insects and diseases. Studies have established that weeds cause more damage than insects and fungi combined (FAO 2002).

11.3 Genetic Resources and Conservation

Date palm cultivation in Niger is characterized by a lack of technical knowledge and of high-quality germplasm. There is a need to produce fruits of high quality at reasonable prices, to replace the low-yielding date palms, and to facilitate the expansion of date palm cultivation in Niger. An effective extension service in the date-producing regions would go a long way toward the improvement of date cultivation in the country.

11.3.1 Zones of Date Palm Production

Except for some isolated and small groups of date palm trees spread throughout the country, there are two main zones of date palm cultivation in Niger (Fig. 11.3). First, the traditional zones: They are located in the north (Aïr Mountains, Ingal valley) and northeast (Bilma plateau: Djado-Kawar-Agram) of Niger belonging to the Agadez region. Date palm has been cultivated on the Bilma plateau since the twelfth century (Munier 1963). Second, the marginal zones, located in the southeast part of the country and covering the Zinder (Damagaram) and Diffa (Manga) regions. French colonialists first introduced dates from the Aïr Mountains to this latter region in 1917–1918 (Jahiel 1989). Date planting continued throughout the first half of the twentieth century.

The potential zones are located in the south from west to east along perennial rivers (Niger and Komadougou Yobé), Lake Chad, the Dallols (Bosso, Maori) with a shallow water table, and intermittent rivers (Goulbi and Maggia). The system of rivers comprising the Niger River crosses the country in its western part (550 km) and the Komadougou River in its southeastern part (150 km). Adding to these two rivers, intermittent rivers provide important surface water resources estimated to about 30 billion m^3 in an average per year. Less than 1 % of this water is used for irrigation.

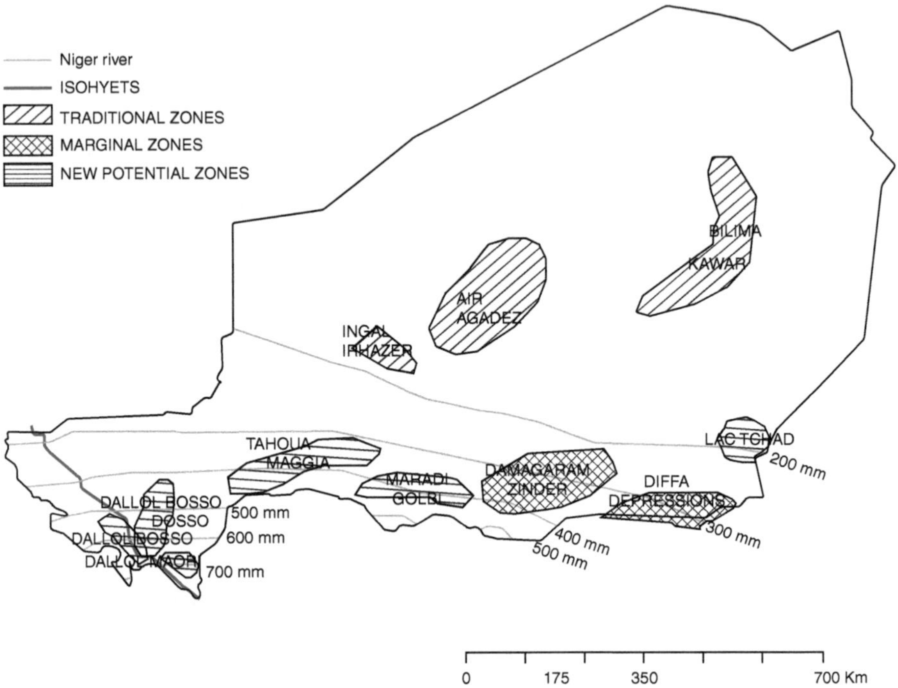

Fig. 11.3 Niger date palm cultivated and potential zones

Renewable ground water resources are about 2.5 billion m^3/year of which less than 20 % are used for human and animal consumption as well as for small-scale irrigation. The most important water resource potential in Niger remains the huge nonrenewable ground water reserves of more than 200 billion m^3 in the north of the country, of which a very small portion is used for mining activities. The major constraint of these resources is accessibility caused by their relatively great depth.

11.3.1.1 Aïr

Situated between 17° and 19° N lat., the Aïr palm groves are located in the Aïr Mountains and Ingal valley (Fig. 11.4). Dates have been cultivated since the sixteenth century (Munier 1963) in oases situated on high plateaus some 1,000–2,500 m above sea level. The number of trees was estimated by Lenormand (1986) to be 50,000–70,000, and the average yield is 11 kg per tree. Since 1986 many new groves of good quality date palms have been planted, but there is no record of the current number of date palms in Aïr.

11.3.1.2 Bilma Plateau

The Bilma plateau is situated between 18° and 22° N lat., east of the Aïr Mountains. The two regions are separated by the sand desert of Ténéré. The

Fig. 11.4 Date palm cultivation in Niger. (**a**) Typical palm groves in the Aïr Mountains: dates are planted over shallow aquifers formed at the banks of large seasonal rivers. (**b**) Ingal valley palm groves: a mix of old and young trees with high planting density

Fig. 11.5 (**a**) Tree with good fruit set, (**b**) mounds of salt

Djado-Kawar-Agram region is a set of oasis in the middle of the Bilma erg, one of the largest in the world. The oases are located under cliff escarpments extending from south to north.

The source of water in Bilma is springs originating from deep aquifers in this region. The ground water table is shallow (less than 4 m deep, usually 1.5 m) rising to the surface in some locations. The climate is arid with an average annual rainfall as low as 15 mm. Evaporation (4,302 mm/year) and daylight (3,700 h/year) are very high. Day to night in temperature amplitude is large. Conditions in the palm groves are, however, relatively moderate because of the microclimate effect created by the palm canopy. In most cases the ground water is slightly saline but the dates can tolerate the salinity.

The Bilma region farmers have a long tradition of date palm cultivation, and it represents the second-ranked economic activity after salt production (Fig. 11.5). Bilma is situated in the path of caravans that cross the Sahara from the Maghreb countries in the north to sub-Saharan Africa. The frequencies of these caravans have diminished during the past 40 years (Lenormand 1989).

Table 11.1 Common date cultivars in Bilma with quality and productivity ranking

Cultivar	Fruit quality rank	Productivity rank
Tilmoiran	1	5
Agous	2	3
Karab	3	6
Gameye	4	1
Tidirchi	5	7
Terseb	6	2
Dilo	7	4

Date cultivation apparently began during the seventh century (Munier 1963) although it was not mentioned until the twelfth century (Jahiel 1993). The number of palm trees is about 250,000–300,000 (Biarnes 1982) of which only 40 % are productive. The average yield per tree is 12 kg.

Lenormand (1989) provides the names of seven date cultivars grown in Bilma and ranks them according to fruit quality and productivity (Table 11.1). Tilmoiran (1) is considered to have the best fruit quality and Dilo (7) the lowest. From our investigations, Karab is not an individual cultivar name but a collective term for all cultivars that are directly consumed after harvest and not commercialized. We found also that names given by Lenormand often are not correctly spelled.

11.3.1.3 The Depressions of Southeast Niger

The soil depressions of southeastern Niger are situated between 13° and 14° N lat., from Zinder along the Nigerian border and reaching all the way to Mainé-Soroa in the Diffa region. The area is a large sandy plateau in which basins are cropped with date palms and vegetables; the basins are extended by low-lying ground. The depressions cover an estimated area of 4,300 km^2. The fertile soil at the bottom of the depressions (Fig. 11.6) and the shallow ground water make this region a very important agricultural resource for Niger. Windstorms throughout the year cause sand dunes to advance and threaten palm groves, gardens, and buildings. The rainy season is short (2–3 months).

The soil depressions of southeastern Niger are an unusual topographic feature. The average area of an individual depression is about 40 ha. and they resemble a volcanic crater. A flat bottom composed of clay is surrounded by a hill of sand, rising 15–30 m above the center. Rainfall over the sandy plateau drains toward the center of the depression and creates a shallow water table. The depth of the water table ranges from 1 to 5 m. The evolution of these depressions is described by Durand and Nicole (1978).

One can identify five distinct sectors surrounding the depressions. These sectors are described in Table 11.2. The number of date palms in the depressions was estimated at 100,000 in Damagaram (Munier 1963) and 200,000 in Manga (Jahiel 1989). The extension of the date palm cultivation beyond its traditional areas in the basin of Mainé-Soroa to the south was attempted but faces sociological, climate, and environmental constraints (Jahiel and Candelier 1991).

Fig. 11.6 Typical soil depression in Damagaram region: valley of date palms mixed with other tree species and surrounded by sand dunes

Table 11.2 Characteristics of ecological sections in a soil depression unit

Section no.	Type of landscape	Soil texture	Depth of water table	Soil cover
1	Live dunes	Sand	>15 m	Sparse vegetation
2	Pastures	Sandy	15 m	Woody and herbaceous plants
3	Top of depression	Sandy	10 m	Rainfed crops and woody plants
4	Middle of depression (deep water)	Sandy loam	3–5 m	Woody plants and horticulture crops
5	Middle of depression (shallow water)	Silt clay	0–4 m	Woody plants and horticulture crops

Source: Jahiel (1989)

11.3.1.4 New Date Production Zones of the Sahelian Region

Annual precipitation in the Sahelian region ranges from 400 to 600 mm/year. The rainy season begins in June and ends in September. Relative humidity during the months of July–August can be very high; this humidity combined with the relatively high temperatures can be harmful to date fruits that start to mature in the month of June. For this reason in the Sahel it is recommended to plant cultivars such as Barhi which can be harvested at the rutab stage, when the fruit color has turned yellow. ICRISAT has been introducing and testing improved tissue culture cultivars in the Sahel since 2001. It constructed a plant hardening-off facility (Fig. 11.6) and has been importing tissue culture dates from laboratories in the UK and in the UAE. To date ICRISAT and FAO have introduced 6,500 tissue culture plants to Niger, 1,200 plants to Mali, 1,500 plants to Burkina Faso, and 1,030 to Senegal. The establishment of a tissue culture laboratory for dates in the Sahel would go a long way toward the promotion of date palms in this region.

11.3.2 Research and Development

Past work on date palm in Niger was initially based mainly on surveys and evaluations of palm groves in all production zones (Gaillard 1988; Girard 1980; Girard and Munier 1980; Lenormand 1986). A project was funded by the European Development Fund with technical support from CIRAD (Centre de Cooperation Internationale en Recherche Agronomique pour le Développement) scientists. Important information on cultivation techniques, number of trees per production zone, production and yield, and genetic resources were provided by the different missions. Recommendations for future research and development work were made for the following regions.

11.3.2.1 Aïr and Ingal

Main constraints to date production in the Aïr Mountains and Ingal valley are fluvial erosion, use of traditional wells, lack of improved cropping techniques, lack of improved cultivars, lack of skills on offshoot propagation, lack of training on fertilization techniques, lack of date palm grove protection against animals, and pest control. Physical and/or biological measures should be undertaken to restore and protect seasonal riverbanks against erosion that reduces areas of date palm plantations. Traditional wells should be replaced by cement ones to provide sustainable sources of irrigation water. Training farmers and technicians on offshoot propagation and replacing old trees by young ones are key issues to expand date palm production areas and improve productivity through the introduction of improved cropping systems. Technicians and farmers should be trained on fertilization (use of manure or compost, mineral fertilizers, and legume forage) and water management techniques. Date plantations should be fenced and mechanisms developed to control white aphids and spiders.

11.3.2.2 Djado-Kawar-Agram

In this region wind erosion is the major agricultural problem as sand is invading date plantations (Fig. 11.7). Bilma is the only region where date palm cropping techniques such as offshoot propagation and manual pollination are known. However, there is a need for capacity building to improve date palm cropping techniques including irrigation and water management techniques, pest control, and

Fig. 11.7 Djado-Kawar-Agram. (**a**) Sand dunes and (**b**) sand invading date palm groves

fertilization: (use of manure, composting, and mineral fertilizers). A means to control white aphids and spiders should be developed. It is important to design a scheme for the development of an irrigated program around boreholes and springs. Lack of roads is a major constraint to date fruit trade shipments to other cities of the country, such as Agadez, Zinder, Maradi, and Niamey. There are a high percentage of old trees in the palm groves. A program for renewing palm groves should be designed.

11.3.2.3 Damagaram and Manga

Introduction of the date palm into this region is recent, compared to the north and northeast zones. Depressions being surrounded by sand dunes, wind erosion is a major constraint to date palm production. Development of date palm production can be enhanced by the following actions: training technicians and farmers on date palm cropping techniques, promotion of oasis cropping systems (intercropping dates with fruits and local food crops (vegetables and cereals)), promotion of improved date palm cultivation techniques, promotion of good water resources management, creation of new palm groves by introducing better-performing cultivars from Aïr and Kawar, and biological control of wind erosion through sand dune stabilization. For these reasons, the Kojemeri research station and the Agadez plant protection center, with an entomology laboratory for biological control of white aphids (*Parlatoria blanchardi* Targ.), were created.

The National Agricultural Research Institute of Niger (INRAN), in collaboration with CIRAD, carried out some research on date palm during the period 1975–1986 at Bonkoukou research station in western Niger and in 1988–1994 at Kojemeri in the southeast (Abdoussalam and Chégou 2000). In Agadez, in collaboration with CIRAD, INRAN carried out research on biological control of white aphids as well as chemical control of spiders. Likewise, ICRISAT has conducted some studies at its regional research station south of Niamey. The results of this research are described below.

11.3.2.4 Bonkoukou Research Station

The aim of research in Bonkoukou was to test the behavior of date palm under Sahelian conditions. Offshoots from Aïr and Bilma in the north and Damagaram in the southeast were introduced. From observations between 1975 and 1986 made on characteristics and behavior of young date palms of different provenance, the following were determined:

(a) The rate of offshoot regeneration for all provenances is better when offshoots are planted during the dry season (January–March) rather than during rainy season. Offshoots weighing 15–25 kg are more successfully established.
(b) Flowering occurred after 4 years and material from Damagaram grew better than that from Aïr.
(c) Double flowering has been observed on plants from Damagaram. This phenomenon, detailed below, is already known from parent materials.
(d) The July production season is the most important. Material from Damagaram is more productive than that from Aïr but fruit does not mature on the tree.

Based on these results, recommendations were made to plant early cultivars and to give more attention to those maturing in March (Haury 1982). However, there was no assessment based on clearly defined descriptors of the diversified date palm material potential in Niger. A decrease in the level of the phreatic aquifer was observed from 1987 to 1992 (Lenormand 1992). Since then, palm trees have not been regularly irrigated and management has been inadequate. Most of the palms propagated from offshoots died and have been replaced by tissue culture plants of international improved cultivars, including Medjool, Barhi, and Boufeggous, provided by an ICRISAT/FAO project.

11.3.2.5 Research at Kojemeri

Southeastern Niger is characterized by a Sahelian climatic type with annual rainfall between 200 and 500 mm. Research on date palm in the Manga depressions was mainly based on propagation methods, flowering, and fruiting cycles with specific attention to the double-flowering phenomenon.

The transfer of date palm in the Sahel led to changes of faster vegetative growth and to flowering twice annually (Jahiel 1996). Date palm flowering and fruiting is known to occur once a year in the traditional date palm production zones. However in the southeastern Niger basin, two flowering cycles were observed: an atypical cycle which starts after the rainy season and a normal cycle starting at the end of the cooler season.

Jahiel and Blay (1994) pointed out that the two annual cycles of flowering and fruiting observed in this area involve distinct groups of axillary buds and that the transition to the stage of flowering is subject to the same conditions of induction. Through dissections on adult date palms, they evaluated the period between the entry into the elongation phase of flower buds and their visual appearance in the canopy to approximately 50 days. This observation coupled with monitoring of weather conditions established a relationship between climatic parameters and the flowering cycles. Research by Jahiel (1996) concluded:

(a) On most trees, two flowering periods per year were observed. The first flowering (September–October to March–April) is less productive than the second (February–March to June–July).
(b) Harmattan or hot desert winds and rainfall might have some effect on date palm flowering cycles in this zone. However it is difficult to conclude that climatic factors alone are responsible for this phenomenon. Genotypic characteristics may also play an important role.

According to Gaillard (1988), the date palm's ability to flower and produce fruit twice a year is influenced by temperature, age of plant, genotype, and environment. Dates in the Manga depressions rarely complete maturation on the tree. The first fruiting gives a low yield (6.2 kg/tree) and good quality of dates, while in the second, the yield is higher (17.8 kg/tree), but due to the high humidity, the quality of the dates is poor (Jahiel 1996). Insect pests such as aphids, spiders, and locusts and predatory birds are not limiting factors to date palm development in the depression zone.

11.3.2.6 Agadez Biological Pest Control Center

Biological pest control consists of reducing and destroying harmful insects by using their natural enemies. Successful results were obtained on white aphids (*Parlatoria blanchardi* Targ.) by using the coccinellid or lady beetle (*Chilocorus bipustulatus* var. *iraniensis*), according to research by the French Research Institute on Fruit Trees (IRFA) in Mauritania (Tourneur 1970). Based on these results, in 1972 INRAN, in collaboration with IRFA, initiated a similar operation in northern Niger (Mounkaila 1981b; Tourneur et al. 1976). An entomology field laboratory was established in Agadez. Mauritanian lady beetles were multiplied and introduced in Al Arses and Ingal palm groves. Direct lady beetle releases were made in 1973 in Aïr, Kawar, Djado, and Fachi palm groves with great success. Despite this initial success, however, permanent establishment of a laboratory for continuous lady beetle multiplication failed.

Spiders are present in all Niger palm groves (INRAN 1983). For example, acarids (*Oligonychus afrasiaticus* Mac Gr.) attack fruits, causing acariosis disease. Immediately after fruit set, spider eggs are deposited to produce larvae which will feed on the fruits and later cover them with a spider web which captures sand particles. Spiders multiply rapidly causing fruit drop. Spider control was based on chemical spraying using a mix of micronized sulfur and clay dust as inert or neutral matter (Herisse 1976; Mounkaila 1981a). Experimental spraying trials were undertaken in Ingal, Tabelot, and Agadez. By weight, the product used contained 14 % micronized sulfur and 86 % of clay dust. Other inert materials which can be mixed with micronized sulfur are talcum, wood ash, or lime. Damage from spiders is severe in neglected plantations.

11.3.2.7 Observations at ICRISAT Research Station

In 2001, ICRISAT planted in its Sadoré regional research station three plots of drip-irrigated Barhi and Medjool dates. The demonstration plot size was 2 ha. Three date palm-based cropping systems were put in place (Fig. 11.8a, b). Planting trials were (a) sole plot, only cv. Medjool at 10×8 m without any intercropping, (b) cv. Barhi at 10×8 m intercropped with vegetables, and (c) cv. Medjool at 10×10 m intercropped with citrus trees.

Both vegetables and fruit trees were also drip irrigated; however, data in relation to yields are not available. It was observed that date palms had a positive effect on the growth of the vegetables. Date palms intercropped with vegetables grew faster than dates without intercropping or than dates intercropped with fruit trees. Apparently dates benefited from increased soil fertility brought about by vegetables production and the additional quantity of water.

Dates were planted in December 2001 and the first flowering of cv. Barhi was observed in 2005. Growth abnormalities, such as dwarfism, characterized by an important growth restriction of date palms derived from tissue culture (Oihabi and De Wet 2006), were observed on Medjools obtained from the PALMDAT Namibia laboratory and were replanted in August 2003 with plants obtained from Al Wathba Marionnet in the UAE.

Fruit yield was recorded only on Barhi cv. as fresh fruit harvested in its rutab stage (Fig. 11.8c, d). Mean fruit yield of Barhi was 35 kg per tree in 2007 and 52 kg in 2008. Medjool cv. fruit never reached maturity under Sadoré conditions.

Intercropping of dates with irrigated crops is a common practice in the Sahara where maximum air temperatures can be as high as 50 °C. This type of oasis agriculture can be an effective technology to combat climate change in the Sahel that will be typified by very high temperatures.

11.4 Plant Tissue Culture

The date palm is a good income-generating crop as well as a source of employment. A rapid increase of production areas and productivity in Niger can be better ensured by tissue culture propagation. As a mass propagation technique, it can provide needed quantities of the high-quality plants. The general commercial upscaling of dates in Niger can ideally be carried out through the use of tissue culture-propagated plants.

Through the dates for the Sahel project, a regional promotion initiative, ICRISAT and FAO introduced international improved date palm tissue culture cultivars into Niger and other Sahelian countries: Burkina Faso, Mali, Senegal, and Cameroon. Plants were obtained from commercial tissue culture laboratories including Date Palm Developments Ltd (DPD) in the UK and Al Wathba Marionnet LLC in the UAE.

Fig. 11.8 Date palm-based cropping systems. (**a**) Dates intercropped with vegetables and citrus trees in the background. (**b**) Dates in pure stand. (**c**) Harvested Barhi fruits. (**d**) Six-year-old Barhi plant with bunch protection

All plants were hardened off at a special facility (Fig. 11.9) constructed at the ICRISAT research station in Niger. In total 6,500 tissue culture plants of improved date palm cultivars were introduced to Niger by ICRISAT and the Technical Cooperation Program of FAO. An attempt by ICRISAT to establish a date tissue culture laboratory in Niger was unsuccessful.

11.5 Cultivar Identification and Description

It is likely that all planting material for the traditional Sahelo-Saharan date palm groves is from the same origins: Libyan and Egyptian oases (Lenormand 1986). Propagation of date palm in these zones started from east to west. Bilma is considered to be the center of date palm cultivation in Niger and the reservoir of all existing cultivars from which others are derived.

Lenormand (1986) based his classification of date palms in Aïr on local names but recognized the difficulty for two reasons: (a) contradictory information from producers and (b) natural and biological mutations or induced mutations. In relation to this classification, there are three main cultivar types based on color: (a) Tazouza ((green color) small green-yellow fruit, good yield, average quality, semisoft fruit), (b) Taorak ((yellow color) elongated fruit, nice and sweet, semisoft), and (c) Zugar ((red color) nice elongated fruit, semisoft and sweet, good conservation).

Fig. 11.9 ICRISAT tissue culture seedlings hardening-off facility

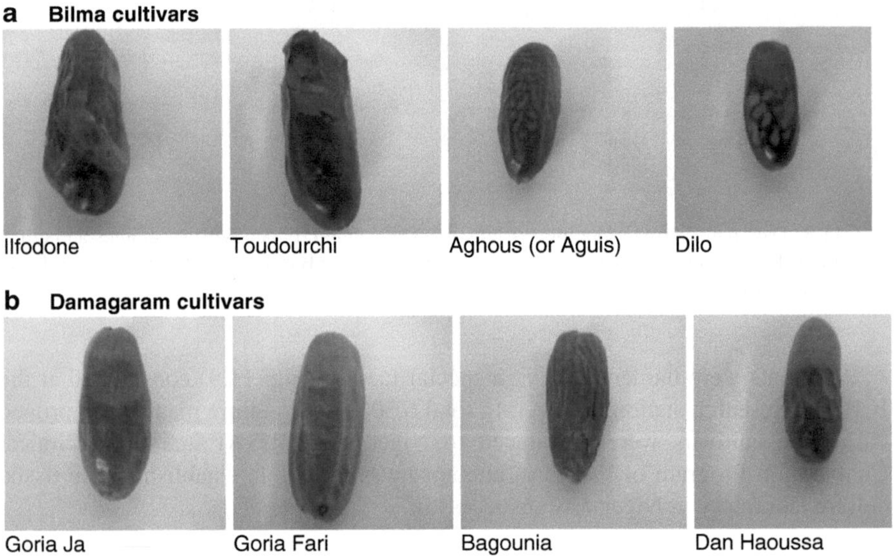

Fig. 11.10 Fruit morphology of some known cultivars. (**a**) Bilma and (**b**) Damagaram

Jelbadane is a cultivar with large soft fruits of good quality but it is not common. Other cultivar names according to localities exist: Guewass, Aria, and Alaka from Taorak; Intie-Sadane, Edjemel, and Taroudoum from Zugar; and Guewass and Racha from Tazouza. Representative date samples of the most common cultivars grown in Niger are shown in Fig. 11.10.

Date palm cultivar identification (Table 11.3) is based on fruit color and characteristics, provenance, and social considerations. Bilma cultivars are not described in detail because of the lack of personal contact in the region.

Names given in this table are not exhaustive, and there is a need for thorough investigations to help identify other local cultivars and to conduct a survey of their characterization based on clearly defined descriptors.

Table 11.3 Date palm cultivars in different regions of Niger

Cultivar	Description	Cultivar name	Description
Aïr Mountains local cultivars			
Touzouwzaw	Strong green color	Tawragh	Common cultivar having a yellow fruit
Tazwagh	Immature fruit is red and turns black at maturity	Guewass	Name of the village
Taghayat	Good quality and easy to conserve	Injar ibadan	Easy to harvest
Tanibott	Name meaning: *without father*		
Ingal valley local cultivars			
Almadeyna group			
Almadeyna	Large fruit, best quality, and most popular	Tanghal	Yellow fruit, black when mature, good quality, and similar to Almadeyna
Madebdebe	Large green fruit, consumed green, sweet, low number of bunches (7–8) compared to other cultivars which can bear 15–20 bunches	Talharma	Fruit are red but become black at maturity
Toumbay group			
Kaanihery	Red fruits	Baghaberha	Yellow fruit with round shape
Atratinna	Black fruit at maturity	Tanzirakh	Fruit with 2 colors (top is yellow and bottom is black)
Touwila	Small fruit	Talittat	Resembles Kaanihery
Teneykogo	Hard when dry	Tassaragrag	Slim
Bossou-Bossou	Gray fruit		
Bilma local cultivars			
Dilo	Semisoft date, fruits are black at maturity, consumed at early maturity stage, poor conservation capacity and not commercialized, very good date for direct consumption, mainly produced in Fachi district. Average sugar content, good pulp/seed ratio (small seed)	Gameye or Ngamaya	Very good quality, specific to Bilma district area, semisoft date, better than Toudourchi but presents same characteristics
Tidirchi or Toudourchi	Oval fruits, semisoft, from green color fruits become violet and then take on a chocolate color at maturity. The cultivar is produced all over Kawar but in small quantity	Aghous or Aguis	Good geographical distribution mainly intended for sale and for preparation of a special local dessert called *arsa*. It is a dry date with good adaptation to climatic conditions, high yield

(continued)

Table 11.3 (continued)

Cultivar	Description	Cultivar name	Description
Karab	A collective name for all delicious dates intended for consumption (Dilo, Hadib, Ilfodone, and Krouss Krouss). Only Aghous is excluded as commercial date	Gilfodon or Ilfodone	Idem Hadib
Tilmoiran or Cliyarom	Mature fruits while green (*cli* means *green* in Kanouri, local dialect), very early date	Akanirom	Early date, very delicious, short period between green and yellow stages. Highly consumed, exceptional taste, fruits are elongated oval shaped. Mainly intended for subsistence consumption and not commercialized
Krouskrous	Identical to Hadib, round fruit, hard when dry. Good cultivar but not common	Hadib	Intended for consumption, round fruits, looks like grapes when dry
Depression zone cultivars			
Bagounia	Red fruit	Goria Ja	Red large fruit, sweet
Dan Haoussa	Small fruit, low quality	Goria Fari	White large fruit, sweet
Maiwa	Very sweet green fruit	Koukouma	Yellow fruit, not sweet
ICRISAT-introduced varieties			
Barhi	From Iraq, good appearance, pale brown color, semisoft fruits, and early cultivar. Fruit small to medium, ovate to nearly round, golden yellow. Hard ripe doka (khalal) fruits very sweet and suitable for eating raw	Zamli	From Saudi Arabia. Very high-quality date palm cultivar. Good alternative to Barhi. Good adaptation to severe environmental conditions (humid regions and it endures high temperatures and hot summer storms). Tree is strong and gives very high yield. Fruits are golden yellow and large. It could be consumed in rutab, bisr, and tamar stages. It is a soft cultivar. Fruits are very sweet in fresh hard yellow stage (bisr). Zamli rutabs have an excellent and long shelf life

Cultivar	Description	Cultivar name	Description
Medjool	Origin from Morocco in Tafilalet and Ziz regions. Excellent appearance, brown color, semisoft date, late-maturing cultivar. A late-ripening cultivar, it is suitable for preparation of dry dates. The fruit is large and broadly oblong-oval to somewhat ovate, orange yellow with a fine reddish-brown stippling and highly astringent at doka stage	Um Ed Dahan	An Iraqi cultivar, mid-season fruiting, bright yellow fresh fruits, size is medium oblong shape. Resistant to high humidity. Has great resemblance to the Tunisian Deglet Noor but much softer in texture. Could be consumed in all three stages of ripening: bisr (balah), rutab, and tamar. The date is characterized by its soft skin and the complete absence of fiber
Khalas	A mid-season cultivar. Fruit small to medium, oblong-oval, yellow and sweet at doka stage, has an oblique base and irregular outline. It is suitable for eating fresh and for processing as soft dates	Dayri	Derrie or Dayri, the monastery date, from southern Iraq; these are long, slender, nearly black, and soft
Shishi	A renowned Saudi cultivar. Mid-season ripening. Greenish color that turns brown while ripening. Large fruits, high productivity. Soft in texture. Dates are characterized by their oblong shape, large size, greenish color, and soft skin. It has an excellent fruit setting rate, very good adaptation in wide range of climates. Widely planted in the Gulf Region and Iran and considered among the high-priced dates	Saggai	A well-known Saudi cultivar. One of the most important dates in the Gulf Region. Tree is strong with high productivity. Fruits are large elongated. Saggai rutab is golden yellow in color. Mostly consumed as tamar (fully ripened dry fruits). It is a semidry cultivar. Fruits are characterized by the ivory ring at the base. Saggai has excellent shelf life for storing in tamar stage
Ashaal	Origin: Saudi Arabia. Red yellow, khalal stage. Round/oval shape, small to medium fruit size. 40–60 kg per palm annually. Mid- to late ripening. Eaten soft	Nemeilshi	Yellow khalal, oval shape, 40–60 kg per palm annually. Mid-season ripening. Eaten either soft or dried

(continued)

Table 11.3 (continued)

Cultivar	Description	Cultivar name	Description
Khadrawy	A comparatively dwarf and mid-season cultivar. Fruit matures slightly later than Halawy. Small to medium, oblong-ovate, and greenish yellow at doka stage	Boufeggous	From Morocco (all regions), good appearance, brown color, soft date, early season maturity
Male pollinizers			
Jarvis	Early flowering. Heavy pollen yield	Fard	From Oman
Ghannami			

11.6 Date Production and Marketing

Annual date fruit production in Niger is about 8,000 mt, from a total of 720,000 date palm trees (Jahiel 1996). The average fruit yield is 10 kg/plant, only about 10 % of potential. Date exports are estimated to be 15,000 mt and imports 20,000 mt. More than 90 % of imports originate from Algeria. Expenditures relating to date imports represent approximately 2 % of total agricultural imports in Niger. In the Sahelian region, low- to middle-quality cultivars have a market value in excess of USD 1 per kg. Thus, one productive date palm can potentially generate about USD 50 or more, solely from fruit sales. Dates are also exported to Nigeria where prices are USD 1.6/kg. The Niger date industry suffers from competition by the Maghreb countries that export low-quality dry dates particularly, particularly during Ramadan. Imports from Algeria compete with dates produced in Bilma and in the Aïr region. Competition from Maghreb counties can be counteracted by the cultivation of high-quality cultivars. However, there is no competition to the Barhi fresh dates that are sold in July.

There is no organized marketing of dates in Niger, resulting in lower prices for the fruit, and no data are collected on the quantity sold or of the prices received. There are no effective sorting and packaging facilities.

An effective extension service for date cultivation will result in increased production and productivity and in improved postharvest handling of the fruit. Fruit quality is good but yield is very low in Bilma (Jahiel 1993). Fruits are consumed locally or sold to markets in other parts of the country (west and southeast) as well as in Nigeria. With regard to dates produced in Niger, Girard and Munier (1980) stated that:

(a) Ingal dates are of excellent quality and have a good reputation in the country. Fruits are semisoft but the conservation period is short. They are sold in traditionally made small containers woven from date palm leaves. A 250 g quantity sells for USD 0.5; sold in bulk 1 kg fetches USD 1.25.
(b) Dry dates in Aïr are pounded with millet or cheese. Soft fruits are consumed as they are.

Fig. 11.11 Date market in southeastern Niger. (**a**) Damagaram depressions local dates in Droum village market: fruit are immature and quality is very low. (**b**, **c**) Dry dates from north traditional zones sold in the southeast at Droum village market

(c) Dates in Bilma are of excellent quality. Because of limited trade, prices are very low: USD 0.05–0.15 per container of less than 1 kg. Transportation costs from Bilma to Agadez are about USD 4 per 100 kg bag. Dates produced in Bilma are usually traded for millet, sorghum, or wheat.

Dates in the depression zone are a precious source of cash (Abdoussalam and Chégou 2000). Income from date sales is used to buy millet which is the most important food item. Usually mature dates are not sold in local markets or in Nigerian markets (Fig. 11.11a). Selling is done in different ways: on the tree, in sacks, or in bulk. In any case the producer must sell quickly in order to avoid high loss. Dry dates from the north traditional zones are also sold in southeast village markets (Fig. 11.11b, c).

11.7 Processing and Novel Products

In zones of established date palm cultivation (e.g., north and northeast zones), dates and by-products have diverse uses (Girard and Munier 1980). Date palm tree products are used to build tents (local houses) and make woven tools such as rope, fans, baskets, food tools, mats, and rope made from the petiole; the petiole also is used for building, as fences, and as firewood and the seeds as animal feed as the nutritive value of date palm seeds is equivalent to barley forage. Most of these uses have not spread in the southeast zone.

Dates contribute to enhancing the income of farmers. During the harvesting period, in Ingal, for example, all fruits are sold to traders coming from Agadez and Niamey (Luxereau 2002). It is believed that at harvest time producers are wealthier than the merchants.

In the depression zones of northeast, dates are sold on the tree to traders coming from Nigeria. The lack of good organization of a date value chain is a problem all over the country.

11.8 Conclusions and Recommendations

Date palm cultivation could become a major source of income and provide food security for Niger. The hot and dry climate is suitable for date palm production. There is a very large internal market for dates and a huge potential regional market. Dates are an ideal food security product providing a great source of energy. Dates have more than 3,000 cal/kg compared to cooked rice with 1,800 cal/kg. The average yield of a date palm in Niger in its three traditional production regions is only 10 kg/tree, only 10 % of the potential. There is an opportunity to increase yield through creation of an effective extension service and the introduction of high-quality, high-yielding cultivars.

Niger has large underutilized water resources that could be used for date palm irrigation. Research should be carried out to identify cultivars producing fruit in the dry season for the Sahelian region of Niger. Intercropping dates with vegetables and other irrigated crops will provide a solution for anticipated climate change characterized by high temperatures (the so-called oasis agriculture), since date palms markedly improve the microclimate for vegetable growing. Establishment of a tissue culture laboratory for date palm plant production for Niger and the rest of the Sahel will increase the accessibility of growers to superior plant material and reduce the prices of currently imported tissue culture material. The proliferation of date palm cultivation in Niger will be achieved only after the government takes responsibility for this undertaking.

References

Abdoussalam S, Chégou M (2000) Date palm Cultivation in Niger. In: Proceedings international symposium on date palm. Windhoek, Namibia, pp 281–287

Alassane I (1988) The double flowering pattern of date palms in southeast Niger. A report submitted to the Ministry of Agriculture and Animal Husbandry of Niger. INRAN Library, Niamey, Niger, 11 p. In French

Biarnes A (1982) Systèmes de production et de commercialisation des oasis du Kawar. Mémoire 3ème Cycle Agronomie, Paris-Grignon

Cook RE (1956) A study of the relationship of heat units to the ripening times of dates. Date Grow Inst Ann Rpt 33:13

Durand N, Nicole M (1978) Rapport de mission au Sud-Est du Niger. Soumis au Ministère de l'agriculture du Niger

FAO (2002) Date palm cultivation, FAO Plant production and protection. Paper no. 156, Rev.1, FAO, Rome

Gaillard JP (1988) Compte rendu de mission au Niger, CIRAD, Département "Fruitiers" IRFA

Girard F (1980) Les palmeraies et la phoeniciculture dans les massifs de l'Aïr (Nord Niger). Rev Fruits 35(6):383–391

Girard F, Munier P (1980). Mission Palmeraies en République du Niger, CIRAD, Département Fruitiers, IRFA

Haury M (1982) Etude de comportement du palmier dattier au Niger: Bonkoukou, Dallol-Bosso. Rev Fruits 37(10):627–633

Herisse D (1976) Programme générale de lutte contre *Oligonychus afrasisticus*. Centre de lutte biologique d'Agadez. Documentation INRAN

Institut National de Recherche Agronomique du Niger INRAN (1983) Fiches techniques de vulgarisation: Actualisation des résultats de recherches, 2nd edn, INRAN Library, Niamey, Niger

Jahiel M (1989) Intérêt et particularités du palmier dattier dans les zones en cours de désertifications: Exemple du Sud-est du Niger. Mémoire de DEA en Botanique Tropicale Appliquée. Institut de botanique appliquée de Montpellier, INRAN Library, Niamey, Niger

Jahiel M (1993) Effet des récentes perturbations climatiques sur la phoeniciculture dans le sud-est du Niger. Rev Sècher 4(1):7–16

Jahiel M (1996) Phénologie d'un arbre méditerranéen acclimaté en région tropicale: le dattier au Sud du Niger et son appropriation par la société Manga. Thèse de Doctorat, Montpellier II

Jahiel M, Blay JC (1994) La double floraison du palmier dattier dans le Sud-est du Niger. http://publications.cirad.fr/auteur.php?mat=2178

Jahiel M, Candelier P (1991) Le palmier dattier dans le Sud-Est du Niger (cas des cuvettes de Maïné-Soroa). Fruits 46(1):83–92

Laborie A (1999) Réalisation d'une cartographie des sites d'implantation favorables à la culture du palmier dattier au Niger – Projet "Promotion de la culture du palmier dattier au Sahel", Desert Margins Program, ICRISAT-Niamey, Niger

Lenormand C (1986) Mission de reconnaissance des palmeraies du Nord Niger: Identification de la production dattiere (Projet FED "Vallées de l'Aïr" No 51300-33-40-34, INRAN, Section Recherches Fruitières, DRA Niamey

Lenormand C (1989) Reconnaissance des Palmeraies du Nord Niger Phase II. Projet F. E. D. "vallées de l'Aïr No. 5100.33.40.034

Lenormand C (1992) Section Recherches Fruitières: Synthèse des activités 1988–1989–1990 et suggestions de programmes 1991–1992. INRAN Library, Niamey, Niger

Luxereau, A (2002) Les dates d'In Galle, un produit patrimoine. http://www.ingall-niger.org/index.php/flore/199-les-dattes-d-in-gall-un-produit-patrimoine

Mounkaila MM (1981a) Rapport de synthèse des résultats de recherches du programme de lutte contre l'Acariose du palmier dattier. Programme proposé à la vulgarisation. Documentation INRAN

Mounkaila MM (1981b) Rapport de synthèse sur la lutte biologique contre la cochenille blanche du palmier dattier au Niger. Documentation INRAN

Munier P (1963) Prospection phoenicicole du territoire de la République du Niger. IFAC, Paris

Oihabi A, de Wet PF (2006) Date palm abnormalities in Namibia and Niger. In: Proceedings of the international workshop on true-to-typeness of date palm tissue culture-derived plants, Morocco, 23–25 May 2005

Tourneur JC (1970) L'utilisation des coccinelles prédatrices en lutte biologique. Rev Fruits 25(2):97–107

Tourneur JC, Lenormand C, Mounkaila MM et al (1976) Intervention écologique au Niger destinée à lutter contre la cochenille blanche du palmier dattier par l'introduction de *Chilocorus bipustulatus*. Rev Fruits 31(12):763–773

Chapter 12
Date Palm Status and Perspective in Cameroon

Sali Bourou, Noé Woin, and Mohammed Aziz Elhoumaizi

Abstract Date palm tree is an economic crop which is grown in the Sudan Savannah and Sahel regions of North Cameroon. Its role in food production, foreign exchange earnings, raw materials for industries, and income and employment generation makes it a crucial asset for national economic development. Yet, Cameroon does not figure among the date-producing nations in Central Africa and the world. World date production is led by Egypt with 1,350,000 mt, followed by Iran with 1,088,040 mt, and Saudi Arabia with 1,052,400 mt. Data available for 2009 placed Niger as the highest producer in West Africa with 37,794 mt. No data exist on date palm production in Central Africa. Actions have been taken to increase date palm production in Cameroon. A number of elite cultivars were introduced in 2005–2006, thanks to a technical cooperation program with the FAO. Seven well-known cultivars (Ghanami, Khalas, Khadrawi, Medjool, Barhi, Ambarah, and Zahidi) have been introduced in rural areas. At present, the Cameroonian government, through its agricultural diversification policy, encourages the production of date palm by making available improved planting material, training, and the creation of an actual production sector. These will likely result in the future to the classification of Cameroon among date palm-producing countries.

Keywords Agriculture • Cameroon • Date palm • Diversification • Extension • Local cultivars

S. Bourou (✉) • N. Woin
Fruits Plants Program, Agricultural Research Institute of Cameroon, Garoua Station, P.O. Box 415, Garoua, North Cameroon
e-mail: sali.bourou@gmail.com; woin57@yahoo.com

M.A. Elhoumaizi
Department of Biology, Faculty of Sciences, University of Oujda, P.O. Box 717, Oujda, Morroco
e-mail: elhoumaizi@yahoo.fr

J.M. Al-Khayri et al. (eds.), *Date Palm Genetic Resources and Utilization: Volume 1: Africa and the Americas*, DOI 10.1007/978-94-017-9694-1_12

12.1 Introduction

The socioeconomic life of the entire area of the African savannah depends upon the production of annual food crops and cotton. But in the hot and humid tropical zones, cultivation by plowing necessarily leads to the mineralization of organic matter and soil degradation after a period of 15–30 years, affecting the soil's texture, fertility, and management system (Boli et al. 1991).

As recently as 40 years ago, when farmers felt that the soils were depleted, they abandoned their farmed sites for a long-fallow period (10–50 years), and the village moved to clear new land; this slash and burn shifting cultivation system was the main important cultivation system (Boli et al. 1997).

Now, shifting agriculture is no longer acceptable in some parts of Africa (Roose 1994) because of the low proportion of suitable farmland (often <30 %), pressure on land (we have reached the limit of local farmland), population pressure (the population doubles every 25 years), and the development of permanent village structures like tracks, mills, wells, churches, clinics, and sheds for agricultural equipment and supplies.

It is in this context that agricultural diversification policy was deemed important to boost agricultural production and improve the income of the rural population. Agriculture in northern Cameroon is not generally very productive (Boli et al. 1997); yields of major crops remain low and the majority of farmers live below the poverty line. However, we cannot say that these producers have remained inactive. They showed, on the contrary, great innovation by highlighting new land by reclamation, using more intensive techniques, adopting new crops, implementing the original institutional arrangements, and inserting themselves into the regional economy.

If the diversification of agricultural production is primarily based on annual crops, the current systems are not sustainable. Nevertheless, in northern Cameroon, the introduction of several annual cash crops, including soybeans (*Glycine max*) and sunflower (*Helianthus annuus*), has taken place. However, to enhance the diversification initiative, fruit production, such as the production of mango, guava, lemon, etc., which have a strong socioeconomic importance. The production of date palm in northern Cameroon remains little known, but it has great potential importance. It is within this context that the Government of Cameroon, with the support of its bilateral and multilateral partners, is committed to boost date palm production in northern Cameroon and consider the date palm as a cash crop. This chapter focuses on promoting date palm in North Cameroon and also to identifying the main strategies for date palm production. Very little general information or economic data exist on date palm in Cameroon. However, in sub-Saharan African areas such as Cameroon, date palm is expected to contribute to the goal of agricultural diversification. Realistically, we think that by 2035, Cameroon could expect date palm to be among the top five major crops in the Sahelian zone of the country.

12.1.1 Overview of Cameroon

The Republic of Cameroon is located in Central Africa; it extends from the Gulf of Guinea to Lake Chad, between 2° and 13° north latitude and 8°30′ and 16°10′ east longitude. The country has a surface area of 475,650 km^2 with a long coastline of 402 km. Triangular in shape, its length from north to south is 1,400 km and from east to west is about 800 km. It is bordered to the south by Equatorial Guinea, Gabon, and Congo, to the west by the Atlantic Ocean and Nigeria, to the north by Lake Chad, and to the east by Chad and the Central African Republic. Cameroon has many physical assets, including considerable hydroelectric potential, an impressive variety of agricultural activities, and important mineral resources such as bauxite and cobalt. It is also provided with potential natural areas that remain to be developed (Atlas de l'Afrique 2000).

The foundation of current Cameroonian agriculture originates from the German colonial era of the late nineteenth century. The Germans chose to develop crops used as raw materials for European industries. At that time, opportunities were provided for these promising products. Farm work was poorly paid and the use of forced labor common, but commercial agriculture led to the emergence of the first railroads, ports, and other infrastructure. The first crops developed were rubber, oil palm, coffee, and cocoa. Production areas, where the first settlers planted these crops, began in the coastal zone, progressing towards the interior. Agriculture was marked by two types of operations: large plantations established by settlers using hired labor and small family farms. The first large plantations were established in the south and west along the Atlantic Coast.

Infrastructure was developed around the first economic centers. The initial goal was to export the production. There was a dominant trend in the development of some crops to supply Western markets. The subsequent French colonization reaffirmed this belief and added new crops. Thus cotton was grown in the northern part of the country, an area also favorable to growing date palm. At the time of independence in 1960, Cameroon was a prosperous country with a diversified agriculture which was export and/or industrially oriented.

The period since independence was one of strong economic growth. Based on the results of agriculture due to the nationalization of major industries and major currency gains realized, the country launched the development of other sectors with the state as the main actor. National strategic choices focused on the development of import substitution industries and the establishment of large parastatals to compensate for the absence of a private sector. The emphasis was on cash crops inherited from the colonial era with the encouragement of smallholdings by improving public services. Despite these efforts, agricultural production remained stagnant except for sugarcane whose development expanded rapidly after the creation of the CAMSUCO (Cameroon Sugar Company).

The progress which took place with food crops was somewhat different. These crops do not directly benefit from government actions. Development efforts appear to be related to population growth and the goal of food self-sufficiency. The crops receiving attention are extremely varied, but the major crop groups consist of carbohydrates (i.e., roots, tubers, and plantains), cereals, and oilseeds. Although there is an apparent increase in production, there is a marked deterioration in output per capita. This is due to lower yields related to soil depletion, inadequate production techniques, aging plantations of perennial crops, aging farmers, and the rural to urban exodus (MINADER 2008).

The high price of cash crops has long hidden the structural weaknesses of Cameroonian agriculture. In the early 1980s, a drastic fall in prices of major raw materials plunged the country into a deep recession that halted the reforms. This manifested itself in the suspension of implementation of the 5-year plans and a return to austerity policies with structural adjustment. In agriculture, specific policies had been developed for the advancement in support of a partnership policy. A liberalized environment involved required divestitures and accountability of actors who were then called upon to organize and take over management of the activities.

Crop diversification is a new model to overcome the uncertainty of farm income related to the volatility of exports and to ensure an acceptable income for producers. Administrative structures have also evolved to have limited missions when they do not disappear completely.

Unofficially and informally, the introduction of date palm goes back hundreds of years to the days of slavery. However, Cameroon only became actively interested in the date palm from the 2000s. In 2006 Cameroon introduced the first elite date palms, through an FAO technical cooperative program. It involved importation of seven cultivars from the UAE: Ghanami, Khalas, Khadrawi, Medjool, Barhi, Ambarah, and Zahidi. The main objective was food security, improved food quality, and also in response to drought.

12.1.2 Importance of Agriculture to Cameroon

Cameroon's economy is quite diversified. It is mainly based on the agriculture and forestry and petroleum sectors (MINADER 2008). Agriculture and forestry are key sectors of the national economy and account for 42 % of gross domestic product (GDP). This justifies the importance of reviving agricultural research as concerns cacao, coffee, cotton, bananas, and rubber. This cash-crop farming contributes 40 % to agricultural exports and provides 72 % of food products (MINADER 2008). The agricultural contributions are important to the development of the rural sector and to poverty reduction through better integration of agricultural research (Atlas de l'Afrique 2000).

Cameroon has considerable agricultural potential, both in terms of food production and export crops. In the southern part of the country, family farms exist, devoted to food production (tubers, plantains, maize) and cash crops (cacao, coffee) in a mixture of small farms, along with large industrial plantations specialized in oil palm, rubber, or bananas. Agricultural systems in the northern part of Cameroon combine cereal crops (sorghum, maize), vegetables, and cash crops (cotton, in particular). Cameroon is also a livestock-raising country. Cattle are herded predominantly in the Adamawa Region. In the west and around urban centers, poultry and pig production has increased considerably (Atlas de l'Afrique 2000).

Cameroon has a diverse agricultural sector as a result of its various agroecological zones. For this reason, there is a wide range of crops and livestock species. The importance of the agricultural sector in Cameroon's economy is quite large. Despite a gradual decline in importance since political independence in 1960, agriculture has continued to play a key role in the national economy as far as GDP of the rural sector is concerned. Indeed, in 2005, the agricultural sector contributed 20 % to the national GDP with an annual growth rate of 4.1 %, as compared to 3.4 % growth reported for the rest of the economy (MINADER 2008). Agriculture provides employment to more than 60 % of the active population and has the lowest unemployment rate. In terms of export taxes, the rural sector contributed 54.5 % in 2003. The sector is also important in terms of tax collection. The importance of the rural sector in the PRSPs (Poverty Reduction Strategy Papers) is pronounced. The part of the budget used for food consumption is very high at the household level; this percentage is 43 % for all households and 53.4 % in poor households. Within the African subregion, the Cameroonian agriculture plays a central role. A large proportion of the Cameroonian production is in fact exported to neighboring countries (Atlas de l'Afrique 2000; MINADER 2008). Since independence, the rural sector has played a leading role in the national economy. The golden age of the 1970s was built on the strength of cash crops, such as coffee, cacao, banana, tea, and cotton, and their rising world prices. Agriculture still occupies almost 75 % of Cameroon's population and urban centers depend on it for the supply of food products (MINADER 2008).

However, the development of Cameroon's agriculture faces many obstacles. These include farms suffering from poor access to credit for both investment and inputs, like fertilizers and pesticides, and a degraded natural environment. The isolation of some production areas makes market access difficult. Despite some progress, related professional organizations in the country still lack structure. In this context, despite the great potential of its agriculture, Cameroon is still importing food. Rice and wheat are the largest agricultural imports by volume. The production of cane sugar and vegetable oils is insufficient, which requires the country to fulfill its requirements on the international market. Despite the existence of an expanding domestic market, which is an immediate outlet for domestic production, food crop production continues to decline because of poor cultivation practices and soil degradation.

In Cameroon, the date palm can play a role in strengthening food security, providing environmental protection, and generating income for farmers and to improve the international balance of payments by reducing imports and providing potential exports. However, the promotion of the date palm cultivation in Cameroon requires that a national strategy be developed.

12.1.3 Cameroon Environmental Conditions

Cameroon is divided into four primary agroecological zones: Sudano-Sahelian, high-Guinean savannas, highland, and rainforest; the latter can be further subdivided into areas of monomodal and bimodal rainfall. Within each agroecological zone, several cropping systems are being practiced.

Fruit tree cultivation is practical and appropriate to all the agroecological zones; however, some fruit crops are specific to certain of them. These include coconut in the humid forest zone and date palm in the Sudano-Sahelian zone.

Rainfall varies from one area to another within the country. Annual precipitation reaches 4,000 mm in the coastal area but amounts to only about 300–500 mm in the area of Kousseri in the Far North. However, in North Cameroon where natural stands of date palm are found, there is an annual rainfall of 300–1,000 mm per year.

Depending on the climate, there are areas of high temperatures (30–35 °C) and high relative humidity (>80) prevailing throughout the year (the coastal zone) but also areas with average annual temperatures of 20–25 °C, as in the case of the forest area in the center of Cameroon. In the northern zone, the annual range of temperature is characterized by two periods: one of high temperatures (40–45 °C) from late February to late June and another of dry cold conditions marked by low temperatures (18–25 °C) from late November to January. During the remainder of the year, the temperatures are stable with an average around 30 °C.

Physical characteristics of the land in Cameroon can be classified into: (a) highland regions such as the Mandara in the north, the Adamawa Plateau in the center, and the highlands to the west which extend southwestward to the Atlantic Ocean in a mountain chain with the highest peak of Mount Cameroon at 4,070 m and (b) plains mainly in the coastal area and extending inland to include the plains of Chad, Diamare, and Benue.

There are many soil types with the lateritic clays most common in southern Cameroon. In the north, mainly ferruginous tropical soils are found. These soils characterize the tropical areas (1,000 mm/year) with a pronounced dry season with rainfall for a period of 7–8 months and include the Sahelian and Sudanian zones. In northern Cameroon, tropical ferruginous soils represent about 1.9 million ha and constitute 60 % of cultivated land (Boli et al. 1997; Roose 1994). They are characterized by low clay content, significant leaching, and undeveloped surface structure, with a sandy clay horizon at depth. The organic matter content is low at the surface (1–2.5 %), decreasing rapidly with depth. They are acidic, with a pH of 5–6, and possess a low cation exchange capacity (CEC = 5–10 meq/100 g). The available water holding capacity (AWHC) is low (73 mm of water/80 cm of soil) (Roose 1994). These tropical ferruginous soils serve for date palm cultivation. Indeed, a good development of date palm already exists on these soils in the north. Production systems involve intercropping with cotton, peanuts, maize, soybeans, and cowpeas.

12.1.4 Evolution of Scientific Research in Cameroon

The Institute of Agricultural Research for Development (IRAD) is the research institution through which the Government of Cameroon has given new impetus to date palm cultivation. IRAD is in charge of testing production systems based on date palm cultivation and suitable cultivars. The complex history of IRAD needs to be explained in summary form.

Even before Cameroon's independence, scientific and technical research activities were undertaken by certain colonial institutions. In 1935, the Cameroonian Society Studies (SECAM) was created with the mission to investigate all matters relating to the social and human sciences, geology, ecology, and hydrology. The scientific work of SECAM led to the publication of the first scientific journal in Cameroon, The Bulletin of the Cameroonian Society Studies.

In the early years of independence, Cameroonian authorities expressed the political objective to make science and technology an engine of the development economy. The Federal University of Cameroon was founded in 1962, followed a year later by the Office of the National Scientific and Technical Research (ONAREST). Continued consolidation of new governmental systems led to the absorption of ONAREST into the General Delegation for Scientific and Technical Research (DGRST) in 1979 and the subsequent replacement of DGRST in 1984 by the new Ministry of Higher Education and Scientific Research (MESRES). The Ministry's Department of Scientific Affairs was given authority over the country's research institutes, including the Agricultural Research Institute (IRA) and the Institute of Animal Research (IRZ). The objective of the Department was to provide a collaborative framework allowing for the development of a truly national science and technology policy and to monitor its implementation through the Board of Higher Education and Scientific Research and Technology. However, activities were curtailed sharply during the economic recession in the period 1986–1997. As a result of decreased research funding, there was a decline of national scientific activity, and this was compounded by the cessation of publication of the journal Science and Technology, suspension of recruitment of researchers, and the reduction of nearly 50 % in the salaries of existing researchers. These circumstances led to a major demobilization of researchers resulting in a loss of credibility of the National System of Research, both domestically and internationally. It was not until a decade after the economic recession that the agriculture sector was given needed attention, with creation of the Institute of Agricultural Research for Development (IRAD), in 1996. IRAD's mission was to ensure the implementation of research in rural areas by assuming responsibility for the activities of the former IRA and the IRZ.

12.2 Cultivation Practices

Several problems hinder farming systems in the region of North Cameroon. We thus note soil degradation, variable rainfall, and also the socioeconomic context of the environment. We will highlight the soil and climatic parameters (Bourou et al. 2006).

In the savannah of northern Cameroon, where date palm is cultivated, intensive cropping of cotton/maize rotation in sandy areas has contributed to soil degradation over the past 15 years. However, shifting cultivation is no longer acceptable because of recent permanent village structures (houses, roads, boring wells, etc.) and the limited areas suitable for such agricultural activities. Research has been carried out to assess erosion hazards, factors of soil degradation, and restoration and to select cultural practices allowing intensive and sustainable production, without degrading the rural environment (Boli et al. 1991, 1997). Because of these reasons, a mixed cropping system was advised using date palm with some annual crops such as cotton, maize, cowpea, peanuts, and even millets or sorghum.

Over a period of 4 years, conventional tillage gave the best yields, but water and soil losses (loss in organic matter, clay, and loam in suspension) were critical and explained the rapid soil degradation. Reduced tillage in plantations under light litter reduced water and soil losses down to these of savanna. But during high-rainfall years, maize yields decreased 20–40 %; in response to increased nutrient leaching, 20 kg ha^{-1} of nitrogen must be added after heavy rainfall weeks (Boli et al. 1997). Scouring 50 mm of topsoil is involved in 30 % reduction of maize production. But after 4 years of cropping on runoff plots, it was observed that a difference of 4 mm of sheet erosion involved a 40 % reduction of maize yield.

Tropical soil restoration is very difficult because of the low storage capacity of kaolinite clay, the high rate of organic matter mineralization, and significant losses through erosion and leaching. After 5 years, analytical characteristics of the topsoil were not improved. But within 3 years, a high level of production was obtained on degraded sandy soils considered suitable for date palm cultivation in this region.

The combination of trees and annual crops is very old in agricultural landscapes. Today, new forms of agroforestry are emerging to meet the constraints of current agricultural systems. The main changes compared to traditional agroforestry are the choice of species, the arrangement of trees, and their density. Introduced agroforestry based on date palms increases the value of the operation, without significant reduction in farm income. Date palms in agroforestry also allow the operator to offer a different picture of the farming profession by having a positive visible impact on the landscape. Agroforestry systems produce more biomass per hectare, particularly compared to rotation, where there are separate trees and agricultural crops. It helps to improve the natural soil fertility and thus offers the possibility of reducing the need for inputs and preserves the soil against erosion and protects groundwater.

Cultural practices are characterized by the use of appropriate tools to increase labor productivity while maintaining soil quality. In this context it is essential to use, within different ecological zones of the country, the best-adapted agricultural tools. Such is the case of the hand planter, a manual tool able to sow one hectare by hand. The establishment of a process of production and maintenance of these types of tools at the village level is needed to ensure the sustainability of their use. The development of appropriate farming techniques, suitable extension services, making available farming equipment, and support for the creation of centers for multiplication of planting material are all needed (PNUD 1996).

The fight against crop pests is a major concern to the authorities in the countries of the region. Due to the high cost of chemical pesticides and the human and environmental risks their use entails, each country has recognized the need at a regional level. Indeed, the problem of crop protection can only be resolved through cooperation between, for example, member countries of the Lake Chad Basin (LCB). To do this, the LCB has opted for Integrated Pest Management (IPM), which favors non-chemical methods, protects the environment, and is very inexpensive for the farmer. It combines the use of resistant cultivars, agronomic techniques, biological control, effective phytosanitary measures, and direct participation of the farmers.

12.3 Genetic Resources and Conservation

Biodiversity of the date palm refers to variation within the species, while genetic diversity represents the heritable variation that can be found within and between oases, populations, and cultivars of the date palm throughout its distributional range (Dada et al. 2012). Conservation of genetic diversity is essential for present and future functioning of oasis agroecosystems and the well-being of their human communities. Recently, there has been increasing awareness of and interest in the need to adopt a holistic view of dynamic conservation and sustainable utilization of date palm within its natural habitat, especially oases (Eneh and Nwawe 2000). A thorough understanding of date palm genetic diversity and how it is structured in different oases, populations, and cultivars is essential for its dynamic conservation and sustainable use. It will help farmers, scientists, and policymakers determine what to conserve, as well as where to conserve, and will improve the understanding of the taxonomy, origin, and evolution of this unique fruit tree. Additionally, this knowledge is essential for germplasm collecting and the efficient use of the cultivated species and their wild relatives for crop improvement. The vast array of adaptive genetic variation present in date palm, which is generally quantitative and responsive to habitat differences, often reacts to biotic, abiotic, and anthropogenic factors. Genetic diversity studies on date palm have clearly demonstrated that there is a direct association between population characteristics and the environments (i.e., oases) in which they grow, whereas ecological factors largely determine the extent and distribution of genetic diversity in wild populations (Dada et al. 2012).

Local seedling date palms (*dibinojeh naoura*) with outstanding adaptation to climatic, edaphic, and management factors have been observed in Cameroon, especially in arid areas of the north (Elhoumaizi 2007). They are used essentially for ornamental purpose (Fig. 12.1). No particular data have been collected about these local date palms in relation to vernacular name, cultivar identification, and morphological and molecular characterization.

These local cultivars are the product of centuries of interaction between farmers, the genetic and breeding systems of the date palm, and the environment. The breeding systems of date palm, as well as several ecological pressures, affect the distribution of intrapopulation variability and determine the genetic composition of cultivars. In addition, selection for ecological adaptation may have resulted in the accumulation of genotypic and/or phenotypic differences in plants derived originally from identical clones but grown under different environmental conditions and/or management practices. The natural breeding system of the date palm is still determined to an extent by climatic factors such as wind. In Cameroon, apart from recent introductions (2005–2006) by the Technical Cooperation Programme (TCP/FAO) of elite cultivars (Ghanami, Khalas, Khadrawi, Medjool, Barhi, Ambarah, and Zahidi), no other improved cultivars have been introduced;

Fig. 12.1 Local date palm used for ornamental purpose

all those observed in northern Cameroon are considered local (Bourou et al. 2006; Elhoumaizi 2007). Wild palm species have breeding systems somewhat distinct from those of the domesticated species and, therefore, present different problems with respect to their biodiversity, collection, maintenance, and regeneration in the field. These cultivars are mostly associated with annual crops, such as cotton, maize, or groundnut (Fig. 12.2).

No special technical conservation of in vitro date palm plants exists in Cameroon; what conservation is done is in situ. There is no plan at present to set up the technology to identify trait-specific molecular markers.

12.4 Plant Tissue Culture

Plant tissue culture techniques are used to regenerate a whole plant from cells or plant tissue culture medium, using modern techniques of cell culture (McCubbin et al. 2004). It produces plants free of viruses and any other infections; in addition it has the capacity to quickly produce a large amount of plantlets. It is used for the creation of new plants, the multiplication of commercial plants producing few or no offshoots, or the conservation and propagation of rare species. Typically, plant cells (in fact only one or a few cells are enough) are placed on a sterile growth medium including sucrose as a source of energy, vitamins, and nutrients and placed in an environment with temperature, humidity, and light carefully controlled. Tissue

Fig. 12.2 Barhi cultivar in an associated system with cotton

culture operates on the principle that each cell possesses the necessary and sufficient potential to multiply and especially to organize differentiated tissues to reconstruct a plant with its DNA, the totipotency of plant cells (McCubbin et al. 2004).

In Cameroon, up to now, tissue culture exists only for the banana plant. Some private organizations are engaged in collaboration with scientific research institutes. These include the Centre for Agricultural Research for Development of Ekona at Nkolbisson and the Agricultural Research Station of Njombe (Bourou et al. 2006). In the private companies, banana plantlets are transferred to soil (vermicompost) from plant media. This process is done for acclimatization of plantlets to the soil as they were previously grown in plant media. After growing for some days, the plantlets are transferred to the field.

In Cameroon, three methods of tissue culture can be considered for date palm: (a) meristem culture or stem apex, which remains the most generalized technique and ensures the reproduction of true-to-type plantlets, (b) regeneration of plants by formation of new buds and roots on callus, and (c) somatic embryogenesis, which allows obtaining embryos and plant regeneration from somatic cells. These three methods are well known among Cameroonian researchers but little practiced. Tissue culture plants often have improved physiology and morphology, giving them the benefits of starting force, better than rooting cuttings or layering (NIFOR 2008). No direct application has been made on the date palm in Cameroon; however, in vitro date palm seedlings were introduced into Cameroon in 2005–2006. These came from the Marionnet Laboratory in the UAE and DPD Laboratory in the UK (Bourou et al. 2006; Elhoumaizi 2007). Plans exist for future tissue culture work on date palm in Cameroon.

12.5 Cultivars Description

Up to now in Cameroon, no local date palm cultivars have been named, identified, or described in terms of morphological or molecular characteristics and so on. Date palms are found in the North and Far North but are limited to scattered individual trees and recent plantings from seeds obtained from imported date palm fruits. Isolated date palm plants are common at residences around villages in the North and Far North of Cameroon. Date palm plantations (of more than 60 palms) also can be seen around the walls and villages of Garoua, Guider, and Maroua (Bourou et al. 2006; Elhoumaizi 2007). Some plantations were established with seeds from fruits imported from Saudi Arabia, some ranging in area between 0.4 and 1.0 ha. Several farms with older plants, 10–16 years of age, exist in the region of the Far North, for example, Diamare, Mayo-Sava (Far North), Mayo-Louti, Benue, and Mayo-Rey (North) (Bourou et al. 2006; Elhoumaizi 2007). Besides these private initiatives, the Regional Delegation of Agriculture and Rural Development of the Far North began in 2001, as part of crop diversification and environmental protection and promotion of date palm growing through farmer organizations (FOs).

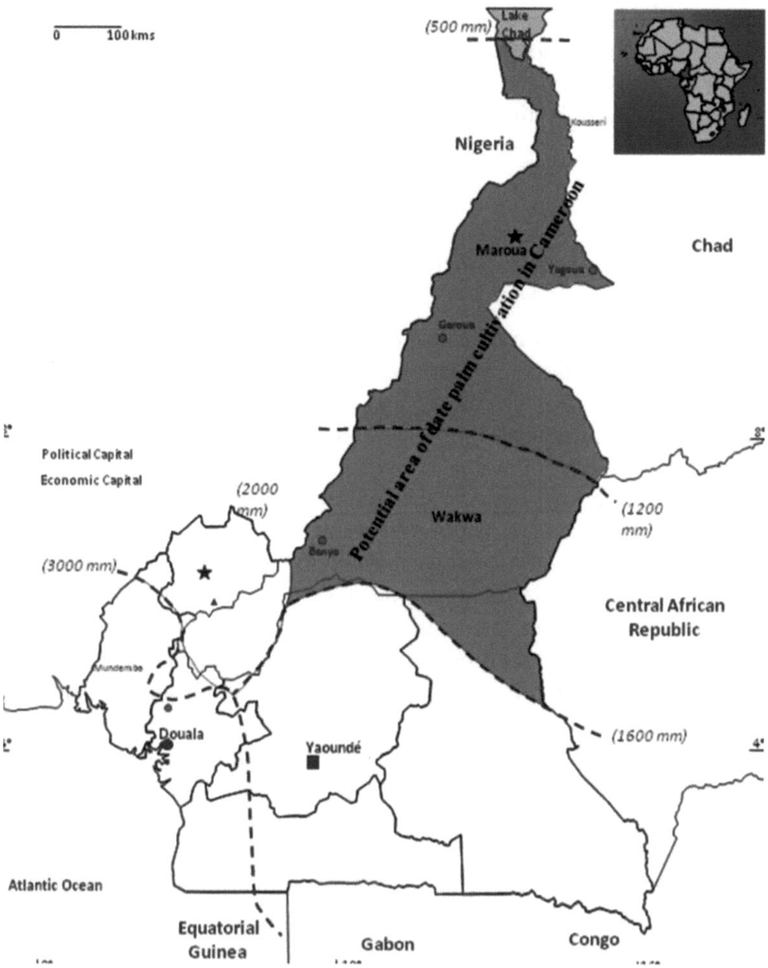

Fig. 12.3 A map of Cameroon showing the potential area of date palm cultivation

Climatic data indicate that conditions are suitable for date palm growing, especially in the Far North. The areas of Cameroon potentially suitable for date palm cultivation are shown in Fig. 12.3.

Overall date palm yields in Cameroon are low with fruit of very poor quality. This local production can only be used for home consumption. All initiatives thus far have been limited by two factors: first, the inadequate supply of improved planting materials and, second, lack of trained supervisors in techniques of date palm cultivation. Moreover, it should be noted that the date fruits are consumed by people of the northern part of the country and in major cities. The date palm is a prestigious plant in chiefdoms, the concessions held by some important community leaders (Bourou et al. 2006; Elhoumaizi 2007). In the north, a few land developers have begun to establish date palm orchards. All these date palm plantations are located in the northern zone with an annual rainfall of around 700 mm. Despite the enthusiasm of the people of the North for the date palm in particular, research on the palm is limited to cultivar collections (Bourou et al. 2006; Elhoumaizi 2007).

12.6 Date Palm Production and Marketing

Currently no Cameroonian company produces and distributes young date palms. However, the Garoua Research Station's experimental orchard at Kismatari and IRAD is in the process of producing and releasing date palm offshoots in rural areas, derived from introduced improved cultivars in 2005–2006. This production technique is described as follows.

Two types of offshoots occur on the date palm in general: the lower which are older and the upper which are considered younger, the latter are not yet of concern in Cameroon. It is believed that lower offshoots are more active physiologically than the higher ones; they probably grow faster, the number of leaves produced increasing with age. In fact, the high offshoots have less carbohydrate than low offshoots, resulting in low root production and consequently a lower survival rate. It is also suspected that high offshoots develop when no fruit is on the palm (Eneh and Nwawe 2000; Mahmoudi et al. 2008).

Early offshoot removal is desirable because (a) it allows easy access to the palm, (b) it improves the development and fruit production of the parent tree, and (c) planting young offshoots is advantageous as they will produce a greater number of offshoots than older palms.

Numerous factors must be considered when rooting offshoots, including the size of an offshoot (often expressed in weight), type (upper or lower), origin of the offshoot, method of removal and preparation for planting, and treatment of the offshoots after planting. To stimulate and promote rooting, the base of the offshoot should be in contact with moist soil for at least 12 months before removal. Production of high offshoots is primarily related to a particular cultivar character but also in some cases related to a damp climate. For these high offshoots, boxes or plastic bag materials can be fastened around the base of the offshoot. Another technique is to leave them on the mother palm until they are more mature. They are then removed and rooted in a nursery as in the case of the experimental orchard of Kismatari. Date palm offshoots in Cameroon are expensive, selling for about XAF 15,000 (EUR 22.68).

In Cameroon dates are appreciated by the people, but the date palm's future is uncertain. Dates are sold in bags of 100 kg by wholesalers and in various quantities by retailers. The price per bag of 100 kg in Kousseri fluctuates from XAF 40,000 to 45,000 (EUR 60.97–68.60) from September to May and XAF 45,000 to 50,000 (EUR 68.60–76.22) from June to August (Elhoumaizi 2007). The dates sold are dry

Fig. 12.4 Date palm price and quality in the North area of Cameroon

and of poor quality. The primary retail measure used is a 2.5 l container called a *goro*. The price range of a *goro* of dates is XAF 1,500–2,500 (EUR 2.28–3.81).

The human populations of the northern regions of Cameroon are all present or potential consumers of dates, along with about 6 million people in the large cities such as Yaounde and Douala. Dates of poor quality (Fig. 12.4) are sold at very high prices of about XAF 1,000 (EUR 1.52) per ½kg (Elhoumaizi 2007). During the period 1998–2003, about 34,000 kg dates were imported per year. A large amount of informal importation of dates also occurs across the borders from Chad and Nigeria. The date market in Cameroon is entirely in the hands of the informal sector. Dates consumed locally in Cameroon mostly come from Kousseri, through the northern areas from Chad (Faya - Largeau), Libya, Sudan, and Saudi Arabia (Bourou et al. 2006; Elhoumaizi 2007).

The main issues impeding development of a profitable date palm industry in Cameroon are the lack of (a) new cultivars and programs to encourage producers to plant offshoots rather than seeds, (b) skilled national date palm specialists, and (c) modern farming techniques for the date palm. Date palm could adapt well to the adverse conditions of the dry, warm areas of the Far North, providing they are supplied with adequate quantities of irrigation water.

The date palm can play a key role in the protection of the environment with the creation of favorable microclimatic conditions for the development of other understory fruit crops as well as functioning as oasis windbreaks. Date palm can also contribute to the generation of significant income by marketing the fruits and palm by-products. It should be noted that local date palm cultivars have an annual yield of only 15–20 kg of fruits/tree; therefore, considerable potential exists for much higher yields with the growth of improved cultivars. Enhanced quantity and quality of dates in Cameroon will also improve the international trade balance of the country through the export of dates to international (Europe) and regional markets such as the Sahelian countries, and finally it will generate jobs through the development of agriculture.

12.7 Conclusion and Recommendations

Date palm production can contribute to Cameroon's economy in the next few years. Current consumption of date fruit in Cameroon is mostly based on imports from the Middle East and neighboring countries, thus depleting foreign exchange reserves. Date imports could be eliminated with improved high-yielding cultivars adapted to Cameroonian ecological conditions. Cameroon has the capacity to produce enough dates to meet local demand and to earn foreign exchange through export of any surplus production into the international market. Perhaps the greatest potential in this respect lies in the commercial development of date products from naturally occurring cultivars with a ready market.

Cameroon is not yet established as a commercial date palm-producing country but has important potential in terms of human resources, ecological advantages, market, etc. Creating date palm plantations will not only add to the total economic GDP of the country but will undoubtedly check the increasing rate of desertification, and it will also make a positive impact on the diet of the people, thereby contributing to improving the standard of living of not only the farmers but the entire population. Cameroon has two harvesting seasons, wet and dry, which is an advantage for the country to compete on the international market when other countries that produce dates are out of season. More research is still needed to add to what has already been achieved. In order for the development of date products trade to result in tangible improvement in sustainable management of the date resources in the country, partnerships between rural producers, national policymakers, and the private sector are essential.

References

Atlas de l'Afrique (2000) 2éd. Ed. du Jaguar, Paris

Boli Z, Bep Aziem B, Roose E (1991) Enquête sur l'érosion en région cotonnière du Nord-Cameroun. Bull Réseau Eros 11:127–138

Boli Z, Roose E, Bep Aziem B et al (1997) Recherche de systèmes de culture intensifs et durables en parcelles d'érosion à Mbissiri (1991–1994), Nord Cameroun. In: Seyni Boukar L, Poulain JF, Faure G (eds) Agricultures des savanes du Nord, Actes de l'Atelier de Garoua. CIRAD-CA, Montpellier, pp 255–272

Bourou S, Woin N, Elhoumaizi AM (2006) Synthèse des rapports de projet: Projet FAO/CMR/3003, Promotion de production du palmier dattier dans la zone du Nord-Cameroun

Dada M, Nwawe CN, Okere RA, Uwubanmwen IO (2012) Potentials of date palm trees to the Nigerian economy. World J Agri Sci 8:309–315

Elhoumaizi AM (2007) Promotion de la production du palmier dattier dans la zone Nord du Cameroun. Rapport de projet TCP/CMR/3003

Eneh FK, Nwawe CN (2000) Towards effective utilization of research results: commercializable technologies developed by the Nigerian Institute for Oil Research (NIFOR)

Mahmoudi H, Hosseininia G, Azadi H, Fatami M (2008) Enhancing date palm processing, marketing and pest control through organic culture. J Organ Syst 3(2):29–39

McCubbin MJ, Zaid A, Van Staden J (2004) A Southern African survey conducted for off-types on date palms produced using somatic embryogenesis. Emir J Agric Sci 16:8–14
MINADER (2008) Annuaire des statistiques agricoles
NIFOR (2008) Nigerian Institute for Oil Palm Research. House Res Rev 165–204
PNUD (1996) Plan de gestion de l'environnement volume II. Analyse des secteurs d'intervention au Cameroun. Rapport de synthèse
Roose E (1994) Introduction à la GCES. Bulletin Pédologique FAO, Rome, No 70

Chapter 13
Date Palm Status and Perspective in Djibouti

Abdourahman Daher, Nabil Mohamed, and Frederique Aberlenc-Bertossi

Abstract Djibouti has acquired date palm genetic resources by successive introductions of cultivars originating from neighboring countries. The most common cultivars grown are Beladi, Harissa, Ambabo, Sheeri, Oumo-Assala, Medjool, Zahidi, and Khadrawy. Four different categories of groves are found in the country. In the oasis gardens developed on the edges of the wadis in the southeastern and northern coastal plains, date palms are propagated from seeds and offshoots. In the inland continental plains, an agropastoral system combining the growing of crops and pasturage is practiced. Date palm groves at higher elevations are spread along the wadis close to volcanic structures in the mountainous zones of the North. Lastly, pilot date palm farms were established a few years ago and tissue culture-derived date palms were planted in these sites where water and good alluvial soil are available. There is an increasing demand for dates within the country. Domestic production estimated at 541 mt nationwide cannot cover the needs of the local market. To meet this demand, Djibouti imports about 400 mt of dates per year. Despite numerous constraints which prevent the development of date palm cultivation in Djibouti, the country is rich in genetic resource. The expanding date palm sector should be aided by complementary training for farmers to improve date production.

Keywords Agropastoral system • Date palm groves • Djibouti • In vitro • Oasis gardens • Seedling • Wadi

---

A. Daher • N. Mohamed
Institut des Sciences de la Vie/CERD-Centre d'Etude et de Recherche de Djibouti, Ministère de l'Enseignement Supérieur et de la Recherche, BP 486 Djibouti, Djibouti
e-mail: abd_daher@yahoo.fr; nabillahgui@yahoo.fr

F. Aberlenc-Bertossi (✉)
IRD-Institut de Recherche pour le Développement, Palm Developmental Biology Group, UMR DIADE, 911 avenue Agropolis, Montpellier 34394, France
e-mail: frederique.aberlenc@ird.fr

J.M. Al-Khayri et al. (eds.), *Date Palm Genetic Resources and Utilization: Volume 1: Africa and the Americas*, DOI 10.1007/978-94-017-9694-1_13

13.1 Introduction

With a total area of 23,200 km^2, Djibouti is located on the Horn of Africa at the confluence of the Red Sea and the Indian Ocean. This particular biogeographic position gives Djibouti a strategic role as a commercial hub and is of ecological importance with respect to climate change and food security. The country has an arid and hot climate. Average temperatures range between 23 °C in January (cool season) and 39 °C in August (summer season). The climate is characterized by low and irregular rainfall (average annual rainfall is 150 mm), high humidity in summer, hot and dry winds (*khamsin*) blowing during the summer for around 50 days, and high evapotranspiration (2,800 mm/year) (Müller 1982). These special conditions result in the absence of perennial streams across the country. Thus, the groundwater resources for sustainable food and irrigation are accumulated in volcanic and sedimentary subsoil. The Djibouti landscape is mostly dominated by a volcanic terrain characterized by mountain chains with altitudes of 500–2,010 m alternating with uplands and sedimentary plains of alluvial or lacustrine origin. The soils are generally rocky since they are derived from the weathering of rocks (basalts and rhyolites). However, the potential arable land is estimated at 1,400 km^2 which represents 6 % of the total area. These harsh environmental conditions, combined with the pastoral origin of the rural population, make agriculture difficult to practice in Djibouti. Current agricultural production is estimated at about 6,000 mt, fulfilling about 10 % of food needs. Most food needs are imported from abroad, and the country's economy primarily depends on the service sector, which represents 80 % of activities. Recognizing this, the government has embarked on a policy to fight against poverty and has developed a strategy for food security. Within this strategy, date palm cultivation was chosen as the driving force of agricultural development in Djibouti. The date palm agriculture system, which consists of the date palm oasis and food crops, is well suited to Djibouti. This initiative aims to boost the primary sector and reduce the rural to urban exodus by increasing the creation and expansion of oases. Indeed, in these arid and hot environments, date palm cultivation is an important socioeconomic factor, because it ensures the survival of many families whose livelihood primarily depends on revenues generated directly and indirectly by date palms.

13.1.1 History of Date Palm Introductions

The timeline of the earliest introductions of date palm trees in Djibouti is unknown. The creation of the first palm groves was in the mid-nineteenth century and was related to the occupation of the western shores of the Red Sea by the Egyptian army of the Ottoman Empire (Awaleh 1990). The remains of these old date palm plantations are found in the coastal plain of Tadjourah. During World War II and the French occupation of the territory, a second generation of date palms was brought in from the Arabian Peninsula. Yemeni migrants established palm groves in Ambouli Wadi, as well as other vegetable crops and aromatic herbs. Furthermore date palms were planted in principal towns such as Ali-Sabieh, Obock, Tadjourah, and Dikhil. A third introduction of imported offshoots and vitro-plants occurred during the post-independence period and under the Djibouti date palm development project carried out with French cooperation in 1990 (Peyron 2000). Date palm groves were set up with the introduction of offshoots obtained from Abu Dhabi, Yemen, Oman, Sudan, and Iraq and vitro-plants produced by a French team for date palm research (Peyron 2000; Toutain 1990). In 2004, new date palm plantations distributed in five regions within the country were established by the government. Thus, more than 18,000 high-quality vitro-plants (Nabut Sultan, Khalas, Rzizi, Kenezi, Barhi cvs.) from Saudi Arabia were planted over more than 36 ha. Since their inception, these palm groves have become a laboratory to test how date palm trees adapt to the Djibouti environment, and palm tree behavior is monitored by specially trained teams of the Center of Study and Research of Djibouti (CERD) (Nabil 2010). The successive introductions of date palms from various sources have allowed Djibouti to acquire a genetically rich date palm population adapted to local soil and climatic conditions.

13.1.2 Main Date Palm Groves

There are around 1,700 private oasis gardens in the entire country and date palms are found in each of them. The total number of date palms is estimated to be more than 50,000; about one-third of them are very old. Date palm may reach an age of 70–150 years and height of 15–20 m. Djibouti can be divided into four different categories of date palm groves (Figs. 13.1 and 13.2).

13.1.2.1 Coastal Date Palm Groves

The oasis gardens in the coastal plain are divided into two groups according to their geographical location. On the Southeast plain, date plantations have been developed on the edges of the wadis close to the sea, such as those at Ambouli, Douda, Damerjog, and Atar (Figs. 13.1 and 13.2a). Soils are essentially composed of recent alluvia transported by intermittent stream flow within the wadi. The oasis gardens are all more or less of the similar size. Date palms in these oasis gardens are propagated from seeds and offshoots. The majority originates from Yemen and has low productivity. The dates are large in size but their organoleptic properties are low.

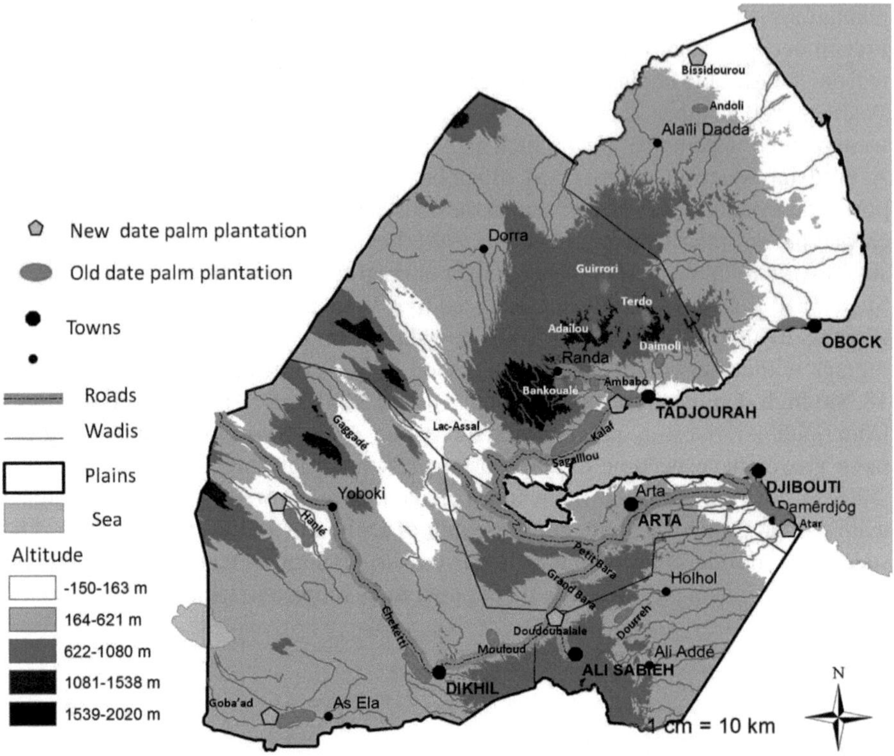

Fig. 13.1 Map of Djibouti date palm grove locations

The garden space is used harmoniously and date palm plantations are associated with vegetables, forage, and fruits trees (guava, mango, and citrus). In some gardens, a sedentary farming method including cattle rearing is practiced. In the Northern coastal plain, date palms are found in Sagallou, Kalaf, Ambabo, Tadjourah, and Obock plantations (Figs. 13.1 and 13.2b). Each of these plantations contains more than 100 palms originating in general from Iraq (Peyron 2000). Soil texture is generally sandy. Date palms were essentially propagated by young offshoots except in Sagallou zone where the small plantations originated from seed and contain a high proportion of male trees. Date production is low (3–5 mt/ha). The dates are small in size but much appreciated by the local and nomad communities. Beneath these date palms, different crops are grown such as vegetables, forage (sorghum, durra, a new strain of sorghum, and nitrogen-fixing *Leucaena* shrubs), and fruit trees (doum palm). Vegetable production is often observed further from the sea. Livestock is kept in some gardens especially goats. It is observed that the date groves near the sea show yellow leaves because of the high salinity in the coastal zone. In addition, the coastal area has high humidity, an ideal condition for fungal diseases to damage date palm flowers and fruits. This excessive moisture could be responsible for incomplete fertilization and the development of parthenocarpic fruits.

Fig. 13.2 Main date palm groves in Djibouti. (**a–b**) Oasis gardens in coastal zone (**a** in the southeast such as Damerjog and **b** in the north such as Ambabo and Kalaf). (**c–d**) Date palm groves in inland plains (**c** in Dikhil city and **d** in As-Eyla, Chekeyti). (**e**) Date palm groves at higher elevations. (**f**) Pilot farms of vitro-plants such as at Doudoubalala (Source: ISV/CERD)

13.1.2.2 Continental Palm Groves

The inland plains are the most intensely cultivated areas in the country. These include Hanle, Goba'ad, Grand and Petit Bara, Gagadé, and Doda. Surrounded by mountain terraces, a majority of the plains are transformed into ephemeral lakes after heavy rains. Soils are derived from lake sediments composed of silt and clay that are deposited on a basaltic substratum. Oasis gardens are located on the borders of the wadis which cross these plains. These zones have sandy alluvial soils. The subsoil water is relatively abundant. The constraints related to excessive moisture are not found in continental palm groves where the humidity is low. In contrast, it is the case in Hanle, Mouloud, Chekeyti, Doudoubalala, As-Eyla, Yoboki, and Dourreh oasis that gardens of 0.5–2 ha in size are found. In the best maintained oasis gardens such as in Dikhil, Mouloud, Hanle, and As-Eyla, the upper canopy is occupied by date palm trees associated with fruit trees (doum palm and guava), and vegetables and forage crops are cultivated in the understory (Figs. 13.1 and 13.2c). A small space is reserved for nursery seedlings of tomatoes, onions, lettuce, peppers, etc. Therefore, a farming system with a three-layer cultivation system has been created in these oases. Indeed, an agropastoral system combining the growing of crops and pasturage is practiced. Food crops for livestock are cultivated in the shade provided by the planted trees. Organic fertilizer used to increase agriculture productivity consists of livestock manure and crop residue. Usually there are more than 100 date palm trees per garden, offering a relatively satisfactory date production (5–7 mt/ha).

Dikhil and As-Eyla cities have the oldest date palm plantations in the region with long histories. Today, they are used for tourism. In contrast, some gardens in Chekeyti and As-Eyla localities have been abandoned (Figs. 13.1 and 13.2d). The only remaining dates probably survive by having roots in deep groundwater. Male date palms are high in number and pollination occurs naturally by wind action. It should also be noted that there is a high proportion of female date palms bearing parthenocarpic fruits due to low fertilization rates.

13.1.2.3 Date Palm Groves at Higher Elevations

In the mountainous zones of the North, oasis gardens are spread along the wadis close to volcanic structures. These include the Guirrori, Rippaleh, Weima, and Bankoualé date groves. Most oases are small plantations organized in terraces (Figs.13.1 and 13.2e). Their size varies from 600 to 4,000 m^2 according to the possibilities offered by volcanic formations. These date plantations originated from seed. Environmental conditions are optimal because of permanent water availability and good climate. The main difficulty is to mobilize the water because of the difficulty of the terrain. Furthermore, oasis gardens located at higher elevations suffer from a lack of potential and extendable areas and from their distance from the market. Date production in most of these gardens is better as compared to the coastal area. Many types of vegetables are intensively cultivated under date palm trees.

13.1.2.4 Pilot Farms

Six pilots date palm farms (Bissidourou, Ambabo, Hanlé, Goba'ad, Doudoubalala, and Atar) were established in five regions of the country a few years ago (2006) (Figs.13.1 and 13.2f). Tissue culture-derived date palms from Saudi Arabia were planted in these pilot sites where water and good alluvial soils exist. The pilot farms represent a total area of 36 ha with 125 vitro-plants per ha at a tree spacing of 8×10 m. Water collected in a large tank is used exclusively for irrigation purposes and drip irrigation has been adopted on all pilot sites. Cultivars Medjool, Barhi,

Khalas, and Nabut Sultan were selected based on their commercial value and their ecological plasticity. Early flowering has been observed, from the fourth year after planting. Date production is currently low because not all female palms have reached flowering stage. Between the palms, no vegetable growing is practiced.

13.1.3 Cultural Importance and Potential Uses

In Islam, the date palm has a well-respected position and remains a blessed tree. It has been feeding human beings through good and bad years for centuries. When all other resources fail, due to perturbations within the harsh environment, the date palm becomes the farmer's companion at all times. Dates are staple foods for Muslims, specifically during the month of Ramadan. Thus dates are the most coveted commodity because of their sacredness and nutritional value. Palm leaves are used in traditional marriages as a decoration on doors to symbolize celebrations. Moreover, date palm leaves have many other uses; for example, they are used to feed livestock, and also mats and baskets are made from the rachis and leaflets. Dry palm leaves are used as fences, roofs, and walls for rural houses and shops. They are also used to cook food when others fuels are unavailable.

13.2 Cultivation Practices

13.2.1 Description of Cultivation Techniques

In Djibouti, old date groves are composed mainly of local genotypes. Some such as Harissa are highly appreciated by local people in the North and Southwest regions while Beladi is mostly valued as a fresh date and has a good demand in the Arta and Djibouti regions. Therefore, farmers propagate selected trees according to their productivity and the quality of their dates by offshoots and seeds. However, propagation by seed leads to a proportion of males in the next generation. The main cultivation period of date palm is the winter season from October to April. In new plantations, date palm trees are planted at a high density because farmers are accustomed to plant the young palm trees close to the old ones. Therefore, they are rarely organized in a way that allows associated crops. Cultivation practices are limited to pollination, irrigation, pruning, cutting dry leaves, and removal of offshoots. However, annual date production never exceeds 10–20 kg/tree. This is the consequence of deficiencies in farmers' knowledge and skills in the following areas: pollination, pruning, thinning, bunching, and irrigation. In recent decades, knowledge and experience have not been capitalized upon. One of the reasons is that the majority of oasis garden owners hire Ethiopian workers to maintain their palm groves. Unfortunately, these workers rarely stay very long because they generally immigrate to Arabian countries. For example, in several oasis gardens, date palm trees are too densely planted. This high density does not allow female date palms to fully express their productive capacities. Farmers also report lower quality and precocious abscission of dates because of a lack of pollination practices. Pollination know-how has been lost as Yemeni immigrants left Djibouti over the last few decades. Also, there are many unwanted offshoots at the base of the trees, as well as dead or infected palms caused by *Parlatoria blanchardi* in most of the cases. Unwanted offshoots and dead palms influence water and mineral absorption by the mother palm and its date production.

13.2.2 Phenology

Phenological observations show that date palms exhibit an annual season of flowering that begins in Djibouti in December with the precocious flowering palms until April for late-flowering individuals. Male flowering usually starts a few weeks before the female phase and occurs in a continuous pattern during the flowering period. Female and male flowering has been observed to exhibit a bimodal pattern with a first peak in mid-January and a second peak at the end of February. Flowering intensity at the individual level varies according to gender, genotype, and environment. Female date palms produce more inflorescences per year (10–40) than males (5–25). All individuals produce inflorescences that are fertile and unisexual the majority of the time. However, incomplete hermaphrodite flowers have been very occasionally observed in the staminate inflorescence of male plants under particular environmental conditions. This flowering abnormality is more a condition of parthenocarpic development of the carpels rather than a perfect hermaphroditism. Under local conditions, flowering lasts 60–120 days during cold periods with rainfall. Date palms in the coastal area begin flowering before the inland ones. Fruit maturation takes place usually from April to August and requires low humidity and high temperatures.

13.2.3 Pollination

Under natural conditions, date palm is wind pollinated; however, palms are pollinated artificially in most groves. The commonest method of date pollination practiced is based on hand pollination. The farmer climbs into the female tree and places 6–8 male spikelets on each of the female inflorescences just after opening and for several days. The best time for pollination is early morning or afternoon to prevent high temperature and high humidity which have been observed to damage pollen germination. Farmers use fresh male inflorescences of the season because conservation techniques for pollen are not perfectly mastered. Male inflorescences are highly variable in pollen production. They differ greatly in vigor and pollen quality. As a result, fruit set differs from one individual palm to another. Date palm growers are now beginning to realize the need for the selection of superior males.

13.2.4 Fruit Harvest

During fruit maturation, the number of pollinated strands on every female bunch is reduced by thinning (pruning) the excess branches. This practice is generally used for cultivars with dense inflorescences or long spikelets to avoid damaging dates and to facilitate fruit growth. The date harvest takes place when the majority of dates are ripe. The picking method for dates also remains rudimentary. The farmer carries a basket and climbs the female trees using ropes to harvest the ripe dates. Most of the date production is consumed locally and any surplus dates are transported to the Djibouti city market.

13.2.5 Irrigation Systems and Assessment of Water Requirement

The method used to supply water to date palm trees is submersion. Solar or gas-powered pumps bring water from the shallow well to an open water tank of about 20–70 m^3. Tanks are equipped with taps. The depth of the wells is 4–12 m depending on the water table. Water impounded in the tank is distributed through a pipeline by gravity to the date palm grove. Furrow irrigation has been commonly used in farms in Djibouti. Water fills the basins surrounding the date palms and extra water goes to associated crops. The diameter of basins varies according to the extent of the canopy. Water is in short supply in the continental plantations but it is of good quality. However, coastal groves have lesser water quality containing excessive dissolved salts. There are two aquifer systems in the coastal area, namely, freshwater occurring above the salted water table. When freshwater is pumped out, the saline water invades the wells, which causes water salinization up to 10 g/l and leads to land abandonment. The traditional irrigation system based on basins demands more water for the farmers. For example, the annual water requirement for 1 ha of date palm grove ranges from 20,000 to 30,000 m^3. Although each farmer has its own shallow well, the Mouloud date palm groves have one large water tank to provide for irrigation. The common system is managed by the Mouloud association which organizes a calendar for water usage. Therefore, they ensure that water availability and cultivated surface area are balanced.

To improve date productivity, pilot date palm farms have adopted modern irrigation systems (drip irrigation) and therefore contribute to water usage optimization and reduction in water loss. To further develop optimized water management methods in date palm oases, CERD researchers are carrying out studies on the amounts

Fig. 13.3 (**a–b**) Experimental farm to estimate the optimum irrigation for date palm grove. (**b**) Neutron probes and tensiometers installed around a stipe of date palm tree. (**c–e**) In vitro culture of date palm (**c**) calluses, (**d**) cluster of buds with elongated leaves, and (**e**) plantlet with roots ready for acclimatization (Source: ISV/CERD)

of water required. Researchers use tensiometers and neutron probes which are installed around the trunk of a date palm at different depths (Fig. 13.3a–b). Different scenarios of irrigation such as water infiltration and drying tests are implemented. Results obtained have revealed that the roots of date palm absorb water only from 40 to 60 cm in depth. The above experiments will be complemented by studies of date palm transpiration. These data will collectively allow the CERD researchers to determine the water balance between the soil and the plant and to develop a water requirement model for date palm groves.

13.2.6 Pest Status

The existing knowledge of insect biodiversity in Djibouti is poor. Thus, a preliminary study to characterize insect fauna has been carried out in eight oases by the Plant Protection Laboratory of Life Science in CERD. The species collected belong to four classes of arthropod with insects being the most important. Among specimens, *Parlatoria blanchardi* (date palm scale insect) and *Ommatissus binotatus* (dubas bug) were the most abundant species. They were found in all oasis ecosystems. This unpublished study revealed a high insect infestation in date groves which therefore is a threat to date production, particularly the coastal date palm groves with high relative humidity. However, among the most important zoophagous species collected in date palm groves, there are two predators, *Micraspis lineola* and *Pharoscymnus ovoïdeus* (both ladybird beetles), which play an important role in biological control. A significant reduction in insect infestations is therefore observed in oases where these predators are present. Finally, this study is the first description of insects affecting Djibouti date palm groves and will allow establishment of an alternative biological control approach for the pests of date palm.

13.3 Genetic Resources and Conservation

From the nineteenth century onwards, Djibouti acquired date palm genetic resources from various origins which have been enriched by successive introductions of cultivars. However, the lack of precise data makes it difficult to identify date palm cultivars and their geographical origin. In addition, the recent date palm development program, which involves the creation and expansion of date palm groves, requires reliable knowledge of the genetic diversity of local genotypes. For this reason, researchers in CERD and IRD are carrying out studies to characterize the genetic diversity of palms in Djibouti using microsatellite markers (Aberlenc-Bertossi et al. 2014). The preliminary results show a relatively low genetic variability in different date palm groves. Thus, there is also low differentiation observed between the Chekeyti, As-Eyla, Sagallou, Ambabo, and Ambouli groups because there are genetic exchanges between the date palm plantations. Finally, the genetic analyses reveal that many of the date palm trees have close genetic relations with cultivars from the Arabian Peninsula.

Since the introduction of plantations of high-quality vitro-plants, rural people have shown more interest in agriculture, especially in the oasis system. When there is insufficient availability of female vitro-plants, farmers select female date palms from seedlings which provide good fruits and high-quality dates. They propagate these selected female by seeds, but there remains a high proportion of male trees which monopolizes agricultural space in the oasis gardens. However, for the first time, a Franco-Tunisian team led by IRD in collaboration with CERD researchers has identified molecular markers for the early sex determination of date palm (from 6 months of germination) (Chérif et al. 2013). These markers will provide farmers with the opportunity to select female plants, thus avoiding the cultivation of unnecessary male trees. Thus the CERD biotechnology laboratory is engaged in the identification of local female genotypes/individuals with desirable agroeconomic characters, their multiplication via seedlings, and the selection of female plants in the progeny.

13.4 Plant Tissue Culture

Since 2006, the Government of Djibouti has strengthened research and development in the date palm sector. As a result, a plant biotechnology laboratory has been created in CERD. The objective of this laboratory is to produce plants free of diseases, from elite cultivars of date palm trees that are well adapted to the local environment. These studies have been conducted by CERD researchers with the expertise of IRD French team (Aberlenc-Bertossi et al. 1999; Sané et al. 2006). The selection of these high-quality date palms requires good phenological, organoleptic, and genetic characterization. The tissue culture laboratory established in the Life Sciences Institute of CERD is fully equipped with facilities for tissue culture in an aseptic environment. The plant biotechnology laboratory used for somatic embryogenesis processes has obtained thousands of rooting plantlets such as cvs. Medjool, Zahidi, Nabut Sultan, Barhi, and Khalas (Fig. 13.3c–e). In late 2009, the organogenesis technique developed by the Phoenix station in Spain was optimized for the multiplication of several cvs. such as Medjool and Khalas (Ferry et al. 2003). All in vitro-regenerated plantlets are currently at the acclimatization phase, a very critical and final step of the tissue culture process.

13.5 Cultivar Identification

Date palm groves in Djibouti are composed of several cultivars. Most of them originated from neighboring countries (Saudi Arabia, Yemen, Iraq, and Egypt). The most common cultivars are Beladi, Harissa, "Ambabo," Sheeri, Oumo-Assala, Zahidi, and Khadrawy (Fig. 13.4a–c). These palms are propagated from offshoots. The main cvs. produced by tissue culture are Khalas, Barhi, Nabut Sultan, Rzizi, Zahidi, Khadrawy, and Medjool. The tissue culture plantlets were acquired within

Fig 13.4 Examples of Djibouti Dates: "Ambabo" (**a**), Beladi (**b**), and Sheeri (**c**) successfully cultivated in many oasis gardens. They are soft and well appreciated by the communities (Source: ISV/CERD)

the framework of Djibouti-France cooperation in 1991 and more recently in 2006–2009 by the Ministry of Agriculture. Phenological observations have revealed that cultivars show significant differences in the date weights at the tamar stage. Cultivars Harissa, Imri, Oumo-Assala, Medjool, Khadrawy, and Kisba have large fruits (9–14 g), while the fruits of "Ambabo", Beladi, and Zahidi are small (3–6 g), although all cultivars show similar changes in date color during the stages of maturation. Sheeri cv. is an exception with regard to fruit color; it is red at the khalal and rutab stages and becomes dark red at the tamar stage. Most dates are soft and semisoft (Medjool, Beladi, Khadrawy, Oumo-Assala, Sheeri, Zahidi cvs.), but a few produce dry dates such as the Harissa cv. of the Goba'ad Plain and Kisba cv. of the Arta and Djibouti regions.

13.6 Cultivar Descriptions

Some traits of the most important date palm cultivars grown in Djibouti including statistics data and growth regions are listed in Table 13.1.

Table 13.1 Most important date palm cultivars grown in Djibouti

Name of cultivar[a]	Yield/ tree (kg)	No. planted trees	Fruit quality and characteristics	Growing region
"Ambabo"	10–20	More than 2,000	Good, soft	Tadjourah
Barhi[b]	40	More than 3,000	Good, soft	Ali-Sabieh, Arta, Dikhil, Tadjourah
Beladi	10–20	More than 4,000	Good, semisoft	Arta, Djibouti, Tadjourah
Harissa	10–20	More than 3,000	Good, dry	Dikhil
Imri	10–20	More than 50		Arta, Dikhil
Khadrawy[b]	70	Around 50	Good, soft	Ali Sabieh
Khalas[b]	40	More than 4,000	Good, soft	Ali-Sabieh, Arta, Dikhil, Tadjourah
Kisba	10–20		Good, dry	Arta, Djibouti
Medjool[b]	90	More than 500	Good, semisoft	Ali-Sabieh
Nabut Sultan[b]	40	More than 3,000	Good, soft	Ali-Sabieh, Arta, Dikhil
Oumo-Assala	10–20	More than 50	Good, soft	Arta
Rzizi[b]	30	More than 1,500	Good, soft	Ali-Sabieh, Arta, Dikhil
Sheeri	10–20	Around 100	Good, soft	Arta, Dikhil, Tadjourah
Zahidi	80	Around 30	Good, soft	Ali-Sabieh, Dikhil

[a]In alphabetical order
[b]Vitro-plants at their first years of production stage

13.7 Date Production and Marketing

In 2004, date production in Djibouti totaled about 541 mt. The Tadjourah area had the highest number of trees with 57 % of the national production. Overall, date palms are not very productive and the annual fruit yield may be as little as 20–30 kg per tree. The entire date production is consumed locally. Peak consumption is during festive periods such as Ramadan, wedding days, and other religious festivals. Usually dates are eaten fresh. However, some consumers have a particular attraction for yellow-colored dates, i.e., at their crisp (khalal) stage. There is an increasing demand for dates and for better-yielding vitro-plants. Therefore, domestic production cannot cover the needs of the locally marketed dates. To satisfy the local market, Djibouti imports about 400 mt of dates per year. The total national date demand is about 1,000 mt. Statistical data reveal that imports of dates have doubled in the last 10 years from 240 to 405 mt, between 2002 and 2013 (Fig. 13.5). Dates may be imported in different ways:

(a) Official imports are made by licensed companies which import large quantities of dates during the Ramadan and Eid periods.
(b) Informal imports are made by *charchari* (self-employed women) who import dates in bags of 15–30 kg from Saudi Arabia, Yemen, Oman, Qatar, and Kuwait. The *charchari* respond quickly to the needs of the market. Dates are sold at around USD 1.7 per kilogram except during the month of Ramadan where the price of dates may reach USD 3.4 per kilogram.

The government has funded many development projects to provide sustainable solutions to the major challenge of food insecurity which threatens more than 210,000 rural people (one-fourth of the national population) by promoting a policy aimed to improve agricultural production and productivity. One of the strategic axes consists of the development of intensive oasis agriculture and the enhancement of the quality of the date palm genetic resources. A goal is to make the country self-sufficient in dates. Different main issues were covered:

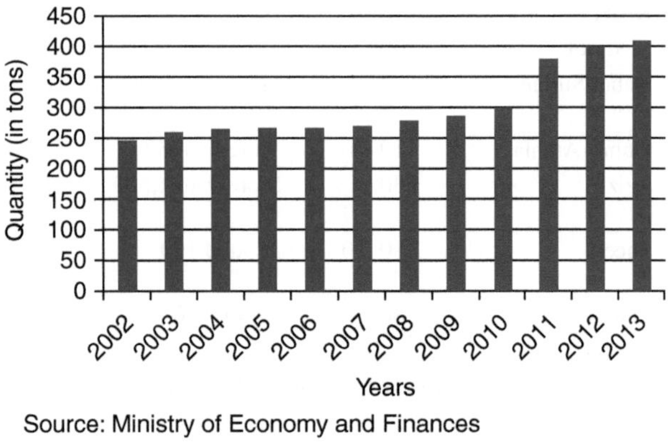

Fig. 13.5 Djibouti date imports from Arabian countries 2002–2013

(a) Developing and improving date palm plantations in the country by using new technologies and capacity building:

- Rehabilitation of existing date palm plantations
- Introducing elite cvs. like Medjool and Barhi and establishment of five date palm pilot farms for demonstration
- Introducing modern irrigation system (drip irrigation)
- Use of renewable energy solar pumping
- Provide adequate training for technicians and growers on date palm production

(b) Developing a research and development program which aims to:

- Characterize local genotypes and select the best performing of them that correspond to the demand of farmers and consumers
- Improve productivity of tissue culture laboratory
- Monitor and test behavior of vitro-plants imported from foreign countries

13.8 Processing and Novel Products

In Djibouti, most date palm groves are traditional plantations. Date production is marginal and does not meet the local demand. A portion of dates produced is consumed during maturation stage. The other part is consumed fresh at harvest stage. Spoiled dates are fed to livestock. Furthermore, the dates are not used for food preparations and for conservation. In addition, due to low production and high demand, no date processing is present in the country.

13.9 Conclusion and Recommendations

Djibouti has a rich and original date-growing heritage which covers a wide genetic diversity and has given rise to preferred and locally adapted date palm genotypes. Both foreign and local cultivars show promising production and good values in the local market. The date palm sector is expanding as a result of agriculture policies and a renewal of interest among the rural population for date palms. However, the development of date palm cultivation in Djibouti faces some constraints. First, some are hydroclimatic in nature. Date growing requires the availability of sufficient quantities of water of good quality. However, seasonal variations in groundwater availability and increasing salinity are the main limiting factors. The second main obstacle to the oasis development is the lack of know-how in cultivation practices and the instability of the workforce. The growing date palm sector should be aided by complementary training activities for farmers and measures to improve awareness of the general public of the importance of date palms, thus allowing the country to obtain sufficient numbers of productive palms and to respond to the high domestic demand for dates.

Acknowledgements We thank Sabira Abdoulkader, Adwa Abdou, Hami Said and Souleiman Hassan for their support in the writing of the manuscript. The authors would like to thank James Tregear for English corrections.

References

Aberlenc-Bertossi F, Noirot M, Duval Y (1999) BA enhances the germination of oil palm somatic embryos derived from embryogenic suspension cultures. Plant Cell Tis Org Cult 56(1):53–57

Aberlenc-Bertossi F, Castillo K, Tranchant-Dubreuil C et al (2014) *In silico* mining of microsatellites in coding sequences of the date palm (Arecaceae) genome, characterization, and transferability. Appl Plant Sci 2(1):1300058. http://www.bioone.org/doi/full/10.3732/apps.1300058

Awaleh M (1990) Les caractéristiques de la phoeniciculture djiboutienne et ses perspectives de développement. Options Méditerranéennes; Sér. A 1 O 1, – Les systèmes agricoles oasiens

Chérif E, Zehdi S, Castillo K et al (2013) Male specific DNA markers provide genetic evidence of an XY chromosome system, a recombination arrest and allow the tracing of paternal lineages in date palm. New Phytol 197(2):409–415

Ferry M, Ruiperez E, Navarro J (2003) Date palm axillary productions: morphogenesis and in vitro propagation. In: The date palm: from traditional resource to green wealth. The Emirates Center for Strategic Study and Research, Abu Dhabi, pp 87–102

Müller W (1982) Inventaire et mise en valeur des ressources en eau de la République de Djibouti. Bundesanstalt für Geowissenschaften und Rohstoffe. Project N° 78.2233.1. German hydrogeological cooperation, Hanover

Nabil M (2010) Le développement du palmier dattier en République de Djibouti. Actes du IIIème Séminaire du réseau AUF–BIOVEG, Biotechnologies du palmier dattier. F. Aberlenc-Bertossi (Ed), Editions IRD, France, pp 55–64

Peyron G (2000) Cultiver le palmier dattier. Groupe de Recherche et d'Information pour le Développement de l'Agriculture d'Oasis, France

Sané D, Aberlenc-Bertossi F, Gassama-Dia YK et al (2006) Histocytological analysis of callogenesis and somatic embryogenesis from cell suspensions of date palm (*Phoenix dactylifera* L.). Ann Bot 98:301–308

Toutain G (1990) Phoeniciculture et agriculture d'oasis à Djibouti. Situation actuelle et perspectives, Montpellier Gridao, Cirad-dsa

Part III
The Americas

Chapter 14
Date Palm Status and Perspective in the United States

Robert R. Krueger

Abstract The date industry in the United States is small in comparison with that of many other date-producing countries. The United States has no indigenous date palm (*Phoenix dactylifera* L.) germplasm, so the industry developed using imported cultivars in the early years of the twentieth century. Due to its specific climatic requirements, date production in the United States is confined to the low-elevation desert areas in California and Arizona. Date production has centered in the Coachella Valley of Riverside County, California, but has recently expanded into the Bard Valley of Imperial County, California, and Yuma County, California. The dominant cultivar is Deglet Noor but recent plantings are primarily Medjool, and this is becoming the dominant economic cultivar. The US industry is highly mechanized compared to the date industry in most other countries. The primary use of dates produced in the United States is for fresh consumption, but small amounts of processed products are also produced. Production practices, handling, and research activities are also discussed.

Keywords Backcross • Bard • Breeding • Coachella • Indio • Processing • Production • Tissue culture • Yuma

14.1 Introduction

The date palm (*Phoenix dactylifera* L.) is native to the Middle East, most probably originating in the area of present-day Iraq and Iran, and has been cultivated in that region since the third or fourth millennium BC (Zohary and Hopf 2000). Cultivation of the date palm spread into adjacent areas in North Africa, the Mediterranean region, and South Asia having suitable climates starting in that period and extending into the early centuries of the present era (Munier 1973; Zohary and Hopf 2000). During this period, it became established in southern Spain, and it was the Spanish

R.R. Krueger
USDA-ARS National Clonal Germplasm Repository for Citrus and Dates,
1060 Martin Luther King Blvd, Riverside, CA 92507-5437, USA
e-mail: robert.krueger@ars.usda.gov

J.M. Al-Khayri et al. (eds.), *Date Palm Genetic Resources and Utilization: Volume 1: Africa and the Americas*, DOI 10.1007/978-94-017-9694-1_14

who originally imported dates into what is now the United States (Nixon 1971; Rivera et al. 2013).

Dates were introduced into the Americas as seeds very soon after the initial European contact in 1492. Historical documentation exists that date palms were established in the Caribbean area and the mainland, extending to the Pacific coasts of Peru and Chile by the late sixteenth century. Spanish colonization of what is now northwestern Mexico and southern California did not occur until a later date; date palms were apparently introduced into Baja California and Alta California by the Jesuit and Franciscan missionaries in the late eighteenth century (Rivera et al. 2013). Date palm trees in the Californias were established at the missions. In southern California, the fruit was of minor interest due to the coastal climates in which the missions were established; nevertheless, some trees survived into the mid-twentieth century (Trent and Seymour 2010). In Baja California, date palms became naturalized in the desert oases and are still cultivated (Rivera et al. 2013).

These early date palm plantings never became established as agricultural industries after Alta California was annexed to the United States in 1850. However, dates as a food commodity were known to residents of the United States at least by the early nineteenth century, but the closer proximity of the European market to the production areas in the Middle East and North Africa resulted in a lack of supply of dates in the United States (Colley 1967a). This led the US Department of Agriculture (USDA) to investigate the possibility of producing dates in the Southwestern United States. Klee (1883) made detailed studies of various places in California possibly suitable for date production from Winters in the Sacramento Valley to San Diego in the south. Most areas were not suitable for date production due to low temperatures and humidity. However, the Central Valley was considered to have potential for date production. Unfortunately, Klee (1883) did not consider the desert regions of California, and it was not until 1904 that a detailed report on the possibilities of date production in the Salton Basin of the Coachella Valley was made by W. T. Swingle. Swingle (1904) found that the soils, climate (high temperature and low humidity), and water quality of the Coachella Valley were very similar to that of areas in the Middle East producing high-quality dates. Indeed, dates produced in the area were so superior to those produced in other areas of California that commercial date production was not further considered outside the low desert areas (Colley 1967a).

Early attempts at growing dates from seed were not completely successful due to the variability of date seedlings, and this led to attempts to import offshoots of desirable cultivars from the Old World (Colley 1967a; Nixon 1971). The first attempts to import offshoots were unsuccessful, and the first importation of offshoots from which palms survived occurred in 1890 under the direction of H. E. Van Deman of the USDA (Nixon 1971). The palms that survived were planted in areas generally not suited for date production; however, some palms planted in southern Arizona fruited, and this led to requests for additional importations of offshoots from the Old World. The USDA began importing offshoots successfully in 1900, the first such successful importation being from Algeria under the direction of W. T. Swingle. Additional USDA importations were made, the most important being in 1902 from Iraq, Baluchistan, and Egypt by D. Fairchild and in 1905 from Algeria and Tunisia by T. Kearney (Nixon 1947a, 1950, 1971). The last experimental importation of offshoots by the USDA was done by R. W. Nixon from Iraq in 1929 (Nixon 1971). In total, the USDA introduced 1,076 lots (more than 20,000 offshoots) between 1890 and 1929 (Nixon 1971; Traub and Robinson 1937). Additional importations were made by commercial interests from about 1911 to 1922, totaling over 40,000 offshoots. These importations are elaborated in Nixon (1947a, 1950, 1971).

Evaluation of the imported dates was initiated by W. T. Swingle in 1904 with the establishment of an experimental date garden at Mecca, California. In 1907, due to the threat of flooding of Mecca from the newly formed Salton Sea, a second research site was established near Indio, California. The Mecca Station was operated until about 1929 (Nixon 1971). The Indio facility was originally designated the US Date Garden until 1934, when its name was changed to US Date Field Station. In 1950, the name was changed again to US Date and Citrus Station (Cooper 1995). The Date Station was the principal site of date-related research in the United States until it was closed in 1982. Since 1982, date-related research in the United States has continued at a lower level, chiefly by the USDA, University of California, and University of Arizona personnel.

Commercial date production in the Coachella Valley began in the first decade of the twentieth century (Colley 1967b; Nixon 1971). One of the pioneers of Coachella Valley date production was Fred Johnson, who established a planting that included cv. Deglet Noor in 1905. Johnson's success with Deglet Noor was one of the main factors encouraging the planting of that cultivar (Nixon 1971). Various other growers established plantings during the period 1905–1915, and the first date packinghouse (the American Date Co.) was opened in 1912 (Colley 1967b; Nixon 1971). During this same time period and extending into the 1920s, various cooperative associations were established to promote and market dates. These included the Coachella Valley Date Growers Association (established 1913), the Coachella Valley Date Association (established 1919), and the California Date Growers Association (established 1919 as the Deglet Noor Date Growers Association) (Nixon 1971). Various packinghouses were also established during this period to support these associations as well as handle imported dates (Colley 1967b; Nixon 1971). It was in the early years of the industry that the Deglet Noor cultivar became predominant in the Coachella Valley, proving superior to the mostly seedling dates produced earlier.

One of the most serious threats to date palm production in the Coachella Valley was the *Parlatoria* scale (*Parlatoria blanchardi* (Targ.)). This insect pest was apparently introduced to the United States along with some of the offshoot importations. The offshoots were fumigated, but some scale apparently escaped death and became established in the Coachella Valley. Due to a lack of natural enemies, *Parlatoria* scale early on became a serious economic pest of date palms in the Coachella Valley. Eradication efforts were initiated by the USDA in 1914. These efforts were fairly successful but new infestations discovered in 1927 in California and Arizona made more intensive efforts necessary. A large-scale, concerted effort of inspection and treatment by the Federal Government and the State Governments of California and Arizona resulted in the successful eradication of the *Parlatoria* scale by 1934; the program was discontinued in 1936. The eradication of the *Parlatoria* scale was one of the somewhat rare examples of the complete eradication of an economic pest (Boyden 1941; Colley 1967a; Nixon 1971).

During the Depression Era of the 1930s in the United States, date growers endured hard times as did the remainder of the country. Fruit production actually expanded, but the local markets that had utilized most of the crop were not expanding. Fruit quality was variable and this resulted in low prices when it could be sold. In the mid-1930s, efforts were made to regulate the industry to eliminate low-grade dates in various manners, but these were unsuccessful. In 1938, a state marketing order was established that made it illegal to market inferior dates. It was only with this marketing order that high-quality dates were made the norm for commerce in the United States (Colley 1967b; Nixon 1971).

Date production in California has centered in the Coachella Valley, with a few smaller production areas in Imperial County, San Diego County, and Death Valley. Surface area devoted to date production in the Coachella Valley increased from slightly less than 700 ha (1,300 mt) in 1930 (Riverside County Agr Comm 1930) to a peak of approximately 1,977 ha in 1951 (Nixon 1971). (The production area was 1,959 ha (12,196 mt) in 1950.) Thereafter, the planted surface area decreased to a low of 1,628 ha (11,928 mt) in 1980 (Riverside County Agr Comm 1980). After 1980, the area around Indio began to urbanize rapidly, and date production shifted to the southwest of Indio and is now centered near the towns of Thermal and Mecca. Date production area increased to 2,217 ha (14,219 mt) in 2000 (Riverside Agr Comm 2000). In 2011, the last year for which production statistics were available at the time of writing, date production area in the Coachella Valley was 3,775 ha (24,002 mt) (Riverside County Agr Comm 2011). The shift in production area to the southeast portion of the Coachella Valley has resulted in many plantings that contain immature trees. Consequently, production per unit area has decreased somewhat despite advancements in production technology. The abandonment of mature date plantations has lead to the development of a satellite industry of selling mature palms for landscaping purposes. A grower may sell mature palms, sell the property for real estate purposes, and establish a young planting in a more rural area. Palms used for landscaping are valuable, bringing several thousand dollars per tree. This has resulted in some growers planting trees with the goal of eventually selling them for landscaping with the production of dates being a side venture until the trees are large enough for landscaping use.

Although the Coachella Valley is the largest date growing district in California, there is another important date growing area located in the Bard Valley to the northwest of Yuma, Arizona. Dates are mentioned in early reports from the Imperial County Agricultural Commissioner (Imperial County Agr Comm 1930). However, those reports do not report production surface and include early date plantings in the Imperial Valley. Date production in the Imperial Valley never achieved the importance it did in the Coachella Valley due to having heavier soils and more humid summer conditions, both of which are unfavorable for the production of the Deglet Noor cv. For this reason, the Medjool cultivar, which is more tolerant of these conditions, began to be planted in the Bard Valley in the 1940s. The Imperial

County Agricultural Commissioner (1970) reported 20 ha (158 mt), at least some apparently in the Imperial Valley since Nixon (1971) reported that there were 101 ha of cv. Medjool dates planted in the Bard Valley. This might be an error as the 1971 report (Imperial County Agr Comm 1971) reported 49 ha (414 mt) and the 1972 report (Imperial County Agricultural Commissioner 1972) reported 57 ha (481 mt). Possibly, Nixon (1971) misprinted *250 acres* (=101 ha) instead of *50 acres* (=20 ha), the size presented in the 1970 report. In any case, the early 1970s were the beginning of larger-scale plantings of Medjool in the Bard Valley. The planted surface increased steadily to 416 ha (2,573 mt) in 2000 (Imperial County Agr Comm 2000). In 2011, the last year for which production statistics were available at the time of writing, date production area in the Bard Valley was 514 ha (4,128 mt) (Imperial County Agr Comm 2011). The Bard Valley plantings are almost exclusively Medjool.

Across the Colorado River from the Bard Valley is Yuma, Arizona. The University of Arizona has maintained an experimental planting of dates in Yuma since 1905. Early plantings of the Deglet Noor cv. near Yuma were unsuccessful due to the climate. Consequently, the date industry of the Yuma area remained very small until recently (Nixon 1971). However, beginning in the late 1990s, the Medjool cv. began to be planted in large quantities in Yuma County due to the suitability of the climate to this cultivar and the lower cost of land and other inputs as compared to California (Dale 1997). As of 2013, there were approximately 2,000 ha of date palms in Yuma County (and less than 40 ha in the remainder of Arizona), of which approximately 70 % are considered mature trees; production is estimated at 9,000 mt (Wright G, 2013, personal communication). Date culture had also been tried in the Salt River Valley in Central Arizona beginning at approximately the same time as in the Coachella Valley. However, the climate of the Salt River Valley is not suitable for date culture due to the amount and pattern of rainfall. Consequently, date plantings in the Salt River Valley peaked at about 220 ha in 1947. Urbanization after World War II resulted in the subdivision of most date gardens in the Salt River Valley (Nixon 1971). The distinctive Sphinx cultivar associated with the Salt River Valley was cultivated in the backyards of suburbanites after World War II and harvested and packed by commercial interests. This has largely ceased, but a few homeowners still harvest and market Sphinx (Johnson et al. 2002).

Date palm culture was also attempted in South Texas starting in 1926. However, it soon became apparent that the conditions in South Texas were not conducive to date culture, and by the time of World War II, these experimental plantings had been abandoned. Similarly, the climate of Florida is not suited for date production, and no experimental plantings were made nor was a date industry established (Nixon 1971).

Date production in the United States is confined to small areas in the low desert areas of California and Arizona due to the specific climatic requirements for date production. Consequently, the date industry is small in comparison with most other agricultural industries. In 2011, the latest year for which statistics were available at the time of writing, dates were valued at USD 41,116,000 in the Coachella Valley and USD 24,001,000 in the Bard Valley. Total agricultural production in Riverside and Imperial Counties was USD 1.282 billion and USD 1.964 billion, respectively.

Thus, date production accounted for only a small percentage of agricultural receipts in the counties in which dates are produced. Agricultural production in the Coachella Valley was approximately USD 526.3 million, however, and dates thus represented slightly less than 8 % of receipts. A separate compilation of Bard Valley production was not available. Dates ranked ninth in crop value in Riverside County but did not appear in the top 10 crops in Imperial County (Imperial County Agr Comm 2011, Riverside County Agr Comm 2011). Note that the total value of agricultural production in California in 2011 was approximately USD 43.5 billion (California Department of Food and Agriculture 2012). Date production in Arizona is not as well documented. Dates are not shown separately in agricultural statistics for Arizona but are grouped in with Miscellaneous Fruits and Nuts, the value of which was USD 32,980,000 in 2011; total agricultural receipts for the state of Arizona were USD 4,372,000,000 (National Agricultural Statistics Service 2012). According to Arizona Cooperative Extension personnel, gross receipts for dates in Yuma County are approximately USD 79,000 per ha for mature date palms. Annual receipts might be as high as USD 60 million, most trees in Yuma County being fairly young (Wright G, 2013, personal communication).

Although the date industry is small in the scope of American or California agriculture, it is locally important in the desert valleys in which the dates are produced. The sale of date palms for landscaping purposes has resulted in additional income for date farmers, but the income from this source is not documented in the agricultural statistics. Date palms produced in the desert valleys are protected from disease by state quarantines and thus are considered of *high health* and thus more valuable to landscapers. This also allows movement into other areas of California and to other states.

The country of Mexico has a small date industry that is almost exclusively Medjool. There are approximately 400 ha in northern Sonora and Baja California. Production statistics are not known to the writer, but it has been suggested that the United States, Israel, and Mexico are currently the largest producers of Medjool in the world (Glasner B, 2012, personal communication). The traditional oases in Baja California Sur produce mainly *criollo* dates of low quality and value for local consumption. There is one commercial planting of various Old World varieties east of San Ignacio. There has also been an FAO project to promote date growing in Baja California Sur, but it does not appear to have resulted in any recent commercial plantings. Temperature and precipitation patterns in Baja California Sur might not be optimal for date production (personal observation of the author).

The US date palm industry currently faces several challenges. These may be divided into economic and cultural. Economic challenges include market development. Americans as a whole do not consume many dates. Therefore, as production increases with the newer plantings in Arizona and California, additional domestic and international markets must be developed. Development of export markets may be difficult as American-produced dates are high quality but are also costly to produce. Date consumers in other countries tend to be poorer and could not afford a high-priced date from the United States shipped across the Atlantic Ocean. Lower-quality dates from the Middle East or North Africa would be less costly to produce and ship, for example, to Europe. The additional economic constraints are basically those that cause American dates to be of high cost. These include input costs (labor, cultural inputs) and fixed costs (taxes, equipment, facilities). Cultural challenges include soil and water quality. Soils in the Coachella Valley are low in organic matter and very stratified. The strata can restrict root development, resulting in a less efficient tree. Equipment usage is high in the United States, and this can further compact soils. Water tends to be saline, resulting in buildup of salt and other production problems. Flood irrigation can result in high mosquito populations. Some of these problems can be addressed with improved soil and water management, such as tiling, slip plowing, and use of mini-sprinklers for irrigation (Adbul-Baki et al. 1997).

14.2 Cultivation Practices

The development of the US date industry involved a substantial input of research aimed at identifying appropriate cultural practices. The area of production in the United States was selected based upon climatic similarities with Old World areas producing superior-quality dates, and knowledge of production practices in those areas was available (Swingle 1904). However, adaptation of traditional date-producing practices to specific US conditions as well as the more industrial or mechanized society in the United States as compared to Old World date production areas was necessary. The majority of research concerned with improving US date production was performed by personnel associated with the US Date Garden, maintained by the USDA. In particular, the career of Roy Nixon from 1925 to 1965 resulted in many advances in date culture and its dissemination to growers (Cooper 1995). His work, and that of others, was consolidated into the publication *Date Culture in the Unites States*, first published in 1945, with a final edition appearing in 1978 (Nixon 1945, 1959, 1966; Nixon and Carpenter 1978). Reference will be made to the final (1978) edition of this document. Contributions to date production technology were also made by personnel from the University of California and University of Arizona. Some more recent advances have been summarized by Aslan et al. (1991) and Adbul-Baki et al. (2002).

The United States has always relied almost exclusively on propagation by offshoots (see Sect. 14.4). Early studies (Aldrich et al. 1945; Nixon and Carpenter 1978) on the propagation of date palms by offshoots indicated that the best results were achieved with offshoots that have reached a sufficient size (15–35 cm diameter, 10 kg) and have good root development. Best results are achieved when soil temperatures are sufficient for root growth (generally May through July). More recent studies (Hodel and Pittenger 2003a, b) confirmed these requirements and quantized the best offshoots for propagation as having 25 roots and 20 leaves. Planting density for date palms in the United States has traditionally been 250 trees per ha (approximately 9 m between and within row). More recently, more dense plantings have been utilized in some cases as growers sell mature palms for landscaping use and do not have large, mature palms as in the past.

Date palms require high inputs of water for acceptable production. Pillsbury (1937) estimated that Deglet Noor cv. date palms growing in a sandy soil consumed approximately 210 cm of water per year out of a total of approximately 270 cm applied. Application efficiency was therefore 75 %. Furr and Armstrong (1956) observed that the small-statured cv. Khadrawy extracted 130 cm of water from the upper 2.1 m of the soil profile. About 80–85 % of the date palms' feeder roots occurred in this portion of the soil profile. They estimated that the actual water usage was approximately 150 cm per year. Combining this figure with an estimated application efficiency of 75 % and taking into account Pillsbury's (1937) findings, they concluded that date production in the Coachella Valley would require 270–300 cm per year, including a leaching fraction. This figure has been used since that time without much comment or further investigation (Adbul-Baki et al. 2002).

Historically, date palm irrigation in the United States has been by flood or furrow irrigation and done on a calendar schedule. Bearing date gardens on light soils were irrigated every 7–14 days during the summer and every 20–30 days during the winter, with the goal of keeping the upper 2–2.5 m of the soil profile moist (Nixon and Carpenter 1978). This would have to be adjusted if cover crops are used, with young trees, or in heavier textured soils. In some cases, irrigations are reduced during the harvest season to reduce the possibility of physiological disorders such as blacknose that reduce fruit quality (Nixon and Carpenter 1978). More recently, the use of micro-sprinklers has become more common, as has the use of evapotranspiration data and soil moisture measuring devices for irrigation scheduling (Adbul-Baki et al. 2002). However, flood irrigation is still the most common method of irrigation application in date palms (Hodel and Johnson 2007).

Date palms are salt tolerant but excessive salt intake can cause reductions in yield and fruit quality (Adbul-Baki et al. 2002; Aslan et al. 1991). Most date palms are irrigated with water originating from the Colorado River. This generally has about 750 ppm total salt concentration. This should not cause problems with date production; however, occasionally, problems due to salinity or boron toxicity are encountered in US date production (Adbul-Baki et al. 2002). This might necessitate the application of excess water in order to leach the salts from the soil. In some cases, soil amendment is necessary to counteract the effects of excessive sodium buildup (Hodel and Johnson 2007). Soil modification is also sometimes necessary when soil conditions restrict water uptake or penetration (Adbul-Baki et al. 2002; Aslan et al. 1991).

Most investigations into the mineral nutrition of date palms in the United States has centered on nitrogen. Under the conditions of the various studies, responses to nitrogen fertilization have been variable. Investigations into the fertilization of date palms began in the 1920s (Winslow 1928). Furr and Barber (1950) reported increased yields from fertilized as compared to unfertilized dates, even though soil nitrogen levels were very low. The use of high levels of fertilization in date palms was attributed to experience with citrus and said to be of *questionable economy*. Furr and Cook (1952) observed nitrogen application of 2.7–3.6 kg N per tree to significantly increase the yield of Deglet Noor but not of Khadrawy. This was attributed to the presence of a cover crop in the Khadrawy and its overall low nutritional status. In contrast, Furr et al. (1951) did not observe differences in yield between fertilized and unfertilized Deglet Noor in the initial years of the study, although in subsequent years the unfertilized trees did decline in yield as compared to the fertilized trees. Similarly, Furr et al. (1952) did not observe a difference in yield between fertilized and unfertilized Khadrawy, but this might have been due to the presence of a cover crop of sweet clover. Furr and Armstrong (1957) reported little to no response to nitrogen applied to Deglet Noor for the first 7 years after planting and in Medjool cv. for 5 years after achieving full production. Furr and Armstrong (1960) also observed no response to fertilization for the first 5 years after planting Medjool.

These observations led to the recommendation for application of 1.8–2.7 kg actual N per palm per year from all sources (Nixon and Carpenter 1978). This might be organic or inorganic nitrogen, depending upon grower preference. Some growers prefer the use of manure, particularly if they wish to be organic certified. Manure rates of 11–33 mt per ha are generally used. Manure is applied in the late fall or early winter unless a cover crop is present, in which case it is applied in the spring after the cover crop is turned under (Nixon and Carpenter 1978). Date palms in the United States and other arid areas have generally not responded to applications of potassium or phosphorous (Nixon and Carpenter 1978).

Postlethwaite (1925) reported the use of a legume cover crop to increase soil nitrogen. Swann (1925), reporting from the heavier-soiled (as compared to Coachella Valley) Imperial Valley, noted that declining yields had been increased by the use of leguminous cover crops. Schoonover (1935) noted that "it may be possible to meet the requirements almost completely through cover crops" and also noted their beneficial effects on soil structure. Thackery and Leach (1936) also suggested the use of cover crops to improve soil organic matter and date production, with the emphasis on the use of leguminous crops. These early reports lead to the use of cover crops of legumes, particularly in young plantings. Cover crops can sometimes supply all the nitrogen needed by smaller palms. Summer cover crops, on the other hand, have not proven useful in US date production (Nixon and Carpenter 1978).

In addition to the use of cover crops as noted in the previous paragraph, Schoonover (1935) also discussed the importance of other aspects of soil management in date production. Many of the problems and concepts noted in Schoonover (1935) and other early reports, and efforts to eliminate these problems with slip plowing, cover crops, etc., are reiterated in the more recent reports of Aslan et al. (1991) and Adbul-Baki et al. (2002). Adbul-Baki et al. (2002) surveyed date palm producers in the Coachella Valley and observed soils with pH generally high (limiting micronutrient uptake) and organic matter levels generally low. Although no micronutrient deficiencies were observed, macronutrients were often unbalanced. The authors recommend various methods of soil modification (amendment, physical modification, cover cropping) and the use of leaf and soil analysis to deal with these problems.

Under US conditions, it was observed that yield was generally in proportion to the number of green leaves in the crown (Nixon 1940, 1943). Hence, pruning of functional leaves is not generally necessary except to facilitate cultural practices such as tying up bunches or pollinating. In a few cases, at least with the Deglet Noor cv., a palm may retain too many leaves, with many being below the fruiting level. This can increase the humidity of the microclimate around the bunches, thus promoting checking and other physiological problems. In this case, removal of leaves below the fruit zone has proven beneficial; however, with short-stalked cultivars, this can lead to reductions in yield (Nixon 1947b). If it is necessary to remove green leaves, this should be done in June so as to decrease humidity in the fruiting zone. In addition, if mature functional leaves are removed, bunches should also be thinned to maintain an adequate leaf-to-bunch ratio. Deglet Noor in full production can be pruned to 100–125 leaves (4–5 years of growth) and carry one moderately thinned bunch for each eight to nine leaves. Dead or partially dead leaves can be removed at any time, but this is best done before the leaf base hardens. Basal spines are generally removed from leaves during the winter to facilitate cultural care the following season (Hodel and Johnson 2007; Nixon and Carpenter 1978).

Depending upon climatic conditions for a specific year, date palms in the United States generally start blooming in February and bloom lasts until late March or early April. Pollen is generally collected, stored, and diluted before use in pollination. Dried pollen in sealed containers may be held over from one season to the next at temperatures of approximately 5 °C (Nixon and Carpenter 1978). Collection and drying is generally done by hand, although mechanical pollen extractors are available (Burkner and Perkins 1975). Pollinations are generally done in March and April (Nixon and Carpenter 1978). Under US conditions, pollination early in the season under cool temperatures can result in poor fruit set (Brown et al. 1969). Individual growers have various methods of performing pollinations. Mechanical pollinators are available (Perkins and Burkner 1973), but this can result in lower fruit set than with hand pollination (Nixon and Carpenter 1978). The use of personnel on lift trucks enables the use of small-scale pollinators combined with the advantages of rapid movement from tree to tree. For best fruit set, pollen is applied at 4–7-day intervals during the growing season (Nixon and Carpenter 1978). Early in the history of US date production, male or staminate trees were used more or less indiscriminately for pollen collection. Eventually, growers identified superior males based upon time of bloom, number and size of inflorescence, flower and pollen characteristics, and compatibility with female or pistillate cultivars. These male trees are generally maintained at the grower level although a few have become germplasm accessions (see Sect. 14.3). Metaxenic effects are also discussed in Sect. 14.3.

Under US conditions, fruit thinning was early on shown to be necessary for increasing fruit size, improving fruit quality, and preventing delayed ripening of date fruit; reducing the weight and compactness of the bunches; and ensuring adequate flowering the following growing season (Nixon and Crawford 1937). Fruit thinning is done by reducing the number of fruit per bunch (bunch thinning), reducing the number of bunches per palm (bunch removal), or a combination of both. Bunch thinning is done by either removing individual fruit from strands, removing entire strands, or both so that 50–75 % of the entire fruit load of the bunch is removed. Removal of individual fruit from the strands is labor intensive and expensive, so it is usually done only on the more profitable Medjool cv., which depends upon its large size for some of its economic value (Nixon 1951, 1956). As stated previously, entire bunches are sometimes removed in order to reduce the entire crop. This is done based upon the number of green leaves present on the palm and also to reduce alternate bearing tendencies. In some cultivars, the first bunches to appear are smaller than the later ones and are removed preferentially so the more robust later bunches develop the fruit crop. In addition, bunches that do not set well or are damaged should be removed. Information in this paragraph is mostly from Nixon and Carpenter (1978), which contains a more complete description of thinning methods.

After the pollination season, bunches are generally pulled down through the leaf blades and the fruitstalks are tied to leaf midribs. This is done to prevent scarring of the fruit and also provides support for the crop as fruit weight increases during the season. After the fruit reaches the khalal stage, the bunches are generally covered with brown craft paper or mesh bags. Some cultivars with open crows (e.g., Khadrawy, Halawy) are less sunburned when white rather than brown paper is used; Medjool growers generally use mesh bags. This practice is done to reduce the effects of moisture on the fruit. Increasing the ventilation of the bunch by the use of spreader rings placed in the center of the bunch also has this effect (Nixon and Carpenter 1978).

Date production in the United States has experienced relatively low levels of pest and disease pressure as compared with many other date-producing areas. Cultural practices and varietal composition influence disease and arthropod pest pressure. For instance, thinning can reduce humidity in the bunch, which can reduce rot and improve pesticide coverage, whereas bagging can increase humidity, reduce pesticide coverage, and allow the accumulation of dehisced fruit in the bottom of the bag, which can attract nitidulid beetles. Irrigation practices can increase pest pressure. Moisture stress can increase mite damage. Flood irrigation can result in additional pest pressure due to the presence of standing water but can also reduce pest pressure by hastening the decomposition of fallen dates, which can harbor insect pests. Orchard sanitation is important in pest control in dates, as are good management practices in general (Davies and Mauk 1997; Warner et al. 1990a).

Various arthropod pests have been reported in dates in the United States, the most important are the Banks grass mite (*Oligonychus pratensis* (Banks)), although the carob moth (*Ectomyelois ceratoniae* (Zeller)) and beetles of the Nitidulidae. Other arthropod pests are occasionally reported in dates but are not considered major economic pests.

The Banks grass mite (BGM) can cause scarring and drying of the date fruit surface during fruit expansion (Elmer 1965; Stickney et al. 1950). Damage from the BGM can result in extensive economic damage if not controlled. Traditionally, BGM was controlled using sulfur (Elmer 1966). However, during the 1990s, it was observed that sulfur had become ineffective. This was thought to be due to either resistance to sulfur developing in BGM or the sulfur reducing populations of beneficial insects and predators (Gispert et al. 2001). Various predators have been reported and help control the BGM, and newer pesticides have also been shown to be effective (Gispert et al. 2001; Mauk et al. 2005).

The carob moth was not reported in early investigations on date pests in the United States (Elmer 1966), as it was first reported in the Coachella Valley in 1982 (Davies and Mauk 1997). The carob moth can cause economic damage in dates from the feeding of the larvae, which leave webbing and deposit frass. The carob moth is usually controlled using malathion (Davies and Mauk 1997; Warner et al. 1990b). Current research is centered on pheromone trapping.

Nitidulid beetles are not considered to cause as much general economic damage as the BGM or the carob moth but can be as damaging to individual growers (Davies and Mauk 1997). The nitidulid beetles attack ripe fruit, especially soft and sour dates. Eggs are laid on the fruit, and the larvae feed on the fruit until they are grown, when they pupate in the soil. Control measures include fumigation in the packinghouse, field treatment with malathion, and orchard sanitation (Nixon and Carpenter 1978; Warner et al. 1990a, b).

The red palm weevil (*Rhynchophorus ferrugineus* (Olivier)) (RPW) is considered the most destructive insect pest of palms worldwide (Faleiro 2006). This pest was reported in a local planting of *Phoenix canariensis* Hort. ex Chab. in Laguna Beach, California, in October 2010 (USDA-APHIS 2010). This area is approximately 2 h west of the date production area and separated by mountain ranges. This detection is considered a severe threat to the date industry in the United States. However, actions by the California Department of Food and Agriculture and the USDA Animal and Plant Health Inspection Service have apparently eradicated RPW or confined it to the small section of Laguna Beach. A report of the related but less destructive South American palm weevil (SAPW) (*R. palmarum* (Linnaeus)) was reported in San Diego County, California, in May 2011 (USDA-APHIS 2011). Like the RPW, this has apparently been confined to a limited area separated from the date production area. Both of these introductions are being monitored, and it is hoped that they will be either eradicated or confined to the coastal area in the case of the SAPW.

The most important nonarthropod pests of date palms found in the United States are the various root-knot nematodes (*Meloidogyne* spp.). These are widely distributed in date palm orchard soils and can attack and kill date palm seedlings; however, the extent of their damage to producing stands is unknown and probably minor (Carpenter 1964).

Disease pressure is low in the date palm-producing areas of the United States. Minor losses from various pathogens have been reported but are not generally encountered. California state quarantines (see Sect. 14.3 below) keep the most potentially damaging pathogens out of the desert areas where date palms are produced.

In the coastal areas of California, *Fusarium oxysporum* Schlecht pv. *canariensis*, in association with *Gliocladium vermoeseni* (Biourge) Thom, causes wilt, dieback, and sometimes death in plantings of *Phoenix canariensis*. This pathovar of *F. oxysporum* also affects *P. dactylifera* (Feather et al. 1989). Consequently, a state quarantine prevents movement of palms from the coastal areas to the desert production areas, preventing the introduction of the wilt pathogens and protecting the date industry. This is considered the most serious potential disease threat to date production in the United States.

Pathogens occasionally reported in US date orchard but not considered to present economic threats include *Thielaviopsis paradoxa* (De Seyn) Hoehn. (= *Ceratocystis paradoxa* (Dade) C. Moreau), causing black scorch; *Graphiola phoenicis* (Moug.) Poit., causing Graphiola leaf spot; *Omphalia pigmentata* Bliss and *O. tralucida* Bliss, causing *Omphalia* root rot; and *Diplodia phoenicum* (Sacc.) Fawc., causing *Diplodia* disease. A minor disease of unknown etiology, rhizosis or rapid decline, has also been reported. Fruit rots caused by various pathogens can cause serious losses under humid conditions, causing spotting, rotting, and dropping. These are generally controlled by bunch management and other cultural practices but can occasionally cause considerable economic losses. Additional information on these pathogens and diseases can be found in Nixon and Carpenter (1978) and Carpenter and Elmer (1978).

14.3 Genetic Resources and Conservation

Date palms are not indigenous to the New World. Consequently, all genetic resources utilized in the development of the date industry were imported from the Old World (see Sect. 14.1 above). The most recent importation of date palm germplasm to the United States occurred as long ago as 1929 (Nixon 1947a, 1950, 1971). Although the number of cultivars currently used commercially is small (see Sect. 14.6 below), during the early years of date palm culture in the United States, many different cultivars were imported for experimental observation. Most developmental work on existing cultivars as well as breeding was done at the US Date Garden (USDG) in Indio, California. At its peak, the USDG probably had several hundred cultivars in its collection (Carpenter 1979b; Nixon 1950).

Date palm germplasm activities in the United States have been largely the work of the USDA. There were three main areas of germplasm-related activities carried out in the United States: varietal selection, metaxenia studies, and a breeding program.

The early research on date palm reproduction and breeding focused on the selection of seedlings from known cultivars (Swingle 1904). These were generally not successful. Most of the selected seedlings were inferior to the named Old World cultivars from which they were derived. By the 1930s, a few growers produced seedlings that "may have merit" (Traub and Robinson 1937). By the 1950s, there were many different cultivars that had been produced and named by local date growers (Hodel and Johnson 2007; Nixon 1955). However, most of these were grown only on a small scale by the originators and never became commercially important. Since the 1950s, a few more selections have been made by local growers, mostly in the Coachella Valley as production in the Bard Valley and Yuma County centers on Medjool. Generation of new cultivars of date palms by bud mutations has not been reported in the United States; bud mutations in date palms are "comparatively rare" (Traub and Robinson 1937).

The most important date palm germplasm-related activities that occurred in the United States were cultivar evaluations of the offshoots imported early in the twentieth century. The identification of suitable cultivars for the Coachella Valley and other areas led directly to the development of the US date industry as it exists today. The earliest evaluation of date palm cultivars was done by B. Drummond and S. C. Mason at the Date Station (Nixon 1971). Much of the later work was performed by Roy Nixon, who was instrumental in the development of many of the cultural practices used by the US date industry as well as the cultivar work. In addition to the straightforward characteristics shown, the cultivars were evaluated for such characteristics as yield, flavor, tolerance to humidity, and other characteristics which affect performance and profitability. The results of these investigations are summarized in Nixon (1950).

Swingle (1926) coined the term *metaxenia* to describe the direct effects of pollen from different male trees on the somatic tissue of the date fruit. Nixon (1928, 1934) showed that pollination of Deglet Noor with pollen from a male seedling of the Fard cv. produced larger dates which matured about two weeks earlier than pollination with pollen from the cv. Mosque. Pollen from *Phoenix reclinata*, *P. canariensis*, *P. roebelenii*, and *P. rupicola* produced smaller, later fruit than pollen from *P. dactylifera*, whereas pollen from *P. sylvestris* produced slightly larger fruit than mixed *P. dactylifera* pollen (although not as large as some date palm male selections) (Crawford 1935; Nixon 1928, 1934, 1935). Pollen from some male selections produced larger fruit in conjunction with increased size due to thinning. However, the increased size of the fruit produced from pollen from male selections did not lead to an increase in checking as did the increased size due to thinning (Nixon 1956).

These observations lent support to the common belief that some males are better than others for pollinating certain cultivars. This in turn suggests the identification and selection of superior males and the development of clones thereof to produce a desirable type of pollen in quantity. These considerations led to the identification of several seedling male palms of local origin with valuable characteristics. These were acquired before 1954 and are currently included in the Date Palm Germplasm Repository holdings (Carpenter 1979b; Krueger 2001).

The overall objective of date breeding is to achieve the highest fruit quality and yield (and profitability) consistent with local requirements (Carpenter and Ream 1976). The latter might include:

(a) Tolerance or resistance to cold, extreme heat, high humidity, rain damage, insect attacks, diseases, saline soil or water, poor drainage, and other soil-related problems
(b) Adaptation to mechanical harvesting, processing, and pest control
(c) Modification of growth habit to reduce the rate of vertical growth, reduce the number and size of spines, increase the length and flexibility of fruitstalks, improve distribution and numbers of fruits per bunch to increase size, and reduce thinning operations

(d) Modification of fruit quality, seed size, uniformity, and time of ripening and reduction of skin separation in soft cultivars
(e) Development of male palms with metaxenic characters that could be used to manipulate fruit production
(f) Development of inbred lines to produce seed with sufficiently uniform characters to permit propagation of palms from seed
(g) Some other factors which should be considered (Barrett 1973) are:

  (i) Identification or discovery of hermaphroditic flowers or monoecious lines
  (ii) Identification, discovery, or production of precocious lines

The most serious drawback in date breeding is time. The average time from seed to flowering is about 6 years (Nixon and Furr 1965). An additional drawback is the time required to produce enough offshoots for trials (5 years minimum or more if a large number are needed which would entail several generations of offshoot production). Finally, date palms do not reach full production until they are 10–15 years of age. All these factors make date palm breeding a long-term project. This is especially true when backcrosses are made.

Initial attempts at breeding date palms in the United States were made by the University of Arizona in 1912. These attempts were made to study the inheritance of fruit characteristics by inbreeding of Deglet Noor cv. Some observations of the inheritance of fruit color were made, but none of the progeny produced fruit as good as that of Deglet Noor and the project was discontinued after three generations (Nixon and Furr 1965).

In 1948, J. R. Furr and R. W. Nixon of the US Date and Citrus Station began a comprehensive date improvement program. Other participants over the years included C. L. Ream, H. Barrett, and J. B. Carpenter. The overall aims of the program (Barrett 1973; Carpenter 1979a; Carpenter and Ream 1976; Ream 1975; Nixon and Furr 1965) were:

(a) Production by backcrossing of male palms that approach the female parent in genetic composition
(b) Production of new and superior females by use of advanced backcross males in intervarietal hybridization
(c) Selection of superior male and female seedlings with the potential for commercial development

The initial phase of this breeding program lasted from 1948 to 1970 and was concerned primarily with production of backcrossed males. Towards the end of this period, intervarietal hybrids began to be made. The intervarietal crossing was intensified in the early to mid-1970s. However, it was during this period that US governmental policy with respect to the Date and Citrus Station changed. Support levels were cut, and the breeding program terminated. The station was closed in 1982, and the breeding lines were incorporated into the National Date Palm Germplasm Repository (Carpenter 1979b; Krueger 2001).

The cultivars used in making the backcrosses were the same as would be used as female parents in later intervarietal crosses. These cultivars were selected on the basis of possessing one or more outstanding characters (as expressed in Coachella Valley conditions) that might be desirable in a new cultivar. Primary considerations were fruit characteristics, such as large size, attractive color and appearance, good texture, and flavor; good shipping and storing quality; time of ripening; high yield; and rain tolerance (Nixon and Furr 1965).

Males resulting from the BC project were selected on the basis of good flower characters and resemblance of leaf characters to the female parent in the hope that there was sine linkage with fruit characteristics. Any male considered desirable for intervarietal crosses was also a candidate for saving. By 1965, 38 BC lines had been established. Some cultivars had advanced to the third BC generation (BC3), and many of them strongly resembled the female parents based upon vegetative characteristics (Nixon and Furr 1965). By 1976, the number of lines had been reduced to 22 lines representing 14 cultivars, as a result of genetic weaknesses in some lines and changes in breeding objectives (Carpenter and Ream 1976; Ream 1975). Some lines had advanced as far as the BC5 level. The more recently developed BC males had not been used in any intervarietal hybrids at the time the program was discontinued (Carpenter 1979b). There are currently 27 lines of BC males representing 11 cultivars in the Date Palm Germplasm Repository. These represent the lines considered most valuable at the time that the Date Station was closed.

The males produced above were to be screened for desirable male characteristics (metaxenia), but the ultimate objective was to utilize them in making intervarietal crosses to develop female cultivars with the following characteristics (Carpenter and Ream 1976):

(a) Quality and yield equal to, or superior to, those of Deglet Noor
(b) Freedom from blacknose, a serious physiological disorder of ripening Deglet Noor fruit
(c) Adaptation to mechanical harvesting and processing
(d) A fruitstalk at least equal in length and flexibility to that of Deglet Noor to facilitate handling of fruit bunches
(e) A moderate rate of vertical growth

Although production of BC males commenced in 1948, it was not until 1961 that males considered appropriate for intervarietal crosses were available. The first intervarietal crosses were made in 1961, and at the time of the first report on the breeding program, little information was available on the progeny (Nixon and Furr 1965). However, many advanced BC males flowered in 1970 and 1971, and 62 intervarietal crosses were made between 11 female and 13 BC male parents. From 1971 to 1974, about 3,000 hybrid seedlings, of which 1,200 were females, were planted, and evaluation of fruit characteristics of the females began in the following years (Carpenter and Ream 1976; Ream 1975). At the time of the report cited, nearly 70 % of the seedlings had flowered, considered a good percentage for trees less than 4 years old from seed. Limited data suggested that cvs. Empress, Khadrawy, and Thoory females and Khadrawy BC3, Tadala BC1, and Thoory BC3 males induced early flowering in a high proportion of the crosses in which they are used (Carpenter and Ream 1976; Ream 1975).

Limited observations of the characteristics of the crosses were summarized by Carpenter (1979b). Most of the crosses yielded inferior female seedlings. Many progenies were discarded after initial evaluation. The characteristics considered most deficient in the discarded progeny were fruit size, tolerance to high humidity and/or rain during the khalal stage, texture, flavor, appearance, and storage characteristics. The crosses made in 1971 yielded 9 seedlings considered worth saving. These included dry, semidry, and soft-fruited selections. Some of the soft or dry selections had the potential to compete with commercial cultivars. However, no semidry selection was found that appeared to equal Deglet Noor in quality, although some may have lent themselves to mechanical harvesting and processing. The 9 selections were incorporated into the National Date Germplasm Repository when the Date Station was closed (Carpenter 1979b; Krueger 2001). One selection, Medjool X (Dayri X Deglet Noor BC3) (PI 555427), has recently received some interest due to its large size (approximately 50 % larger than Medjool). It has a mild flavor and probably would not ship well due to its soft skin. However, it may be released to the public in the future if interest persists.

As there are no indigenous date palm or other *Phoenix* species in the New World, germplasm and genetic resource conservation of date palms in the United States has centered on the genotypes obtained from Old World sources. What might be termed the "National Date Palm Germplasm Repository" is currently a part of the USDA-ARS National Clonal Germplasm Repository for Citrus and Dates (NCGRCD). The origins of the date palm collection lie in the offshoots imported from the Middle East in the early part of this century and maintained and evaluated at the Date and Citrus Station. This collection of Old World (Nixon 1950) and local (Nixon 1955) date cultivars grew to large proportions over the years. In 1971, the decision was made to retain only those female cultivars that had commercial uses or had some valuable characteristics for breeding. As US governmental support for the Date and Citrus Station decreased, the date breeding program was terminated, and the most valuable materials were incorporated into the National Date Palm Germplasm Repository in 1977 (Carpenter 1979b). The Date and Citrus Station was completely closed in 1982, and the date palm collection was repropagated as offshoots to the USDA-ARS Irrigated Desert Research Station in Brawley, California, about 100 km south of Indio in Imperial County. The NCGRCD was established by the USDA Agricultural Research Service in 1987 as a part of the US National Plant Germplasm System (Shands 1995). The date palm germplasm collection was incorporated into the NCGRCD in 1989. US governmental support for the Brawley station began to decrease in the early 1990s. The threat of closure prompted the repropagation of the collection to the University of California Coachella Valley Agricultural Research Station (CVARS) in Thermal, California. Thermal is only a few km from Mecca, the original site of the US Date Garden, and is a more suitable area for date culture.

The date palm germplasm collection currently (2013) consists of 28 Old World female cvs., 9 New World female cvs., 30 BC male lines, 17 intervarietal hybrids, 5 superior males, 10 seedling lines from historic Baja California Sur oases, and 18 unverified accessions, for a total of 117 accessions. There has also been an effort to acquire additional *Phoenix* species (sometimes as seed), and currently, 15 accessions representing 7 *Phoenix* species are maintained (*P. acaulis*, *P. canariensis*, *P. hanceana*, *P. paludosa*, *P. reclinata*, *P. roebelenii*, and *P. sylvestris*). These accessions are free of pathogens due to a California State Internal Quarantine of the date production area and are available for distributions to qualified researchers worldwide as offshoots, seed, and pollen.

The referenced California State Internal Quarantine basically states that palm materials may not be moved into the date palm production area in California except under special permit. This is intended to prevent the introduction of pathogens not currently present in the production area, specifically *Fusarium oxysporum* Schldtdl.:Fr.1824 (CCR Title 3 Section 3419). In addition, there exists a California State Interior Quarantine intended to prevent the introduction of *Phymatotrichum [Ozonium] omnivorum* (Shear) Duggar 1916 into the date palm production area (CCR Title 3 Section 3401) and a California State Exterior Quarantine intended to prevent the introduction of palm lethal yellowing from outside the State (CCR Title 3 Section 3282). These regulations at the state level are intended to protect the date industry. They have had the side effect of resulting in high health palms in general, and ornamental palms produced in the California desert are considered desirable by the landscaping industry. At the federal level, date palms are a *prohibited* commodity as per USDA-APHIS (7 CFR Section 319.37-2). This is also intended to prevent the introduction of pests exotic to the United States. The existence of these quarantines makes introduction of new cultivars or germplasm difficult. Specific permits need to be issued based upon protocols designed to minimize risks. This entails facilities and programs that are not currently available. Political differences between the United States and countries that are the center of origin and diversity have also complicated the introduction of new date palm germplasm in the past. Some of these political differences have been resolved but the phytosanitary restrictions remain. The recent emphasis on the "clean plant" concept by USDA-APHIS has resulted in increased restrictions in some areas. For instance, in the past, it was possible to acquire date palm cultivars as tissue-cultured plantlets under a 2-year postentry quarantine. However, even tissue-cultured date palms are now prohibited entry into the United States without a departmental permit.

The conservation of date palm genetic resources is the main effort in the United States in the area of genetics and breeding of date palms. Funding levels from the US Federal Government are such that there is only one person actively working in this area part time. Financial resources from the date palm industry and the various universities are directed towards management and production problems. Nevertheless, there are limited amounts of research activities that have been carried out in the area of date palm genetics and breeding.

The United States has only a relatively low range of date palm genotypes used by the industry. However, those genetic resources that are maintained have been well curated over the decades and remain a valuable resource. It is envisioned that the United States will continue to collaborate in various investigations with researchers in countries in which date palms are more agriculturally and culturally important. Moreover, the US NPGS is at the forefront of global efforts in plant genetic resource conservation. Possible future roles for the United States may include serving as a backup for date palm germplasm resources from other countries with richer date palm genetic resources. This would be dependent upon development of novel quarantine protocols and long-term preservation methodologies, as well as increased international cooperation (Krueger 2011). These issues will most probably be resolved, and the United States will remain an important international resource for specific areas of date palm genetic and breeding research.

14.4 Plant Tissue Culture

In the United States, tissue culture propagation of date palms was first reported by Tisserat (1979) and Tisserat et al. (1979). These scientists reported the production of adventitious plantlets from lateral buds, shoot tips, embryos, and stem and rachilla tissue of date palms on a modified Murashige-Skoog medium. More complete descriptions of the technique were subsequently published by Tisserat (1981, 1982, 1983, 1984, 1988). Tisserat (1988) expanded his technique to other palm species. Tisserat and DeMason (1980) and Tisserat (1985) reported the histological and developmental events associated with the production of date palm vitroplants.

Smith and Aynsley (1995) reported that Barhi cv. date palms produced by a commercial tissue culture laboratory according to the protocol of Tisserat (1981) produced fruit indistinguishable from fruit produced by offshoot-propagated palms. Fruit production began in the sixth year and reached commercial yields in the eighth year, a similar time frame to fruit production by offshoot-propagated date palms. Despite this report, commercial propagation of date palms never became viable in California, and the private laboratory mentioned above went out of business in approximately 2000. Grower acceptance of tissue-cultured date palms in the United States was variable. Some growers reported many offtypes and poor fruit production, while other growers reported that the trees and fruit were acceptable (Ben Laflin, personal communication, ca 1998).

The reason for this is primarily the small size of the US date industry and the stability of the cultivar profile. Tisserat (1983) estimated that utilizing offshoots for propagation would allow only an expansion of only 40–80 ha per year. However, until recently, the extent of the industry was relatively stable or declining slightly. Both of the dominant cultivars, Deglet Noor and Medjool, produce relatively large numbers of offshoots. Most of the recent expansion of date palm cultivation area has been to Medjool, which produces a large number of offshoots. The fact that offshoot production is inexpensive and has proven sufficient for the US industry has reduced the need for a commercial tissue culture industry for date palms in the United States. Recently, there has been renewed interest in commercial tissue culture production of date palms. However, efforts in this area are oriented towards external markets more than the United States.

14.5 Cultivar Identification

As the United States has no indigenous date palm germplasm, there were no native varieties or genotypes to describe. Consequently, efforts in morphological descriptors were aimed at cultivars imported from the Old World, which often had attached some sort of description. These descriptions were sometimes derived from the native sources and recorded by early travelers and plant explorers and, in other cases, were made directly by the travelers and collectors. As mentioned in Sect. 14.1, these plant explorers were primarily USDA personnel, although in some cases they were working for private interests; however, other plant explorers were in fact private citizens. The accounts of their travels and explorations are interesting and contain important information on date varieties and culture. These published accounts include Fairchild (1903), Kearney (1906), Milne (1913), Popenoe (1913a, b, 1914, 1926, 1973), Mason (1915, 1923, 1927), Dowson (1921–1923), Brown (1924), Nixon (1934), Brown and Bahgat (1938), and Swingle (1945). A more complete list, including more obscure documents, is found in Nixon (1950).

After importation of Old World date cultivars into the United States, it was necessary to validate their identity by morphological characterization. Morphological characteristics of specific cultivars in the US might vary somewhat from their characteristics in the Old World due to climate and cultural practices. However, since the area chosen for date production in the United States has a climate similar to that in date-producing countries (Swingle 1904), morphological characteristics should be more or less the same in both areas. Obvious offtypes would be identified and either removed or, if possessing superior or interesting characteristics, retained as a new line. The morphological characterization of imported cultivars and documentation of which cultivars were best suited for production in the United States were important steps in the development of the American date industry. Similarly, the discovery of interesting and potentially useful seedlings required observation and documentation.

The vast majority of this morphological characterization was performed at the USDA Date Station, particularly by Roy Nixon. American workers had early on realized the importance and utility of using leaf characteristics for identification purposes (Mason 1915), and the descriptor system developed by the Date Station incorporated vegetative as well as fruit characteristics (Nixon 1950). Vegetative characteristics documented included general appearance (stature, stoutness), leaf characteristics (color, curvature, length), leaf base characteristics, spine characteristics, pinna characteristics (size, droopiness, angles, spacing), and fruitstalk characteristics. Fruit characteristics included color, shape, size, calyx, skin and flesh characteristics, flavor, season, and seed characteristics.

By 1950, over 200 Old World cultivars had been characterized (Nixon 1950). Of these, 16 were considered major cultivars, 28 minor cultivars, and the remainder obscure. Not all cultivars described were known to be accurately named, and in some cases, reidentifications were made based upon characteristics or other observations. Due to the date of publication, Nixon (1950) lists cv. Medjool as a minor cultivar, whereas today (2013), it is tending towards dominance. Nixon (1955) made similar observations of seedling selections made in the United States. Hodel and Johnson (2007) include Nixon's observations with some newer commentary and color photographs. Due to the contraction in the number of cultivars of dates grown in the United States, the lack of importations of new cultivars since 1929, and the proprietary nature of newer grower selections, there has been little descriptive work done on dates in recent decades. Some internal characterization has been made at NCGRCD for specific purposes.

The first marker system developed for use in date palm genetic analysis was based on isozymes. Tisserat and Torres (1979) and Torres and Tisserat (1980) used isozymes to study inheritance in seedling populations with known parents from the USDA breeding program. Five isozymes were coded by 7 polymorphic genes with 14 alleles. Additional polymorphism was detected in other *Phoenix* spp. As expected, inheritance followed classical Mendelian genetics, and the genetic uniformity of offshoots compared to mother palms was confirmed.

Although various marker systems have been used to assess date palm genetics (Meerow et al. 2012), there has been only a small amount done in the United States. The first widely disseminated report on their use in date palm genetic analysis was by Cao and Chao (2002). Four primer sets were used to detect polymorphisms in 23 date cultivars maintained by NCGRCD. Based upon the UPGMA-cluster analysis of 328 polymorphic bands, the majority of the cultivars fell into two major groups. The same group (Devanand and Chao 2003a, b) used AFLP to determine genetic consistency between commercial plantings of Deglet Noor and Medjool. They reported little variation within cultivated Deglet Noor but considerable variation within cultivated Medjool. This led the authors to postulate that Medjool represents a landrace in its native Morocco. Further analysis of Moroccan-derived Medjool showed only 79 % genetic similarity within 66 collections of Medjool from three locations in Morocco, thus supporting the contention that Medjool is a landrace (Elhoumaizi et al. 2006). The same group used AFLP to analyze Egyptian date palm cultivars (El-Assar et al. 2005). The majority of the accessions fell into one major cluster with only a few cultivars showing large divergence. It was postulated that this main cluster represented the major group of date palm germplasm in North Africa, which could have been derived from introductions from the center of diversity in the Middle East that had only low levels of genetic diversity. Chao's group also reported variation in the methylation status of mother plants and offshoots (Fang and Chao 2007). Changes in DNA methylation on deoxycytidine residues have been shown to be involved in the regulation of plant development by way of gene expression at the transcriptional level and in response to environmental stresses (Finnegan et al. 2000).

Johnson et al. (2009) analyzed 18 cultivars maintained at the NCGRCD using SSR markers originally developed for oil palm (*Elaeis guineensis* Jacq). Three primer pairs were identified that unambiguously identified each of the date palm cultivars. The polymorphic bands were used to design new internal primers. However, all amplifications with the new primers yielded only polymorphic bands, indicating that the variation among these date palm cultivars was at or near the original primer sites, with internal sequences preserved.

Zhao et al. (2013) analyzed 28,889 EST sequences from the date palm genome database to identify simple sequence repeats (SSRs) and to develop gene-based markers, i.e., expressed sequence tag-SSRs (EST-SSRs). We identified 4,609 ESTs as containing SSRs, among which, trinucleotide motifs (69.7 %) were the most common, followed by tetranucleotide (10.4 %) and dinucleotide motifs (9.6 %). The motif AG (85.7 %) was most abundant in dinucleotides, while motifs AGG (26.8 %), AAG (19.3 %), and AGC (16.1 %) were most common among trinucleotides. A total of 4,967 primer pairs were designed for EST-SSR markers from the computational data. In a follow-up laboratory study, we tested a sample of 20 randomly selected primer pairs for amplification and polymorphism detection using genomic DNA from date palm cultivars. Nearly one-third of these primer pairs detected DNA polymorphism to differentiate the 12 date palm cultivars used (maintained at NCGRCD). Functional categorization of EST sequences containing SSRs revealed that 3,108 (67.4 %) of such ESTs had homology with known proteins.

Although there is not any active genomic work being done with date palms in the United States, there has been an important involvement with the genomic work being done by the Weill Cornell Medical College – Qatar Date Palm Genome Project (http://qatar-weill.cornell.edu/research/datepalmGenome/index.html). This group recently submitted a complete genome generated by shotgun sequencing using massively parallel sequencing of a Khalas cv. female date palm (Al-Dous et al. 2011; Malek 2010). The date palm genome was estimated to be 658 Mbp in size. In addition, 8 additional genomes were sequenced, including Deglet Noor and Medjool females and backcrossed males maintained at NCGRCD. A total of 28,890 gene models were predicted, of which over 85 % of the protein-encoding genes showed homology with sequences in the NR database at NCBI. Approximately 1.7 M SNPs were called in 381 MB of sequence for a heterozygosity rate of 0.46 %. More detailed information and annotations are available at the website. The Qatar group attributes a large amount of the success of this project to the availability of genotypes maintained at NCGRCD, in particular Deglet Noor BC5 (PI 555432). Collaboration between NCGRCD and the Qatar group has continued.

14.6 Cultivar Description

Because the United States is not a center of origin or diversity for dates, the industry developed using imported cultivars (see Sect. 14.1). Due to the pattern of industry development, two cultivars currently dominate US date production: Deglet Noor and Medjool. Deglet Noor was originally imported in 1900, and large-scale importation occurred during the years 1911–1921 (Nixon 1950). Due to the characteristics well adapted to production in the Coachella Valley, Deglet Noor became the predominant cultivar produced there. Importation of Medjool did not occur until 1927, and it was not released from quarantine until 1936 and released to growers until 1944 (Swingle 1945). By this time, Deglet Noor was established as the leading cultivar in the United States. However, it quickly became apparent that Medjool was superior under US production conditions and it became the second cultivar. The later establishment of date production in the Bard Valley and the recent developments in Yuma County rely exclusively on Medjool. Even in the Coachella Valley, most recent date plantations have been established using Medjool, and the predominance of Deglet Noor has decreased. The current (2013) season's cultivar mix in the Coachella Valley is 70 % Deglet Noor, 29 % Medjool, and 1 % minor cultivars (Cooper L, 2013, personal communication). With the Bard Valley being nearly 100 % Medjool, it would appear that in California as a whole, Medjool now accounts for approximately 38 % of production surface. With the Yuma County plantings being almost entirely Medjool, the production surface of Medjool in the United States probably exceeds that of Deglet Noor. Many of the Arizona plantings are not yet mature, so production of Deglet Noor might still be higher than that of Medjool. However, this can only be an estimate since there are no official figures available for Arizona date production.

The following descriptions were adapted from Nixon (1950, 1955) and Hodel and Johnson (2007). More complete descriptions of these cultivars as well as many other genotypes are available in those references.

The Deglet Noor cultivar is said to have originated early in the seventeenth century near Touggourt, Algeria. It was early on recognized as a superior cultivar and was established in many other oases in Algeria and Tunisia by the end of the century, and production there increased greatly during the French colonial period. It became apparent during the initial years of its importation that Deglet Noor did not do well in Arizona or in Imperial County, California, due to the heavier soils in those areas and more humid conditions during the summer months (Deglet Noor is susceptible to fruit damage from humid conditions). Consequently, the Coachella Valley became the center of Deglet Noor culture and of date production. Deglet Noor fruit are attractive in appearance and possess a delicate, distinctive flavor. They are firm in texture, shrink less in curing, and hold their shape better in packing, handling, and storage than softer cultivars. In storage, Deglet Noor fruits are less marred by "sugar spots" (white or light-colored areas on the skin) than other cultivars. However, fruit quality is very sensitive to environmental conditions, particularly heavy or alkaline soils and humid conditions during ripening. Deglet Noor fruit checks (develops small, linear scars near the apex) and develops blacknose (blackening and shriveling of the tip) more readily than most other cultivars when grown in humid conditions. Losses due to rot and fermentation under these conditions also occur. Deglet Noor is usually a semidry date under US conditions. However, as with other semidry types, a relatively high proportion of dates will fall into the dry category when grown under very dry conditions or seasonal weather. Deglet Noor grows and yields best on relatively light soils that retain water but still allow sufficient drainage. Growing on such soils and managed appropriately, Deglet Noor comes into full bearing in 12–15 years, at which times yields are generally in the range of 90–140 kg per tree. Under less than optimum conditions, Deglet Noor yields and quality are often disappointing. Deglet Noor also appears to be more susceptible to root rots caused by *Omphalia pigmentata* Bliss and *O. tralucida* Bliss. Offshoot production of Deglet Noor is usually in the range of 8–12 relatively

Fig. 14.1 Mature Deglet Noor cv. date palm orchard, Coachella Valley, California (Photo by D. Karp)

slender offshoots. Since Deglet Noor has long been the most commonly grown date palm, most of the date palms transplanted for ornamental use are Deglet Noor, even though Medjool and Zahidi are considered more attractive and valuable.

Deglet Noor palms have long (Fig. 14.1), slightly arched leaves with relatively stiff pinnae, the longest and oldest of which show slight drooping and occasional bending. The olive green of the foliage, the long spine area, and the numerous spines, a few of which occur in groups of three, are a combination of characteristics usually sufficient for characterization. The light, coral red color of the khalal fruit is distinctive. The trunk of Deglet Noor is slender to medium heavy. Leaf curvature is slight and fairly uniform with a blade length of 300–500 cm. Leaf bases are narrow to medium broad, green, and somewhat glaucous, with older ones having a little maroon. Leaf spines are 10–60 in number and occupy one-quarter to one-half of the blade length. Pinnae are slight to moderate, with irregular drooping and occasional bending in the pinnae lower blade; the longest pinnae are 70–90 × 1.4–2.3 cm just above the spine. Fruitstalks are up to 1.6 m long, slender to medium heavy, with 61 rachillae. The color of Deglet Noor fruit is light red at khalal, amber at soft fruit rutab, brown or straw colored at dry fruit rutab, and slight deeper at tamar (unless subjected to high temperatures or held a long time to storage), with a light bloom. The shape is oblong to ovate, 40–50 × 20–25 mm; skin is medium thick, adhering to the flesh and forming coarse wrinkles and folds in curing; flesh 4–5 mm thick, firm, soft, and amber except for paler inner zone. Deglet Noor is a late ripening cultivar. Seeds are medium brown, narrowly elliptical, 23–30 × 7–9 mm, with central germ pore and furrow usually closed through the middle and continuing as a slight depression towards the apex and base, with ventral surface more or less flattened.

The Medjool cultivar has become the most important and desired date worldwide due its large size, attractive appearance, soft flesh, and excellent taste. Medjool was early on exported to European markets and was highly esteemed. However, production in Morocco was nearly wiped out by bayoud disease. All current Medjool in the United States originated from a single bayoud-free palm in the Boudenib oasis in Morocco, from which offshoots were taken to the United States in 1927 by W. T. Swingle. When released from quarantine, the desirability of Medjool for production in the United States became apparent. Although the first plantings of Medjool were in the Coachella Valley, the Bard Valley and later Yuma County, Arizona, became the center of Medjool production. Medjool is better adapted to soils and seasonal humidity in those areas than is Deglet Noor. In particular, it shows little checking when exposed to humid summer conditions. Fruit of Medjool can be quite large and can become crowded and damaged on the rachillae. This makes thinning necessary. Thinning of both individual fruit and rachillae is done, and bunches opened up with spreader rings at the khalal stage. This also helps increase airflow through the bunch and reduces blacknose. Inflorescence removal is also practiced to increase fruit size and stabilize yields from year to year. Production of Medjool is thus more labor intensive, but the higher price commanded by the large, high-quality fruit more than makes up for this. Medjool generally yields 70–90 kg per tree when mature, although higher yields are probably possible. Offshoot production is prolific, and this has contributed to the expansion in production of Medjool in the United States and Mexico.

The size, shape, color, and texture of Medjool fruit are distinctive, and seeds are generally winged. The trunk of Medjool is medium heavy. Leaves are short to medium long, with slight, uniform curvature. Leaf bases are medium broad, green, slightly glaucous at first, later becoming yellowish with maroon mottling or streaking in the center, very sparse scurf on leaves. Spines number 30–38 and occupy approximately one-quarter of the blade length. Pinnae show slight to moderate drooping, the largest being 70–82 × 2.4–3.0 cm, in groups of three or four. Fruitstalks are orange-yellow and are short to medium long and heavy. Fruit are orange-yellow with a reddish-brown stippling at khalal, ripening to amber at rutab, and reddish brown, more or less translucent, at tamar. The fruit has moderate to pronounced bloom. The shape is broadly oblong-oval to somewhat ovate, 48–48 × 26–32 mm, with medium-thick skin that adheres to the flesh in curing and forms coarse, irregular wrinkles. Flesh is 5–7 mm thick, moderately soft, with very little rag. This cultivar is early ripening. Seeds are dark brown, oblong or oblong-elliptical, usually winged, 18–24 × 8–9 mm, with germ pore below middle and furrow variable, closed in the middle or narrow and shallow to deep, a little wider at the base and apex.

While Medjool and Deglet Noor cvs. account for nearly all the dates grown in California, there are a few Old World cultivars that are occasionally sold on site or at other small venues such as farmers' markets. These are generally only available in small quantities.

Barhi cv. is a soft date originally from Basra, Iraq. The fruit is small to medium, ovate to nearly round, and yellow at khalal. It becomes amber on ripening and deep brown when cured. The fruit is low in astringency at the khalal stage, and there is a small niche market for khalal-stage fruit among residents of Middle Eastern or

North African heritage. Barhi is late ripening and has heavy yields of up to 140 kg per tree. Mature fruits of Barhi are of high quality, but quality deteriorates gradually during storage. The fruit is somewhat susceptible to checking, but losses from fruit rot and souring are low.

Dayri cv. is a semidry date from Iraq producing medium to large fruit, oblong to oblong-elliptical, and dull rose over a deep chrome yellow. It turns into dark reddish brown as it matures. The softer fruit is attractive but the drier fruit usually proves disappointing unless softened commercially. Dayri ripens mid-season. Yields are variable, usually being in the range of 70–90 kg per tree. Dayri leaves (*lulav*) are the most demanded for use in Jewish *succot* rituals, and rabbis often come to the Coachella Valley to purchase them for use in other areas.

Halawy cv. is a soft date from Iraq, where it is regarded as one of the better cultivars. It is damaged relatively little by humid conditions but has a tendency to shrivel during ripening. This is usually not enough to be considered a market defect, unless the season is dry and it is grown on light soils with low water holding capacity. The best production of Halawy is on heavy soils with adequate irrigation, where yields of 90–115 kg per tree are obtained. The fruit is small to medium, oblong with a rounded apex, and yellow, becoming light amber upon ripening and translucent golden brown when cured. Halawy ripens early.

Khadrawy cv. is a soft date from Basra, Iraq and is well adapted to a range of conditions. The fruit is small to medium, oblong-ovate, and light yellowish, becoming greenish amber on ripening and reddish brown when cured. The fruit is slightly less susceptible to humid conditions than many other varieties and is not seriously affected by rot. The fruit cures and keeps well with less shrinkage than most other cultivars. The trees are distinctively small in stature as compared with any other cultivar, and yields are also light, rarely exceeding 55 kg per tree. Khadrawy is an early ripening cultivar.

Thoory cv. is a date originating in Algeria. It is the best dry date grown in the United States due to the fruit's large size, attractive appearance, and good flavor, but it is rarely planted due to the market's preference for softer dates. The fruit is only slightly damaged by humid conditions. The fruit is medium to large, oblong with a rounded apex, and yellow, ripening and curing to a light grayish brown or straw color. When fully cured, astringency is low in this late ripening cultivar. However, there is a tendency to harvest it early since dry dates do not show much contrast between the fresh-ripe and cured stages, and when this occurs, the flavor may have more tannin content. Yields are usually 90–115 kg per tree.

Zahidi cv. is a semidry date from Iraq, where it was one of the major cultivars in the Basra area. The fruit is less tolerant of humidity than cvs. Halawy and Khadrawy. The fruit are small to medium, obovate, and yellow, turning amber when ripe and reddish brown when cured. The flavor is considered good but not outstanding, and it ripens mid-season. Zahidi is considered to handle well; growers do not need to harvest individual fruit but can remove the entire bunch and manipulate early and late fruit in maturation chambers and humidifiers, respectively. Yields are high at 90–135 kg per tree. Zahidi is considered one of the most desirable landscaping palms.

In addition to Old World cultivars, some growers have made proprietary selections, usually seedling selections of Deglet Noor or Medjool. Some of the older selections are described in Nixon (1955) and Hodel and Johnson (2007) and some of the newer ones in the latter reference only. Mention will be made of the distinctive Sphinx cv. of Arizona (also known as Black Sphinx). The origin of Sphinx is unknown, but it is possible that it is a seedling of Hayany discovered near Phoenix, Arizona, around 1920. Sphinx is well adapted to conditions in the Salt River Valley, and its distinctive culture was mentioned in Sect. 14.1. The fruit is large, oblong to oval, medium red at khalal, and very dark brown to nearly black at rutab and tamar. The flesh with little rag and the flavor are considered to be good. Sphinx is a late ripening cultivar with high yields of up to 135 kg per tree. Vertical growth is slow, Sphinx being only somewhat taller than Khadrawy.

14.7 Dates Production and Marketing

The extent of the date industry in the United States is discussed in Sect. 14.1 above. Date producers in the United States are generally relatively large industrial producers utilizing high levels of production inputs. Production inputs include contract labor or company personnel, equipment (tractors, lifts, etc.), fertilizer, and water. Date production is labor-intensive compared to most contemporary farming activities. Although mechanization has made date farming easier than in the past, the cultural practices described in Sect. 14.2 above still involve much manual activity. Most of the cultural practices (pollination, thinning, etc.) require access to the crown of the tree, which may be 10 m or more above the ground in a mature date garden. In the past, ladders, climbing belts, or a combination were used. In the remaining mature date gardens in the Coachella Valley, it is common to see 3–6 m ladders permanently fixed to trees just below the crown. This was accessed by a ladder moved from palm to palm. After reaching the crown of a tree, a climbing belt was used, and a chain (sometimes with a hook) was used to attach the belt to three or four leaf bases for support while working in the crown. In some cases, only climbing belts were used (Nixon and Carpenter 1978). Working in the crown of a tall tree in this manner is hazardous. Currently, most growers access the crowns of the trees using various types of mechanical lifts. Although there are still hazards associated with working in the crown of a tall tree, they are reduced using mechanical lifts.

As stated in Sect. 14.1, there is only a small amount of khalal harvesting done in the United States. The main cultivar harvested at the khalal stage is Barhi, which typically reaches this stage in July or August. There is also only a small production of dry-type dates. These are generally harvested after fully ripening on the tree, at which time all fruit is removed at once (Nixon and Carpenter 1978; Rygg 1977).

Semidry dates dominate US production. The harvest season usually lasts 3–4 weeks for early-maturing cultivars and 3–4 months for later-maturing cultivars. In addition to a cultivar's earliness or lateness, the harvest date is influenced by local weather conditions and consumer preference as well as production factors such as labor availability. The threat of an early rain can move harvest dates up in order to avoid fruit yield and quality losses due to the humid conditions. The cost of grading dates that have suffered moisture or insect damage is higher than it would otherwise be. However, the cost of handling dates harvested before full maturity is

Fig. 14.2 Harvesting Medjool cv. dates using a mechanized lift and platform. Note mesh bags protecting fruit bunches (Photo by D. Karp)

also higher than if allowed to ripen more fully on the tree, since artificial methods must be used to complete the ripening process and in many cases dehydration is also necessary. Dates begin to lose both astringency and color after the khalal stage is reached. Moisture also begins to be lost at khalal and this continues up to the tamar stage unless conditions are humid in the date garden or storage facility. In some regards, date palatability is highest for many people immediately after khalal when the fruit is at its juiciest and plumpest condition. However, the fruit is also most fragile and difficult to handle at this time. Most people prefer to consume dates at a slightly later stage, when they have lost some moisture and become more pliable but not yet tough. Proper harvesting of dates requires coordination between the grower and packer in order to assure a successful harvest (Nixon and Carpenter 1978; Rygg 1977).

Currently, dates in the United States are harvested mostly using human labor. Workers are raised into the crown of the tree on lifts (Fig. 14.2), and dates are harvested into various types of containers (Fig. 14.3). Generally, these are either plastic bins or specialized wire baskets. Due to the tenderness of the dates, these containers cannot be more than two or three layers of fruit in depth (Nixon and Carpenter

Fig. 14.3 Harvested Medjool cv. dates being loaded into trays for transport to packing facility, Coachella Valley, California

1978). These containers are usually lowered to the ground from the lift's platform, where they are placed in shallow trays to dry or taken to the packinghouse after being shaken off the stalks by manual or mechanical shaking. The valuable soft cultivar Medjool is usually harvested sequentially in order to obtain the largest number of large, high-quality fruit. However, each harvesting event is an additional expense, and with less valuable cultivars, the entire bunches is generally harvested. Deglet Noor is well adapted to this technique, and generally, an entire block is harvested in one or two bunches harvested. The semidry Deglet Noor can also be *mechanically* harvested (Peightal 1962; Perkins and Brown 1964). This is somewhat of a misnomer, since it does involve human labor. A worker in a boom harvests bunches into the boom's basket. The basket is then lowered to the ground and the dates are mechanically shaken by a trailer pulled by the boom truck. Most Deglet Noor fruits are currently harvested in this manner, and labor input is reduced by approximately 75 % (Rygg 1977) with no loss in quality (Nixon 1963; Nixon and Furr 1961, 1962).

Fig. 14.4 State-of-the-art Medjool cv. packing facility, Yuma, Arizona (Photo by D. Karp)

Dates in the United States are generally packed by the individual grower, by a cooperative, or by custom packing operations. Since date production is a small industry in the United States, packing equipment designed specifically for date handling is rare. Equipment in packinghouses is generally adapted from that used for handling other types of fruit. Pack lines for dates are generally fairly simple, particularly those used by small independent producers. Recently, a large cooperative packing operation was opened in Yuma, Arizona (Fig. 14.4). This operation is oriented entirely towards Medjool and represents the current state of the art in date packing in the United States.

Upon arrival at the packinghouse, the first operation to be performed has traditionally been fumigation for disinfestation of insects. This requires specialized facilities that are well sealed to prevent the escape of noxious gases and are temperature controlled to ensure the death of the insects. Some packers fumigate outside under plastic tarps. Fumigation has traditionally been done with methyl bromide (Nixon and Carpenter 1978; Rygg 1977). Since methyl bromide is being phased out of agricultural use, investigations are underway to identify the most efficient and safest alternate method to disinfest dates and other crops. Currently (2013), methyl bromide use in date packing has remained constant for the last several years. This is due to the fact that the date industry is small, and only small amounts of methyl bromide are used in comparison with other crops. It is expected that methyl bromide use will be restricted in 2014 or 2015. The methyl bromide alternative sulfuryl fluoride is less useful since it does not kill insect eggs as methyl bromide does. Another alternative, phosphide, is not economically feasible since it would require the modification of the existing fumigation facilities (Cooper L, 2013, personal communication).

After fumigation, dates are cleaned and culls removed and sorted into lots of uniform ripeness, size, consistency, and appearance for further handling. This is generally done on tables in small packinghouses and belts in the larger houses (Nixon and Carpenter 1978; Rygg 1977). More advanced packinghouses have recently started using electronic devices to automate the various sorting functions. In the United States, date standards are defined in 7CFR.B.IX.987. These standards actually apply only to Deglet Noor, Zahidi, Halawy, and Khadrawy produced in Riverside County (i.e., Coachella Valley). This marketing order is administered by the California Date Administrative Committee, located in Indio, California (http://www.datesaregreat.com). Bard Valley and Yuma County have a separate association, Bard Valley Medjool Date Growers Association (http://www.naturaldelights.com).

In some instances, dates are picked when not completely mature and artificial ripening is necessary. This is generally done in maturation rooms that have controlled temperature and humidity and a means of circulating air. Temperatures for maturation of Deglet Noor should not exceed 35 °C. Whether or not artificial maturation is necessary and how much is a matter of judgment based upon observations of fruit quality made by the packer. Chemically induced maturation has not proven useful with dates (Nixon and Carpenter 1978; Rygg 1977). Dates with very high moisture content must be dried before they can be marketed. This is generally accomplished either in the maturation room or a similar facility at temperatures of 29–23 °C. Deglet Noor is dried to a moisture content of 23–25 % and Medjool to 25–30 %. Conversely, some dates are picked when too dry due to delayed harvest, inadequate irrigation, or seasonal weather patterns. In the United States, dates are generally hydrated by exposing them to steam at 60 °C for 4–8 h. In some cases, particularly with small lots, dates can be hydrated by placing them in water under a partial vacuum and then returning them to atmospheric pressure. This causes the uptake of water into the flesh (Rygg 1977). In some cases, dates are pasteurized to prevent spoilage.

Dates intended for fresh consumption are sometimes pitted. Pitting is also sometimes necessary for processed products. Mechanical pitting machines are available. Dates for fresh consumption are generally packed in 2.3 kg, 4.6 kg, or 6.9 kg cardboard boxes. Smaller lots are often packaged in various sizes and shapes of plastic boxes or bags. Dates are generally stored in cold facilities after being packed as described previously. The dates are generally in the cardboard boxes or sometimes in bulk bins. The storage life of dates is affected by temperature and humidity, with lower temperature and humidity enabling a longer storage life. Deglet Noor can be stored for up to a year at 0 °C, whereas Medjool and other soft types maintain better quality when stored at −18 °C. Refrigerated transportation is also necessary for successful distribution to retailers (Rygg 1977).

Most dates produced in the United States are consumed fresh. The scope of the industry estimated from state-level statistics is outlined in Sect. 14.1 above. According to the US National Agricultural Statistics Service (NASS) (http://www.nass.usda.gov/Statistics_by_Subject/index.php), in 2011, the total production of dates in the United States was 30,000 mt with a value of USD 37,356,000 at the farm gate and USD 45,347,000 as utilized. According to these NASS statistics, none of the production was processed.

According to the USDA Economic Research Service (http://www.ers.usda.gov/data-products/fruit-and-tree-nut-data/) in market year 2011/2012, total US exports of dates were 5,400 mt at a value of USD 35 million. Date exports in the 2011/2012 market year went primarily to Canada (35 %, USD 10,800,800) and Australia (29.4 %, USD 10.7 million). According to the California Department of Food and Agriculture (2012) statistical review, 19 % of California date production (5,600 mt) was exported for a value of USD 31 million. This was said to be 100 % of US date exports. This figure may include Arizona production as many producers in Yuma County, Arizona, are based in the Bard Valley, California. However, as stated in Sect. 14.1, date production in Arizona is not well documented and the California share may be overestimated. This may be reflected in the slight difference between US and California statistics. Date exports in 2011 went primarily to Canada (34 %, USD 10.3 million), Australia (28 %, USD 8.4 million), and the European Union (17 %, USD 5.1 million). US dates are of high quality but also expensive, so exports tend to be to developed countries.

In market year 2011/2012, the United States imported 22,000 mt of fresh dates valued at USD 30 million and only 157 mt of processed dates valued at USD 320,000. The main countries from which fresh dates were imported into the United States in 2011/2012 were Israel (39 %, USD 9.3 million), Pakistan (23 %, USD 8 million), Tunisia (20 %, USD 9.3 million), and Mexico (6 %, USD 2.2 million). Thus, the United States is a net importer of dates when volume is considered. However, when value is considered, the United States is a net exporter of dates due to the higher returns from premium US-grown dates.

The per capita date consumption in the United States is generally characterized as low. However, this really depends upon the country used in the comparison. The per capita date consumption in the United States is low compared to countries in the Middle East or North Africa where dates are traditionally an important part of the diet. However, US per capita date consumption is probably not particularly low compared to countries where dates are not traditionally grown and consumed. The fact that date production in the United States has increased in the last several decades and that the United States imported 22,000 mt in comparison with 30,000 mt domestic production in 2011/2012 suggests that date consumption in the United States is sufficient both to support the domestic date industry and a substantial amount of imports. The large population of the United States may compensate somewhat for low per capita consumption.

In any case, the date packing industry actively promotes the use of dates both for fresh consumption and cooking. As for fresh consumption, the emphasis is on the health benefits of dates. Promotion of the use of dates for cooking has recently involved their use in gourmet or *foodie* uses as compared with traditional uses in baking. The California Date Administrative Committee has in recent years sponsored a competition aimed at promoting the use of dates in haute cuisine. In addition, as more people from traditionally date-consuming countries immigrate into the United States, the use of dates in ethnic restaurants will also increase. In terms of consumption, the future of the US date industry looks good.

However, the US date industry also faces challenges. In the Coachella Valley, the traditional heart of date production in the United States, the industry has already moved its center southeast of its original area in response to urbanization pressures. Urbanization results in upward pressure on land values (and thus taxes) and increased competition for water and sometimes generates conflicts between urban residents and production agriculture. Urbanization pressures are less intense in the Bard Valley and Yuma County, but the latter is also urbanizing. All date-producing areas in the United States face the usual internal pressures from rising input costs, labor shortages, as well as state and federal regulations. How the US date industry balances its advantages against these challenges remains to be seen. However, the recent expansion in area, although not indefinitely sustainable, points towards an optimistic industry with a secure future.

14.8 Processing and Novel Products

Most dates produced in the United States are consumed fresh. However, a few processed products are produced from lower-grade dates, mostly Deglet Noor. The processed date industry is a small segment of the industry as a whole but does provide a way to utilize low-grade dates that would otherwise be without value. Date processed products fall into two categories: those aimed at industrial use and those aimed at direct consumer use.

Industrial products are used in baked foods, dairy products, cereals, or confections (http://www.datesaregreat.com/food-professionals/). These products are said to offer the advantages of being cost-effective, easily incorporated ingredients. They are less acidic than some other dried fruits and consequently will not interfere with chemical leavening systems or greatly alter yeast-leavened dough characteristics. Their natural sweetness compliments a wide range of dairy product formulations. Industrial products include:

(a) Dehydrated date pieces (also known as date sugar). Used to add flavor and sweeten very firm dough where fruit must remain intact; visual identity for dry bakery mixes and fillings
(b) Date paste, finely ground, extruded through a fine screen. Used as filling for baked goods, confections, and cereals; formed with enhancers; natural, substitute sweeteners. Maintains product freshness
(c) Extruded date pieces, coated with oat flour, rice flour, dextrose, or other anticaking material. Used in applications that require uniform date product based on specific size needs

(d) Diced dates, coated with oat flour, rice flour, dextrose, or other anticaking material. Used in cereal and baked products that require the particles separate and are free-flowing
(e) Date concentrate, a concentrated and liquefied date by-product. Used in liquid/beverage applications such as dairy shakes; fills, baking
(f) Macerated dates, which contain no pits and have excellent flavor and texture. This is the most versatile date product. Hygroscopic properties help maintain freshness where large pieces of fruit are desired such as breads and cakes. Used in confections or made into paste

Processed products aimed at direct consumer use include products similar to the industrial products listed above but are used at consumer-level cooking. In addition, some processed products are used directly in the consumption of food. These include date butter and date syrup. In addition to these uses for lower-grade dates, good-quality dates are processed into various directly consumable products such as stuffed dates, dates covered with powdered sugar or mixed with nuts, etc.

In contrast with many other countries, which utilize other plant parts of the date palm in addition to the fruit, in the United States, the date fruit is the main product. There is no other industrial use of date palm materials. The use of mature date palm trees in landscaping was mentioned in Sect. 14.1 above. This is the main economic use of date palms other than fruit production.

14.9 Conclusions and Recommendations

Date production in the United States is a small, niche enterprise possible only in a few geographical areas having a suitable climate. The industry has expanded in recent decades, but future expansion will probably be limited by decreasing amounts of land with a suitable climate. The recent expansion into Yuma County, Arizona, has greatly increased the area of Medjool plantations, and Medjool will probably be the preferred cultivar for any future expansion of the industry. Urbanization of the traditional Coachella Valley growing area has pushed date production to the southeast. Continued urbanization and its attendant effects (higher land prices, more expensive labor, etc.) will potentially constrain expansion of US date production. US date production is already more expensive than in other countries, but the high quality of US dates has so far allowed the industry to remain profitable.

Date production in the United States will remain a small industry in the future, even if some expansion occurs. Date production, although not a native enterprise in the desert valleys, has become naturalized and contributes not only to the local economy but also to the local culture and society of the US low desert communities. Its survival as a profitable enterprise is dependent upon continued high quality and increasing the efficiency of production. If increased efficiency enables the continued profitability of date production, it will remain an important part of local agricultural enterprises, culture, and society.

References

Adbul-Baki AA, Aslan S, Cobb S, Aguiar J (1997) The future of date production in the US. Chron Hort 37(1):18–19

Adbul-Baki A, Aslan S, Linderman R et al (2002) Soil, water and nutritional management of date orchards in the Coachella Valley and Bard, 2nd edn. California Date Commission, Indio

Al-Dous EK, George B, Al-Mahmoud ME et al (2011) De novo genome sequencing and comparative genomics of date palm (*Phoenix dactylifera*). Nat Biotech 29:521–527

Aldrich WW, Leach GH, Dollins WA (1945) Some factors influencing the growth of date offshoots in the nursery row. Proc Am Soc Hort Sci 46:215–221

Aslan S, Neja R, Estrada D et al (1991) Soil, water, and climatic considerations in selecting date palm planting sites in the Coachella Valley. Coachella Valley Resource Conservation District, Coachella Water District, and Riverside County Farm Bureau, Indio

Barrett HC (1973) Date breeding and improvement in North America. Fruit Var J 27:50–55

Boyden BL (1941) Eradication of the parlatoria date scale in the United States. US Department of Agriculture Miscellaneous Publications 433, Washington, DC

Brown TW (1924) Date palm in Egypt. Egypt Min Agr Tech Sci Bull 43, Min of Agr, Cairo

Brown TW, Bahgat M (1938) Date-palm in Egypt. Egypt Min Agr Hort Sect Booklet 24, Min of Agr, Cairo

Brown GK, Perkins RM, Vis EG (1969) Temperature and heat unit occurrence during date pollination in the Coachella Valley of California. Date Grow Inst Ann Rep 46:21–24

Burkner PF, Perkins RM (1975) Mechanical extraction of date pollen. Date Grow Inst Ann Rep 52:3–7

California Department of Food and Agriculture (2012) Calif Agr Stat Rev 2012–2013. California Department of Food and Agriculture, Sacramento

Cao BR, Chao C-CT (2002) Identification of date cultivars in California using AFLP markers. HortScience 37:966–968

Carpenter JB (1964) Root-knot nematode damage to date palm seedlings in relation to germination and stage of development. Date Grow Inst Ann Rep 41:10–14

Carpenter JB (1979a) Breeding date palms in California. Date Grow Inst Rep 54:13–16

Carpenter JB (1979b) The national date palm germplasm repository. Date Grow Inst Ann Rep 54:29–32

Carpenter JB, Elmer HS (1978) Pests and diseases of the date palm, Agriculture handbook 527. US Department of Agriculture, Washington, DC

Carpenter JB, Ream CL (1976) Date palm breeding, a review. Date Grow Inst Ann Rep 53: 25–33

Colley CC (1967a) The California date growing industry, 1890–1939. Part I. South Calif Quart 49:47–63

Colley CC (1967b) The California date growing industry, 1890–1939. Part II. South Calif Quart 49:167–191

Cooper WC (1995) The US Horticultural Research Laboratory: a century of USDA subtropical-horticultural research. Florida Citrus Research Foundation, Orlando

Crawford CL (1935) Growth rate of Deglet Noor dates in metaxenia. Proc Am Soc Hort Sci 33:51–54

Dale D (1997) A growing market. Calif Farm (September):35–38

Davies J, Mauk PA (1997) PMA pest management evaluation. Proposal to the California Date Commission. California Date Commission, Indio

Devanand PS, Chao CT (2003a) Identification of genetic strains of 'Medjool' and 'Deglet Noor' date palm cultivars in California using Amplified Fragment Length Polymorphism (AFLP) markers. Acta Hort 623:333–340

Devanand PS, Chao CT (2003b) Genetic variations within 'Medjool' and 'Deglet Noor' date (*Phoenix dactylifera* L.) cultivars in California detected by fluorescent-AFLP markers. J Hort Sci Biotech 78:405–409

Dowson VHW (1921–1923) Dates and date cultivation of the Iraq. Agricultural Directorate, Ministry of the Interior, Mesopotamia
El-Assar AM, Krueger RR, Devanand PS, Chao C-CT (2005) Genetic analysis of Egyptian date (*Phoenix dactylifera* L.) accessions using AFLP markers. Genet Res Crop Evol 52:601–607
Elhoumaizi MA, Devanand PS, Fang J, Chao C-CT (2006) Confirmation of 'Medjool' date as a landrace variety through genetic analysis of 'Medjool' accessions in Morocco. J Amer Soc Hort Sci 131:403–407
Elmer HS (1965) Banks grass mite, *Oligonychus pratensis*, on dates in California. J Econ Ent 58:531–534
Elmer HS (1966) Date palm insect and mite pests in the United States. Date Grow Inst Ann Rep 43:9–14
Fairchild DG (1903) Persian Gulf dates and their introduction into America. Bull No 54. US Department of Agriculture, Washington, DC
Faleiro JR (2006) A review of the issues and management of the red palm weevil *Rhynchophorus ferrugineus* (Coleoptera: Rhynchophoridae) in coconut and date palm during the last one hundred years. Int J Trop Insect Sci 26:135–154
Fang J-G, Chao CT (2007) Methylation-sensitive amplification polymorphism in date palms (*Phoenix dactylifera* L.) and their off-shoots. Plant Biol 9:526–533
Feather TV, Ohr HD, Munnecke DE, Carpenter JB (1989) The occurrence of *Fusarium oxysporum* on *Phoenix canariensis*, a potential danger to date production in California. Plant Dis 73:78–80
Finnegan EJ, Peacock WJ, Dennis ES (2000) DNA methylation, a key regulator of plant development and other processes. Curr Opin Genet Develop 10:21723
Furr JR, Armstrong WW (1956) The seasonal use of water by Khadrawy date palms. Date Grow Inst Ann Rep 33:5–7
Furr JR, Armstrong WW (1957) Nitrogen fertilization of dates – a review and progress report. Date Grow Inst Ann Rep 34:6–9
Furr JR, Armstrong WW (1960) Effects of source of nitrogen and rate of application on young non-bearing Medjool date palms. Date Grow Inst Ann Rep 37:13–15
Furr JR, Barber HD (1950) The nitrogen content of some date garden soils in relation to soil management practices. Date Grow Inst Ann Rep 27:26–30
Furr JR, Cook JA (1952) Nitrogen content of pinnae, fruit, and seed of Deglet Noor and Khadrawy date palms as related to nitrogen fertilization. Date Grow Inst Ann Rep 29:13–14
Furr JR, Currline EC, Hilgeman RH, Reuther W (1951) An irrigation and fertilization experiment with Deglet Noor dates. Date Grow Inst Ann Rep 28:17–20
Furr JR, Currlin EC, Armstrong WW (1952) Effects of water shortage during ripening and of nitrogen fertilization on yield and quality of Khadrawy dates. Date Grow Inst Ann Rep 29:10–12
Gispert C, Farrar C, Perring TM (2001) Seasonal abundance of the Banks grass mite Oligonychus pratensis (Banks) (Prostigmata: Tetranychidae) and a predatory mite, and their response to sulfur treatment on commercial date palms *Phoenix dactylifera* L in *Southern California. Proceedings of the 10th International Conference of Acarology*, pp 403–408
Hodel DR, Johnson DV (2007) Imported and American varieties of dates (*Phoenix dactylifera*) in the United States. ANR Publication 3498. University of California, Oakland
Hodel DR, Pittenger DR (2003a) Studies on the establishment of date palm (*Phoenix dactylifera* 'Deglet Noor') offshoots. I. Observations on root development and leaf growth. Palms 47:191–200
Hodel DR, Pittenger DR (2003b) Studies on the establishment of date palm (*Phoenix dactylifera* 'Deglet Noor') offshoots. II. Size of offshoots. Palms 47:201–205
Imperial County Agricultural Commissioner (1972) Imperial County Agricultural Commissioner Report. Imperial County Agricultural Commissioner, El Centro
Johnson DV, Joyal E, Harris RK (2002) Date palm varieties in Arizona. Fruit Gard 34(5):6–9, 26
Johnson C, Cullis TA, Cullis MA, Cullis CA (2009) DNA markers for variety identification in date palm (*Phoenix dactylifera* L.). J Hort Sci Biotech 84:591–594

Kearney TH (1906) Date varieties and date culture in Tunis. Bull No 1125. US Department of Agriculture, Washington, DC
Klee WG (1883) Culture of the date. US Department of Agriculture Department Report 24. Washington, DC
Krueger RR (2001) Date palm germplasm: overview and utilization in USA. In: Proceedings the first international conference on date palms. Al-Ain, March, 1998, pp 2–37
Krueger RR (2011) Date palm germplasm. In: Jain SM, Al-Khayri JM, Johnson DV (eds) Date palm biotechnology. Springer, New York, pp 313–336
Malek JA (2010) Next generation DNA sequencing applied to the date palm tree (*Phoenix dactylifera* L.). Acta Hort 882:249–252
Mason SC (1915) Dates of Egypt and the Sudan. Bull No 271. US Department of Agriculture, Washington, DC
Mason SC (1923) The Saidy date of Egypt: a variety of the first rank adapted to commercial culture in the United States. Bull No 1125. US Department of Agriculture, Washington, DC
Mason SC (1927) Date culture in Egypt and the Sudan. Bull No 1457. US Department of Agriculture, Washington, DC
Mauk PA, Perring TM, Shea T (2005) Management of Banks Grass mite in dates. Topics Subtrop 3(4):2
Meerow AW, Krueger RR, Singh R et al (2012) Coconut, date, and oil palm genomics. In: Schnell RJ, Priyadarshan PM (eds) Genomics of tree crops. Springer, New York, pp 299–351
Milne D (1913) The date palm and its cultivation in the Punjab. Punjab Government, Layallpur
Munier P (1973) Le palmier-dattier. G-P Maisonneuve & Larose, Paris
National Agricultural Statistics Service (2012) 2011 Arizona Agricultural Statistics Bulletin. Arizona Agricultural Statistics, Phoenix
Nixon RW (1928) The direct effect of pollen on the fruit of the date palm. J Agr Res 36: 97–128
Nixon RW (1934) The Dairee date, a promising variety for testing in the Southwest. Circ No 300. US Department of Agriculture, Washington, DC
Nixon RW (1935) Metaxenia in dates. Proc Am Soc Hort Sci 32:221–226
Nixon RW (1940) Fruit thinning of dates in relation to size and quality. Date Grow Inst Ann Rep 17:27–29
Nixon RW (1943) Flower and fruit production in the date palm in relation to the retention of older leaves. Date Grow Inst Ann Rep 20:7–8
Nixon RW (1945) Date culture in the United States. Circ 728, US Department of Agriculture, Washington, DC
Nixon RW (1947a) Importations of date offshoots and the men who made them. Date Grow Inst Ann Rep 24:20
Nixon RW (1947b) Can a date palm carry too many leaves? Date Grow Inst Ann Rep 24:23–27
Nixon RW (1950) Imported varieties of dates in the United States. US Department of Agriculture Circ 834, Washington, DC
Nixon RW (1951) Fruit thinning experiments with the Medjool and Barhee varieties of dates. Date Grow Inst Ann Rep 28:14–17
Nixon RW (1955) American varieties of dates. Unpublished manuscript on file at USDA-ARS-NCGRCD
Nixon RW (1956) How many fruit per strand should be left in thinning the Medjool date? Date Grow Inst Ann Rep 33:14
Nixon RW (1959) Date culture in the United States. Agr Inf Bull 207, US Department of Agriculture, Washington, DC
Nixon RW, Furr Jr. (1962) Frequency of picking in relation to quality of Deglet Noor dates in 1961. *Report. 39th annu. Date Grs' Inst., Coachella.* 16–18
Nixon RW (1963) Quality of Deglet Noor dates in relation to method of harvesting. Date Grow Inst Ann Rep 40:10
Nixon RW (1966) Date culture in the United States. Agr Inf Bull 207, US Department of Agriculture, Washington, DC

Nixon RW (1971) Early history of the date industry in the United States. Date Grow Inst Ann Rep 48:26–30
Nixon RW, Carpenter JB (1978) Date growing in the United States. Agr Info Bull 207, US Department of Agriculture, Washington, DC
Nixon RW, Crawford CL (1937) Fruit thinning experiments with Deglet Noor. Date Grow Inst Rep 34:107–115
Nixon RW, Furr JR (1961) Quality of Deglet Noor dates in relation to frequency of picking. Date Grow Inst Ann Rep 38:14–15
Nixon RW, Furr JR (1962) Frequency of picking in relation to quality of Deglet Noor dates in 1961. Date Grow Inst Ann Rep 39:16–18
Nixon RW, Furr JR (1965) Problems and progress in date breeding. Date Grow Inst Ann Rep 42:2–5
Peightal BJ (1962) Remarks on the mechanization of date harvesting. Date Grow Inst Ann Rep 39:18
Perkins RW, Brown GK (1964) Progress in mechanization of date harvesting. Date Grow Inst Ann Rep 41:19–23
Perkins RW, Burkner PF (1973) Mechanical pollination of date palms. Date Grow Inst Ann Rep 50:4–6
Pillsbury AF (1937) How much water does a date palm use? Date Grow Inst Ann Rep 14:13–16
Popenoe PB (1913a) Babylonian dates for California. Pomona College J Econ Bot III(2): 459–477
Popenoe PB (1913b) Date growing in the old and new worlds. West India Gardens, Altadena
Popenoe PB (1914) The home of the Fardh date. Calif State Comm Hort Monthly Bull 3(1):9–19
Popenoe PB (1926) Le dattier au Maroc. Rev Bot Appliquee D'Agr Colon 6(55):129–136
Popenoe PB (1973) The date palm. Fields Research Projects, Miami
Postlethwaite RH (1925) *Melilotus indicus* as a cover crop. Date Grow Inst Ann Rep 2:7–8
Ream CL (1975) Date palm breeding – a progress report. Date Grow Inst Ann Rep 52:8–9
Rivera D, Johnson D, Delgadillo J et al (2013) Historical evidence of the Spanish introduction of date palm (*Phoenix dactylifera* L, Arecaceae) into the Americas. Genet Res Crop Evol 60:1433–1452. doi:10.1007/s10722-012-9932-5
Riverside County Agricultural Commissioner (various years) Riverside County Agricultural Commissioner Report. Riverside County Agricultural Commissioner, Riverside
Rygg GL (1977) Date development, handling, and packing in the United States. Agricultural Handbook 482, US Department of Agriculture, Agricultural Research Service, Washington, DC
Schoonover WR (1935) Some suggestions on soil management in date gardens. Date Grow Inst Ann Rep 12:4–6
Shands HL (1995) The US National plant Germplasm system. Can J Plant Sci 75:9–15
Smith RJ, Aynsley JS (1995) Field performance of tissue cultured date palm (*Phoenix dactylifera*) clonally produced by somatic embryogenesis. Principes 39:47–52
Stickney FS, Barnes DF, Simmons P (1950) Date palm insects in the United States. Agr Circ 846, US Department of Agriculture, Washington, DC
Swann G (1925) Result of cover cropping. Date Grow Inst Ann Rep 2:6
Swingle WT (1904) The date palm and its utilization in the southwestern states. US Department of Agriculture Bureau of Plant Industry Bulletin 53, Washington, DC
Swingle WT (1926) Metaxenia or the influence of the male parent on the tissues of the mother plant outside of the embryo and endosperm, especially as exemplified in the date palm. *Third Pan Pacific Sci.* Cong, 1164–5
Swingle WT (1945) Introduction of the Medjool date from Africa into the United States. Date Grow Inst Ann Rep 22:15–16
Thackery FA, Leach GH (1936) Progress report on preliminary cover crop trials. Date Grow Inst Ann Rep 13:17
Tisserat B (1979) Propagation of date palm (*Phoenix dactylifera* L). J Exp Bot 30:275–1283

Tisserat B (1981) Date palm tissue culture. Advances in Agricultural Technology AAT-W-17. USDA-ARS, Oakland
Tisserat B (1982) Factors involved in the production of plantlets from date palm callus cultures. Euphy 31:201–214
Tisserat B (1983) Tissue culture of date palms – a new method to propagate an ancient crop – and a short discussion of the California date industry. Principes 27:105–117
Tisserat B (1984) Propagation of date palms by shoot tip cultures. HortScience 19:230–231
Tisserat B (1985) Occurrence and histological structure of offshoots and inflorescences produced from *Phoenix dactylifera* L plantlets *in vitro*. Bull Torrey Bot Club 112(1):35–42
Tisserat B (1988) Palm tissue culture. Publication ARS-55. USDA-ARS, Washington, DC
Tisserat B, DeMason DA (1980) A histological study of development of adventive embryo in organ culture of *Phoenix dactylifera*. Ann Bot 46:465–472
Tisserat B, Torres AM (1979) Isozymes as genetic indicators in date palms. Date Grow Inst Ann Rep 54:24–28
Tisserat B, Foster G, DeMason D (1979) Plantlet production in vitro from *Phoenix dactylifera* L. Date Grow Inst Ann Rep 54:19–23
Torres AM, Tisserat B (1980) Leaf isozymes as genetic markers in date palms. Am J Bot 67:162–167
Traub HP, Robinson TR (1937) Improvement of subtropical fruits other than citrus. USDA Yearbook of Agriculture, Washington, DC, pp 1–77
Trent H, Seymour J (2010) Examining California's first palm tree: the Serra palm. J San Diego Hist 56:105–120
USDA-APHIS (2010) First detection of Red Palm Weevil, *Rhynchophorus ferrugineus*, in California. DA-2010-55, USDA-APHIS, Washington, DC
USDA-APHIS (2011) First detection of South American Palm Weevil (*Rhynchophorus palmarum*) in California. DA-2011-45. USDA-APHIS, Washington, DC
Warner RL, Barnes MM, Laird EF (1990a) Reduction of insect infestation and fungal infection by cultural practices in date gardens. Envir Entom 19:1618–1623
Warner RL, Barnes MM, Laird EF (1990b) Chemical control of a carob moth, *Ectomyelois ceratoniae* (Lepidoptera: Pyralidae) and various Nitidulid beetles (Coleoptera) in 'Deglet Noor' dates in California. J Econ Entom 83:2357–2361
Winslow MM (1928) Date fertilizer trials in the Coachella Valley. Date Grow Inst Rep 5:4–5
Zhao Y, Williams R, Prakash CS, He G (2013) Identification and characterization of gene-based SSR markers in date palm (*Phoenix dactylifera* L.). BMC Plant Biol 12:237
Zohary D, Hopf M (2000) Domestication of plants in the old world: the origin and spread of cultivated plants in West Asia, Europe, and the Nile Valley, 3rd edn. Oxford University Press, Oxford, UK

Chapter 15
Date Palm Status and Perspective in South American Countries: Chile and Peru

Hugo A. Escobar and Rafael G.J. Valdivia

Abstract Date palm was introduced to South America, probably from Morocco, by Spanish colonialists to the central coast of Peru, from where it spread to Mexico and North America. Currently, date palm is little known as an agronomic crop; its cultivation in South America is limited to specific areas of Peru, Chile, Argentina, and Brazil. Excellent climatic conditions for date palm cultivation are present in northern Chile from Arica to Copiapó, as well as in areas of central and northern Peru such as Ica, Zaña, and Pisco. Important germplasm has developed from seed propagation which has produced all the named cultivars such as Medjool, Zahidi, and Deglet Noor. The importance of seedling-derived material lies in its adaptation to edaphic and climatic conditions different from those found in its area of origin, including acquisition of tolerance to high levels of salinity and boron in soil and irrigation water. It is interesting that there are date palms cultivated in locations at 1,500 m elevation such as the valleys of Pisco and Ingenio in Peru and Codpa in Chile. Currently, date production in Chile and Peru is all consumed locally. As well as having areas with optimal climate for the cultivation of date palms, Chile and Peru have important strengths and opportunities for the development of this crop, such as internationally recognized prestige in the production and exportation of fruit, government support of innovation, and multiple free trade agreements. There are also weaknesses, mainly the absence of local agronomic researchers familiar with the crop, lack of trained workers, insufficient knowledge about this fruit, and limited consumption. Date palm has an important potential in South America due to favorable agricultural and economic conditions as well as an expanding international demand.

Keywords Arica • Azapa • Chile • Date palm • Esmeralda • Ica • Lluta • Peru • Pica • Pisco • Zaña

H.A. Escobar (✉)
Facultad de Ciencias Agronómicas, Universidad de Tarapacá, Arica, Chile
e-mail: hescobar@uta.cl

R.G.J. Valdivia
Agronomist, Private Consultant, Santiago, Chile
e-mail: rafaelvaldiviaurra@gmail.com

J.M. Al-Khayri et al. (eds.), *Date Palm Genetic Resources and Utilization: Volume 1: Africa and the Americas*, DOI 10.1007/978-94-017-9694-1_15

15.1 Introduction

Although the date palm is known as a fruit tree linked to Arabian countries, its cultivation has spread into the Americas, introduced and disseminated during Spanish colonization. In certain American countries, the date palm adapted to favorable climates; however, it has been little developed as a crop. Chile and Peru have agroclimatic conditions which would allow date palm cultivation with high-quality fruit, such as those currently produced in family orchards but not in commercial quantities. In locations such as Ica, Zaña, and Pisco in Peru, and in the oasis of Pica in Chile, dates are produced which are highly sought after locally. The initial efforts to develop this fruit in Chile are still present, represented mainly by those in the former Esmeralda Experimental Station inland from Iquique, in a part of the Atacama Desert currently dedicated to growing tropical and subtropical fruit. There are larger plantations in Peru, from where it is thought that date palms were spread to North America and other South American countries. The importance that dates have attained in countries of North Africa and the Middle East, coupled with the similarity of their climate to the warm deserts of Chile and Peru, opens the possibility of developing this fruit in South America, where there are large unexploited desert areas of saline soils which only support cultivation of the most tolerant species such as the date palm. There is almost no precipitation in these deserts, an important factor for the cultivation of date palms, especially in the ripening period. Although there is limited published information available, the objective of this study is to present what is known about the history of the date palm in Chile and Peru, along with the geographic locations where plantations have been established, the agricultural and climatic characteristics, and an analysis of future possibilities of date palm cultivation in these two countries.

15.2 Historical and Current Agricultural Aspects

The first introduction of the date palm to the New World has not been documented, although it is likely that it was Spanish colonizers who carried seeds to South America and Spanish religious congregations who propagated it in the sixteenth and seventeenth centuries. Date palm seeds presumably reached America from Morocco; their initial propagation appears to have been on the coast of central Peru. The Jesuit Bernabé Cobo wrote in 1612 about the date palms in the Viceroyalty of Peru (Mateos 1956), indicating that "…the dates from palms born in the same valley (Pisco) mature as well as those which are brought from Berberia" (Barbary Coast, North Africa).

During the Spanish colonial period in Peru, the Jesuits mention the establishment of date palms in Peruvian farms. The interest in establishing date palms may not have been just for their fruit, but also associated with traditional Spanish religious ceremonies such as Palm Sunday, which begins the Catholic Holy Week and which requires palm fronds, ideally from the date palm. The production of dates

Fig. 15.1 *Phoenix dactylifera* in Esmeralda, Iquique, Chile

Fig. 15.2 Date palms introduced to Arica, Chile, from California in 1965–1970

as edible fruit was only important in specific areas where the fruit can mature, such as Zaña, Pisco, and Ica in Peru. This crop was an important activity for at least 200 years; however, currently these palms are surviving without appropriate crop management. The importance that Peru may have had in this crop is also reflected in its being an important possible distribution center of date palm seeds to Mexico and the Baja California Peninsula in North America, due to the abundant commerce between these regions in the seventeenth and eighteenth centuries; this also generated an interchange of crop management practices (Aschmann 1957). Date palms were present during the colonial epoch; however, the commercial interest in this crop only began in 1965–1970, when the Chilean government introduced offshoots from California to the Tarapacá Region. Most of the 240 plants brought in were planted in the Esmeralda Experimental Station (Fig. 15.1), inland from Iquique, and a few in the Lluta Valley in the Arica and Parinacota Region (Fig. 15.2). The cultivars introduced were Medjool, Zahidi, and Deglet Noor, which have been the progenitors of an abundant quantity of plants, mainly

Fig. 15.3 Date palm ecotype in the Lluta Valley, Arica, Chile (*Source*: Universidad de Tarapacá)

Fig. 15.4 Date palm ecotype in the Lluta Valley, Arica, Chile (*Source*: Universidad de Tarapacá)

propagated by seed (Figs. 15.3, 15.4, and 15.5), which have been distributed in the cities of Arica, Pica, and Iquique. However, the project originated from the Public Works Ministry (Fernández 2006), beginning in 1957 with a study of the ground waters in Esmeralda and Canchones, inland from Iquique, followed in 1968 by the establishment of the Esmeralda Experimental Station of CORFO (Corporación de Fomento de la Producción), on which 6 ha of date palms was established (Pavez et al. 2007). This station has been under private ownership since 1997. In 1995, the University of Tarapacá established a small plantation of 25 date palms and a row of approximately 50 of the plants in the Lluta Valley, Arica, all from seed. In the Villa Frontera section of the city of Arica, at the mouth of the Lluta River, there are two orchards of 1,400 and 380 palms, of 3 and 2 ha, respectively. These were planted between 1995 and 1998 and constitute the plantations closest to the coast, at a distance of about 6 km from the ocean.

Fig. 15.5 Date palm ecotype in the Lluta Valley, Arica, Chile (*Source*: Universidad de Tarapacá)

Fig. 15.6 Date palm ecotype in the Azapa Valley, Arica, Chile

15.3 Agricultural Importance

Although date production is concentrated mainly in the Northern Hemisphere, it is important to consider as well the presence of date palms in the Southern Hemisphere. Here, Peru is important; it has 50,000–80,000 plants of productive age, with a potential to produce about 4,000 mt of fruit annually (Robles 2006). These are possibly the descendants of oldest plantings in South America; however, the industry so far has not had been efficiently developed and fruit is only commercialized locally (Azañero et al. 2000). In Chile, date palms are mostly found in the extreme north of the country and do not constitute an important industry; most are isolated family plantings. The most important concentration of date palms in Chile is the orchard in Esmeralda, community of Pica, with 6 ha. Other important localities with date palms are the valleys of Lluta and Azapa (Fig. 15.6) in the community of Arica.

Fig. 15.7 Date-growing locations in Chile and Peru

Both in Chile and Peru, date palms are an important potential agricultural crop in regions where irrigation water and soil have high concentrations of salt and boron and an advantageous climate for its cultivation.

Along the Pacific Coast in the desert of Peru and Chile, date palms are found from Zaña in Peru (6°55′ S) to Pica in Chile (20°30′ S), a distance of 2,300 km (Fig. 15.7). This north-south latitudinal distribution is the most extended in the world for *Phoenix dactylifera*. It is important to note that in this zone, flowering and fruiting of date palms may be obtained at elevations from sea level to 1,500 m in the valleys of the western slope of the Andes (Pisco and Ingenio in Peru and Codpa, Pica, and Tarapacá in Chile). This distribution provides the date palm with one of the largest ranges of geographic adaptation to elevation in the world.

An increase in the area planted with date palms in Chile would allow expanding the use of this plant not only for fruit production, but also to use its leaves as a subproduct for the preparation of biochar. This would minimize the liberation of CO_2 to the atmosphere and serve as a product to improve the physical and chemical conditions of the soil and make possible increased horticultural production. The first experimental trial of this technique in Chile was published by Lara (2011), with biochar from date palm leaves used to cultivate beets in highly saline soils with excess boron. In this study, the yield of edible roots and the fresh weight and area of leaves were greater when plant biochar was added to the soil at a concentration of 1 % (v/v).

Table 15.1 Reported locations of date palm growth in Peru and Chile

Country	Location/city	Latitude (°S)	Elevation (m)	Distance from the ocean (km)
Peru	Zaña/Chiclayo	6° 55′	120	16
Peru	Paracas/Pisco	13° 51′	5	0
Peru	Cachiche/Ica	14° 10′	400	45
Peru	Ocucaje/Ica	14° 22′	330	60
Peru	Río Grande/Nazca	14° 32′	265	55
Peru	Camaná/Arequipa	16° 40′	5	5
Peru	Ilo /Moquegua	17° 42′	35	8
Chile	Azapa/Arica	18° 32′	250	15
Chile	Lluta/Arica	18° 32′	300	20
Chile	Villa Frontera/Arica	18° 32′	20	6
Chile	Tarapacá/Iquique	19° 55′	1,500	70
Chile	Esmeralda/Pica	20° 30′	1,180	110

Source: Modified from Pavez et al. (2007)

As well as in the zones indicated in Table 15.1, date palms could be cultivated in Chile up to 1,000 km south of the city of Iquique; climatic conditions would allow the establishment of date palms in the Taltal and El Salado valleys (26° 30′ S), the Copiapó valley (28° S), and San Félix in the community of Vallenar (29° S, at 1,600 m elevation and 125 km from the ocean).

The main date production zone in Peru is the Department of Ica, 300 km south of Lima, including the Paracas district in Pisco, Villacurí, Cachiche, and Ocucaje in Ica and Río Grande in Nazca. This area is recognized as the most important in South America due to the density of date palms established there; its peak was reached in the eighteenth and nineteenth centuries. In this zone, there are 50,000 adult female palms, although 90 % of them are abandoned and surviving in a natural state. Only about 4,000 palms are managed under rustic conditions, while another 1,000 are in industrial plantations with adequate technical management and irrigation. The potential for new plant production by offshoot separation is approximately 300,000 trees; however, the parent plants were propagated from seed and provide little incentive to increase the planted surface (Letts and Pavez 2000). Earlier information (INIA-CONAFRUT 1998) reported 60,000–80,000 productive date palms in all of Peru, which makes it the largest in South America. These palms are located in the coastal valleys of Zaña, Chilca, Palpa-Río Grande, Nazca, Acarí, Yauca, Camaná, Tambo, Ilo, and La Yarada/Tacna.

The low yields obtained in Peru, equivalent to 3,000 kg/ha and 20–30 kg/plant, are the result of deficient management of the plantations, most of which do not receive irrigation, pollination, or sanitary treatments. Thus, as Table 15.2 illustrates, there has been a decrease of commercial date production in Peru, mainly affected by the decrease in the cultivated area. This production is consumed almost completely in the area of cultivation; there is only one company in Peru (Huerto Alamein S.A.) which packages and sells dates in Lima. There are also some locally important artisanal preparations such as candied dates at Zaña in Chiclayo and Camaná in

Table 15.2 Commercial date production in Peru

Year	Surface (ha)	Production (mt)	Yield (kg/ha)
1966	150	808	5,386
1986	270	777	2,877
2004	97	260	2,680
2011	104	379	3,644

Source: FAOSTAT (2014), Pavez et al. (2007)

Table 15.3 Average monthly temperatures (°C) at locations where date palms are grown

Location and country	Maximum (warmest month)	Minimum (coolest month)	Annual average
Paracas, Peru	29.4	9.8	18.7
Ica, Peru	33.3	6.6	21.2
Ocucaje, Peru	35.6	5.9	20.6
Azapa, Chile	28.5	11.5	19.0
Esmeralda, Chile	34.1	1.0	18.8

Source: Pavez et al. (2007)

Arequipa. The fruit and its possibilities of processing and incorporation into the food industry are almost completely unknown in the internal market.

15.4 Climatic Conditions

15.4.1 Temperature

In Chile, the optimum isotherm of 18 °C for cultivation of the date palm occurs only around the city of Arica; however, there are other locations in the country where good-quality dates are produced with the 15 °C isotherm, such as Pica and Esmeralda. Other localities with the same characteristics extend along the entire littoral from Arica (18° S lat.) to Carrizal Bajo (28° S lat.) and inland of Iquique and Arica (Valdivia 2012).

The subtropical climate of this area is modified by marine influence, with thermal seasonality. The highest temperatures (28.5–35.6 °C) occur from December to March, the lowest (1.0–11.5 °C) from June to August (Table 15.3). Mean minimum temperatures are rarely below 8 °C near the coast, and it never freezes. However, temperature differences increase in the interior of the continent at intermediate elevations; day-night oscillation may reach 30 °C, as in Esmeralda, at 1,180 m elevation, 110 km from the coast. Table 15.4 presents the average maximum daily temperatures in some localities of Chile and Peru.

Earlier research by Munier (1973) and Rebour (1971) indicated that date palm needs an isotherm of at least 18 °C to produce quality dates; this isotherm is present in Chile only round the city of Arica, and there are no higher isotherms in the country. However, inland from Iquique, specifically in Pica, high-quality dates are produced

Table 15.4 Average maximum daily temperatures during the year (°C)

Month	Paracas, Peru	Ica, Peru	Ocucaje, Peru	Azapa, Chile	Esmeralda, Chile
January	28.9	31.7	33.1	28.0	34.0
February	28.8	33.3	35.4	28.5	34.0
March	29.4	32.6	35.6	27.7	33.3
April	26.9	30.9	34.8	24.1	31.7
May	26.1	28.2	33.7	22.7	29.9
June	23.5	24.0	27.0	20.3	28.5
July	22.7	23.6	24.1	19.3	29.5
August	21.9	24.1	26.3	19.5	31.1
September	22.4	24.5	29.2	20.7	32.3
October	22.4	27.1	29.1	22.1	34.0
November	23.5	27.8	31.3	23.9	34.1
December	24.9	30.4	32.4	26.0	33.9
Average	24.2	28.3	31.1	23.7	32.2

Source: Pavez et al. (2007)

Table 15.5 Heat units at various date-growing areas in Chile and Peru (base 18 °C)

Months	Paracas, Peru	Ica, Peru	Ocucaje, Peru	Azapa, Chile	Esmeralda, Chile
January	337.9	424.7	468.1	310.0	496.0
February	302.4	428.4	487.2	294.0	449.2
March	353.4	452.6	545.6	300.7	474.3
April	267.0	387.0	504.0	213.0	411.0
May	251.1	316.2	486.7	145.7	368.9
June	165.0	210.0	270.0	69.0	315.0
July	145.7	173.6	220.1	40.3	356.5
August	120.9	189.1	257.3	46.5	406.1
September	132.0	224.0	336.0	81.0	429.0
October	136.4	282.1	344.1	127.1	496.0
November	165.0	294.0	399.0	177.0	483.0
December	244.9	384.4	446.4	248.0	492.9
Annual total	2,621.7	3,767.1	4,764.5	2,052.3	5,177.9

Source: Pavez et al. (2007)

although the annual isotherm is below 15 °C. Thus, in spite of the opinion of Rebour (1971) that there are few places in Chile where date palms could produce quality fruit, early-bearing cultivars such as Medjool should have no trouble producing fruit, at least in areas with an annual isotherm of 15 °C or greater, which includes the entire littoral from Arica to Carrizal Bajo. This concurs with the opinion of Pavez et al. (2007) who indicated that the date palm may be cultivated in the coastal area from Arica to 1,000 km south of Iquique and only inland of Iquique and Arica up to 1,500 m.

Table 15.5 presents the accumulated heat units in certain localities of Chile and Peru, confirming the feasibility of date production since the heat units are considerably above

Table 15.6 Flowering period and heat units of tree date palm cultivars in Esmeralda, Chile

Cultivar	Flowering period[a]	Heat units (base 18 °C)
Deglet Noor	July–August	356.5–406.1
Medjool	September–October	429.0–496.0
Zahidi	July–August	356.5–406.1

[a]Alexander Cáceres, administrator of the Fundo Esmeralda (personal communication 2011) and Enrique Arroyo, date producer of Pica, Region I (personal communication 2011)

Table 15.7 Rainfall at various date-growing areas in Chile and Peru

Location	Rainfall (total annual, mm)	Monthly rainfall (rainiest month)	Rainiest month
Paracas, Peru	1.6	0.5	August
Ica, Peru	4.4	1.2	January
Ocucaje, Peru	0.3	0.2	December
Azapa, Chile	0.5	0.3	January
Esmeralda, Chile	0.6	0.3	February

Source: Pavez et al. (2007)

1,000 °C (base temperature 18 °C) which is the lower limit for quality date production established by Munier (1973) cited by Zaid and Wet (2002). Table 15.6 shows the heat units and flowering periods of three date palm cultivars in the Esmeralda Station in Chile.

15.4.2 Precipitation

Precipitation amounts are minimal and rainfall sporadic in the zones where the temperature would allow date production in Chile. According to the Dirección Meteorológica de Chile (2001), rainfall is usually around 1 mm per year, although inland from the cities of Arica and Iquique totals reach nearly 20 mm/year.

The coastal desert of Peru and Chile is considered the driest area of the world; it almost never rains. Only short (hours) and very sporadic rains occur, followed by long, dry sunny periods. However, in northern Peru in Zaña, Chiclayo, summer precipitation is influenced by the El Niño effect, a periodic climatic event associated with warmer than normal offshore waters, which alters stability and produces rainfall of considerable magnitude and duration in the months of fruit formation, between December and April (Table 15.7).

15.4.3 Relative Humidity

Humidity is another climatic factor which significantly affects the date palm; according to Klein and Zaid (2002), in the areas where the relative humidity is high, date palm plantations should adopt a lower planting density (125 palms or less per ha), and they should not be intercropped with other fruit trees since this increases the effect of high humidity on the dates. High relative humidity occurs in the

Table 15.8 Average daily relative humidity at selected locations in Peru and Chile

Location	Maximum relative humidity (%)	Minimum relative humidity (%)	Annual average (%)
Paracas, Peru	87.0	77.0	81.5
Ica, Peru	99.0	39.0	69.3
Azapa, Chile	97.0	41.2	72.0
Esmeralda, Chile	85.0	20.0	31.4

Source: Pavez et al. (2007)

Table 15.9 Average monthly relative humidity (%) at selected locations

Months	Paracas, Peru	Ica, Peru	Azapa, Chile	Esmeralda, Chile
January	82.0	67.5	68.9	40.8
February	80.5	67.0	70.1	42.6
March	80.5	68.0	72.1	43.5
April	80.5	68.0	72.9	36.9
May	82.5	69.5	74.1	28.1
June	83.0	73.0	75.8	28.0
July	81.0	74.5	75.2	23.7
August	81.0	71.5	74.2	25.0
September	82.5	69.5	72.3	23.7
October	81.5	67.0	70.4	23.3
November	81.5	68.5	69.4	28.6
December	81.0	68.0	68.4	32.7
Annual average	81.5	69.3	72.0	31.4

Source: Pavez et al. (2007)

northern coastal desert of Chile (Dirección Meteorológica de Chile 2001) during the night and early morning in the coastal desert of Chile and Peru (Tables 15.8 and 15.9), but decreases during the day due to high solar radiation. Above 1,200 m elevation, humidity is less because of the dry warm air which descends from the Andes. This low humidity may dry dates excessively; plants also transpire more and thus have higher water requirements (Pavez et al. 2007).

High relative humidity and high cloud cover reduce the commercial production of quality dates in a large part of the littoral area of central Peru between Chiclayo in the north and Chincha to the south, including the capital, Lima. Thus, it must be mentioned that the ideal conditions of relative humidity for *Phoenix dactylifera* are found in the hot, dry sunny valleys in the departments of Ica, Arequipa, Moquegua, and Tacna, and in interior valleys north of Lima, but at greater elevation and distance from the ocean (Pavez et al. 2007).

15.4.4 Hours of Sunlight

The coastal zone of Peru and Chile enjoys many hours of sunlight during the year, which reaches a maximum in the interior desert and at higher elevations such as in Esmeralda, Chile (Table 15.10). The coasts of Peru and northern Chile are located

Table 15.10 Average hours of sunshine in selected locations

Months	Paracas, Peru	Ica, Peru	Azapa, Chile	Esmeralda, Chile
January	217	223	261.8	350.3
February	199	177	252.8	299.6
March	213	217	265.4	322.4
April	211	223	247.8	312.0
May	177	235	215.4	257.3
June	136	196	177.0	243.0
July	119	184	198.8	313.1
August	153	205	193.7	313.1
September	168	221	217.5	333.0
October	198	260	246.7	334.8
November	197	235	249.5	348.0
December	201	218	269.4	356.5
Annual total	2,189	2,594	2,795.8	3,783.1

Source: Pavez et al. (2007)

between 7° and 20° S latitude. For this reason they receive more solar radiation than the traditional areas of date production in the Northern Hemisphere, which are located in subtropical areas between 18° and 39° N latitude.

Considering the exceptional levels of heat accumulation, an absence of cold temperatures and precipitation, there are exceptional natural conditions for the development of date palm cultivation.

15.4.5 Soil Conditions

The soils in the zone with isotherms of 15 °C or greater usually have depths greater than 2 m; however, they have high salt and boron concentrations which would affect the productivity of dates when the electrical conductivity (EC) is above 5.3 dS*m^{-1}. According to Pavez et al. (2007), in the valleys of Lluta, Azapa, Chiza, and Miñi-Miñe in Chile, the production of this species would not be limited by this factor, while in the valleys of Suca, Taltal, and Copiapó, the plants would not produce dates, since these areas have EC above 27 dS*m^{-1} (Table 15.11). However, this study also concluded that the levels of boron in the water and soil in these valleys would not affect the establishment or productivity of the species. Date palms adapt well to soil with salt and boron concentrations, which favor its introduction in a wide variety of soils. In Peru, the main concentrations of date palms in the *pampas* (plains) of Pisco and Ica valley are located in desert areas with an elevated concentration of salts, but have good-quality ground water at depths of 2–10 m, originating from the rivers descending the western slopes of the Andes.

The locations in Table 15.11 represent a variety of local agricultural and ecological conditions found in the Atacama Desert. The valleys of Copiapó, Taltal, Liga, and Suca have extreme salinity, while Taltal, Suca, and Lluta have extreme levels of

Table 15.11 Chemical analysis of the soil in selected valleys of northern Chile (in saturation extracts)

Parameter	Lluta	Azapa	Chiza	Suca	Liga	Miñi-Miñe	Taltal	Copiapó
pH	7.6	7.6	8.5	7.4	6.9	8.1	7.1	8.2
EC (dS/m)	3.07	5.25	2.38	40.6	16.19	0.84	27.1	62.1
Ca (meq/l)	10.1	30.4	6.3	70.2	75.3	3.7	59.0	41.6
Mg (meq/l)	4.8	4.6	2.0	26.2	19.2	1.0	39.8	135.0
Na (meq/l)	14.1	14.63	14.5	315.2	66.4	3.4	157.0	265.2
K (meq/l)	1.5	1.05	0.8	4.2	1.4	0.3	13.0	7.0
HCO_3 (meq/l)	2.8	2.7	9.4	7.5	4.2	2.8	3.8	10.8
Cl (meq/l)	18.3	29.63	11.0	317.2	140.3	5.2	228.8	388.9
SO_4 (meq/l)	6.9	27.5	3.7	52.4	14.8	0	0	165.0
B (mg/l)	13.2	7.82	3.1	17.6	8.8	1.8	24.5	6.4

Source: Escobar et al. (1995), Figueroa et al. (1993)

boron. The Lluta and Camarones valleys are typically saline, with serious limitations for agriculture. The main crops in these valleys are local cultivars of alfalfa and maize which are highly resistant to salinity and boron. The Azapa, Chaca, and Miñi-Miñe valleys have the best soil and water conditions, almost without restrictions for cultivation of a number of fruits and vegetables, including olives, citrus, guavas, mangoes, tomatoes, and green beans. The Miñi-Miñe valley has ideal conditions for all types of subtropical agriculture, thanks to a microclimate at 2,000 m elevation which allows the growth of mangoes, bananas, and citrus fruits.

15.4.6 Water Conditions

Irrigation water in this zone presents limitations for many crops, due to its high concentration of salts and boron. However, Pavez et al. (2007) indicated that irrigation water usually has EC values which allow the production of dates, since in the Lluta, Azapa, Chiza, Suca, Miñi-Miñe, and Copiapó valleys, EC values are below 3.5 $dS*m^{-1}$ (maximum value in which dates do not lose productive potential); in Taltal, EC reaches 5.34 $dS*m^{-1}$, which would produce a decrease in the allowable yield of the plant (Table 15.12).

Thus, we conclude that the date palm may be cultivated in any zone of the littoral from Arica to Carrizal Bajo, extending 120 km inland from Iquique and Arica especially in the transverse valleys where other fruits which are much less tolerant to salts than date palm (e.g., olives, pomegranates, citrus, mangoes) have been cultivated for many years. Dates can also be grown in areas with isotherms greater than 15 °C where other crops are not cultivated due to the high concentration of salts and boron in the water and soil, keeping in mind that the date palm may grow and establish, but will decrease its potential production according to the differential concentration of these elements in each locality and availability of water; this is the situation in Suca, Taltal, and Copiapó, which have high concentrations of salts in the soil.

Table 15.12 Chemical analysis of irrigation water in different valleys in northern Chile

	Lluta	Azapa	Chiza	Suca	Miñi-Miñe	Taltal	Copiapó
pH	7.0	8.6	7.6	6.8	8.2	7.0	7.0
EC (mS/cm)	3.15	0.66	1.83	2.01	0.94	5.43	2.06
Ca (meq/l)	5.4	0.9	1.3	16.6	3.2	24.0	9.0
Mg (meq/l)	5.3	0.5	1.5	2.2	1.1	10.7	5.5
Na (meq/l)	20.4	3.1	13.0	4.1	5.3	18.2	6.1
K (meq/l)	1.1	0.3	0.2	0.3	0.2	1.0	0.4
HCO_3 (meq/l)	2.3	3.38	7.7	3.2	4.9	3.5	4.5
Cl (meq/l)	22.0	1.62	8.5	6.7	0.94	45.4	3.7
SO_4 (meq/l)	6.5	3.7	2.8	11.8	0.8	5.2	11.2
B (mg/l)	16.6	0.8	2.2	1.9	1.3	1.2	1.4
R.A.S.	8.8	2.9	11.1	1.3	3.6	4.4	2.3

Source: Escobar et al. (1995), Figueroa et al. (1993)

15.5 Propagation and Crop Management

The small plantations of date palms in northern Chile suffer from traditional management practices. The technicians and laborers of the plantations work without specific knowledge about this species, acquiring information as the plants developed. To the present, professional consultants on irrigation, nutrition, harvest, and postharvest for the most advanced plantations must be contracted from outside the country.

The main center of propagation and dispersion of seeds and plantlets of date palms in Chile has been the Esmeralda Station, inland from Iquique, which has the main group of cultivars Medjool, Zahidi, and Deglet Noor. Its seeds have been used to establish small plantations in Pica, the Azapa and Lluta Valleys and Villa Frontera. Later, other plants were propagated from these materials. Palms of these cultivars have been maintained in the area of Pica, adjacent to Esmeralda. In northern Chile, there is currently the potential to develop in vitro propagation in the Plant Tissue Culture Laboratory, University of Tarapacá, Arica. This modern laboratory would also allow importation of plants from other producer countries, as well as the micropropagation of local genotypes.

These small date palm plantations are generally drip irrigated with fertilizers added to the water or to the wells. Irrigation flow is applied without a specific methodology with respect to the variables involved, such as soil type, climate, and the plant needs. The irrigation criterion is mainly determined by the available water. The presence of sandy soils and salts is an important factor to consider in northern Chile. Other practices are unknown or there is no record of their application; thus, there is haphazard management of pollination, leaf and stem trimming, sanitary management, fruit thinning, and harvest, which are important factors that limit the production of dates. It is also important to note the maturation problem of the Deglet Noor cv., whose fruits do not fully mature; they are left on the tree until the next epoch of flowering and fruit set of the following year in the hope that they will finish ripening. This results in dehydrated and extremely hard fruits with almost no commercial value and generates cycles of alternate bearing in these plants.

The lack of industry in northern Chile does not allow a greater level of investment in these small family orchards and plantations. This affects processes such as the harvest, which is inefficient for the lack of appropriate machinery; harvesting is done by ladder, which increases production costs due to increased labor cost. The main postharvest problems are due to the lack of efficient storage, resulting in fruit damage by insects, birds, and rodents.

15.6 Production, Consumption, Importation, and Exportation of Dates in Chile

According to Valdivia (2012), Chile has no reported date production and no statistics of production, yield, or area harvested, according to FAOSTAT. However, according to the records of the Fundo Esmeralda for 2011, approximately 5 mt was harvested, with a yield of at least 10 kg per palm. Most of the production was of the Deglet Noor cultivar; local prices were Deglet Noor USD 3/kg, Medjool USD 6/kg, and Zahidi USD 4/kg (Cáceres A, Administrator of the Fundo Esmeralda, 2011, personal communication).

Date consumption in Chile between 2001 and 2007 averaged about 11 mt, less than 1 g per person per year, but increased 5 % per year (Valdivia 2012). Importations between 2001 and 2009 averaged USD 28,000 (CIF); this was variable, but increased by 32 % per year. The mean volume of fruit imported of 20 mt per year had a mean value of USD 1,462 (CIF)/mt. Most of the imports came from Iran and the USA (69 and 26 %, respectively), with small amounts from Tunisia and Syria. The monthly distribution of importation shows none in the months of March, April, October, and November and peak values in December and January. The unitary value followed almost the same pattern as the total of importations, while the volume was variable, increasing frequently in the summer months and with a small increase in May.

Reexportation of dates from Chile in the period 2001–2009 averaged USD 14,000 (FOB) per year in value, decreasing at a 12 % annual rate. The volume of reexportation declined even more rapidly (21 %/year) with an average of 10 mt/year; the mean unitary value of this fruit was USD 1,911 (FOB)/mt, decreasing by 8.3 %/year. Chile reexported dates mainly to Brazil, Colombia, and Costa Rica, 28, 27, and 20 %, respectively, with occasional shipments to Japan, the USA, and Venezuela. Reexportation decreased from October to December, revealing two patterns; one, dates were not exported in January and, two, the value and volume of reexportations showed similar behavior, reaching their maximum in June. The unitary value decreased continuously during the year beginning in February; it was above USD 3/kg in October and December, but less than USD 2/kg in March, April, July, and August.

Table 15.13 Detail of variable cost and variable unit cost

Component	Variable cost (USD/ha/year)	Variable unit cost (USD/mt)	Relative importance (%)
Consultants	774	62	3.08
Fertilizer	804	64	3.20
Electricity	1,586	127	6.31
Maintenance of irrigation system	95	8	0.38
Rent of equipment	534	43	2.12
Total labor costs	2,134	171	8.49
Agricultural and security implements	342	27	1.36
Storage	193	15	0.77
Packing Materials	6,250	500	24.85
Subtotal (Preshipment)	12,712	1,017	50.55
Shipment to port	1,934	155	7.69
Shipment (Maritime Europe)	10,500	840	41.76
Total variable costs	25,146	2,012	100.00

Source: Valdivia (2012)

15.6.1 Costs of Date Production in Chile

There are no official statistics on the production costs of dates in Chile, because it is a little-cultivated species, and because the producers do not have complete records. Thus, in order to estimate the variable unit costs and get an idea of the competitiveness of Chile in the international market for dates, it was necessary to use price quotes and earlier experience (Fernández 2006; SACOR 1986), decoupling the variation in fixed costs in date palm cultivation in different places. Table 15.13 shows that among the component of this variable cost, the most important are preshipment items that were packing costs, labor, and electricity, which represented 49, 16, and 12 %, respectively, of the variable unit cost before shipping. This cost is half of the total variable unit cost, almost all the remainder was the cost of surface shipment to Europe (USD 840/mt).

Thus, the total variable unit cost of exporting dates to Europe was USD 2,012 (CIF)/mt, noting that the unit variable FOB to ship dates to the closest port (Iquique) was USD 1,172/mt. Thus, we may conclude that according to the calculated total variable unit cost, Chile would have the opportunity to insert itself into the world market for this fruit; however, the success of individual producers will depend on their own fixed costs and the market price obtained.

15.6.2 SWOT Analysis for Chile

To understand and evaluate the real possibility of Chile entering the world date market, we performed an analysis of the strengths, weaknesses, opportunities, and threats (SWOT) of the present and future development of this area, supported by the opinions of qualified informants and previous experience, ordering them as appropriate for this tool, which is shown below.

15.6.2.1 Strengths

The absence of serious date palm pests and diseases in Chile creates a great advantage both in the production and commercialization of this fruit, since it means almost zero cost to control of outbreaks except for rodent and bird control, minimizing the entrance requirements of many markets.

According to qualified informants and Fernández (2006), the dates produced in northern Chile are larger than those in the world market, which would generate a competitive advantage. According to Pavez et al. (2007), some climatic conditions in Chile favor a reduction in the cost of production, since, for example, the absence of rainfall avoids the repetition of manual pollination which is necessary after a rain, increasing the labor requirement for the development of this crop.

The proximity of an embarkation port to the production zone would also generate a cost reduction, giving Chile an advantage compared to other producers. Chile has a very good reputation as a fruit exporter in many markets where dates fetch a good price, as in most European countries; it also has a good exportation infrastructure for foods and fruits, along with many free trade agreements which facilitate the entrance of Chilean products into European, US, and Latin American markets.

The promotion of the introduction and diversification of new alternatives for agricultural development in northern Chile, the generation of new jobs, and control of desertification may generate benefits appreciated by the government and/or private institutions, facilitating the initial development of this crop in Chile. The sale of offshoots and larger palms for ornamental use in the cities of northern Chile should also be considered, which would generate a scrap value of senescent date palm that may aid in the formation of new plantations or recovery of old ones.

Of course, the possibility exists of intercropping date palms with other fruits, vegetables, and fodder, increasing the productivity of the fields in northern Chile with an acceptable availability of workers, machinery, and agricultural inputs.

Finally, the establishment of immigrant Moslem communities in Chile and neighboring countries such as Brazil may help to promote the initial sales and exportation of dates.

15.6.2.2 Weaknesses

The small number of qualified workers with technical knowledge; the lack of specialized equipment; the limited experience with packing, processing, and commercialization of dates; the low local availability of offshoots; and an absence of micropropagation techniques in date palms are all factors which limit the development of this product in Chile and increase production costs. The greater distance to the main markets and/or consumer countries also limits the insertion of Chile into them.

Lack of knowledge of the date palm may influence the farmers of northern Chile to continue to grow crops better known to them. Also, the level of investment necessary to establish new plantations may limit the development of this crop, since the cost of water rights and electricity and their accessibility often implies high costs and investment which farmers cannot or do not want to incur without the certainty that the returns will justify the investment. This is especially true in the areas which have suitable conditions to cultivate this crop, since agriculture must compete with local mining activities for water and labor.

The lack of accumulation of heat units, in some areas, for the maturation of the late-bearing cultivars, could limit cultivation to early-bearing cultivars, such as Medjool and other selected local ecotypes.

The high concentration of salts and boron in water and soil is common in northern Chile, which may limit the development and yield of date palm; this may increase production costs, losing competitiveness in the market and to alternative crop productions. However, it must be taken into account that the date palm is the most tolerant fruit to these conditions.

15.6.2.3 Opportunities

The scarcity of pests and serious disease which affect this crop in Chile represents a real possibility of producing organic dates, generating higher prices and better acceptance by consumers. The deep cultural and religious association between dates and the Moslem people insures consistent commerce and consumption of this fruit, which extends to new Islamic communities established in other countries, indirectly making the fruit known to the world. Chile has greater political and economic stability than that shown by some of the large producers of this fruit in the Middle East, fostering greater consumer confidence. Also, there are government grants and other economic incentives supporting agricultural research and development and innovative initiatives for production. Between 2001 and 2009, both the unit value and the volume of importations of this fruit grew at an annual rate close to 6 %.

15.6.2.4 Threats

One threat, which may be interpreted as competition, is that represented by other South American countries having natural potential for date palm cultivation and the production of good-quality dates. Also, there may be poor acceptance of dates produced in Chile by Moslem consumers, due to an emotional attachment linked to the origin or place of production of the fruit.

15.7 Conclusions and Recommendations

Based on the analyses in this chapter, it is concluded that the coastal desert from Zaña in Peru (7° S) to Pica in Chile (20° S), a distance of 2,300 km, has optimum agroclimatic conditions for date palm development; it could also be grown up to 1,000 km farther south in specific areas, to the city of Vallenar (29° S). It is important to note that the presence of old and new date palm plantings in Chile and Peru demonstrates the technical feasibility of producing quality fruit. Most of the palms in these countries are propagated from seed, representing valuable germplasm which has been grown in and adapted to areas with high levels of salinity and boron. These genotypes are important to keep in mind for the improvement and selection of palms for an agriculture limited by an excess of salts and desert characteristics such as high solar radiation and extreme drought. The strengths described for this crop in the warm deserts of the Chilean and Peruvian coasts, as well as their high prestige as fruit producers and exporters, government support for agricultural research and innovation, free trade agreements, and growing demand for dates, combine to create the basis to predict a high probability of success for the development of the date palm. However, the weaknesses should also be considered, mainly the minimal technical experience with this crop in South America, lack of familiarity with and low consumption of dates, and the current small area of cultivation which does not yet produce enough to export. Conventional and in vitro propagation, agronomic management, importation of the main cultivars, and the establishment of agreements and support networks with the major producing countries are the main needs to provide incentives for the cultivation of date palm in South America.

References

Aschmann H (1957) The introduction of date palms into Baja California. Econ Bot 11(2):174–177

Azañero M, Figueroa U, Mendoza N (2000) Dátiles. Seminario de Agro Negocios. Sección Administración. Universidad del Pacifico, Facultad de Administración y Contabilidad, Lima

Dirección Meteorológica de Chile (2001) Climatología regional. Departamento de Climatología y Meteorología. http://164.77.222.61/climatologia/

Escobar H, Bastías E, Tapia L et al (1995) Análisis del comportamiento de frutales tropicales y subtropicales en el valle de Azapa; selección y micropropagación de clones sobresalientes. Informe Final de Proyecto. Fondo Nacional de Ciencias, Universidad de Tarapacá, Arica

FAOSTAT (2014) Date crop production. FAO, Rome

Fernández F (2006) Prospección económica de la palma datilera, *Phoenix dactylifera,* y evaluación de localidades de la región de Tarapacá para su cultivo. Taller de Licenciatura, Pontificia Universidad Católica de Valparaíso, Facultad de Agronomía, Quillota

Figueroa L, Tapia L, Bastías E et al (1993) Selección y evaluación de portainjertos de olivos (*Olea europaea* L.) resistentes a las condiciones de salinidad y toxicidad específica para el norte de Chile. Informe Final de Proyecto, Fondo Nacional de Ciencias, Universidad de Tarapacá, Arica

INIA-CONAFRUT (1998) Cultivo del datilero. Bol Téc 17. Comisión Nacional de Fruticultura, Ministerio de Agricultura, Lima

Klein P, Zaid A (2002) Land preparation, planting operation and fertilisation requirements. In: Zaid A (ed) Date palm cultivation. FAO Plant Production and Protection Paper 156. Food and Agriculture Organization of the United Nations, Rome, pp 107–129

Lara R (2011) Efecto del biochar de maíz y palma datilera en el crecimiento, relaciones hídricas, contenidos iónicos foliares y rendimiento de la planta de betarraga (*Beta vulgaris* var. *Detroit Dark Red*) en el Valle de Lluta, XV Región de Arica y Parinacota. Tesis Ingeniero Agrónomo, Universidad de Tarapacá, Arica

Letts R, Pavez A (2000) Peru in South America: excellent conditions for commercial date cultivation. In: Proceedings of the Date Palm international symposium, Windhoek, pp 349–350

Mateos F (1956) Obras del P. Bernabé Cobo (1580-1657). Historia del Nuevo Mundo. Biblioteca de autores españoles. Madrid

Munier P (1973) Le palmier dattier. Maison Neuve et Larose, París

Pavez A, Escobar H, Johnson D (2007) Date palm cultivation in Chile and Peru (South America): current status and future prospects for development. Acta Hort 736:71–85

Rebour H (1971) Frutales mediterráneos. Mundi Prensa, Madrid

Robles F (2006) Dátil, tesoro del desierto. http://www.a-campo.com.ar/espanol/fruticultura/fruticult10.htm

SACOR (1986) Evaluación técnico económica de la hortofruticultura del área de Esmeralda. I Región. Sociedad Agrícola CORFO Ltda, Santiago

Valdivia R (2012) El mercado mundial del dátil: evaluación de posibilidades de inserción para Chile. Memoria de Título. Facultad de Ciencias Agronómicas, Universidad de Chile, Santiago

Zaid A, de Wet P (2002) Climatic requirements of date palm. In: Zaid A (ed) Date palm cultivation. FAO Plant Production and Protection Paper 156. Food and Agriculture Organization of the United Nations, Rome, pp 58–73

Appendixes

J.M. Al-Khayri et al. (eds.), *Date Palm Genetic Resources and Utilization: Volume 1: Africa and the Americas*, DOI 10.1007/978-94-017-9694-1

Appendix A

Fruit characteristics of major date palm cultivars grown in various locations along with estimated tree yield and number of trees per cultivar of some date-producing countries in Africa and the Americas

Country	Cultivar	Yield per tree (kg)	Number of planted trees	Fruit characteristics	Location
Algeria	Bent Qbala	50–90	na	Ovoid, small to medium. Weight of 20 fruits: 135–213 g. Yellow khalal and amber tamar. Soft, fibrous, perfumed date	Frequent: M'zab. Infrequent: Ouargla. Rare: Oued Righ, Zibans
Algeria	Degla Beida	41–60	8,266,244	Ovoid or narrowly oblong, small to medium. Weight of 20 fruits: 70–165 g. Yellow khalal and yellow or amber tamar. Dry, variable, tangy date	Abundant: Zibans, Souf, Oued Righ, El Meniaa. Infrequent: M'zab, Ouargla, Aurès
Algeria	Deglet Noor	64–150	6,845,205	Ovoid and sometimes oblong-ovate, very small to medium. Weight of 20 fruits: 82–230 g. Red khalal and variable tamar, semidry to dry, often fibrous, perfumed date	Abundant: Zibans, Souf, M'zab, Oued Righ, Ouargla, Aurès. Frequent: Metlili, El Meniaa. Rare: Tidikelt Gourara, Tassili
Algeria	Ghars	52–100	2,710,047	Narrowly oblong-obovate, medium. Weight of 20 fruits: 94–340 g. Yellow khalal and brown or amber tamar. Soft to semidry, fibrous, perfumed date	Abundant: Zibans, Souf, M'zab, Oued Righ, Ouargla, Aurès, Metlili. Frequent: El Meniaa. Rare Gourara, Tassili, Tidikelt
Algeria	Halwa	40–60	na	Oblong, medium. Weight of 20 fruits: 80–160 g. Yellow khalal and amber tamar, semidry, often farinaceous use, perfumed date	Frequent: El Meniaa, Zibans Infrequent: Aurès. Rare: Oued Righ, Oued Souf
Algeria	Kentichi	40–60	na	Oblong, small. Weight of 20 fruits: 74–136 g. Yellow khalal and brown at tamar. Dry, floury, perfumed date	Frequent: Zibans. Rare: Oued Righ, Oued Souf, Ouargla

Algeria	Litima	50–120	na	Ovoid, medium. Weight of 20 fruits: 132–135 g. Yellow khalal and amber or red tamar. Soft, fibrous, perfumed date	Frequent: Ouargla. Infrequent: Zibans, Oued Righ. Rare: M'zab, Oued Souf
Algeria	Mech Degla	40–60	na	Ovoid or oblong, small. Weight of 20 fruits: 100–130 g. Red khalal and amber tamar, dry, floury, tangy	Aurès, Zibans
Algeria	Tafezwin	50–80	na	Narrowly oblong, medium. Weight of 20 fruits: 100–205 g. Yellow khalal and amber or red tamar, semidry, fibrous, perfumed date	Abundant: M'zab, Ouargla. Frequent: Metlili, Oued Righ, Oued Souf. Rare: Zibans
Algeria	Takarmust	35–50	na	Round, small. Weight of 20 fruits: 84–250 g. Yellow khalal and black or brown tamar. Soft or semidry, fibrous, perfumed date	Abundant: Ouargla, M'zab Infrequent: Tidikelt, Oued Righ
Algeria	Tantbucht	40–80	na	Round, small, Weight of 20 fruits: 110–277 g. Yellow khalal and brown tamar. Variable, fibrous, perfumed date	Abundant: Oued Righ Infrequent: Ouargla, M'zab Rare: Tidikelt, Zibans, Aurès
Algeria	Taqerbucht	35–55	na	Round, small. Weight of 20 fruits: 110–235 g. Yellow khalal and amber tamar. Semidry or dry, fibrous, flattened	Abundant: Tidikelt. Frequent: Touat, Gourara. Infrequent: Saoura. Rare: M'zab
Algeria	Timjuhart	50–90	na	Ovoid, medium. Weight of 20 fruits: 65–250 g. Red khalal and black tamar. Dry or semidry, fibrous perfumed date	Abundant: El Meniaa, M'zab Frequent: Touat, Gourara. Infrequent: Ouargla, Metlili. Rare: Tidikelt, Oued Righ, Oued Souf
Cameroon	Dibinojeh naoura	15–20	na	Dibinojeh naoura, general term for seedling dates; no specific cultivars are named. Poor quality	Northern Cameroon
Djibouti	Barhi	40	3,000	Good and soft	Ali-Sabieh, Arta, Dikhil, Tadjoura
Djibouti	Beladi	10–20	4,000	Good and semisoft	Arta, Djibouti, Tadjourah

(continued)

Country	Cultivar	Yield per tree (kg)	Number of planted trees	Fruit characteristics	Location
Djibouti	Harissa	10–20	3,000	Good and dry	Dikhil
Djibouti	Harissa	10–20	2,000	Good and soft	Tadjourah
Djibouti	Imri	10–20	50	na	Arta, Dikhil
Djibouti	Khadrawy	70	50	Good and soft	Ali-Sabieh
Djibouti	Khalas	40	4,000	Good and soft	Ali-Sabieh, Arta, Dikhil, Tadjourah
Djibouti	Kisba	10–20	na	na	Arta, Djibouti
Djibouti	Medjool	90	500	Good and semisoft	Ali-Sabieh
Djibouti	Nabut Sultan	40	3,000	Good and soft	Ali-Sabieh, Arta, Dikhil
Djibouti	Oumo-Assala	10–20	50	na	Arta
Djibouti	Rzizi	30	1,500	Good and soft	Ali-Sabieh, Arta, Dikhil
Djibouti	Sheeri	10–20	100	na	Arta, Dikhil, Tadjourah
Djibouti	Zahidi	80	30	Good and soft	Ali-Sabieh, Dikhil
Egypt	Aglany	163	167,730	Fruit has poor taste at khalal and becomes sweet at rutab stage	Sharkia, Ismailia Governorate
Egypt	Amhat	107	253,498	Fruit taste is not suitable for consumption at khalal stage although it becomes sweet with little fiber at rutab stage	Giza, Fayoum Governorate
Egypt	Amry	175	79,860	Fruit medium-sweet taste with moderate flesh thickness	Sharkia, Ismailia Governorate
Egypt	Bartamoda	50	13,307	Fruit has good sweet taste at tamar stage with high content of fiber	Upper Egypt especially Aswan
Egypt	Bent-Eisha	128	340,774	Taste is sweet with a thin skin and little fiber	Nile Delta, Sharkia Governorate
Egypt	Gargoda	30	3,000	Flesh texture is fibrous with moderate sweet taste	Upper Egypt especially Aswan

Egypt	Gondaila	35	17,220	Fruit tastes deliciously sweet and has moderate flesh thickness	Upper Egypt especially Aswan
Egypt	Hayany	122	2,538,131	Fruit has acceptable taste at khalal and good at rutab stage with moderate fiber flesh	Sharkia, Domiat, Ismailia, Beheira Governorate
Egypt	Malkaby	40	4,410	Taste is extremely sweet and the fruit skin is smooth and united with flesh	Upper Egypt especially Aswan
Egypt	Oraiby	120	224,440	Fruit has acceptable taste at khalal and is moderately sweet with little fiber flesh at rutab stage	Nile Delta Governorates
Egypt	Sakkoty	55	99,016	Fruit flesh is thick and deliciously sweet and skin is smooth and loose from flesh	Upper Egypt especially Aswan
Egypt	Samany	130	648,411	Skin is smooth, thick flesh, moderately sweet and succulent	Mediterranean Coast, Sharkia Governorate
Egypt	Shamiya	35	2,000	Excellent dry cultivar, medium-sweet taste with little fiber	Upper Egypt especially Aswan
Egypt	Siwy	98	1,822,419	Acceptable taste at khalal and tamar is very sweet with little fiber	Oases of the West Desert, Giza, Fayoum Governorate
Egypt	Zaghloul	116	1,043,061	Softest quality cultivar, extremely sweet, very crunchy when served fresh, skin is smooth and crisp flesh and sweet taste	Nile Delta, the Mediterranean Coast
Libya	Abel	75–100	na	Slow growth. High productivity, good market value, easy to harvest, store, and transport. Fruit quality very good	Oasis of Al Jufrah
Libya	Bamour	60	na	Trade and commercial value for daily consumption and when dried to tamar stage, used for animal feed. Fruit quality good	Oasis of Al Jufrah
Libya	Berni	50	na	Fresh fruit consumption and cake and paste. Fruit quality very good	Oasis of Al Jufrah

(continued)

Country	Cultivar	Yield per tree (kg)	Number of planted trees	Fruit characteristics	Location
Libya	Bestian	65	na	Rapid growth. High productivity, date cultivar most recommended for diabetes sufferers. Good for postharvest processing and animal feed. Fruit quality good	Oasis of Al Jufrah
Libya	Halima	60	na	Poor rooting ability. Low productivity with high storability. Represents top quality Libyan date with an excellent market value. Fruit quality excellent	Oasis of Al Jufrah
Libya	Hamria	50	na	Suitable for dry farming system and also for animal feed and other uses. Fruit quality good	Oasis of Al Jufrah
Libya	Kathari	70–80	na	Slow growth. Resistant to soil and water salinity. Fruit quality very good. Highly esteemed. High quality of offshoots. Reasonable market value. Fruit quality very good	Oasis of Al Jufrah
Libya	Libyan Deglet	80–100	na	Rapid growth. Very suitable for long-period conservation and highly acceptable for consumption. Fruit quality very good excellent	Oasis of Al Jufrah
Libya	Noyet Meka	50	na	Very good offshoot capacity. Daily human consumption and when dried to tamar stage, used for animal feed. Fruit quality good	Oasis of Al Jufrah
Libya	Omglaib	50	na	Well adapted for both dry farming and irrigated farming. Reasonable offshoot capacity. Daily consumption. Fruit quality suitable for cake or paste and storage. Fruit quality good	Oasis of Al Jufrah
Libya	Saiedi	70	na	Rapid growth. Resistant to pests and very productive, regular productivity. Good market value suitable for storage and transport. Fruit quality very good	Oasis of Al Jufrah

Libya	Saila	50	na	Not highly commercial and use for daily fresh consumption. Reasonable offshoot capacity. Fruit quality sufficient	Oasis of Al Jufrah
Libya	Sokeri	50	na	Reasonable offshoot capacity, one of the best cvs. to produce vinegar and date syrup. Fruit quality good	Oasis of Al Jufrah
Libya	Tagiat	80	na	Rapid growth. High and constant production every year. The best cultivar to prepare cake and syrup using tamar stage. A popular old saying claims that hunting dogs run fast because they eat Tagiat dates. Fruit quality very good	Oasis of Al Jufrah
Libya	Talis	50	na	Fresh consumption and for cake. Very adaptable cultivar under both dry farming and irrigated farming. Very good offshoot capacity. Fruit quality good	Oasis of Al Jufrah
Libya	Tameg	50	na	Daily consumption. Best cultivar to obtain fiber for rope and daily fruit consumption. Good offshoot capacity and fruit quality	Oasis of Al Jufrah
Libya	Trasferit	40	na	Good for storing. Well adapted to the area and used for swapping commodities during the caravan period. Very good offshoot capacity. Fruit quality good	Oasis of Al Jufrah
Libya	Zebur	60	na	Because of its hard skin, it is used for livestock feed. Reasonable offshoot capacity. Fruit quality sufficient	Oasis of Al Jufrah
Mauritania	Ahmar Danca	45–60	na	Largest fruit size, high percentage of fruit pulp, date of mid-season, good fruit and moderate to good storability	States of Adrar, Laassaba, and Tagant

(continued)

Country	Cultivar	Yield per tree (kg)	Number of planted trees	Fruit characteristics	Location
Mauritania	Ahmar Dli	45–60	na	Relatively large fruit size, especially high fruit pulp, very late producer, good fruit and moderate to good storability	States of Adrar, Laassaba, and Tagant
Mauritania	Al-Falha	50–65	na	Date of mid-season, good fruit, and moderate to good storability, longest fruiting bunch, relatively high fruit weight and percentage of seed on date and small protuberances rarely appear on the seed surface, low percentage of fruit pulp	State of Adrar (Oued Seguelil),
Mauritania	Alfat Al-Bahoua	40–50	na	Very precocious, average quality of fruit and moderate storability, moderate weight of date and percentage of fruit pulp	States of Adrar, Laassaba, and Tagant
Mauritania	Alfat Foum Agadir	45–60	na	Date of mid-season, good fruit but the seed is large, moderate storability, low weight of date and percentage of fruit pulp, fruit often pear shaped	States of Laassaba and Tagant
Mauritania	Al-Hanaouia	45–60	na	Date of mid-season, good fruit and storability, moderate weight of fruit, high percentage of fruit pulp	State of Laassaba
Mauritania	Amchekhssi	45–60	na	Late maturing, average quality of fruit and moderate to good storability, moderate weight of date and percentage of fruit pulp	State of Adrar
Mauritania	Basbrik	45–60	na	Date of mid-season, good fruit relatively and moderate storability, oval shape of the fruit in lower part, moderate weight of date, high fruit pulp content	States of Tagant and Laassaba
Mauritania	Boujira	45–60	na	Late maturing, relatively good fruit and moderate storability, long fruiting bunch (part free of fruits), relatively high weight of seed, moderate weight of date and, percentage of fruit pulp	State of Adrar

Mauritania	Lakhdira	55–70	na	Very precocious, good fruit and good storability; shortest fruiting bunch, color of date brown to greenish, long fruit and seed, moderate weight of date, somewhat high percentage of fruit pulp	States of Adrar, Laassaba, and Tagant
Mauritania	Lamdina	50–70	na	Late maturing, good fruit and weak to moderate storability, long fruit bunch and high weight of date and very long and skinny seed in addition to *wade* form of seed slot that is relatively very wide, moderate percentage of pulp fruit	States of Adrar, Laassaba, and Tagant
Mauritania	Louted	65–75	na	Late maturing, good fruit relatively and moderate storability, high fruit-bunch numbers and length of fruiting bunch, yellow to violet color in rutab stage and brown in mature tamar stage and often the presence of a horn on the fruit, very low weight of date and moderate percentage of pulp fruit	State of Tagant
Mauritania	Mahboula	55–70	na	Very precocious production, good fruit and low to moderate storability, high percentage of fruit pulp, low seed weight	States of Adrar, Laassaba, and Tagant
Mauritania	Mriziga	40–50	na	Late maturing, good fruit, moderate to good storability, moderate weight of date, very high proportion of fruit pulp	States of Tagant and Laassaba
Mauritania	Oum Arich	45–60	na	Late maturing, good fruit, weak to moderate storability, very large number fruiting bunches, low date weight, shortest in date length, moderate proportion of fruit pulp, occasionally oval bumps appear on seeds	States of Tagant and Laassaba
Mauritania	Salmadina	50–60	na	Late maturing, good fruit relatively and weak to moderate storability, fruiting bunch relatively long, high weight of date and moderate percentage of pulp fruit	State of Adrar

(continued)

Country	Cultivar	Yield per tree (kg)	Number of planted trees	Fruit characteristics	Location
Mauritania	Soukani	60–75	na	Late maturing, good fruit relatively and moderate storability, moderate date weight and seed and fruit pulp percentages, cylindrical shape of fruit with occasionally appearance of a kind of ring in the fruit center	States of Adrar, Laassaba, and Tagant
Mauritania	Tifred	50–60	na	Late maturing, good fruit relatively, good storability, light fruit weight and shorter fruit compared with other cultivars, oval-opposite rectangular fruit shape, moderate percentage of fruit pulp	States of Adrar and Tagant
Mauritania	Tiguedert	55–70	na	Late maturing, good fruit, good storability, different values of proportions of pulp fruit and seed observed in different oases, low fruit and seed weight	State of Adrar
Mauritania	Tijeb	50–60	na	Date of mid-season, relatively good fruit and moderate storability, violet color in immature fruit *balh*, different values of date weights and percentages of fruit pulp observed in different oases	States of Adrar, Laassaba, and Tagant
Mauritania	Tinterguel	55–70	na	Late maturing, good fruit, good storability, short fruiting bunch, lowest fruit weight and long seed, moderate date weights and percentages of fruit pulp	States of Adrar, Laassaba, and Tagant
Mauritania	Tinwazid	45–60	na	Date of mid-season, good fruit and moderate to good storability, thicker fruit compared to other cultivars, great length and weight of fruit, great length and thickness of seed, moderate percentage of fruit pulp	States of Adrar, Laassaba, and Tagant

Morocco	Ademou	70–100	<50,000	Yellow at khalal stage, light brown at tamar stage, mid-season, high percentage of pulp in fruit, appreciated locally	Drâa, Tafilalet
Morocco	Aguelid	70–100	<100,000	Orange at khalal stage, light brown at tamar stage, very early, moderate percentage of pulp in fruit, valued commercially for its earliness	Drâa, Bani, Saghro, Anti-Atlas
Morocco	Ahardane	70–95	<100,000	Yellow at khalal stage, brown at tamar stage, very early, high percentage of pulp in fruit, valued commercially for its earliness	Drâa, Anti-Atlas, Saghro, Bani, Oriental
Morocco	Aïssa-Iyoub	73–90	<50,000	Yellow to pinkish at khalal stage, light brown at tamar stage, moderately late, moderate percentage of pulp in fruit, appreciated locally	Oriental, Drâa
Morocco	Azigzao	76–90	<100,000	Yellow orange to violet-reddish at khalal stage, light brown at tamar stage, moderately early, moderate percentage of pulp in fruit, locally appreciated	Tafilalet, between Saghro, High-Atlas areas, Ferkla, Gheris, Saghro
Morocco	Aziza Bouzid	65–90	<100,000	Orange to reddish at khalal stage, light brown at tamar stage, mid-season, moderate percentage of pulp in fruit, very appreciated in its native area	Oriental (east)
Morocco	Belhazit	71–95	<50,000	Orange to violet-reddish at khalal stage, brown at tamar stage, mid-season, high percentage of pulp in fruit, locally appreciated	Tafilalet
Morocco	Black Bousthami	92–130	<100,000	Reddish yellow at khalal, black at tamar, moderately late, high pulp content in fruit, appreciated for own consumption, excellent taste, good for processing	Bani, Drâa, Tafilalet, Saghro, Anti-Atlas
Morocco	Boucerdoune	75–95	<100,000	Reddish yellow at khalal stage, light to little dark brown at tamar stage, mid-season, very high percentage of pulp in fruit	Tafilalet, between Saghro, High-Atlas areas, Gheris, Saghro, Oriental

(continued)

Country	Cultivar	Yield per tree (kg)	Number of planted trees	Fruit characteristics	Location
Morocco	Boufegous	70–120	>400,000	Yellow orange at khalal stage, light brown at tamar stage, mid-season, very high percentage of pulp in fruit, known in all oases	All areas
Morocco	Boufegous ou Moussa	72–100	<50,000	Yellow at khalal stage, brown at tamar stage, moderately early, low percentage of pulp in fruit locally appreciated	Bani
Morocco	Bouijjou	70–95	<100,000	Yellow orange at khalal stage, dark brown at tamar stage, mid-season, very high percentage of pulp in fruit	Tafilalet, Oriental, Guir
Morocco	Bouittob	70–100	<100,000	Light orange at khalal stage, light brown at tamar stage, moderately late, high percentage of pulp in fruit, very appreciated in Bani area	Anti-Atlas, Bani
Morocco	Boukhanni	75–100	<50,000	Yellow at khalal stage, light brown to reddish at tamar stage, moderately early, high percentage of pulp in fruit, locally appreciated	Drâa
Morocco	Bourar	75–90	<100,000	Yellow at khalal stage, dark brown at tamar stage, moderately late, very high percentage of pulp in fruit, very appreciated in Drâa Valley	Drâa, Bani, Saghro, Tafilalet
Morocco	Bouskri	78–95	<200,000	Yellow at khalal stage, dark brown at tamar stage, moderately late, moderate percentage of pulp in fruit, very appreciated due to its high level of sugar and good storage quality	Bani, Drâa, Saghro, Todra, Oriental, Tafilalet, between Saghro, High-Atlas areas, Anti-Atlas
Morocco	Bouslikhene	70–100	<100,000	Yellow orange at khalal stage, dark brown at tamar stage, moderately early, moderate percentage of pulp in fruit, very appreciated in Tafilalet area	Tafilalet, Saghro
Morocco	Boutemda	70–90	<50,000	Yellow orange at khalal stage, dark brown at tamar stage, moderately early, high percentage of pulp in fruit, very appreciated in Bani area	Bani

Morocco	Bouzeggar	76–120	<100,000	Reddish at khalal stage, dark brown to black at tamar stage, late, high percentage of pulp in fruit, very appreciated in Drâa Valley	Drâa, between Saghro, High-Atlas areas, Ferkala, Gheris
Morocco	Hafs	56–80	<100,000	Yellow at khalal stage, light brown at tamar stage, mid-season, high percentage of pulp in fruit, appreciated locally	Drâa, Oriental, Tafilalet, between Saghro, High-Atlas areas
Morocco	Haoua	70–95	<50,000	Yellow orange at khalal stage, brown at tamar stage, mid-season, high percentage of pulp in fruit, appreciated locally	Drâa
Morocco	Iklane	90–130	<100,000	Pink to reddish at khalal stage, black at tamar stage, late, very high percentage of pulp in fruit, appreciated in Anti-Atlas area, good for processing	Anti-Atlas, Bani, Drâa, Saghro
Morocco	Jihel	80–120	<100,000	Yellow at khalal stage, brown to dark brown at tamar stage, late, high percentage of pulp in fruit, very appreciated commercially mostly in Drâa Valley	Drâa, Bani, Anti-Atlas, Tafilalet, Saghro, between Saghro, High-Atlas areas
Morocco	Mah-Lbaïd	75–100	<50,000	Yellow orange at khalal stage, dark brown at tamar stage, moderately late, high percentage of pulp in fruit, appreciated locally	Anti-Atlas, Bani
Morocco	Medjool	80–150	>160,000	Orange at khalal stage, brown to dark brown at tamar stage, late in Drâa Valley and moderately late upstream of Ziz Valley and mid-season in its homeland (Tafilalet), very high percentage of pulp, high value, high demand in the market and in international trade	Ziz, Tafilalet, extended to Drâa, Oriental, Bani areas
Morocco	Mekt	70–100	<50,000	Reddish at khalal stage, black at tamar stage, mid-season, very high percentage of pulp in fruit, appreciated locally	Drâa, Bani, Saghro, Anti-Atlas

(continued)

Country	Cultivar	Yield per tree (kg)	Number of planted trees	Fruit characteristics	Location
Morocco	Mestali	66–90	<50,000	Pinkish at khalal stage, dark brown at tamar stage, moderately early, high percentage of pulp in fruit, appreciated locally	Drâa
Morocco	Najda (INRA-3014) selected variety	90–130	>800,000 distributed vitroplants	Pink to reddish at khalal stage, light brown to brown at tamar stage, date rich in sugars, mid-season, very high percentage of pulp in fruit, good storage fruit ability, valued commercially for its resistance to bayoud and good fruit quality. Recommended for reestablishing the devastated orchards by bayoud	Originally from Drâa, massively multiplied, diffused in all oases, avoid planting in trays in altitude or marginal areas
Morocco	Oum-N'hal	66–100	<50,000	Yellow at khalal stage, light brown at tamar stage, moderately late, weak percentage of pulp in fruit, locally appreciated	Anti-Atlas, Drâa
Morocco	Outoukdime	70–96	<100,000	Pink to reddish at khalal stage, brown to dark brown at tamar stage, moderately late, very high percentage of pulp, cultivar grown at average elevations in the mountains, locally appreciated	Between Saghro, High-Atlas areas, Todra
Morocco	Ras Lahmer	72–100	<100,000	Yellow at khalal stage, light brown at tamar stage, mid-season, moderate percentage of pulp in fruit, appreciated locally	Bani, Drâa, Saghro, Tafilalet, between Saghro, High-Atlas areas
Morocco	Saïrlayalate	74–100	<50,000	Yellow at khalal stage, light brown at tamar stage, moderately late, high percentage of pulp in fruit, appreciated commercially in its crib	Bani
Morocco	Tadmainte	70–120	<50,000	Light orange to pinkish at khalal stage, dark brown at tamar stage, mid-season, high percentage of pulp, appreciated locally	Anti-Atlas, Bani, Drâa, Oriental, Saghro

Morocco	White Bousthami	72–120	<50,000	Yellow at khalal stage, dark brown at tamar stage, mid-season, moderate percentage of pulp in fruit, appreciated locally	Anti-Atlas, Bani
Niger	Aghous (Aguis)	na	na	Good geographical distribution over the region. Dry dates with good adaptation to climatic conditions, high yield	Bilma
Niger	Akanirom	na	na	Early date, very delicious, short period between green and yellow stages. Fruits are elongated oval shaped. Mainly intended for subsistence consumption and not commercialized	Bilma
Niger	Almadeyna	na	na	Large fruit, best quality, and most popular	Ingall
Niger	Atratinna	na	na	Black fruit at maturity	Ingall
Niger	Baghaberha	na	na	Yellow fruit with round form	Ingall
Niger	Bagounia	na	na	Red fruit	Damagaram
Niger	Dan Haoussa	na	na	Small fruit, low quality	Damagaram
Niger	Dilo	na	na	Semisoft date, fruits are black at maturity, mainly produced in Fachi district	Bilma
Niger	Gameye or Ngamaya	na	na	Very good quality, specific to Bilma district area, semisoft date	Bilma
Niger	Goria Fari	na	na	White large fruit, sweet	Damagaram
Niger	Goria Ja	na	na	Red large fruit, sweet	Damagaram
Niger	Guewass	na	na	Name of the village	Air Mountains
Niger	Hadib	na	na	Intended for direct consumption, round fruits, looks like dry grapes when dry	Bilma
Niger	Ilfodone	na	na	Elongated, large red fruit, semisoft	Bilma
Niger	Kaanihery	na	na	Red fruits	Ingall
Niger	Koukouma	na	na	Yellow fruit, not sweet	Damagaram

(continued)

Country	Cultivar	Yield per tree (kg)	Number of planted trees	Fruit characteristics	Location
Niger	Krouskrous	na	na	Oval fruit, hard when dry. Good cultivar but not common	Bilma
Niger	Maiwa	na	na	Very sweet green fruit	Damagaram
Niger	Taghayat	na	na	Good quality and easy to conserve	Air Mountains
Niger	Talharma	na	na	Fruit is red but becomes black at maturity	Ingall
Niger	Talittat	na	na	Looks like Kaanihery	Ingall
Niger	Tanghal	na	na	Yellow fruit, black when mature, good quality and close to Almadeyna	Ingall
Niger	Tawragh	na	na	Common cultivar having a yellow fruit	Air Mountains
Niger	Tidirchi or Toudourchi	na	na	Oval fruits, half soft, from green color fruits become violet and then take chocolate color at maturity. The cultivar is produced all over Kawar but in low quantity	Bilma
Niger	Tilmoiran or Cliyarom	na	na	Very early date	Bilma
Niger	Touzouwzaw	na	na	Strong green color	Air Mountains
South America: Chile	Deglet Noor	na	na	na	Arica to Copiapó
South America: Chile	Medjool	na	na	na	Arica to Copiapó
South America: Chile	Zahidi	na	na	na	Arica to Copiapó
South America: Peru	Deglet Noor	na	na	na	Ica, Zaña, Pisco
South America: Peru	Medjool	na	na	na	Ica, Zaña, Pisco

South America: Peru	Zahidi	na	na	na	Ica, Zaña, Pisco
Sub-Sahelian: Chad	Anagow	25–100	na	Soft	Borkou, Kanem
Sub-Sahelian: Chad	Ardousow	30–100	na	na	Borkou, Kanem
Sub-Sahelian: Chad	Aribo	40–150	na	Semidry	Borkou, Kanem
Sub-Sahelian: Chad	Arsandow	25–65	na	na	Borkou, Kanem
Sub-Sahelian: Chad	Bornow	40–150	na	Dry	Borkou, Kanem
Sub-Sahelian: Chad	Dogordow	60–100	na	na	Borkou, Kanem
Sub-Sahelian: Chad	Koîdi bichanga	60–100	na	na	Borkou, Kanem
Sub-Sahelian: Chad	Koîdi dellémadow	40–100	na	Semidry	Borkou, Kanem
Sub-Sahelian: Chad	Koîdow	40–100	na	Semidry	Borkou, Kanem
Sub-Sahelian: Chad	Kougoudou	50–120	na	Semidry	Borkou, Kanem
Sub-Sahelian: Chad	Kouhi	30–75	na	na	Borkou, Kanem
Sub-Sahelian: Chad	Kourdow	20–50	na	na	Borkou, Kanem
Sub-Sahelian: Chad	Lohandjé	60–100	na	na	Borkou, Kanem

(continued)

Country	Cultivar	Yield per tree (kg)	Number of planted trees	Fruit characteristics	Location
Sub-Sahelian: Chad	Marchiano	40–80	na	na	Borkou, Kanem
Sub-Sahelian: Chad	Méboul	75–125	na	na	Borkou, Kanem
Sub-Sahelian: Chad	Méguirti	40–50	na	na	Borkou, Kanem
Sub-Sahelian: Chad	Mékléya	20–30	na	na	Borkou, Kanem
Sub-Sahelian: Chad	Mékoîdi	60–100	na	na	Borkou, Kanem
Sub-Sahelian: Chad	Météréli	40–50	na	na	Borkou, Kanem
Sub-Sahelian: Chad	Métoukouli	40–60	na	na	Borkou, Kanem
Sub-Sahelian: Chad	Mourudow	30–60	na	na	Borkou, Kanem
Sub-Sahelian: Chad	Tirtidou	40–70	na	na	Borkou, Kanem
Sub-Sahelian: Chad	Wallo	60–90	na	na	Borkou, Kanem
Sub-Sahelian: Chad	Wardanga	50–80	na	na	Borkou, Kanem
Sub-Sahelian: Chad	Waserdow	40–65	na	na	Borkou, Kanem
Sub-Sahelian: Chad	Zalao	40–150	na	na	Borkou, Kanem
Sub-Sahelian: Chad	Faqur	40–60	na	na	Borkou, Kanem

Sub-Sahelian: Somalia	Farad	30–60	na	na	Borkou, Kanem
Sub-Sahelian: Somalia	Masili	40–70	na	na	Borkou, Kanem
Sub-Sahelian: Somalia	Nemahan	60–90	na	na	Borkou, Kanem
Sub-Sahelian: Somalia	Sahcari	50–80	na	na	Borkou, Kanem
Sub-Sahelian: Somalia	Suqadari	40–65	na	na	Borkou, Kanem
Sudan	Barttamoda	60	100,000	High quality. Best cultivar in Sudan. Resembles cv. Deglet Noor	Northern State
Sudan	Abidrahim	80	200,000	Hard	River Nile State
Sudan	Asada	85	2,500	Semidry red fruits, resembles Dayri of Iraq	Northern oases
Sudan	Barakawi	80	4,000,000	Hard date that resists insect attack. Most widely grown cultivar in Sudan.	Northern, River Nile
Sudan	Barhi	200	220,000	Harvested ahead of summer rains and marketed as fresh khalal date	Khartoum State
Sudan	Biraira	100	200,000	Meaty soft date which grows very tall	River Nile State
Sudan	Gargoda	120	200,000	Dry date, outyields Barakawy but harder and inferior to it	Northern State
Sudan	Gondaila	40	500,000	Excellent flavor. Hard but fancy date if harvested at early maturity and well kept	Northern, River Nile
Sudan	Khalas	150	180,000	Soft. High quality. Recent introduction under evaluation	Khartoum
Sudan	Khinaizy	150	10,000	Soft. Early maturing, crisp when harvested ahead of rains	Khartoum

(continued)

Country	Cultivar	Yield per tree (kg)	Number of planted trees	Fruit characteristics	Location
Sudan	Kulma	50	3,000	Largest date fruits in the country. Hard with high fiber content	Northern
Sudan	Madina	90	250,000	Earliest maturing indigenous fresh date. Large yellow fruits	Northern, River Nile, Khartoum
Sudan	Mishrig Wad Khateeb	250	3,000,000	Semidry most widespread cultivar in River Nile State. Edible fresh date with some tannins when harvested ahead of rains	River Nile, Northern, Khartoum
Sudan	Mishrig Wad Laggai	150	450,000	Semidry. Excellent quality but small fruits that need thinning	River Nile, Northern, Khartoum
Sudan	Seedling	100	1,000,000	Heavy bearers. Mostly inferior quality but very few trees with top quality dates	All states
Sudan	Tunisi (Deglet Noor)	100	200	Extremely hard date; good when harvested early	Northern State
Tunisia	Alig	85	300,000	Good, semidry, late	All oases
Tunisia	Ammary	80		Average, soft, early	All oases
Tunisia	Arichty/Rochty	80	400	Good, semidry, late	All oases
Tunisia	Bidh Hamam	70	Rare	Good, soft, mid-season	Continental oases
Tunisia	Bisr Helou	73	88,000	Good, dry, mid-season	All oases
Tunisia	Boufaggouss	60	na	Good, soft, late	Jerid
Tunisia	Bouhattam	75	1,500	Good, soft, mid-season	Littoral oases
Tunisia	Choddakh	60	na	Good, soft, mid-season	Nefzaoua
Tunisia	Degla Bidha	65	na	Average, dry, late	Jerid
Tunisia	Deglet Noor	70	4,000,000	Good, semisoft, late	Continental oases
Tunisia	Fermla	75	na	Average, semidry, mid-season	Nefzaoua
Tunisia	Fezzani	60	na	Good, semidry, mid-season	Continental oases

Tunisia	Gharss Souf (Gharss Mettig)	55	na	Good, soft, mid-season	Continental oases
Tunisia	Gonda	60	na	Good, semisoft, mid-season	Continental oases
Tunisia	Gosbi	60	na	Good, soft, early	Continental oases
Tunisia	Hamra	85	na	Good, semidry, mid-season	Continental oases
Tunisia	Hissa	65	na	Good, soft, early	Continental oases
Tunisia	Hloua	60	na	Good, semidry, late	Nefzaoua
Tunisia	Horra	75	80,500	Average, dry, mid-season	Continental oases
Tunisia	Kinta	80	64,800	Good, dry, mid-season	All oases
Tunisia	Lagou	70	na	Good, soft, mid-season	Continental oases
Tunisia	Lemsi	70	150	Good, soft, mid-season	Littoral oases
Tunisia	Malti	70	na	Good, soft, early	Continental oases
Tunisia	Takermest	50	na	Good, soft, mid-season	Continental oases
Tunisia	Tezerzayet Kahla	65	na	Good, soft, mid-season	Continental oases
Tunisia	Tezerzayet Safra	70	na	Good, soft, early	Continental oases
Tunisia	Tronja	50	na	Good, semidry, late	Continental oases
USA	Deglet Noor	90–140	330,000	Very good	Coachella Valley
USA	Medjool	70–90	450,000	Excellent	Coachella Valley, Bard Valley, California; Yuma County, Arizona

This list is not inclusive of all cultivars grown. All entries were provided by the authors of the corresponding country chapter
na not available

Appendix B

Commercial sources of dates in some date-producing countries in Africa and the Americas

Country	Name	Contact information	Website link
Algeria	Agronima Sarl, Tolga Biskra	Telephone: 0661 35 17 22	na
Algeria	Alim Agrofood Sarl	56, Rue Amrane Hafnaoui Doucen Biskra; agrofood@ ramble.ru	na
Algeria	Biodattes Algérie	1 rue Freres Missouri, Oran Activités Import-export de dattes. contact@biodattes.com.	www.biodattes.com
Algeria	Bionoor	1 rue Auguste Renoir, 93600 Aulnay-sous-Bois France; adj.khelil@bionoor.com	www.bionoor.com
Algeria	Bio-Oasis	Zone d'Activité Oumeche Biskra	na
Algeria	Biscofruits Sarl	Zone d'équipement n° 38 Biskra; biscofruits@yhaoo.fr	www.duna-dattes.com.dz
Algeria	Chennoufi Salah	09, Rue Ben Ameur Aissa, Tolga Biskra; chenoufi-salah@ yahoo.com	na
Algeria	Dana Dattes	Zone d'Activité s Oumache Biskra; www.dana-dattes.com	na
Algeria	Dhaouia Sarl	BP 407 Route de Touggourt El Oued; tpa_sea_daouia@ yahoo.fr	www.daouia.yahoo.fr.gd
Algeria	Ecodatte	Zone d'é quipement BISKRA	na
Algeria	El Biskria Commercialisation, Produits Agricoles	Zone d'équipement BP 457, 07000 Biskra; bicpag@ yahoo.fralim	na
Algeria	El Mokhtar Co.	Lot Benghana, n° 3 Biskra; elmoukhtarco@hotmail.com	www.dattes-moukhtar.com
Algeria	Haddoud Salim Ets	Zone d'Activité Tolga Biskra; haddou_s@yahoo.fr	na
Algeria	Kerbaa Abdelkrim Etablissement	RW N° 03 Lichana Tolga Biskra; kerbaa@yahoo.fr	na
Algeria	Kheireddine Malik Ets	Zone d'équipement Biskra; amarco.kheireddinemalik@ yahoo.fr	na
Algeria	Phenix Sarl	Droh, Commune de Chetma Biskra; sarlphenixdattes@ yahoo.fr	na

Algeria	SODAC	Zone D'Equipement De Biskra, Biskra 7000	na
Algeria	Sud Dattes Eurl	Rue Athmani Mohamed Tahar Tolga Biskra; suddatte@hotmail.fr	na
Algeria	Sudaco Epe Spa	Avenue de la Gare, BP 64 RP; sudaco2002@yahoo.fr	na
Algeria	Utpa Daouia Sarl	Route de Touggourt El Oued; utpa_sea_daouia@yahoo.fr	na
Egypt	El Waha Dates Company	Bahariya Oasis, Giza, Egypt	http://www.elwahaegypt.com/en/index.htm
Egypt	Ganet El-Saharaa Company for Dates	Bahariya Oasis, Giza, Egypt; desertparadisedates@gmail.com	na
Morocco	Riad Tafilalet	Route Boudnib-Errachidia, Boudnib Morocco	na
Morocco	Economic Interest Groups (EIG) Diffate Ziz	Ksar Hannabou, Commune Rurale Ard Sabah Ghriss, Erfoud Morocco	na
Morocco	23 traditional cooperatives (farmer associations) who have names	Both areas of date production	na
Sudan	Kairma Factory, Northern State	Kairma Factory, Northern State, Karima; karimadatates@gmail.com	na
Tunisia	Dar Horchani	Elhamma Tozeur	na
Tunisia	Erribat Douz	Douz; societe-erribat@hotmail.fr	na
Tunisia	Groupement Interprofessionnel Des Dattes Kebili	Route de Gabes BP N° 24, 4200 Kebili	www.gifruit.nat.tn
Tunisia	Palme D'or	commercial.pdor@gmail.com	na
Tunisia	Sodac, Ste Dattes, Chahbani Kebili	Rte de Douz 4200, Kebili	na
Tunisia	Sté Green Fruits SA	Z.I Rte de tozeur Kebili; commercial@green-fruits.com	na
Tunisia	South Organic	Rte de Gabès, El Bayez 4200, Kebili	na
Tunisia	SOVAPO Kebili	Z I BP 226 Kebili 4200	na
Tunisia	VACPA Boudjebel	Route de Menzel Bouzelfa Beni Khalled, Nabeul, Tunisie	www.boudjebeldates.com
USA	Bard Valley Medjool Date Growers Association	PO Box 937, Bard, CA 92222, USA	http://www.naturaldelights.com

na not available

Appendix C

Commercial sources of offshoots in some date-producing countries in Africa and the Americas

Country	Name	Contact information	Website link
Cameroon	Agricultural Research Institute for Development	IRAD, Garoua PO Box: 415 Garoua, Cameroon	www.iradcameroon.org
Egypt	Alex Tarde	496 Al Horreya Avenue, Bolkly, Alexandria, Egypt; alextrade@alex-trade.com	www.alex-trade.com/english.html
Egypt	Nakheel Land	Rashid, Beheira, Egypt	http://www.arabnakhelland.com
Egypt	Al Safa Agricultural Company	Edco, Beheira, Egypt	http://www.aswaqcity.com/thread643374.html
Morocco	Nursery Gheris	Agoumad, Goulmima	na
Morocco	Company Oasis Tafilalet	Ksar Elbouya Jorf, Erfoud	na
Sudan	Zadna International Kadaro, Khartoum North	zadnafarm@hotmail.com	

na not available

Appendix D

Commercial source of in vitro plants in some date-producing countries in Africa and the Americas

Country	Name	Contact information	Website link
Algeria	Institut National de la Recherche Agronomique (INRA)	02, Av. des Frères Ouadek Hacène, BP N°200, EL Harrach, Alger; service.vte@inraa.dz	na
Egypt	Egyptian Hollandia Tissue Culture and Trade	Ahmed Oraby Society, Ismailia Road Line 2, Cairo, Egypt; egyptainhollandiat-issueculture@msn.com	egyptainhollandiatissuecultureand.blogspot.com
Egypt	Egyptian-French Company for Tissue Culture Planting	Km. 4 New Borg El Arab Rd Cairo, Egypt. Fax: 03-4550992	http://www.ameinfo.com/db-3301006.html
Morocco	Les Domaines Agricoles	Domaine El Bassatine, Meknès; bassatine@menara.ma	www.lesdomainesagricoles.com
Morocco	ISSEMGHY Biotechnologies	SN Route 107. Chellalat. Aîn Harrouda Casablanca 20 630, Morocco; contact@issemghybio.com	www.issemghybio.com
Morocco	Laboratories PALMAGRO Biotechnologies	Douar Ourti, Oued Essafa, Biougra, Morocco; contact@palmagro-biotechnologies.com	www.palmagro-biotechnologies.com
Morocco	Invitro Palm Biotechnologie	400, bd Mohamed Zerktouni, 5°ét, 20040 Casablanca Tel.: +212 522474582, 661198237. Production site: Tnin Chtouka, route de Casa/Azemmour, Azemmour	na
Morocco	INVITROPALM Biotechnology	400 Bd Zerktouni 5° Etage N° 26 – Casablanca, abbenj@hotmail.com	na
Morocco	MAGHREB PALM	Inzegane (near Agadir city)	na
Morocco	OASIS Biotechnologie	Erfoud	na
Sudan	Zadna International	Kadaro, Khartoum North; zadnafarm@hotmail.com	
USA	Phoenix Agrotech	16812 Red Hill Ave, Irvine, CA 92606, USA	http://www.phoenixagrotech.com

na not available

Appendix E

Research institutes concerned with date palm research in some date-producing countries in Africa and the Americas

Country	Name	Contact information	Website link
Algeria	CRSTRA (Centre for Scientific and Technical Research In The Dry Areas)	Campus Universitaire Oued Souf, Universite Mohamed Khider, Biskra	www.crstra.dz
Algeria	National Institute of Agronomic Research of Algeria (INRA)	02, Av. des Frères Ouadek Hacen Badi, El Harrach, Alger; service.vte@inraa.dz	www.inraa.dz
Algeria	National Institute of Plant Protection (INPV)	12 Avenue des Frères Ouadek Hacen Badi, El Harrach. Algérie; srpv_alger@hotmail.com	www.inpv.edu.dz
Algeria	Technical Institute of Agricultural Development Saharan (ITDAS)	BP 27 Ain Benoui, Biskra, Algérie	www.multimania.com/itdas
Algeria	Research Laboratory of Arid Areas (LRZA)	BP 32 El Alia Bab Ezzouar, 16111 Alger; lrzadz.boug@gmail.com	www.lrza.usthb.dz
Algeria	National High School For Agriculture (ENSA)	Avenue des Frères Ouadak El Harrach Alger	www.ensa.dz
Cameroon	Agricultural Research Institute for Development	IRAD, Garoua PO Box: 415 Garoua, Cameroon	www.iradcameroon.org
Chad	Université de Djamena, Faculté des Sciences, Environnement et Agriculture, Systématique et Phytoécologie	BP 1117, Avenue Mobutu N'Djamena, Tchad, Chad. Tel.: (235) 251 44 44+(235) 251 46 97, 235-251 45 81; rectorat@intnet.td	www.univ-ndjamena.org
Djibouti	Centre d'Etudes et de Recherche de Djibouti (CERD)	Route de l'Aéroport, BP 486, Djibouti	www.cerd.dj/
Egypt	The Central Laboratory of Date Palm, Research and Development (CLDPRD)	9 Gamma Str., Giza, 12619, Egypt; info@hortinst.com	www.hortinst.com
Egypt	National Research Center (NRC)	El-Tahrir Str., Dokki, 12622, Cairo, Egypt; info@nrc.dci.eg	www.nrc.sci.eg,

Egypt	Desert Research Center (DRC)	1 Mathaf Al-Mataria-Cairo, P. Box: 11753 Mataria, Egypt. Fax: 02-26357858	www.drc-egypt.org
Egypt	Agricultural Genetic Engineering, Research Institute (AGERI)	9 Gamma Str., Giza, 12619, Egypt	www.ageri.sci.eg
Ethiopia	The Ethiopian Institute of Agricultural Research	PO Box 2003, Addis Ababa. Tel.: +251-11-6462633-41. Fax: 251-11-6461294	http://www.eiar.gov.et
Libya	Authorities of Date Palm and Olive Tree Development and Improvement	Ministry of Agriculture	na
Libya	Agriculture Research Centers	Ministry of Agriculture	na
Libya	Faculty of Agriculture	Tripoli University	na
Mauritania	Centre National de Recherche Agronomique et du Développement Agricole (CNRADA)	BP 22 Kaédi, Mauritania; acnrada@maurit.mr	na
Mali	Centre Régional de Recherche Agronomique (CRRA)	CRRA de Gao, Mali. Tel. (223) 820411	na
Mali	Institut du Sahel (IS)	BP 1530, Bamako, République du Mali. Tel: (223) 20 22.30.43/222.47.06/20 23.40.67. Fax: (223) 20 22.78.31. Email: administration@insah.org	www.univ-nkc.mr http://www.insah.org
Mali	Institut d'Economie Rurale (IER) -	Rue Mohamed V, Bamako. Tel.: + (223) 2222606- 223105. Fax: +(223) 223 775; info@ier.ml	www.inra.org.ma
Morocco	National Agronomic Research Institute (INRA)	Avenue de la Victoire BP 415, Rabat, Morocco. Tel.: +212 537770955/+212 537772642/+212 537772639. Fax:+212 537770049	www.inra.org.ma

(continued)

Country	Name	Contact information	Website link
Morocco	Agronomic and Veterinary Institute Hassan II (IAV Hassan II)	Madinat Al Irfane, BP 6202 Rabat Instituts, 10101 Rabat, Morocco. Tel.: +212 537771745, 537771758/59, 537770792,5377743-52. Fax: +212 537778135/537775838	www.iav.ac.ma
Morocco	National School of Agriculture	Km. 10, Route Haj Kaddour, BP S/40, Meknès, 50001, Morocco	www.enameknes.ma
Morocco	Faculty of Sciences Semlalia of Marrakech (FSSM), University Cadi Ayyad	Boulevard Prince My Abdellah, BP 2390, Marrakech 40000, Morocco	www.fssm.ucam.ac.ma
Morocco	Faculty of Sciences and Techniques (FSTG), University Cadi Ayyad	BP 549, Av. Abdelkarim Elkhattabi, Guéliz, Marrakech, Morocco. Email: doyen@fstg-marrakech.ac.ma	www.fstg-marrakech.ac.ma
Morocco	Faculty of Sciences, University Mohammed 1st	Complexe Universitaire Mohamed VI, BP 717, Oujda, Morocco, sciences1.univ-oujda.ac.ma	www.ump.ac.ma
Morocco	Faculty of Sciences and Techniques (FSTE), Université Moulay Ismail	BP 509 Boutalamine, Errachidia, Morocco. Tel.: +212 535574497/84. Fax: +212 535574485, doyen@fste-umi.ac.ma	www.fste-umi.ma
Morocco	Faculty of Sciences (FSA), Université Ibn Zohr	BP: 8106, Cité Dakhla, Agadir. Tel.: +212 528220267/528220268/528220957	www.fsa.ac.ma
Morocco	Complexe d'Agadir, Institut Agronomique et Vétérinaire Hassan II	Km 2 Route de Taroudant Agadir, Maroc. Tel.: +212 48241006/48240155. Fax: +212 48242243, chgadir@mdts.com	www.iavcha.ac.ma
Niger	International Crops Research Institute for the Semi-Arid Tropics (ICRISAT)	BP 12404, Niamey, Niger; icrisatsc@icrisatne.ne	www.icrisat.org
Niger	Institut National de la Recherche Agronomique du Niger	inran@intnet.ne	http://inran.refer.ne

Niger	Faculté des Sciences et Techniques (FST), Université Abdou Moumouni (UAM) de Niamey	Faculté des Sciences et Techniques	http://uam.refer.ne/
Senegal	Institut Sénégalais de Recherche Agricole (ISRA)	Bel Air, Routes des hydrocarbures, BP 3120 Dakar, Senegal. Tel.: 33 859 17 27/33 859 17 25. Fax: 33 832 24 27, dgisra@isra.sn	http://www.isra.sn/siisra/centres/centres.html
Senegal	Conseil Ouest et Centre Africain pour la Recherche et le Développement Agricoles (CORAF)	Address: 7 Avenue Bourguiba, Dakar, BP 48, Dakar RP, Sénégal. Tel.: (+221) 33 869 96 18,(+221) 33, 8699618, infos@coraf.org;secoraf@coraf.org	http://www.coraf.org/
Senegal	Faculté des Sciences et techniques- Université Cheikh Anta Diop	Dakar 5005 Sénégal. Tel.: +221 338248187. Fax: +221 338246318, fst@ucad.edu.sn	http://fst.ucad.sn/
Sudan	Agricultural Research Corporation	ARC, Box 126, Wad Medani, Sudan; arcdg@sudanmail.net	http://www.arcsudan.sd
Sudan	Gezira University	Website: webinfo@uog.edu.sd	www.uog.edu.sd
Sudan	Khartoum University	13314 Faculty of Agriculture, Shambat; agric@uofk.edu.sd	http://www.uofk.edu
Sudan	National Research Council, Khartoum	mgalalma@gmail.com	na
Sudan	Sudan University of Science and Technology	cas@sustech.edu	http://sustech.edu
Tunisia	Centre Régional de Recherche en Agriculture Oasienne de Degueche à Tozeur	2260 Degueche, Tunisie; pdegache@iresa.agrinet.tn	na
Tunisia	Centre Technique des Dattes	Centre Technique des Dattes, Kebili 4200, Tunisia; ctd@topnet.tn	www.ctd.tn

(continued)

Country	Name	Contact information	Website link
Tunisia	Faculté des Sciences de Sfax	1284. Laboratory of Plant Biotechnology, Faculty of Sciences, University of Sfax, 3000, Sfax, Tunisia	na
Tunisia	Institut des Régions Arides	Route du Djorf Km 22.5 Medenine Tunisie; ira.med@ira.rnrt.tn	www.ira.agrinet.tn
Tunisia	Institution de la Recherche et de l'Enseignement Supérieur Agricoles (IRESA)	30 Rue Alain Savary, 1002 Tunis Belvédère, Tunisie; support@iresa.agrinet.tn	www.iresa.tn
USA	USDA-ARS National Clonal Germplasm Repository for Citrus & Dates	1060 Martin Luther King Blvd, Riverside CA 92507, USA; riv@ars-grin.gov	http://www.ars.usda.gov/main/site_main.htm?modecode=53-10-30-00
USA	University of California Riverside Agricultural Experiment Station	1060 Martin Luther King Blvd, Riverside CA 92507, USA; agops@ucr.edu	https://agops.ucr.edu/
USA	United States Department of Agriculture-Natural Resources Conservation Service (USDA-NRCS)	82-901 Bliss St, Indio, CA 92201, USA	http://offices.sc.egov.usda.gov/locator/app?service=page/ServiceCenterSummary&stateCode=06&cnty=065

na not available

Appendix F

Scientific societies concerned with date palm research in some date-producing countries in Africa and the Americas

Country	Name	Contact information	Website link
Algeria	Société d'Histoire Naturelle d'Afrique Du Nord	2 rue Didouche Mourad Alger,16000 (Université d'alger)	www.lrza.usthb.dz
Sudan	Date Palm Society	sudanese.datepalm.society@gmail.com	na
Tunisia	Agricultural Investment Promotion Agency	62, rue Alain Savary 1003, Tunis Cité El Khadra, Tunisie	www.apia.com.tn
Tunisia	Centre Technique de L'agro-Alimentaire	12, Rue de l'usine, ZI Charguia II, 2035 Ariana; CTAA@topnet.tn	www.ctaa.com.tn
Tunisia	Technical Center of Organic Agriculture	BP 54, Chott Mariem 4042 Sousse, Tunisia; ctab@iresa.agrinet.tn	www.ctab.nat.tn

na not available

Index

J.M. Al-Khayri et al. (eds.), *Date Palm Genetic Resources and Utilization: Volume 1: Africa and the Americas*, DOI 10.1007/978-94-017-9694-1

MIX
Papier aus verantwortungsvollen Quellen
Paper from responsible sources
FSC® C105338

If you have any concerns about our products,
you can contact us on
ProductSafety@springernature.com

In case Publisher is established outside the EU,
the EU authorized representative is:
Springer Nature Customer Service Center GmbH
Europaplatz 3, 69115 Heidelberg, Germany

Printed by Libri Plureos GmbH
in Hamburg, Germany